Smart Grid Communications and Networking

The smart grid will transform the way power is delivered, consumed and accounted for. Adding intelligence through the newly networked grid will increase reliability and power quality, improve responsiveness, increase efficiency and provide a platform for new applications. This one-stop reference covers the state-of-the-art theory, key strategies, protocols, applications, deployment aspects and experimental studies of communication and networking technologies for the smart grid. Throughout the book's 20 chapters, a team of expert authors cover topics ranging from architectures and models through to integration of plug-in hybrid vehicles and security. Essential information is provided for researchers to make progress in the field and to allow power systems engineers to optimize communication systems for the smart grid.

Ekram Hossain is a Professor in the Department of Electrical and Computer Engineering at the University of Manitoba, Canada, where his current research interests lie in the design, analysis and optimization of wireless/mobile communications networks, smart grid communications, and cognitive and green radio systems. He has received several awards including the University of Manitoba Merit Award in 2010 (for Research and Scholarly Activities) and the 2011 IEEE Communications Society Fred W. Ellersick Prize Paper Award.

Zhu Han is an Assistant Professor in the Electrical and Computer Engineering Department at the University of Houston, Texas. He received his Ph.D. in electrical engineering from the University of Maryland, College Park, in 2003 and worked for 2 years in industry as an R&D Engineer for JDSD. He is a recipient of the NSF CAREER Award (2010) and the IEEE Communications Society Fred W. Ellersick Prize Paper Award.

H. Vincent Poor is the Michael Henry Strater University Professor at Princeton University, New Jersey, where he is also Dean of the School of Engineering and Applied Science. He is a Fellow of the IEEE, and is a member of the US National Academy of Engineering and of the US National Academy of Sciences. He is also a Fellow of the American Academy of Arts and Sciences, an International Fellow of the Royal Academy of Engineering, and a former Guggenheim Fellow. Recent recognition of his work includes the 2009 Edwin Howard Armstrong Award of the IEEE Communications Society, the 2010 IET Ambrose Fleming Medal, the 2011 IEEE Eric E. Sumner Award, and an honorary doctorate from the University of Edinburgh.

"... an invaluable resource to engineers involved in the design of smart grid ... this book will become an essential reference in the literature of smart grids and smart infrastructures."

Alberto Leon-Garcia, Univeristy of Toronto

Smart Grid Communications and Networking

EKRAM HOSSAIN
University of Manitoba, Canada

ZHU HAN
University of Houston, Texas

H. VINCENT POOR
Princeton University, New Jersey

CAMBRIDGE UNIVERSITY PRESS
Cambridge, New York, Melbourne, Madrid, Cape Town,
Singapore, São Paulo, Delhi, Mexico City

Cambridge University Press
The Edinburgh Building, Cambridge CB2 8RU, UK

Published in the United States of America by Cambridge University Press, New York

www.cambridge.org
Information on this title: www.cambridge.org/9781107014138

© Cambridge University Press 2012

This publication is in copyright. Subject to statutory exception
and to the provisions of relevant collective licensing agreements,
no reproduction of any part may take place without the written
permission of Cambridge University Press.

First published 2012

Printed in the United Kingdom at the University Press, Cambridge

A catalogue record for this publication is available from the British Library

Library of Congress Cataloguing in Publication data
Smart grid communications and networking / [edited by] Ekram Hossain, Zhu Han, H. Vincent Poor.
   p. cm.
  Includes bibliographical references and index.
  ISBN 978-1-107-01413-8 (hardback)
  1. Smart power grids. I. Hossain, Ekram, 1971– II. Han, Zhu, 1974– III. Poor, H. Vincent.
  TK3105.S488   2012
  621.310285′46–dc23   2012008157

ISBN 978-1-107-01413-8 Hardback

Cambridge University Press has no responsibility for the persistence or
accuracy of URLs for external or third-party internet websites referred to in
this publication, and does not guarantee that any content on such websites is,
or will remain, accurate or appropriate.

**For
our families**

Contents

List of contributors		*page* xvii
Preface		xxi

Part I Communication architectures and models for smart grid — 1

1 Communication networks in smart grid: an architectural view — 3

- 1.1 Introduction — 3
- 1.2 Smart grid conceptual model — 5
- 1.3 Smart grid communication infrastructures — 6
  - 1.3.1 Home-area networks (HANs) — 8
  - 1.3.2 Neighbourhood-area networks (NANs) — 8
  - 1.3.3 Wide-area networks (WANs) — 8
  - 1.3.4 Enterprise — 9
  - 1.3.5 External — 9
- 1.4 Interoperability issues — 9
- 1.5 Role of communication infrastructures in smart grid — 12
  - 1.5.1 Customer premises — 12
  - 1.5.2 Core communication network — 15
  - 1.5.3 Last-mile connection — 18
  - 1.5.4 Control centre — 20
  - 1.5.5 Sensor and actuator networks (SANETs) — 21
- 1.6 Security and privacy in the communications infrastructure for smart grid — 23
  - 1.6.1 Component-wise security — 23
  - 1.6.2 Protocol security — 24
  - 1.6.3 Network-wise security — 25
- 1.7 Open issues and future research directions — 26
  - 1.7.1 Cost-aware communication and networking infrastructure — 26
  - 1.7.2 Quality-of-service (QoS) framework — 26
  - 1.7.3 Optimal network design — 27
- 1.8 Conclusion — 27

2 New models for networked control in smart grid — 34

- 2.1 Introduction — 34
- 2.2 Information in today's power system management operations — 35
  - 2.2.1 The management operations in today's power systems — 35
  - 2.2.2 Supervisory control and data acquisition (SCADA) — 37
  - 2.2.3 Basic models for power system controls — 38
  - 2.2.4 Existing power grid controls — 41
  - 2.2.5 The intrinsic difficulties of networked control — 42
- 2.3 Enhanced smart grid measuring functionalities — 43
  - 2.3.1 State estimation — 44
  - 2.3.2 Wide-area measurement system (WAMS) and GridStat — 46
- 2.4 Demand-side management and demand response: the key to distribute cheap and green electrons — 50
  - 2.4.1 The central electricity market — 51
  - 2.4.2 Real-time pricing — 55
  - 2.4.3 Direct load control — 59
  - 2.4.4 Possibilities and challenges at the edge of the network — 60
- 2.5 Conclusion — 61

3 Demand-side management for smart grid: opportunities and challenges — 69

- 3.1 Introduction — 69
- 3.2 System model — 70
- 3.3 Energy-consumption scheduling model — 71
  - 3.3.1 Residential load-scheduling model — 71
  - 3.3.2 Energy-consumption scheduling problem formulation — 72
  - 3.3.3 Energy-consumption scheduling algorithm — 75
  - 3.3.4 Performance evaluation — 76
- 3.4 Energy-consumption control model using utility functions — 77
  - 3.4.1 User preference and utility function — 77
  - 3.4.2 Energy consumption-control problem formulation — 79
  - 3.4.3 Equilibrium among users — 81
  - 3.4.4 The Vickrey–Clarke–Groves (VCG) approach — 84
  - 3.4.5 Performance evaluation of power-level selection algorithms — 86
- 3.5 Conclusion — 88

4 Vehicle-to-grid systems: ancillary services and communications — 91

- 4.1 Introduction — 91
- 4.2 Ancillary services in V2G systems — 92
- 4.3 V2G system architectures — 95
  - 4.3.1 Aggregation scenarios — 97
  - 4.3.2 Charging scenarios — 98
- 4.4 V2G systems communications — 99

		4.4.1	Power-line communications and HomePlug	99
		4.4.2	Wireless personal-area networking and ZigBee	99
		4.4.3	Z-Wave	100
		4.4.4	Cellular networks	100
		4.4.5	Interference management and cognitive radio	101
	4.5	Challenges and open research problems		101
		4.5.1	Fulfilling communications needs	101
		4.5.2	Coordinating charging and discharging	103
	4.6	Conclusion		103

Part II Physical data communications, access, detection, and estimation techniques for smart grid   109

5 Communications and access technologies for smart grid   111

	5.1	Introduction		111
		5.1.1	Legacy grid communications	112
		5.1.2	Smart grid objectives	112
		5.1.3	Data classification	116
	5.2	Communications media		117
		5.2.1	Wired solutions	118
		5.2.2	Wireless solutions	121
	5.3	Power-line communication standards		125
		5.3.1	Broadband power-line communications	126
		5.3.2	Narrowband power-line communications	128
		5.3.3	PLC coexistence	130
	5.4	Wireless standards		131
		5.4.1	Short-range solutions	131
		5.4.2	Long-range solutions	133
	5.5	Networking solutions		136
		5.5.1	Hybrid solutions	136
		5.5.2	Public vs. private networks	137
		5.5.3	Internet and IP-based networking	137
		5.5.4	Wireless sensor networks	139
		5.5.5	Machine-to-machine communications	140
	5.6	Conclusion		142

6 Machine-to-machine communications in smart grid   147

	6.1	Introduction		147
	6.2	M2M communications technologies		150
		6.2.1	Wired vs. wireless	150
		6.2.2	Capillary M2M	152
		6.2.3	Cellular M2M	154
	6.3	M2M applications		156

		6.4	M2M architectural standards bodies	157
			6.4.1 ETSI M2M	158
			6.4.2 3GPP MTC	160
		6.5	M2M application in smart grid	163
			6.5.1 M2M architecture	163
			6.5.2 Transmission and distribution networks	165
			6.5.3 End-user appliances	168
		6.6	Conclusion	171
7		**Bad-data detection in smart grid: a distributed approach**		175
		7.1	Introduction	175
		7.2	Distributed state estimation and bad-data processing: state-of-the-art	176
			7.2.1 Wide-area state-estimation model	176
			7.2.2 Bad-data processing in state estimation	177
			7.2.3 Related work	178
		7.3	Fully distributed bad-data detection	180
			7.3.1 Preliminaries	180
			7.3.2 Proposed algorithm for distributed bad-data detection	181
		7.4	Case study	183
			7.4.1 Case 1	184
			7.4.2 Case 2	187
		7.5	Conclusion	189
8		**Distributed state estimation: a learning-based framework**		191
		8.1	Introduction	191
		8.2	Background	192
		8.3	State estimation model	193
		8.4	Learning-based state estimation	195
			8.4.1 Geographical diversity	195
			8.4.2 Side information	195
			8.4.3 Weighted average estimation	195
			8.4.4 Estimation performance	198
		8.5	Conclusion	198

Part III Smart grid and wide-area networks 203

9		**Networking technologies for wide-area measurement applications**		205
		9.1	Introduction	205
		9.2	Components of a wide-area measurement system	206
			9.2.1 PMU and PDC	206
			9.2.2 Hardware architecture	207

		9.2.3	Software infrastructure	209
	9.3	Communication networks for WAMS		210
		9.3.1	Communication needs	211
		9.3.2	Transmission medium	212
		9.3.3	Communication protocols	213
	9.4	WAMS applications		214
		9.4.1	Power-system monitoring	214
		9.4.2	Power-system protection	217
		9.4.3	Power-system control	221
	9.5	WAMS modelling and network simulations		223
		9.5.1	Software introduction	223
		9.5.2	System infrastructure modelling	223
		9.5.3	Application classification	226
		9.5.4	Monitoring simulation	226
		9.5.5	Protection simulation	228
		9.5.6	Control simulation	229
		9.5.7	Hybrid simulation	230
	9.6	Conclusion		231
10	**Wireless networks for smart grid applications**			**234**
	10.1	Introduction		234
	10.2	Smart grid application requirements		234
		10.2.1	Application types	235
		10.2.2	Quality-of-service (QoS) requirements	235
		10.2.3	Classifying applications by QoS requirements	236
		10.2.4	Traffic requirements	240
	10.3	Network topologies		243
		10.3.1	Communication actors	244
		10.3.2	Connectivity	245
	10.4	Deployment factors		248
		10.4.1	Spectrum	248
		10.4.2	Path-loss	248
		10.4.3	Coverage	249
		10.4.4	Capacity	251
		10.4.5	Resilience	252
		10.4.6	Security	253
		10.4.7	Resource sharing	253
	10.5	Performance metrics and tradeoffs		253
		10.5.1	Coverage area	254
		10.5.2	Capacity	256
		10.5.3	Reliability	258
		10.5.4	Latency	260
	10.6	Conclusion		261

Part IV Sensor and actuator networks for smart grid — 263

11 Wireless sensor networks for smart grid: research challenges and potential applications — 265

11.1 Introduction — 265
11.2 WSN-based smart grid applications — 266
    11.2.1 Consumer side — 267
    11.2.2 Transmission and distribution side — 268
    11.2.3 Generation side — 271
11.3 Research challenges for WSN-based smart grid applications — 272
11.4 Conclusion — 274

12 Sensor techniques and network protocols for smart grid — 279

12.1 Introduction — 279
12.2 Sensors and sensing principles — 280
    12.2.1 Metering and power-quality sensors — 281
    12.2.2 Power system status and health monitoring sensors — 284
12.3 Communication protocols for smart grid — 285
    12.3.1 MAC protocols — 287
    12.3.2 Routing protocols — 290
    12.3.3 Transport protocols — 295
12.4 Challenges for WSN protocol design in smart grid — 297
12.5 Conclusion — 299

13 Potential methods for sensor and actuator networks for smart grid — 303

13.1 Introduction — 303
13.2 Energy and information flow in smart grid — 305
13.3 SANET in smart grid — 306
    13.3.1 Applications of SANET in SG — 307
    13.3.2 Actors of SANET in smart grid — 310
    13.3.3 Challenges for SANET in smart grid — 313
13.4 Proposed mechanisms — 314
    13.4.1 Pervasive service-oriented network (PERSON) — 314
    13.4.2 Context-aware intelligent control — 316
    13.4.3 Compressive sensing (CS) — 316
    13.4.4 Device technologies — 317
13.5 Home energy–management system – case study of SANET in SG — 318
    13.5.1 Energy-management system — 318
    13.5.2 EMS design and implementation — 320
13.6 Conclusion — 321

| 14 | Implementation and performance evaluation of wireless sensor networks for smart grid | 324 |

14.1 Introduction 324
14.2 Constrained protocol stack for smart grid 325
    14.2.1 IEEE 802.15.4 326
    14.2.2 IPv6 over low-power WPANs 327
    14.2.3 Routing protocol for low-power and lossy networks 328
    14.2.4 Constrained application protocol 331
    14.2.5 W3C efficient XML interchange 332
14.3 Implementation 332
    14.3.1 802.15.4 333
    14.3.2 6LoWPAN 333
    14.3.3 RPL 335
    14.3.4 CoAP 336
    14.3.5 EXI 339
14.4 Performance evaluation 339
    14.4.1 Link performance using IEEE 802.15.4 340
    14.4.2 Network throughput with 6LoWPAN 341
    14.4.3 Network throughput with RPL in multihop scenarios 343
    14.4.4 CoAP performance 345
    14.4.5 CoAP multihop performance 347
14.5 Conclusion 348

Part V Security in smart grid communications and networking   351

15 Cyber-attack impact analysis of smart grid   353

15.1 Introduction 353
15.2 Background 354
    15.2.1 Risk management 354
    15.2.2 Prior art 356
15.3 Cyber-attack impact analysis framework 356
    15.3.1 Graphs and dynamical systems 357
    15.3.2 Graph-based dynamical systems model synthesis 358
15.4 Case study 359
    15.4.1 13-node distribution test system 359
    15.4.2 Model synthesis 362
    15.4.3 Attack scenario 1 363
    15.4.4 Attack scenario 2 365
    15.4.5 Attack scenario 3 367
15.5 Conclusion 368

| 16 | **Jamming for manipulating the power market in smart grid** | 373 |

  16.1 Introduction 373
  16.2 Model of power market 375
  16.3 Attack scheme 376
    16.3.1 Attack mechanism 376
    16.3.2 Analysis of the damage 379
  16.4 Defence countermeasures 383
  16.5 Conclusion 384

| 17 | **Power-system state-estimation security: attacks and protection schemes** | 388 |

  17.1 Introduction 388
  17.2 Power-system state estimation and stealth attacks 389
    17.2.1 Power network and measurement models 389
    17.2.2 State estimation and bad-data detection 391
    17.2.3 BDD and stealth attacks 392
  17.3 Stealth attacks over a point-to-point SCADA network 393
    17.3.1 Minimum-cost stealth attacks: problem formulation 394
    17.3.2 Exact computation of minimum-cost stealth attacks 395
    17.3.3 Upper bound on the minimum cost 396
    17.3.4 Numerical results 398
  17.4 Protection against attacks in a point-to-point SCADA network 400
    17.4.1 Perfect protection 400
    17.4.2 Non-perfect protection 401
    17.4.3 Numerical results 401
  17.5 Stealth attacks over a routed SCADA network 403
    17.5.1 Measurement attack cost 404
    17.5.2 Substation attack impact 405
    17.5.3 Numerical results 406
  17.6 Protection against stealth attacks for a routed SCADA network 407
    17.6.1 Single-path and multi-path routing 408
    17.6.2 Data authentication and protection 410
  17.7 Conclusion 410

| 18 | **A hierarchical security architecture for smart grid** | 413 |

  18.1 Introduction 413
  18.2 Hierarchical architecture 415
    18.2.1 Physical layer 418
    18.2.2 Control layer 418
    18.2.3 Communication layer 419
    18.2.4 Network layer 419
    18.2.5 Supervisory layer 419
    18.2.6 Management layer 420
  18.3 Robust and resilient control 420

	18.4	Secure network routing	425
		18.4.1 Hierarchical routing	425
		18.4.2 Centralized vs. decentralized architectures	427
	18.5	Management of information security	429
		18.5.1 Vulnerability management	429
		18.5.2 User patching	430
	18.6	Conclusion	434
19	**Application-driven design for a secured smart grid**		**439**
	19.1	Introduction	439
	19.2	Intrusion detection for advanced metering infrastructures	441
		19.2.1 Smart meters and security issues	442
		19.2.2 Architecture for situational awareness and monitoring solution	443
		19.2.3 Enforcing security policies with specification-based IDS	445
	19.3	Converged networks for SCADA systems	448
		19.3.1 Requirements and challenges for convergence	449
		19.3.2 Architecture with time-critical constraints	450
	19.4	Design principles for authentication	453
		19.4.1 Requirements and challenges in designing secure authentication protocols for smart grid	454
		19.4.2 Design principles for authentication protocols	454
		19.4.3 Use case: secure authentication supplement to DNP3	455
	19.5	Conclusion	458

Part VI Field trials and deployments   463

20	**Case studies and lessons learned from recent smart grid field trials**		**465**
	20.1	Introduction	465
	20.2	Smart power grids	465
		20.2.1 The Jeju smart grid testbed	465
		20.2.2 ADS program for Hydro One	467
		20.2.3 The SmartHouse project	469
	20.3	Smart electricity systems	470
	20.4	Smart consumers	471
		20.4.1 PEPCO	472
		20.4.2 Commonwealth Edison	473
		20.4.3 Connecticut light and power	474
		20.4.4 California statewide pricing pilot	474
	20.5	Lessons learned	475
	20.6	Conclusion	476

Index   478

Contributors

Mahnoosh Alizadeh
University of California Davis, USA

Jesus Alonso-Zarate
CTTC, Barcelona, Spain

Tamer Başar
University of Illinois at Urbana-Champaign, USA

Sara Bavarian
The University of British Columbia, Canada

Robin Berthier
University of Illinois at Urbana-Champaign, USA

Rakesh B. Bobba
University of Illinois at Urbana-Champaign, USA

Nicola Bui
University of Padova, Italy

Karen Butler-Purry
Texas A&M University, USA

Paolo Casari
University of Padova, Italy

Angelo P. Castellani
University of Padova, Italy

Dae-Hyun Choi
Texas A&M University, USA

György Dán
KTH Royal Institute of Technology, Sweden

Yi Deng
Virginia Polytechnic Institute and State University, USA

Mischa Dohler
CTTC, Barcelona, Spain

Nada Golmie
NIST, USA

David Gregoratti
CTTC, Barcelona, Spain

David Griffith
NIST, USA

Vehbi Cagri Gungor
Bahcesehir University, Turkey

Gerhard P. Hancke Jr
Royal Holloway University of London, UK

Gerhard P. Hancke
University of Pretoria, South Africa

Erich Heine
University of Illinois at Urbana-Champaign, USA

Ekram Hossain
University of Manitoba, Canada

Rose Qingyang Hu
Utah State University, USA

Cunqing Hua
Zhejiang University, P. R. China

Jianwei Huang
The Chinese University of Hong Kong, Hong Kong, China

Soummya Kar
Carnegie Mellon University, USA

Nipendra Kayastha
Nanyang Technological University, Singapore

Himanshu Khurana
Honeywell Research Labs, USA

Deepa Kundur
Texas A&M University, USA

Lutz Lampe
The University of British Columbia, Canada

Husheng Li
University of Tennessee, USA

Victor O. K. Li
University of Hong Kong, Hong Kong, China

Hua Lin
Virginia Polytechnic Institute and State University, USA

Salman Mashayehk
Texas A&M University, USA

Javier Matamoros
CTTC, Barcelona, Spain

Amir-Hamed Mohsenian-Rad
Texas Tech University, USA

Dusit Niyato
Nanyang Technological University, Singapore

Arun G. Phadke
Virginia Polytechnic Institute and State University, USA

H. Vincent Poor
Princeton University, USA

Michele Rossi
University of Padova, Italy

Dilan Sahin
Bahcesehir University, Turkey

Pedram Samadi
The University of British Columbia, Canada

Henrik Sandberg
KTH Royal Institute of Technology, Sweden

William H. Sanders
University of Illinois at Urbana-Champaign, USA

Anna Scaglione
University of California Davis, USA

Robert Schober
The University of British Columbia, Canada

Sandeep Shukla
Virginia Polytechnic Institute and State University, USA

Kin Cheong Sou
KTH Royal Institute of Technology, Sweden

Michael Souryal
NIST, USA

Ali Tajer
Princeton University, USA

James S. Thorp
Virginia Polytechnic Institute and State University, USA

Yi Qian
University of Nebraska-Lincoln, USA

Lorenzo Vangelista
University of Padova, Italy

Ping Wang
Nanyang Technological University, Singapore

Zhifang Wang
University of California Davis, USA

Vincent W. S. Wong
The University of British Columbia, Canada

Chenye Wu
Tsinghua University, China

Le Xie
Texas A&M University, USA

Guang-Hua Yang
University of Hong Kong, Hong Kong, China

Tim Yardley
University of Illinois at Urbana-Champaign, USA

Rong Zheng
The University of Houston, USA

Quanyan Zhu
University of Illinois at Urbana-Champaign, USA

Michele Zorzi
University of Padova, Italy

Takis Zourntos
Texas A&M University, USA

Preface

A brief journey through 'Smart Grid Communications and Networking'

A power grid consists of two major parts: the transmission and distribution systems. The transmission system refers to the high-voltage network infrastructure that connects the power generation facilities with the various distribution points. At the distribution points, the electrical carrier is converted to medium and low-voltage signals for the distribution systems that connect the customers. The smart power grid (or *smart grid* in short) refers to the next-generation electrical power grid that aims to provide reliable, efficient, secure, and quality energy generation/distribution/consumption using modern information, communications, and electronics technology. The smart grid will introduce a distributed and user-centric system that will incorporate end-consumers into its decision processes to provide a cost-effective and reliable energy supply. The modern communication infrastructure will play a vital role in managing, controlling, and optimizing different devices and systems in smart grids. Information and communication technologies are at the core of the smart grid vision as they will provide the power grid with the capability to support two-way energy and information flow, isolate and restore power outages more quickly, facilitate the integration of renewable energy sources into the grid and empower the consumer with tools for optimizing their energy consumption.

From an architectural perspective, a smart grid is comprised of three high-level layers: the physical power layer (transmission and distribution), the data transport and control layer (communication and control), and the application layer (applications and services). Each of these high-level layers breaks down further into sub-layers and more detailed market segments. Unlike its predecessor (i.e., the existing electrical power grid), smart grid will use two-way data communication technologies to integrate the utility control system with end-users and consumers, so that intelligent power generation, control, and consumption can be achieved. Moreover, smart grid will allow active participation of users by providing user information related to demand and fault reporting. Many standard bodies and organizations throughout the world are working towards this vision of smart grid. Among many, the Electrical Power Research Institute (EPRI), the National Institute of Standards and Technology (NIST), and European Commission Research (ECR) are working towards developing the most comprehensive frameworks, communication specifications, standards, and roadmaps for the smart grid. However, many issues such as cost, interoperability, cyber and physical security, lack of communication and architectural standards, etc., need to be addressed. Developing the smart grid has become an urgent global priority as its economic, environmental, and societal benefits will be enjoyed by future generations.

The objective of this book is to provide a useful background on advanced data communication and networking mechanisms, models for networked control, and security mechanisms for the smart grid. This book consists of chapters covering different aspects of data communications and networking in the smart grid that include the following: communications architectures and models for smart grid for advanced metering infrastructure (AMI), networked control, demand-side management (DSM), distributed energy resource (DER) management; physical communications, detection, estimation, and access design for smart grid; smart grid and area networks such as home-area networks (HANs), neighbourhood-area networks (NANs), wide-area networks (WANs), wide-area measurement systems (WAMSs); sensor and actuator networks (SANETs) for the smart grid and the related protocol design issues; security in communications infrastructure for the smart grid; and the ongoing projects and field-trials on the smart grid.

This book contains 20 chapters which are organized into six parts. A brief account of each chapter in each of these parts is given next.

Part I: Communication architectures and models for smart grid

A smart grid is a visionary user-centric system that will elevate the conventional power grid system to one that functions more cooperatively, responsively, and economically. In addition to the incumbent function of delivering electricity from suppliers to consumers, smart grids will also provide information and intelligence to the power grid to enable grid automation, active operation, and efficient demand response. A reliable and efficient communication and networking infrastructure will connect the functional elements within the smart grid.

In *Chapter 1*, Kayastha et al. describe the conceptual model for a smart grid adopted by NIST, and describe the interactions among its different domains (e.g., generation, transmission, distribution, customer, service provider, operations, market). In this context, the authors highlight the role and importance of smart grid communications and networking infrastructures, and present an overview of a hierarchical communication infrastructure which spans across the different domains in a smart grid. Such an infrastructure, which is also termed an AMI, comprises many systems and subsystems such as HANs, SANETs, NANs, and WANs. The authors also briefly describe the GridWise Architecture Council (GWAC) framework for interoperability in the integrated smart grid communications infrastructure. In addition, security and privacy issues related to the communications infrastructures in the smart grid are also reviewed.

In *Chapter 2*, Scaglione, Wang, and Alizadeh provide a brief overview of the classical issues of network control and how they relate to the challenges of creating a new architectural model for managing energy distribution in a smart grid that relies on real-time, dependable information gathering and decisions. The authors discuss the important questions that exist in tightening the networked control at the core of the network and at its edges and why these are important parts to unleash innovations in the smart grid.

They discuss how wide-area measurement systems connecting phasor measurement units (PMUs) through novel sensor networking paradigms can help increase the situation awareness in the smart grid. The authors also review the supervisory control and data acquisition (SCADA) model which is currently used for grid monitoring and control. At the edge, the emerging smart metering infrastructure today offers only a glimpse of the possible advantages of having broad consumer participation. The opportunity is to tighten the control of the demand via real-time load scheduling. The authors discuss what are reasonable models for demand and response systems, also referred to as DSM systems, that proactively control smart loads, focusing on the specific example of an electric vehicle, as a compelling case to target for the study of load scheduling.

In *Chapter 3*, Samadi *et al.* present a number of methods for DSM based on smart pricing to improve the efficiency of traditional power grids. Two different objectives for such algorithms are: reducing power consumption and shifting (or scheduling) power consumption. Energy-consumption scheduling can reduce the peak-to-average ratio (PAR) of power consumption as well as minimize the total energy cost in the system. For users, another objective could be to minimize jointly the energy cost and waiting time. The authors consider these design objectives for DSM and present optimization and game-theoretic models to solve the DSM problem. The concept of utility functions is used to model different objectives of users.

In *Chapter 4*, Wu, Mohsenian-Rad, and Huang provide an introduction to vehicle-to-grid (V2G) systems and highlight the role of a reliable and secure communication and networking infrastructure for such systems in the future smart grid. A V2G system can inject power into the grid when required through discharging the batteries of plug-in electric vehicles (PHEVs). Such a system can improve the PAR in the system through a coordinated charging and discharging mechanism for the PHEVs. Also, the V2G power storage mechanism can facilitate integration of renewable energy (RE) sources into the smart grid. In addition, a V2G system can help to regulate frequency and voltage in a power grid. All of these services, which are referred to as ancillary services, can be offered to the power grid efficiently through an advanced communication and networking infrastructure. The authors briefly describe several technologies for V2G system communications which include broadband power-line communication (PLC), ZigBee, Z-Wave, cognitive radio, and cellular wireless technologies. The details of these technologies are discussed in Part II of the book.

Part II: Physical data communications, access, detection, and estimation techniques for smart grid

Different physical data communication technologies for the smart grid will empower the legacy power grid with the capability to support two-way energy and information flow. These technologies will facilitate integration of renewable energy sources into the grid, and empower the consumers with tools to optimize energy consumption. The smart grid will rely on several existing and future wired and wireless communications technologies (e.g., PLC, cellular network, IP networks, ZigBee, Wi-Fi, WiMAX, etc.). Also, advanced techniques for power-system state estimation and data processing (e.g., bad-data detection) will be required for smart grids.

In *Chapter 5*, Bavarian and Lampe provide an exposition on the different communications and access technologies and their applications in smart grid communications. Different wired communications technologies including power-line and optical-fibre technologies, and wireless technologies including cellular, satellite, wireless mesh, and wireless personal-area networking technologies are reviewed. Broadband and narrow-band power-line communications technologies and the related standards (e.g., IEEE 1901, ITU-T G.9960/61, HomePlug) are discussed. Among the wireless technologies, the authors discuss the ZigBee, Wi-Fi, WiMAX, 3GPP LTE, and IEEE 802.22 standards. To this end, the authors also review networking solutions such as Internet and IP-based networks, private networks, wireless sensor and machine-to-machine (M2M) communication networks for smart grids.

In *Chapter 6*, Alonso-Zarate *et al.* review the emerging paradigm of M2M communications, including its definition, historical developments, design drivers, and the status-quo of its standardization efforts. The authors discuss in detail the applicability of the M2M communications to the smart grid and identify open challenges for a symbiotic development of both M2M and smart grid technologies. Different M2M communications technologies including cabled technologies (e.g., PLC, Ethernet), low-power wireless technologies such as ZigBee, Wi-Fi, 6LoWPAN (which are referred to as capillary M2M technologies), and hybrid M2M technologies are discussed. The authors argue that the cellular M2M communications technologies are suitable for smart grid applications such as wide-area situational awareness, interconnection of distributed energy resources, and distribution automation in the transmission and distribution networks. Also, cellular M2M is a technology enabler to build the AMI, and to realize the concept of direct load control (DLC) where intelligent devices can automatically schedule their power loads.

In *Chapter 7*, Xie *et al.* focus on the problem of fast and robust state-estimation techniques for wide-area monitoring, control, and protection in the smart grid. One essential functionality in state estimation is to detect, identify, and eliminate measurement errors, which arise due to the existence of large measurement bias, drifts, or wrong connections. This functionality is referred to as 'bad-data processing', which consists of two steps: bad-data detection and identification. Generally, a chi-square test is used for bad-data detection, and then a normalized residual test is used for bad-data identification. The authors review the state-of-the-art of bad-data processing techniques and present a distributed approach for bad-data detection. The performance of the proposed approach is observed by simulations using the IEEE 14-bus system. The information exchange and communication requirements for the proposed approach are also discussed.

In *Chapter 8*, Tajer, Kar, and Poor also deal with the problem of distributed power-system state estimation taking into account the uncertainties in the underlying physical and sensing models as well as the rapidly varying dynamics of the system. The authors define a learning-based framework for adaptive and distributed power-state estimation. They model the smart grid as a collection of multiple overlapping distributed subnetworks (or clusters) covering the entire network. The subnetworks share their estimates of network state with a central decision-maker entity (central estimator) through a backbone communication network. Then the central estimator combines the local state estimates to obtain the global state of the network. The estimation performance at the central estimator, as well as the estimation quality in each cluster, are modelled analytically using cost functions.

Part III: Smart grid and wide-area networks

Advanced data communication and networking techniques will play a key role in the successful development of the emerging smart grid system. The communication network in the smart grid must be able to support all aspects of generation, transmission, distribution, as well as the requirements of users and utility service providers. The data communication network in the smart grid will be responsible for sensing (i.e., gathering real-time measurements from various locations of the power grid through a WAMS), communication (i.e., bidirectional data exchange between smart meters and control centres), and control (i.e., delivery of control messages to ensure optimal, reliable, and resilient operation of the grid and its subsystems).

In *Chapter 9*, Deng *et al.* focus on the performance evaluation of network architectures and protocols for WAMS applications in the smart grid. The authors review the WAMS architecture (software and hardware) and the different components of WAMS, namely, the PMUs, regional phasor data concentrators (PDCs), centralized super-phasor data concentrator (SPDC), and hierarchically organized communication networks. A WAMS uses a multi-level hierarchical communication network with reliability, real-time responsiveness, scalability, and reliability, to integrate all these components together. The applications of WAMS for power-system monitoring, protection, and control are discussed in detail. A simulation platform based on the OPNET Modeler is designed for a realistic communication system of WAMS and simulation results are obtained for various control, monitoring, and hybrid WAMS applications.

In *Chapter 10*, Griffith, Souryal, and Golmie focus on the use of wireless networks to support the communications quality-of-service (QoS) and traffic requirements of different smart grid applications. These applications include firmware/program update (FPU), field distribution automation maintenance-centralized control (FDAMC) for communications between the distribution management system and various field devices, PHEV messaging, customer information/messaging (CMSG), and meter reading. The QoS requirements (e.g., latency and reliability) and the traffic characteristics of these applications, and also the message flows among the various actors for these applications and the resulting network topologies are discussed. The key factors such as the choice of radio spectrum, wireless channel propagation characteristics, wireless link coverage, and network capacity, resilience and security, which need to be considered for the deployment of wireless networks, are described. In this context, performance metrics such as coverage, capacity, reliability, and latency, which can be used to evaluate different wireless network alternatives, are also discussed.

Part IV: Sensor and actuator networks for smart grid

In a smart grid, wireless SANETs will be deployed in generation systems, transmission and distribution systems, and consumers' premises to monitor and control the functioning of the grid. The existing and potential applications of SANETs in the smart grid include advanced metering, fault diagnosis, demand response and dynamic pricing, energy management, etc. SANETs will be an integral component in future generation smart grids. However, the existing communication protocols for SANETs may need to be modified/optimized taking into consideration the smart grid application requirements.

In *Chapter 11*, Sahin et al. present the potential applications of wireless sensor networks (WSNs) in the smart grid and the related technical challenges. In particular, WSN-based applications have been described for power generation systems, transmission and distribution networks, and consumer facilities. For WSN-based smart grid applications, a number of research challenges exist which involve power, data, and resource management in sensors, interoperability among WSN protocols, QoS provisioning in the network, and system integration.

In *Chapter 12*, Zheng and Hua focus on the sensor technologies and communication protocols for sensor networks in the smart grid. The authors review major types of sensors which are categorized into metering and power-quality sensors and power-system status and health-monitoring sensors. In this context, different sensing principles, which are used to convert the physical parameters into electronic signals, are reviewed. The authors discuss the issues related to designing medium access control (MAC), routing, and transport protocols for WSNs in the smart grid. A brief survey on the existing protocols for general WSNs, along with a qualitative comparison among the different protocols, are also provided. The authors point out that designing sensor networking protocols for the smart grid is challenging due to the unique features of such systems; for example, the complex and heterogeneous nature of the environment, dynamic nature of the system, reliability, availability, and diverse QoS requirements, energy and cost-efficiency, and scalability and security issues.

In *Chapter 13*, Li and Yang focus on addressing the major design challenges of SANETs in smart grids as mentioned before. The authors propose mechanisms such as pervasive service-oriented networking, context-aware intelligent control, compressive sensing, and advanced device technologies (e.g., with low-power, modular, and compact design and power-harvesting mechanisms) to address the challenges. To this end, the effectiveness of the proposed mechanisms is demonstrated with a case study of a home energy-management system (HEMS).

In *Chapter 14*, Bui et al. focus on the implementation and performance evaluation of WSN protocols for smart grid applications in a test-bed built from off-the-shelf wireless sensors. In particular, the authors consider the protocol stack of the 'Internet of things' with IEEE 802.15.4 protocols at the physical (PHY) and MAC layers, 6LoWPAN (IPv6 over low-power wireless personal-area networks) at the routing layer, and CoAP (Constrained Application Protocol) at the application/session layer. The implementation of the test-bed is discussed along with the different optimization techniques used for the network and software implementations. The experimental results for the different layers of the protocol stack are presented. The authors conclude that WSN solutions based on the 'Internet of things' protocol stack are feasible to be integrated with the smart grid.

Part V: Security in smart grid communications and networking

Although the communication infrastructure can considerably improve the efficiency of the power system, it brings significant vulnerability since malicious users can attack the communication system and thus cause various damages to the smart grid, or even result in a large-area blackout. Hence, security is of high priority in the study of smart grids and has attracted substantial attention in industry and academia. We have five chapters which discuss the security issues in the smart grid from different perspectives.

In *Chapter 15*, Kundur *et al.* present a framework for cyber-attack impact analysis in the smart grid. First, background is provided to motivate and introduce fundamental research and development questions on cyber-attack impact analysis. Second, a graph-theoretic dynamical system approach is employed to model the interactions between the cyber and electricity networks in the model synthesis stage. Finally, a test case study is presented to demonstrate the potential for modelling.

In *Chapter 16*, Li proposes a jamming-based attack scheme for manipulating the power market in the smart grid. By intelligently blocking and releasing the information in the power market via jamming the wireless communications, malicious jammers/attackers can manipulate the power price, thus making profit for themselves and causing damage to the power grid. To combat this attack, random frequency hopping can be employed for communication, and a random backoff method is proposed for load adjustment in order to avoid the impulsive impact on the market price and power load due to jamming.

In *Chapter 17*, Dán, Sou, and Sandberg study bad-data injection attacks on state estimation in the smart grid using SCADA systems. State estimation is used to estimate the complete physical state of the power system, and bad-data detection is used to identify faulty equipment and corrupted measurement data. A stealth attack against bad-data detection is investigated, and several algorithms are used to protect the power system against this attack. A realistic model is added for communication of the supervisory control and data acquisition systems. Some new protection mechanisms are also presented.

In *Chapter 18*, Zhu and Başar describe a cross-layer architecture to address security issues in the smart grid. The tradeoff between information assurance and the physical layer system performance is investigated by three security issues at different layers: the resilient control design problem at the physical power plant, the data-routing problem at the network and communication layer, and the information security management at the application layers. The proposed hierarchical model extends the open system interconnection (OSI) and Purdue Reference models for their integration into smart grids.

In *Chapter 19*, Berthier *et al.* discuss an application-driven design approach that builds the large cyber security toolset. A key element is careful enumeration of the control-system-specific aspects of each system and an integrated study of these aspects, cyber security properties, and solutions. Specifically, the following topics are discussed in detail: intrusion detection for advanced metering infrastructure, converged networks for supervisory control and data acquisition, and design principles for authentication of SCADA protocols.

Part VI: Field trials and deployments

The relevance of smart grid is reflected by the increasing number of national and international projects on this topic as well as new initiatives by standardization bodies and organizations such as NIST, EPRI, ECR, and the IEEE. There have been several smart grid field trials in the last few years.

In *Chapter 20*, Hu and Qian provide an overview of several smart grid field trials which are divided into three categories: smart power grids, smart electricity systems, and smart customers. The first category includes the Jeju smart grid testbed in Korea, the advanced distribution system (ADS) programme in Ontario, Canada, and the SmartHouse project in Europe. The second category includes an intelligent protection relay system for smart grids. The third category includes several dynamic pricing schemes. The authors summarize the lessons learned from these pilot projects.

Part I

Communication architectures and models for smart grid

1 Communication networks in smart grid: an architectural view

Nipendra Kayastha, Dusit Niyato, Ping Wang, and Ekram Hossain

1.1 Introduction

The existing electrical grid needs to be smarter in order to provide an economical, reliable, and sustainable supply of electricity [1]. Although the current electrical grid has served well in providing the necessary power supply of electricity, the growing demand, fast depletion of primary energy resources, unreliability, and impact on the environment must be responded to in a vision of the future [2]. This vision is being realized using *smart grid*, which is a user-centric system that will elevate the conventional electrical grid system to one that functions more cooperatively, responsively, economically, and organically [1].

One of the most important features of smart grid technology that makes it smart or smarter than the current grid is the integration of bi-directional flow of information along with electricity, which can be used to provide effective and controlled power generation and consumption [3]. This two-way flow of information in turn enables active participation of consumers, thus empowering them to control and manage their own electricity usage by providing near real-time information on their electric consumption and associated cost. Due to this overlaid communication infrastructure, smart grid will incorporate into the grid the benefits of distributed computing and communications, which would provide the necessary intelligence to instantaneously balance the supply and demand at the device level. Clearly, modern communication and information technology will play an important role in managing, controlling, and optimizing different functional and smart devices and systems in a smart grid. A flexible framework is required to ensure the collection of timely and accurate information from various aspects of generation, transmission, distribution, and user networks to provide continuous and reliable operation [4]. Therefore, the existing and future data communications protocols will have to evolve with the developing smart grid taking into consideration the characteristics of the electrical systems.

The transition to smart grid will require considering and modifying many components and technologies of the current electrical grid. A proper vision could be to understand the specific requirements and characteristics of this future grid. This vision will not only help to identify different inefficiencies in the current electrical grid, but also help to set the required foundation for the transition. The main characteristics and benefits of smart grid are summarized in Table 1.1 [5].

Table 1.1. Smart grid characteristics and benefits

Characteristic	Benefits
1. Self-healing	Capability to rapidly detect, analyse, respond, and restore from fault and failures
2. Consumer friendly	Ability to involve a consumer in the decision process of electrical power grid
3. High reliability and power quality	Ability to supply continuous power to satisfy consumer needs
4. Resistance to cyber attacks	Ability to be immune and to protect the system from any cyber and physical attacks
5. Accommodates all generation and storage options	Ability to adapt to a large number of diverse distributed generation (e.g., wind energy and renewable energy) and storage devices deployed to complement the large power generating plants
6. Optimization of asset and operation	Ability to monitor and optimize the capital assets by minimizing operation and maintenance expenses
7. Enables markets	Offering new consumer choices such as green power products and new generations of electric vehicles, which lead to reduction in transmission congestion

With smart grid, a number of technical and procedural challenges emerge [6]. On the technical side, for example, communication systems must be secured and reliable enough to handle different and new media technologies as they emerge. In addition, smart equipment (e.g., computer-based or microprocessor-based) and data-management techniques must be robust and scalable to handle any existing and future applications. Finally, the new smart grid technologies must be interoperable with the existing electrical grid. On the procedural side, efforts to establish the interoperability framework must consider a broad set of smart grid stakeholders, as every person and business will be affected by this technology. Thus, the shift towards smart grid will be an evolutionary process such that it will follow incremental development. Action plans should be developed to align all the stakeholders of smart grid, from researcher to industry to government, in a direction that ensures an orderly transition to visionary smart grid [7].

In this chapter, we provide a comprehensive study on the importance of communication infrastructure and networking in smart grid. Section 1.2 reviews the conceptual model for smart grid. Section 1.3 describes the importance of communication infrastructure in an attempt to provide better understanding of smart grid hierarchical landscape and its association with the conceptual model. Section 1.4 describes the interoperability issues in smart grid, the GridWise interoperability context-setting framework, and its association with the Open Systems Interconnection (OSI) 7-layer communication model. Section 1.5 outlines the role of communication infrastructure in various stages of smart grid, and provides an overview of the existing communication technologies. The importance of security and privacy is summarized in Section 1.6. Section 1.7 highlights some of the critical issues that still need further investigation. Section 1.8 concludes the chapter.

1.2 Smart grid conceptual model

This chapter considers the smart grid conceptual model adopted by the National Institute of Standards and Technology (NIST) [3], which is followed by many standards such as the Electrical Power Research Institute (EPRI) [8], European Commission Research (ECR) [2], and International Electrotechnical Commission (IEC) [9] as the basis for describing, discussing, and developing the final architecture of the smart grid. The conceptual model not only identifies different smart grid stakeholders, but also provides various electrical and communication interfaces required to understand various interoperability frameworks. For this purpose, NIST has divided the smart grid into seven domains, as shown in Figure 1.1 and described in Table 1.2.

Each domain is comprised of a group of *actors* and *applications*, as shown in Figure 1.1. The actors are typically devices, systems, or programs that make decisions and exchange information through a variety of interfaces in order to perform applications and processes [3]. The applications are various tasks performed by an actor or actors within a certain domain. The domains are able to communicate with one another via communication interfaces, as shown in Figure 1.1 and Table 1.2. This communication

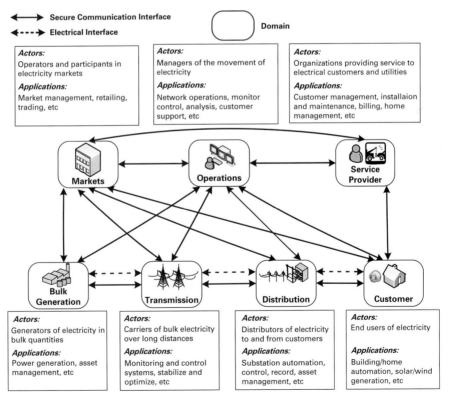

Figure 1.1 Smart grid conceptual model showing interactions among different smart grid domains through secure communication and electrical interfaces.

Table 1.2. Smart grid domains and the associated communication and electrical interfaces with typical applications

Domain	Communication interface	Electrical interface
Customer	Distribution, markets, operations, service provider	Distribution
Distribution	Customer, transmission, market, operations	Customer and transmission
Transmission	Distribution, bulk generation, markets, operators	Bulk generation and distribution
Bulk generation	Transmission, markets, operations	Transmission
Markets	Customer, distribution, transmission, bulk generation, operations, service providers	None
Operations	Customer, distribution, transmission, markets, service providers	None
Service providers	Customer, markets, operations	None

is critical to the overall interoperability of the smart grid, allowing it to collectively generate and distribute electricity efficiently based on the input from all domains.

1.3 Smart grid communication infrastructures

A smart grid can be considered as a network of many systems and subsystems which are interconnected intelligently to provide cost-effective and reliable energy supply for increasing demand response [3]. Moreover, smart grid will be achieved by overlaying the communication infrastructure with an electrical system infrastructure. The application of advanced communication techniques is expected to greatly improve the reliability, security, interoperability, and efficiency of the electrical grid, while reducing environmental impacts and promoting economic growth [3]. Furthermore, in order to achieve enhanced connectivity and interoperability, smart grid will require open-system architecture as an integration platform, and commonly shared technical standards and protocols for communications, and information systems to operate seamlessly among the vast number of smart devices and systems. This makes it hard to realize a single and composite architecture. In fact, smart grid can contain many system architectures developed independently or in association with other systems.

Figure 1.2 shows a hierarchical overview of the smart grid landscape, its relation to NIST domains, and associated examples of components and technologies. Ideally, each of the domains, members, and technologies would interact with each other to provide any of the smart grid's business, technological, and societal goals [10].

This interaction is made possible using *advanced metering infrastructure* (AMI), which will act as the gateway for access, enabling the bi-directional flow of information and power in support of distributed energy resource (DER) management or distributed generation (DG) and consumer participation [1]. There is no standard document defining

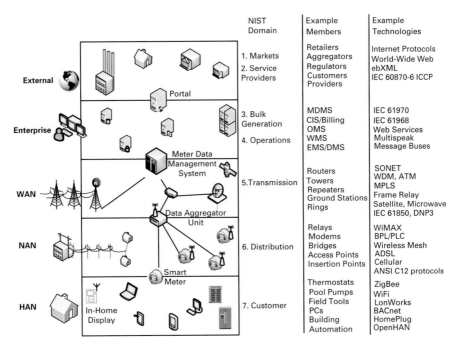

Figure 1.2 Hierarchical overview of smart grid communication infrastructure.

AMI, and it is still open to new development and implementation. Some efforts have been made on behalf of the New York State Electric and Gas Company (NYSEG) [11] to define the system. Following the NYSEG architecture, an AMI consists of several different components such as smart meter, hierarchical networks, and other subsystems shown in Figure 1.2.

The most important requirement of AMI is to provide near real-time metering data including fault and outage to the utility control centre. For this, smart meter will be an integral part of AMI which will support efficient outage management, dynamic rate structures, customer billing, and also demand response for load control [11]. Further, AMI will include a hierarchical network or a multi-tier architecture with star and mesh topologies and a variety of communications technologies such as power-line communication (e.g., broadband over power line (BPL)), cellular network (e.g., GSM and CDMA), other wireless technologies (e.g., Wi-Fi, ZigBee, and worldwide interoperability for microwave access (WiMAX)), and Internet Protocol (IP)-based networks. AMI will comprise local data aggregator units (DAUs) to collect and relay the information from the smart meter to the meter data-management system (MDMS). MDMS will provide storage, management, and processing of meter data for proper usage by other power system applications and services. Also, many systems and subsystems such as wide-area measurement systems (WAMSs), sensor and actuator networks (SANETs) will be grouped under a hierarchical structure based on home-area networks (HANs), neighbourhood-area networks (NANs), and wide-area networks (WANs).

1.3.1 Home-area networks (HANs)

A HAN (sometimes referred to as a premise-area network (PAN) or a building-area network (BAN)) is an important and the smallest subsystem in the hierarchical chain of smart grid as shown in Figure 1.2 [12]. HAN provides a dedicated demand-side management (DSM), including energy efficiency management and demand response by proactive involvement of power users and consumers [3]. HAN consists of smart devices with sensors and actuators, in-home display, smart meter, and home energy-management system (HEMS). HEMS helps to manage the energy consumption in the household.

The HAN communicates with different smart devices using wireline technologies including power-line communication (PLC), or BACnet protocol [13], and wireless technologies (e.g., Wi-Fi and ZigBee [14]). Wireless technology such as ZigBee is becoming a popular choice in contrast to wireline technology due to its low installation cost and better control and flexibility. ZigBee is an open-standard low-power wireless protocol and by far the most popular IEEE 802.15.4 networking standard that meets most of the criteria defined in the OpenHAN system requirement specification (SRS) [15]. The OpenHAN SRS is formed by the OpenAMI task force under the UCA International Users Group (UCAIug) [16] to facilitate interconnection among various appliances in HAN as well as with external networks (e.g., neighbourhood-area network or NAN) using energy service interface (ESI). This ESI acts as a gateway for connecting HAN with other external networks and thus can be referred to as a HAN gateway.

1.3.2 Neighbourhood-area networks (NANs)

A NAN connects multiple HANs and one or more networks between the individual service connections for distribution of electricity and information. As shown in Figure 1.2, all the data from HAN are collected to the data-aggregator unit (DAU). The NAN consists of HAN with smart meters to provide secure and seamless control of different home appliances. DAU consists of a NAN gateway to interface with the HAN and also with the WAN. The DAU communicates with the HAN gateway using network technologies such as PLC, ANSI C12 protocols, WiMAX, or ZigBee. The NAN acts as an access network to forward customer data to the utility local office [15].

1.3.3 Wide-area networks (WANs)

A WAN connects multiple distribution systems together and acts as a bridge between NANs and HANs and the utility network. As shown in Figure 1.2, WAN provides a backhaul for connecting the utility company to the customer premises. In this case, a backhaul can adopt a variety of technologies (e.g., Ethernet, cellular network, or broadband access) to transfer the information extracted from the NAN to the utility local offices [15]. A WAN gateway can use broadband connection (e.g., satellite) or possibly an IP-based network (e.g., MPLS and DNP3) to provide an access for the utility offices to collect the required data. Since information privacy and reliability are the major concerns for the customer, security and fault tolerance of these communication technologies are crucial issues.

1.3.4 Enterprise

Enterprise is the higher-level entity in the smart grid hierarchy which is responsible for processing and analysing all the data collected from various hierarchy levels as shown in Figure 1.2. These distribution systems consist of measurement systems (e.g., supervisory control and data acquisition (SCADA) and wide-area measurement system (WAMS)) to monitor and control the entire electrical grid. WAMS consists of a control centre, phasor measurement units (PMUs), and phasor data collectors (PDCs). The information is acquired synchronously using a global positioning system (GPS)-enabled PMU or synchrophasor which measures the electrical waves on an electrical grid to determine the status of the system. WAMS can be considered as a synchronous version of the conventional SCADA system to collect information about different system components in the power grid. PMU is considered to be one of the most important measuring devices in future power systems [17]. The PDC and PMU are connected in a star network topology, whereas the control centre is connected to the PDC using a wide-area network (e.g., synchronous digital hierarchy (SDH)) [18]. Various applications such as meter data-management system (MDMS), outage-management system (OMS), energy-management system (EMS), distribution-management system (DMS), customer-information system (CIS), and billing are performed at this level.

1.3.5 External

All the retailers, regulators, and providers related to price exchanges, supply and demand support the business process of the electrical system comprising the external level. Communication to and from the external market and service providers should be reliable to match electric production with consumption. Also, various business processes such as billing and customer account management help to enhance customer services, such as the management of energy use and home energy generation. This external level provides new and innovative services and products to meet the new requirements and opportunities provided by the evolving smart grid.

1.4 Interoperability issues

Smart grid will be an interoperable system since it will constitute various communication networks and systems [3]. In this context, interoperability becomes a crucial issue, which allows the information and infrastructure to come together into an interoperable and integrated system for information to flow and be exchanged without user intervention. The most important objective of interoperability is to provide plug-and-play capability, where the component/system automatically configures itself and begins to operate by simply plugging into the main system. Although the concept is simple, achieving plug-and-play capability is not easy and in many situations becomes complex and rather impractical to specify a standard interface between two different systems. For example, consider specifying a customer interface to HEMS. HEMS may use different software tools and protocols to manage the energy consumption in the household. Conventionally, integrating these software tools and protocols will require some manual changes and upgrades so that the interface agreements can be satisfied. Moreover, the interoperability issue increases as the effort to make and test these changes increases. However, standards or best practices such as using a common semantic model (e.g., XML) that a community of system integrators readily understand can decrease the interoperability problems. Thus, by reaching agreements in a specific area of interoperation, a community can improve system integration and effort to achieve interoperation to some extent.

Improving interoperability not only reduces installation and integration costs, but also provides well-defined points in a system in which this interoperability allows for new automation components to connect to the existing system. This can enable substitutability where one automated component can be substituted by other components with a reasonable amount of effort such that the overall integrity of the system is preserved. This substitutability characteristic will provide necessary scalability to the electrical system such that it can evolve to satisfy changing resources, demands, and more efficient technologies. In order to visualize the necessary interoperability issues, this chapter follows the high-level categorization approach developed by the GridWise Architecture Council (GWAC) [19]. Referred to as the *GWAC stack*, the GridWise interoperability context-setting framework identifies eight interoperability categories that are relevant to integration and interoperation of different systems in smart grid. The *GWAC stack* groups these eight categories into three broad types as follows [19]:

- *Organizational* emphasizes the pragmatic (business and policy) aspects of interoperation, those pertaining to the management of electricity. Three layers, namely economic/regulatory policy, business objectives, and business procedures form this category.
- *Informational* emphasizes the semantic aspects of interoperation, focusing on what information is exchanged and its meaning. Business context and semantic understanding layers form this group.
- *Technical* emphasizes the syntax or format of the information, focusing on how information is represented within a message exchange and on the communication medium. Syntactic interoperability, network interoperability, and basic connectivity layers form this group.

The most important feature of the *GWAC stack* is that each layer defines a specific interoperability issue such that establishing interoperability at one layer can enable flexibility at other layers. This means that each layer depends upon, and is enabled by, the layer below it. This chapter focuses on the technical driver and its associated layers, which define the communication networking and syntax issues of smart grid interoperability. Figure 1.3 shows the association of the GWAC stack technical driver with the layers in the Open Systems Interconnection (OSI) 7-layer communication model [20],

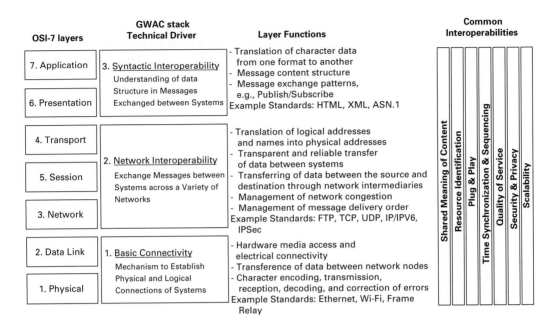

Figure 1.3 Technical interoperability layered category.

the layer functions and common interoperability issues pertaining to overall layers. Note that the common interoperability issues are relevant to more than one interoperability category of the framework and may have concerns that involve aspects at technical levels, informational levels, as well as organizational level.

Based on the GWAC interoperability context-setting framework and common issue areas for interoperability, we highlight the following interoperability issues which are important from a communication point of view [19]:

- *Shared meaning of content.* For effective communication in all interoperability categories, the meaning of content should be interpreted in context both correctly and with clarity. That is, a common semantic understanding is necessary among the content shared between multiple emerging vendors and companies.
- *Resource identification.* This is used to identify resources such as generator, electric appliances, or any functional node so that they can be identified autonomously by all automation components that need to interact. Resource identification is crucial in interoperation since smart grid requires those resources to measure performance metrics and to intelligently control whenever necessary.
- *Plug-and-play.* An important aspect of systems composed of different systems and communication networks such as smart grid is how they are configured so the automation components interact properly once made operational. In this regard, plug-and-play capability is the key to the interoperability issues.
- *Time synchronization and sequencing.* The electrical system is a high-speed, real-time system that reacts very quickly to the disturbances and load shifts.

Thus, systems that monitor and control (e.g., WAMS) must maintain a specific time precision and synchronization for acquiring real-time fault reports. Failure in achieving time synchronization and sequencing may result in a catastrophic effect.

- *Security and privacy.* Since smart grid consists of a number of devices and systems, the point of interface connecting them can be vulnerable to security threats. Security and privacy include aligning security policies such as user, application, and system authentication and authorization. In case of smart grid, defining these policies while allowing multiple systems to interconnect is the main interoperability concern regarding security and privacy.
- *Quality-of-service (QoS).* For any distributed process, QoS signifies the performance and reliability requirements such as response latencies, transaction throughput, etc. In this regard, each automation component and the communication infrastructure should meet the QoS requirement within its portion of the process to provide reliable information exchange. Thus, the implication of QoS requirements should be specified in advance in the collaboration agreements to minimize any faults that may arise.
- *Scalability.* In an electrical system, any upgrades or maintenance should not disturb the overall operation of the system. In this regard, an upgrade path needs to be put forth that allows older (legacy) versions to work with newer (emerging technology) versions of automation interfaces. In addition, as the system evolves, it must have the capability to scale over time to meet the anticipated growth projections.

1.5 Role of communication infrastructures in smart grid

Smart grid requires dynamic architecture, intelligent algorithms, and efficient mechanisms to be developed to meet its new requirements. For this, a wide range of enabling technologies in areas such as integrated communications, sensing and measurement, advanced component, advanced control, and improved interface and decision support must be put into operation [1]. Of these key technology areas, the implementation of integrated communications is the driving factor that will create a dynamic and interactive infrastructure to integrate all upstream (towards the generator) and downstream (towards the consumer) components to work in a unified fashion [22]. A complete communication infrastructure of smart grid highlighting the communication core network and last mile connection is shown in Figure 1.4 [21, 46]. The following subsections focus on the role of data communication in various domains of smart grid. The different research issues and the related approaches in the literature are discussed.

1.5.1 Customer premises

Customer premises refer to home, building, and industry, which are the end users of the electric hierarchical chain. The foremost concern for these customers will be to manage

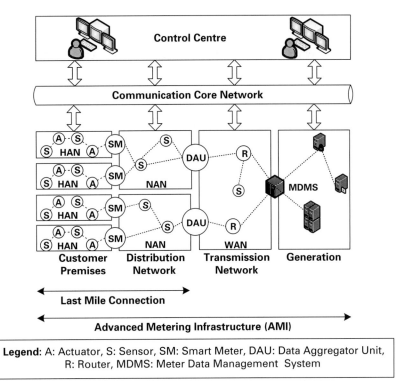

Figure 1.4 The overall communication infrastructure for smart grid.

and reduce their energy usage. The energy-management system (EMS) is emerging as a potential solution for resolving unnecessary electric usage. The EMS acts as a communication gateway to the customer's premises, as it will provide a means of reducing energy consumption by monitoring and controlling different electrical appliances using various sensors, actuators, and communication technologies. Clearly, integrated communication will play an important role in EMS to provide more robust control over different electrical appliances.

The EMS operated using HAN is referred to as a home energy-management system (HEMS). HEMS acts as a subset of EMS and, together with smart meter, provides the necessary interface to the HAN. In fact, there is a growing interest in this approach such as Google (Google powermeter [23]), Microsoft (Microsoft Hohm [24]), and Apple [25], which provide a portal to track users' electricity usage. Generally, HEMS (or EMS) sets certain user limit thresholds based on the information about real-time price-responsive load management and consumption history to control the energy usage of appliances [26]. Therefore, HEMS should be integrated into HAN using available physical communication design which is reliable (i.e., continuous and real-time), cost-effective (e.g., does not require new wiring or high additional cost for installation), interoperable, effective, and dynamic.

Various communications technologies/protocols such as broadband over power-line (BPLC) and wireless technology (e.g., WiMAX and 3G/4G cellular) can be used for backhaul systems. Low-power protocols such as PLC, ZigBee, OpenHAN, or BACnet are used for HANs and also for HEMSs. PLC is already being used in the current electrical power grid. However, narrow-band PLC technology (e.g., FSK and PSK) suffers from noise and also cannot coexist with spread spectrum technology [27]. That is why much effort is made to design efficient low-cost communication protocols to support the HEMS. For instance, dispersed-tone power line communication (DTPLC) [27] that offers plural tones to transmit baseband data by selecting frequencies can be used to avoid narrow-band noise. Similarly, technologies such as KNX [28], which is an OSI-based network communication protocol standard, can be used to perform real-time measurement on the main power socket to provide a physical interface of the energy-management device towards appliances [29]. Also, low-cost, low-power, short-range wireless communications technology such as ZigBee [14] can be used to provide the necessary control and monitoring of various electrical appliances in HAN. The choice of communication technology for HEMS will depend on many factors such as cost, robustness, capacity, and availability.

HEMS has to be integrated into HAN in such a way that it provides efficient energy management within HAN and also offers a channel for consumers to interact with the electrical power grid. Thus, HEMS may reside in the smart meter [26] or may reside in an independent gateway such as the HAN gateway [12] or residential gateway [29] and network adapters [27]. However, incorporating HEMS with/without smart meter needs to be investigated extensively to achieve the optimal operating performance and also the minimum cost. Even though the initial cost of smart meter is high, its implementation will help in standardization of smart grid. However, smart meter with stand-alone HAN gateway would be a feasible choice to implement HEMS into HAN due to the low cost and higher communication capabilities [12].

Another key issue related to HEMS is the granularity of power management in controlling various electrical appliances (e.g., routers, TV, AC, computers, etc.) that provide different services (e.g., wireless access, VoIP calls, ambient temperature control, etc.). These services can be controlled by using different power-control elements (PCEs) such as Ethernet switch, PSTN, and DSL modem [30]. For example, using a broadband Internet service through wireless requires only certain power-control elements (PCEs) such as network adapter, wireless router, DSL modem, and PC to be activated. All other devices can be turned off during this operation. For this, effective power-management techniques that can measure the power consumption to granular level, such as energy-aware plug-and-play (EPnP) devices [30] and universal plug-and-play low-power (UPnP LP) protocol service [31], can be used. Thus, HEMS should provide smart and granular power management that involves all the elements associated with a service for better power-management capability. Also, HEMS should have an open architecture so as to accommodate and configure new devices autonomously without any user intervention [31].

1.5.2 Core communication network

The core communication network of a smart grid will provide all the necessary communication backbone functionalities for effective two-way communication. As mentioned earlier, the AMI will be used to realize this core network. The goal of AMI is to provide better control and management of the electricity consumption [11]. The existence of AMI is more crucial to distribution and transmission networks due to their coverage span to connect both local and remote customer premises. Many communications infrastructures and protocols (e.g., Internet, cellular networks, power-line communication (PLC), etc.) can be used for this purpose. The major issues in AMI are the installation cost and interoperability among different components of AMI.

Since smart meter will be an integral part of the AMI, using a low-cost customized smart meter will help to minimize the cost associated with the AMI deployment. For example, Korea Electric Power Corporation uses PLC for realizing their AMI [32]. The work in [33] presents a detailed design model for such low-cost AMI, that uses customized digital meters. These meters are built using low-cost electronic components instead of custom-made smart meters. Similarly, a cost-effective ZigBee-based smart power meter is introduced in [34]. IEEE 802.15.4 variant ZigBee is becoming more popular in customer-side metering infrastructure due to its low-cost and low-power operation [35]. The proposed smart meter uses a microcontroller (dsPIC30F series microchip) to perform necessary computations. The low-cost and low-power ZigBee system is integrated into the proposed smart meter which is used to transmit the power consumption information and operational statistics to the control centre.

Another important issue related to AMI is interoperability among different components in AMI, which is even more challenging due to the lack of open standards [36]. Moreover, heterogeneity exists even in the same network (e.g., different development languages, communication protocols, deployed hardware, and database [37]). As a result, interaction becomes difficult and there is a need for a flexible, reliable, reusable, and trustable architecture for AMI [38]. One possible solution is the use of a service-oriented architecture (SOA) based on web service [37, 38]. SOA is a software system architecture based on a component model which defines specific functional entities termed services. These services are linked through the interfaces and contracts, which are usually independent of hardware platforms, communications protocols, and programming languages. Therefore, SOA allows uniform interaction among services or systems. However, proper services and interfaces must be defined in the form of web-service description language (WSDL) to ensure interoperability. This SOA not only supports the interaction of services, but also provides good flexibility and scalability for future expansion [37].

The National Institute of Standards and Technology (NIST) [3], International Electrotechnical Commission (IEC), Smart Grid Strategic Group (SG), and European Commission's OPEN meter project [39] are working to define an open standard for AMI for smart grid. The interoperability issue should be addressed first at the system level to leverage any synergy that may be available [40]. For this, a common semantic model such as the common information model (CIM) IEC 61970 [41] and its extension IEC 61968 [42] can be used to map and integrate all the systems [40]. Both IEC

61970 and 61968 define guidelines and specifications for application program interfaces (APIs) for EMS and for information exchanges between electrical distribution systems. Recently, other protocols such as ZigBee [14] also follow this standard to leverage the need for common semantic models. The semantic model helps us to understand the context of data, which is beneficial for better system design and data processing [40]. This in turn helps us to design a specialized protocol for a particular data type, which reduces the computational requirement of a generic protocol. The specialized protocol helps us to reduce the additional overhead in a generic protocol. In the following subsections, we discuss the relevant technologies which can provide the core communication capabilities for AMI.

Internet protocol (IP)-based networks

The AMI should be self-healing, secure, efficient, interoperable, and scalable to handle a variety of applications. Also, it should support open standard with plug-and-play capability. It is necessary to create a convergence layer which would accommodate all the technologies used in smart grid [43]. An effective solution is to utilize an IP-based network (such as public Internet) as the backbone communications infrastructure for smart grid [44]. IP technologies are widely used in data networks and also many network technologies have a convergence layer for IP, for both wired or wireless systems [3, 43].

As smart grid will continue to increase the number of devices connected to its network, the number of addresses needed to uniquely identify these devices in the IP network will also increase substantially. In this regard, the IPv6 protocol is identified to be a better choice than its predecessor IPv4 due to the new features (e.g., larger address space, mobility, and security services [43]). However, not all the existing network technologies have a convergence layer with IPv6. A more flexible approach would be to use a mixed network of IPv4 and IPv6 as mentioned in [43]. However, this will require a proper translation of network boundaries which could act as the initial transition for the shift to a single IPv6 network (e.g., as implemented in the Spanish GAD project [45]).

Some of the variants of IP-based networks that can provide core networking functionality for the envisioned smart grid are discussed below.

- *Internet-based virtual private network (Internet VPN)*: The traditional IP-based network is versatile, but supports only best-effort delivery. That is, it does not ensure the quality-of-service (QoS) of the data being transferred. However, smart grid requires certain QoS guarantees in terms of latency, delay variation (jitter), throughput, and packet loss for efficient and reliable operation. In this context, Internet VPN technology can provide a better alternative to ensure security and QoS requirements of smart grid [46]. Internet VPN technology is a shared communication network architecture for a cost-effective and high-speed core communication network. Internet VPN provides both the functionalities and benefits of a dedicated private network. In other words, Internet VPN can provide reliable, secure, and robust communication with strict QoS guarantee to smart grid over a shared network infrastructure with the same policies and services that the electric utility experiences within its dedicated private communication network. However, Internet VPN still faces several open research issues such as efficient routing, resource management mechanisms, and inter-domain network management, which need to be developed for smart grid applications.

The Internet engineering task force (IETF) has introduced many service models such as differentiated service (DiffServ), multiprotocol label switching (MPLS), integrated sevices (IntServ), and IP security (IPSec) to meet the requirements for high-performance VPN services for smart grid [46, 47]. Among these, DiffServ and MPLS are promising and are used for providing IP QoS. DiffServ is a computer networking architecture. This architecture can be used to provide low-latency, QoS guarantee for critical network traffic while providing simple best-effort traffic guarantees to non-critical services such as web traffic or file transfers. DiffServ can be implemented in IPv4 as well as IPv6. MPLS is also a standard technology for speeding up network traffic flow that integrates layer-2 information about network links (bandwidth and latency) into layer 3 (IP) within a particular autonomous system. This will simplify and improve QoS of IP packet exchange. Both these technologies can provide necessary QoS support for smart grid. For instance, [47] introduces a way to provide QoS guarantee in the IEC 61850 substation automation protocol by using the IPv6 DiffServ model. Also, the integration of MPLS and DiffServ can be used to meet the QoS requirements of data to be transmitted in smart grid [48]. Also, IPSec-based Internet VPN and layer-2 technologies such as *Frame Relay* and *Asynchronous Transfer Mode* can be used to provide secure communications in smart grid. In fact, the actual selection of Internet VPN technologies for smart grid depends upon unique communication requirements and associated capabilities that can be fostered using an available combination of Internet VPN technologies.

- *Hybrid of transmission control protocol (TCP) and user datagram protocol (UDP)*: Another way is to rely on transport layer services of the IP suite, namely transmission control protocol (TCP) and user datagram protocol (UDP). Both protocols provide ubiquitous communication in the transport layer, but in reality complement each other. TCP provides a retransmission mechanism for lost data, but incurs considerable delay in data transmission and high network overheads [4]. Moreover, TCP lacks the multicast ability (to send a single packet to multiple destinations) which might be crucial for sending multiple control signals to many systems and devices in smart grid in case of emergencies. On the other hand, UDP has smaller overheads in the message frame and also supports multicast, but lacks any mechanism to support reliable packet delivery. Thus, a combination of TCP and UDP can provide simple, reliable, and ordered packet service in smart grid [4]. Also, both TCP and UDP can support either IPv4 or IPv6 networks.

WiMAX technology

For most utilities, the use of wireless technologies has always been attractive due to its flexibility and cost-effectiveness. Among the available wireless technologies, WiMAX is emerging as an enabling wireless technology that can address almost all the requirements of smart grid applications [49, 50]. Although WiMAX is not specifically developed for electricity utilities, it provides the necessary features to support smart grid (e.g., the most advanced security protocols, effective traffic management tools, and QoS framework).

With low latency, high throughput, and wider coverage, WiMAX allows electricity utilities to provide a wide range of applications concurrently over the different hierarchical networks but within a unified network core. Another important feature is its interoperable architecture that enables WiMAX to be connected with various other networks such as BPL, ZigBee, Wi-Fi, and even PLC. WiMAX is a versatile technology and recent research has shown that it has the potential to be deployed in various parts of smart grid for last-mile connectivity to backhaul networks [50]. As smart grid is evolving with time, WiMAX technology holds a promising solution to most of the communications problems faced by the current smart grid development. Nevertheless, the issues related to adopting WiMAX technology have to be addressed (e.g., network integration, QoS framework customization, and network deployment).

Communication and networking middleware

Networking middleware provides a higher-level application interface which improves the interoperability, portability, and flexibility of data communication. It provides an interface between high-level application QoS requirements and low-level resource management such as bandwidth requirements and real-time data monitoring. Therefore, it can be used to support interoperability of network protocols to meet the requirements of smart grid. Also, this could be used to augment the capability of both TCP/IP protocols. For instance, a middleware called GridStat [51] is being implemented by Avista Utilities in Washington by incorporating the SCADA system to acquire the necessary data directly from the sensors. GridStat is a simple publish–subscribe middleware which can be used to acquire the status variables and status alerts in the electrical grid.

1.5.3 Last-mile connection

The *last-mile* connection refers to delivering connectivity from the communication provider to a customer. In case of smart grid, it typically means connecting the substation and customer premises to the high-speed communication core network. This adds new challenges to smart grid to provide a cost-effective and reliable communication infrastructure. Many communication technologies (e.g., power-line communication (PLC), wireless communication, satellite communication, optical-fibre communication, etc.) can be used to provide this last-mile connection. In the following, we discuss some of the existing and future technologies.

Power-line communication (PLC)

Due to the ubiquitous presence of power distribution lines and associated advantage of low installation costs, power-line communication (PLC) will be widely adopted in smart grid [53]. PLC refers to transmitting data by modulating the standard 50 or 60 Hz alternating current on the existing electrical power lines [52]. PLC is sometimes referred to as broadband over power line, or BPL, when power lines are used to carry a high-bandwidth data signal. The underlying advantage of PLC is its extensive coverage and low cost due to its ubiquitous presence, which eliminates the need to add extra communication infrastructure for the last-mile connection. PLC can also be used for short-haul communications of home automation and power network management in the smart grid [54]. The most common PLC technology uses a single-carrier system with modulation techniques of FSK and BPSK, which lacks flexibility in selecting the carrier frequency, and hence, results in low throughput and poor reliability.

Studies have shown that the characteristic of a single-carrier PLC varies drastically with different geographical locations or number of network nodes. Also, its performance depends largely on system parameters (e.g., data packet size and response time [53, 55]). Nevertheless, PLC can be improved using different approaches (e.g., single-frequency networking with flooding of message [53], OFDM, and OFDMA modulation [55]). Single-frequency networking with flooding provides transmission of many messages simultaneously using the same frequency [53]. Also, using the OFDMA with PLC provides better resistance to signal distortion and the frequency selectivity problem [55, 56]. However, there are still several technical issues (e.g., high noise over power lines, signal attenuation and distortion, capacity, etc.) and regulatory issues (e.g., lack of regulations for BPL) associated with PLC. As a result, PLC still requires further research effort to become commercially deployable for the last-mile connection.

Wireless communication

Due to the flexibility in installation, wireless communication presents an effective alternative to any wired communication infrastructure. Wireless communication technologies have the potential to provide remote control and monitoring without addition of any cabling cost. Several wireless communication technologies (e.g., cellular network and WiMAX) can be used to provide the last-mile connection in smart grid [46]. Apart from these, wireless technologies such as IEEE 802.11 (Wi-Fi) for local area networks, IEEE 802.15.1 (Bluetooth) for personal area networks, and IEEE 802.15.4 (ZigBee) for small and low-cost networks are also available to provide wireless connection inside HAN or within NAN. All of these use the same carrier-sense multiple access with collision avoidance (CSMA/CA) technique for accessing the media [8]. Further, recent research shows that combining two or more different access technologies with broadband over BPL can improve the data transmission rate and QoS support to the end users. These access technologies could be wireless (e.g., Wi-Fi or WiMAX) or wired network (e.g., fibre or co-axial cable). This is referred to as the hybrid access technology. For example, studies show that the integration of ubiquitous power distribution with Wi-Fi technology can enhance the reliability and reduce the cost of the broadband access network to be used for power distribution networks in smart grid [56, 57]. A BPL unit can act as the wireless interface gateway for converting the medium-voltage line signals to IEEE 802.11a/b/g compatible signals. In this way, the BPL network can be transformed into a hybrid wireless-broadband power line network (W-BPL) [56].

Although wireless communication technologies seem attractive, their high susceptibility to electromagnetic interference, limitation in bandwidth capacity, and maximum distance among communication devices present new challenges for acceptance by the utility industry. Also, wireless communication is more vulnerable to security threats which will require efficient authentication and encryption techniques to be developed.

Satellite communication

Another alternative is to use satellite communication which can offer global coverage to even remote locations without a complete communication infrastructure [8]. Services such as very small aperture terminal (VSAT) are already available to connect remote substations. Also, services such as global positioning system (GPS) can be used for accurate time synchronization and location updates. Satellite communication can be used as a backup for the existing communication network in case of emergency or link failure to send the critical data to the control centre [46]. The main challenges for satellite communication are high costs, long latency, security, and a lack of redundancy. The cost for operating a satellite is still higher than that of other available communication options. Also, the extreme path length of satellite channels contributes to high transmission latency. Since satellite communication uses public airwaves, it is inherently insecure, which will also require additional data encryption techniques.

Optical-fibre

Optical-fibre communication systems are one of the technically promising communication infrastructures providing nearly unlimited bandwidth capacity, strict QoS, fast and reliable data transfer [8, 46]. The advantage of optical-fibre communication lies in its immunity towards electromagnetic interference and radio frequency interference, which makes it an ideal communication medium for high-voltage substations. Furthermore, optical-fibre requires fewer repeaters to amplify data over long distances. The main disadvantage of the optical-fibre is that its initial installation cost might be expensive for an electric utility to install a dedicated network. However, the bandwidth capacity of the optical-fibre can be shared among other users, thus recovering the cost of installation. In this context, optical-fibre might be the most effective, reliable, and cost-effective solution for the last-mile connection in smart grid. Also, new advancements in optical-fibre networks, such as passive optical network (PON), permits a single fibre to be split into 128 sections without any electronic repeaters. Basically, there are three main candidates for optical-fibre technology based on PON, namely broadband PON (BPON), Gigabit PON (GPON), and Ethernet PON (EPON). BPON is used mostly by the telecommunication industry, while GPON and EPON are just in the beginning of deployment. EPON, due to its compatibility with Ethernet, is particularly being used for data communication along the electrical transmission and distribution lines [58].

1.5.4 Control centre

In the area of electrical grid automation and sub-automation, many operation, control, and protection functions need to be optimized to achieve the highest efficiency and supply reliability [59]. The magnitude of the problem faced by these electrical grids varies from simple fluctuations such as voltage, frequency, or angle to major faults such as cascade tripping and blackouts. A control centre is needed to avoid these contingencies. For this, data needs to be collected from various parts of the grid. In fact, the traditional SCADA system already provides steady and non-synchronous data exchange and measurement, where the data is collected by a centralized server. However, SCADA lacks proper integration capability and scalability. Therefore, to provide advanced control, protection, and real-time measurements for smart grid, wide-area measurement system (WAMS) will be used. This WAMS can be integrated with SCADA to provide smooth system migration [60]. In the following, we discuss specific issues in WAMS.

WAMS is being adopted in the power grid by many countries such as the USA, Brazil, China, India, Russia, and Japan [61]. The work in [61] provides a summary of the most advanced stages in WAMS development. The development and implementation functionality of WAMS in Slovenia is presented in [18]. The work in [62] introduces customer involvement in WAMS by providing a simple online monitoring portal to consumers. The system called Campus WAMS is being developed and deployed in Western parts of Japan. Such an initiative decreases the complexity by providing data to the consumer side for further processing. In [63], a decentralization method of data collection is introduced in which the traditional centralized data collection is replaced by a hierarchical network configuration with a simple publish–subscribe structure. In this case, the middleware services such as GridStat [51] can be used to manage and transfer the data to the appropriate application. Also, in [59], a design architecture for flexible and open architecture WAMS is given. From these reports, it is found that hierarchical WAMS design with data concentrator (e.g., phasor data concentrator or collector) that collects and processes the data on the limited scope of operation is the most common structure to be implemented. In particular, a large amount of data is required to be collected and the centralized server can be congested and become bottlenecked easily. Also, there is a need for unified representation of all this information to provide flexible data exchange. Some of the power industry standards such as IEC TC 57 common information model [64] and IEC 61850 standard [65] provide the specifications to promote the interoperability, but at the cost of increased data traffic. Therefore, distributed communications and data architecture are required [63]. The main advantage of distributed communications is that it provides an implementation of the decentralized functionality needed for control and protection of the power system at any computing platform within the distributed control system.

Due to growing interest and the critical nature of WAMS, many researchers are also focusing on the reliability analysis of WAMS. WAMS is an important system in smart grid and its failure may lead to severe consequences. Reliability assessment helps not only to foresee hidden problems and issues, but also to provide necessary guidelines in case of an emergent situation. A comprehensive performance analysis of interdependency among different factors of WAMS is presented in [60]. Similarly, other works such as [66, 67] provide step-by-step reliability analysis of a real WAMS. Generally, analysis tools such as influence diagrams [60], fault-tree analysis, and Markov modelling [66, 67] are used extensively to provide reliable evaluation of WAMS. With advancement of the new technology of communication infrastructure, WAMS will continue to play an important role in the vision of smart grid.

1.5.5 Sensor and actuator networks (SANETs)

A sensor and actuator network (SANET) is a heterogeneous network consisting of a large number of nodes each of which can be a sensor, an actuator, or both [68]. SANETs are used to monitor the operational characteristic and behaviour of smart grid devices so that any outage or disturbance can be prevented. The data gathered by different sensors is sent to a sink which may reside in the network gateway connecting different topologies (e.g., DAU which connects NANs and HANs). The collected data is used either by the actuator locally or sent to the utility control centre via gateways for further processing and action. Specifically, sensors such as PMU are used to measure different system parameters (e.g., light intensity, temperature, voltage and current fluctuation, and power of the line). In contrast to a sensor, an actuator converts the information collected by the sensors and control sent by management system into actions by setting the values of different parameters (e.g., displaying the sensor measurements, or status of a circuit breaker). Both sensors and actuators are installed at multiple sites in smart grid such as at the transformer, distribution substation, or a customer's home.

As the electrical power grid is characterized by its large geographical spread and complex architecture, sensing the status of various electrical power assets presents an important functionality as well as scalability challenges for SANETs [68, 69]. However, the conventional sensors will not be effective for smart grid. In this regard, smart sensors using distributive and cooperative sensing will play an important role in implementing SANETs in smart grid. A distributed sensor network following a hierarchical topology, such as a cluster tree topology, is proposed in [69]. The scalability of the network is improved by organizing the sets of sensors and actuators into a hierarchical topology by forming groups or domains. Similarly, in [70], a cooperative cluster of sensor networks is proposed. Sensor clusters are distributed among the grids such that the sensed data (e.g., voltage waveforms and the voltage quality index) can be shared among each other to provide more reliable global information.

Due to their high flexibility and efficiency, wireless sensor networks (WSNs) will also play a significant role in the success of smart grid. WSNs [71] have several advantages over traditional sensing, including greater fault tolerance (by distributed nature and data redundancy), improved accuracy (by multiple and cooperative sensing), larger coverage (by anonymous and collaborative coverage), and extraction of localized features (by sending necessary information only) [46]. Due to these features, WSN (e.g., the IEEE 802.15.4-compliant [72] sensor network) is considered as one of the enabling technologies to implement wireless automatic meter reading and remote system monitoring, and equipment fault diagnostics [73]. Many utility industries have started implementing WSNs in smart grid. For instance, the utility industry in California, USA, has announced its plans to deploy wireless smart meter by 2013 [74]. Similarly, in Japan, ultra low-power wireless sensor nodes are being deployed to measure the power consumption of home appliances [75]. Also, [73] provides guidelines for developing WSN-based smart grid applications (e.g., remote monitoring and wireless automated meter reading using ZigBee). However, the WSNs have to be optimized considering the unique communications characteristics (e.g., harsh environment of smart grid due to electromagnetic interference) and performance requirements (e.g., time-sensitive latency requirements, limited resources of memory, processing speed, power supply, and the need for variable QoS) [73].

1.6 Security and privacy in the communications infrastructure for smart grid

The credibility of communication infrastructures in smart grid will be the major issue. The smart grid will be expanded to contain more interconnections of systems and subsystems that may become the point of intrusions, malicious attacks, and other cyber threats. The cyber attack on Port of Houston in 2003 and the demonstration of attack on a commercial smart meter at Black Hat Conference in 2009 are some examples of cyber security threats to electrical power grids [76]. Clearly, the integration of the information and communication infrastructure with smart grid introduces new security and privacy-related challenges. However, these technologies also present opportunities to increase the reliability of smart grid by providing proper frameworks and technologies (e.g., improved monitoring and controlling capabilities) to withstand these security attacks.

Many standard bodies, e.g., National Institute of Standards and Technology (NIST), International Society of Automation (ISA), and IEEE standard 1402 group are developing secure and consistent authentication, authorization, and privacy technologies for smart grid [77]. For example, Cyber Security Working Group (CSWG) of the Smart Grid Interoperability Panel (SGIP), a public–private partnership launched by NIST, defines the scope of the cyber security and also highlights different security strategies and requirements across seven commonly accepted smart grid domains (see Figure 1.1). According to CSWG [78], the issue of cyber security in smart grid should also address possible anomalies due to user errors, equipment failures, and natural disasters apart from the possible deliberate attacks. Also, it highlights the importance of a domain-wise approach to security, with one or more security measures and controls implemented at each domain. This will minimize the cascading effect if one component or domain of the smart grid is compromised. Furthermore, there is a need for consumer protection policies against the possible violation of grid security and privacy [79]. Thus, multiple levels of security measures should be implemented.

Different components (e.g., smart meter, IEDs, and WAMS), protocols (e.g., IEC 61850 and DNP3), and networks (e.g., cellular network, fibre-optics, WLAN, and PLC) of smart grid should be secured. Depending upon the structure and operation of smart grid, we classify cyber security for smart grid into the following categories: component-wise security, protocol security, and network-wise security.

1.6.1 Component-wise security

A smart grid consists of many intelligent electronic devices (IEDs) (e.g., smart meters and PMUs), networking infrastructure (e.g., WAMS and AMI), remote terminal units (RTUs), and other controlling devices. Many of these devices will be located at secondary and remote service levels (i.e., outside the facility of the utility company) and will be multi-connected through communication core networks such as the Internet. Due to this, the remote access interface used to access these components might allow any third party to take control over the device, thus creating security vulnerability. Thus, proper authentication and trust management techniques should be implemented to provide the necessary protection for smart grid. For unauthorized access, the security key management systems used for the existing industrial system can be customized specifically for smart grid to minimize the additional cost. Technologies such as public key infrastructure (PKI), advanced encryption standard (AES), and triple data encryption algorithm (3DES) can be used to provide secured and private data exchange [77]. In addition, the security can be further improved by using the device attestation technique which provides a method to check whether a device is tampered or not. The attestation certificates can be installed in the devices by the accredited manufacturers. Further, device authentication can also help to minimize the security risk. This can be achieved if proper trust anchor-management systems are available [77]. For customer-side components such as smart meter, the wired or wireless connection used to transfer meter data might also expose the valuable and private information of a consumer's behaviour to external attack [80]. To minimize these security threats, additional security can be provided using data encryption methods such as pseudo random spreading [81] to smart meters.

Further, for higher-order components such as WAMS, interoperability among solutions is one of the factors which results in security holes for cyber attacks. Due to growing interest in WAMS, various security solutions such as shielding the operation part (e.g., real-time measurement of system operation) from the administrative part (e.g., recording disturbance history and local event recording) [82], domain-wise security [78], and using a dedicated security layer [83] can be used to provide proper security measures in the existing system in a non-intrusive fashion without compromising the grid's reliability.

1.6.2 Protocol security

A smart grid will rely on different communications protocols for providing the necessary bi-directional flow of information needed for reliable control and management of the entire grid. The security of these protocols against cyber attack is a significant issue. The adversity of these cyber attacks can range from sending misleading data to the field device to sending tampered control and command messages to the device which may result in overloading the grid. That is why security is becoming a critical issue for existing and communication protocols for new electric systems. For some of the existing standard automation protocols such as International Electrotechnical Commission Technical Committee 57 (IEC TC 57) protocol suits [84], to be functional, they must use a communications platform for data exchange whose implication can be exposed to a large potential set of vulnerabilities. IEC TC 57 is responsible for development of standards such as DNP3 [85] for process automation, IEC 61850 for IED communications and associated data models in power systems, IEC 61850-7-420 for communications systems for distributed energy resource (DER), IEC 60870-5 for telecontrol and teleprotection standards, and IEC 61970 for energy management systems, IEC 60870-6 (ICCP) to provide data exchange over wide-area networks (WANs) between utility control centres. These standard protocols are mostly open standards and necessary encryption and authentication can be added to provide much needed security, for instance, a privacy-based cyber security pseudo-layer, namely PGP (Pretty Good Privacy) [86], and open-access-compatibility (OAC) security layer [87] for addition of encryption and authentication to DNP3 data-link layer frames below the DNP3 data-link layer. The proposed solution provides authentication capabilities using public key cryptography.

Even though these solutions provide added authentication and encryption to the standard protocol stack, proper protection can only be achieved if these policies are defined within the DNP3 protocol stack. That is why WG15 (IEC 62351) under the IEC TC 57 standard body has been established to undertake the development of security standards for the TC 57 series of protocols including the IEC 60870-5 series, the IEC 60870-6 series, the IEC 61850 series, the IEC 61970 series, and the IEC 61968 series [84]. The WG15 is responsible for defining different security objectives such as authentication of entities through digital signatures, ensuring only authorized access, prevention of eavesdropping, prevention of playback and spoofing, and some degree of intrusion detection. Moreover, security in these protocols can be achieved by using key encryption technologies such as advanced encryption standard (AES) solutions [88]. The popularity of AES is increasing and it has already been integrated into ZigBee protocols. Similarly, the DNP Users Group [85] releases the Secure DNP3 specification which adds five new security-related function codes to the protocol and offers data and source authentication. This certainly makes DNP3 a viable smart grid technology, which leads to standardization of DNP3 for smart grid by the IEEE community [89].

1.6.3 Network-wise security

As mentioned before, the communication core networks will act as the backbone for smart grid which will connect all the components within the smart grid. However, such network topologies are also open to external communications networks for inter-power system operations and transactions of the electricity business [90]. This in turn makes smart grid more vulnerable to potential cyber attacks (e.g., network outage, deliberate sabotage, and authorization violation) that could seriously impact the operation of the electrical power grid.

A simple alternative to minimize security threats would be to use private networks for communication [77]. That is why most of the utility offices still prefer using their own private network, which will greatly minimize the security threats, as there would be no potential access from third parties. Strong authentication and encryption techniques can be applied to private networks. However, in today's highly connected world, having a completely separate network is not economical or efficient. Moreover, interoperability issues arise if all the utility companies use their own private network instead of common or standard network topology. In this regard, technologies such as IP-based VPN, WiMAX, and TCP/UDP provide a better solution in terms of connectivity and also security. Also, security services such as IP security (IPSec) can be used to minimize the vulnerability in the IP-based networks. IPSec helps to authenticate and encrypt each IP packet of a communication session. Also, NIST recommends disjoint protocols for high-risk areas [91]. Further, new network technology such as WiMAX can also be used. WiMAX provides secure communications and also provides support for multiple security standards such as AES, and EAP tunnelled transport layer security.

1.7 Open issues and future research directions

Integrating communications network infrastructure makes smart grid an Internet for energy – an 'energy Internet' [92]. However, at the same time, the use of data communications in smart grid opens many challenges in terms of reliability, stability, cost-effectiveness, and security. The reliability and efficiency of smart grid will depend directly on the performance of the communications infrastructure [93]. Emphasis should be given to develop a standardized distributed communications infrastructure for a specific hierarchy in smart grid [94]. Furthermore, integrating security policies into the design of communications protocols should be focused on both installation and operation phases. Some of the issues related to the data communications and networking are outlined.

1.7.1 Cost-aware communication and networking infrastructure

Many researches (e.g., in [33, 34]) are ongoing for alternative cost-aware and cost-effective networking infrastructures. Cost-aware data protocols are implemented to provide fast and efficient ways of sending/retrieving required data to/from various functional devices in the network in an economical way. The cost of retrieving recent real-time information (e.g., power pricing, metering data, and surveillance data) increases with the increase in frequency of inquiry. As a result, the data bandwidth of the network can be overloaded. However, minimizing the frequency of inquiry results in outdated information which could hamper the decision-making process of the intelligent devices in smart grid. The optimization of information transfer would be required to minimize cost.

In addition, information in smart grid can be sensitive to delay and error, which can result in inefficient decisions. Forecasting models can be used to predict the system state at a given future time. For instance, a probabilistic forecasting model called locational marginal pricing (LMP) is used to predict the distribution of power price in [95]. However, the work in [95] is applied for a simple supplier home model with one power supply and one home appliance. It is hard to model a practical smart grid and it may require other models for power pricing such as reinforcement learning model or using anticipatory control methodology as mentioned in [92]. Therefore, the cost related to data retrieval/sending should be considered when data communications protocols are designed so that the tradeoff between costs of power consumption and communications is balanced under a set of constraints.

1.7.2 Quality-of-service (QoS) framework

In energy networks, real-time data is closely monitored and used to forecast the peak demand of customers so that power generation, transmission, and distribution can be scheduled optimally [92]. Not only the reliability, but also the efficiency are important issues for data communications in smart grid. Therefore, quality-of-service (QoS) is an important factor which ensures the integrity of the data in smart grid. The QoS in smart grid can be defined by the accuracy and effectiveness with which different information such as equipment's state, load information, and power price are delivered in a timely manner to the respective parties [48, 96]. The QoS requirement of smart grid differs from any data communication network where the delay requirement of data is not strict [83]. In addition, since a smart grid will incorporate a large number of devices, the types and quantities of data traffic also increase exponentially. This increased traffic can create a bottleneck in the smart grid communications infrastructure. As a result, delay and loss (e.g., due to congestion) can be severe. Thus, it is important to include the QoS framework while designing a new protocol. Another important factor is to identify the specific QoS requirements and priorities for a specific communication network in smart grid. For instance, power price data should have higher priority and guarantee QoS over the meter data used for summarizing a monthly bill of electric usage.

1.7.3 Optimal network design

For efficient and stable operation, all communication components in smart grid should be installed and deployed optimally to yield a higher productivity. Therefore, networks should be optimized in the design phase. However, optimizing such a large network is a challenging task and requires extensive optimization strategies to be implemented. Most of the current works ignore this issue, although there are a few which suggest using game theory with multi-agent approach [92] or using multi-tier architecture to ensure optimal design [97]. Many factors such as reliability, security, QoS requirements, and cost should be considered while designing an optimal network infrastructure for smart grid.

1.8 Conclusion

In this chapter, we have provided a brief survey of smart grid data communications, which refers to the next generation electrical grid that will revolutionize the way electricity is produced, delivered, and consumed. We have outlined the importance of communication networks in smart grid for achieving its vision. Also, we have presented a hierarchical overview of the smart grid communication infrastructure in an attempt to provide better understanding of the smart grid landscape. Various issues related to interoperability have been summarized. Due to the profound effect of communication infrastructure on smart grid, its role in the development of smart grid has been outlined and discussed. Different approaches proposed for smart grid have been reviewed. At the end, the open issues and further research directions have been discussed.

The scope and awareness of smart grid is increasing, and as a result, many standard bodies (such as NIST, EPRI, IEC, IEEE, European Commission, etc.) are working towards developing frameworks, communications standards, and security policies for smart grid. With the advancement in communications and information technology and active contribution from different research communities, it is only a matter of time until smart grid is a reality.

References

[1] Department of Energy, USA, 'The smart grid: an introduction' [online]. Available at: http://www.oe.energy.gov/SmartGridIntroduction.htm, accessed: April 2011.

[2] European Commission Research, 'European smart grids technology platform vision and strategy for Europe's electricity networks of the future' [online]. Available at: http://ec.europa.eu/research/energy/pdf/smartgrids_en.pdf, accessed: April 2011.

[3] National Institute of Standards and Technology, USA, 'NIST framework and roadmap for smart grid interoperability standards, Release 1.0' [online]. Available at: http://www.nist.gov/smartgrid/, accessed: April 2011.

[4] V.K. Sood, D. Fischer, J. M. Eklund, and T. Brown, 'Developing a communication infrastructure for the smart grid', in *Proceedings of IEEE Electrical Power and Energy Conference (EPEC)*, pp. 1–7, October 2009.

[5] A&B National Energy Technological Laboratory (2007-07-27), 'A vision for the modern grid', *United States Department of Energy*, pp. 5, accessed: May 2010.

[6] Electric Power Research Institute (EPRI), 'Report to NIST on the smart grid interoperability standards roadmap' [online]. Available at: http://www.nist.gov/smartgrid/upload/Report_to_NIST_August10_2.pdf, accessed: April 2010.

[7] E. M. Lightner and S. E. Widergren, 'An orderly transition to a transformed electricity system', *IEEE Transactions on Smart Grid*, vol. 1, no. 1, pp. 3–10, June 2010.

[8] Electric Power Research Institute (EPRI), 'Intelligrid consumer portal telecommunications assessment and specification, technical update' [online]. Available at: http://intelligrid.epri.com/technical_results.html, accessed: April 2011.

[9] International Electrotechnical Commission, 'IEC smart grid standardization roadmap' [online]. Available at: http://www.iec.ch/smartgrid/, accessed: April 2011.

[10] EnerNex Corporation, 'Smart grid standards assessment and recommendations for adoption and development' [online]. Available at: http://collaborate.nist.gov/twiki-sggrid/pub/SmartGrid/H2G/Smart_Grid_Standards_Landscape_White_Paper_v0_8.doc, accessed: April 2011.

[11] New York State Electric and Gas Company, 'Advanced metering infrastructure overview and plan' [online]. Available at: www.dps.state.ny.us/NYSEG_RGE_AMI_Filing.pdf, accessed: April 2011.

[12] N. Ota, 'The home area network: architectural considerations for rapid innovation.' White Paper for Trilliant Inc. [online]. Available at: http://www.trilliantinc.com/library-files/white-papers/HAN_white-paper.pdf, accessed: April 2011.

[13] BACnet, 'A data communication protocol for building automation and control networks' [online]. Available at: http://www.bacnet.org/, accessed: April 2011.

[14] ZigBee Alliance [online]. Available at: http://www.zigbee.org/, accessed: April 2011.

[15] C. Bennett and D. Highfill, 'Networking AMI smart meters', *Proceedings of IEEE Energy 2030*, November 2008.

[16] OpenAMI task force under the UCA International Users Group (UCAIUG), 'UtilityAMI 2008 HAN SRS' [online]. Available at: http://www.utilityami.org/, accessed: March 2008.

[17] J. Giri, D. Sun, and R. Avila-Rosales, 'Wanted: a more intelligent grid', *IEEE Power and Energy Magazine*, vol. 7, pp. 34–40, March–April 2009.

[18] T. Babnik, U. Gabrijel, B. Mahkovec, M. Perko, and G. Sitar, 'Wide area measurement system in action', in *Proceedings of IEEE Lausanne PowerTech*, pp. 1646–1651, July 2007.

[19] The GridWise Architecture Council, 'GridWise interoperability context-setting framework' [online]. Available at: http://www.gridwiseac.org/about/publications.aspx, accessed: April 2011.

[20] H. Zimmermann, 'OSI reference model – the ISO model of architecture for open systems interconnection', *IEEE Transactions on Communications*, vol. 28, no. 4, pp. 425–432, April 1980.

[21] F. Li, W. Qiao, H. Sun, H. Wan, J. Wang, Y. Xia, Z. Xu, and P. Zhang, 'Smart transmission grid: vision and framework', *IEEE Transactions on Smart Grid*, vol. 1, no. 2, pp. 168–177, September 2010.

[22] A. Vojdani, 'Smart integration', *IEEE Power and Energy Magazine*, vol. 6, no. 6, pp. 71–79, November–December 2008.

[23] Google Powermeter [online]. Available at: http://www.google.com/powermeter/about/index.html, accessed: June 2010.

[24] Microsoft Hohm [online]. Available at: http://www.microsoft-hohm.com/, accessed: April 2011.

[25] Apple Smart-Home Energy Management [online]. Available at: http://www.patentlyapple.com/patently-apple/2010/01/apple-reveals-smart-home-energy-management-dashboard-system.html, accessed: April 2011.

[26] Y-S. Son and K-D. Moon, 'Home energy management system based on power line communication', in *Proceedings of international Conference on Consumer Electronics (ICCE)*, pp. 115–116, January 2010.

[27] M. Inoue, T. Higuma, Y. Ito, N. Kushiro, and H. Kubota, 'Network architecture for home energy management system', *IEEE Transactions on Consumer Electronics*, vol. 49, no. 3, pp. 606–613, August 2003.

[28] Konnex Association [online]. Available at: http://www.knx.org/knx/what-is-knx/, accessed: April 2011.

[29] S. Tompros, N. Mouratidis, M. Draaijer, A. Foglar, and H. Hrasnica, 'Enabling applicability of energy saving applications on the appliances of the home environment', *IEEE Network*, vol. 23, no. 6, pp. 8–16, November–December 2009.

[30] Y-K. Jeong, I. Han, and K.-R. Park, 'A network level power management for home network devices', *IEEE Transactions on Consumer Electronics*, vol. 54, no. 2, pp. 487–493, May 2008.

[31] A. Virolainen and M. Saaranen, 'Networked power management for home multimedia', in *Proceedings of IEEE Consumer Communications and Networking Conference (CCNC)*, pp. 331–332, January 2008.

[32] M. Choi, S. Ju, and Y. Lim, 'Design of integrated meter reading system based on power-line communication', in *Proceedings of IEEE International Symposium on Power Line Communications and Its Applications (ISPLC)*, pp. 280–284, April 2008.

[33] S. Rusitschka, C. Gerdes, and K. Eger, 'A low-cost alternative to smart metering infrastructure based on peer-to-peer technologies', in *Proceedings of International Conference on the European Energy Market (EEM)*, pp. 1–6, May 2009.

[34] S.-W. Luan, J.-H. Teng, S.-Y. Chan, and L-C. Hwang, 'Development of a smart power meter for AMI based on ZigBee communication,' in *Proceedings of International Conference on Power Electronics and Drive Systems (PEDS)*, pp. 661–665, November 2009.

[35] C. Bennett and S. B. Wicker, 'Decreased time delay and security enhancement recommendations for AMI smart meter networks', in *Proceedings of Innovative Smart Grid Technologies (ISGT)*, January 2010.

[36] G. Mauri, D. Moneta, and C. Bettoni, 'Energy conservation and smart grids: new challenge for multimetering infrastructures', in *Proceedings of IEEE Bucharest PowerTech*, June–July 2009.

[37] L. Zhang, M. Ge, X. Bi, and S. Chen, 'A SOA-BPM-based architecture for intelligent power dispatching system', in *Proceedings of Power and Energy Engineering Conference (APPEEC)*, pp. 1–4, March 2010.

[38] S. Chen, J. Lukkien, and L. Zhang, 'Service-oriented advanced metering infrastructure for smart grids', in *Proceedings of Power and Energy Engineering Conference (APPEEC)*, pp. 1–4, March 2010.

[39] European Commission, Open Public Extended Networking Metering (OPEN meter) [online]. Available at: http://www.openmeter.com, accessed: April 2011.

[40] D. D. Haynes, 'A case for optimized protocols in the creation of a smarter grid', *IEEE Transactions on Power Delivery*, July 2010.

[41] Energy management system application program interface (EMS-API) part 301: common information model (CIM) base, International Electrotechnical Communication Standard IEC 61970-301 Ed. 2, April 7, 2009.

[42] IEC system interfaces for distribution management part 11: common information model (CIM) extensions for distribution, International Electrotechnical Communication Standard IEC 61968-11 Draft CDV, 2009.

[43] F. Lobo, A. Lopez, A. Cabello, D. Mora, R. Mora, F. Carmona, J. Moreno, D. Roman, A. Sendin, and I. Berganza, 'How to design a communication network over distribution networks', in *Proceedings of International Conference and Exhibition on Electricity Distribution*, June 2009.

[44] F. Lobo, A. Cabello, A. Lopez, D. Mora, and R. Mora, 'Distribution network as communication system', in *Proceedings of CIRED Seminar SmartGrids for Distribution (IET-CIRED)*, June 2008.

[45] Active Demand Management (GAD) Project Under Technological Development Centre of the Ministry of Industry, Tourism and Commerce of Spain (CDTI) [online]. Available at: http://www.gadproject.es/, accessed: April 2011.

[46] V. C. Gungor and F. C. Lambert, 'A survey on communication networks for electric system automation', *Computer Networks Journal (Elsevier)*, vol. 50, pp. 877–897, May 2006.

[47] Y. Ting, Z. Zhidong, W. Jiaowen, and L. Ang, 'Research on transmission data system of smart grid based on IPv6 DiffServ model', in *Proceedings of Power and Energy Engineering Conference (APPEEC)*, pp. 1–4, March 2010.

[48] Y. Ting, Z. Zhidong, L. Angm, W. Jiaowen, and W. Ming, 'New IP QoS algorithm applying for communication sub-networks in smart grid', in *Proceedings of Asia-Pacific Power and Energy Engineering Conference (APPEEC)*, pp. 1–4, March 2010.

[49] X. Zhichao and L. Xiaoming, 'The construction of interconnected communication system among smart grid and a variety of networks', in *Proceedings of Power and Energy Engineering Conference (APPEEC)*, pp. 1–5, March 2010.

[50] M. Paolini, 'Empowering the smart grid with WiMAX', *White Paper presented to Alvarion*, June 2010.

[51] C. H. Hauser, D. E. Bakken, and A. Bose, 'A failure to communicate: next generation communication requirements, technologies, and architecture for the electrical power grid', *IEEE Power and Energy Magazine*, vol. 3, pp. 47–55, 2005.

[52] National Energy Technology Laboratory, 'Integrated communications', *White Paper for the U.S. Department of Energy*, February 2007.

[53] G. Bumiller, L. Lampe, and H. Hrasnica, 'Power line communication networks for large-scale control and automation systems', *IEEE Communications Magazine*, vol. 48, no. 4, pp. 106–113, April 2010.

[54] R. Benato and R. Caldon, 'Application of PLC for the control and the protection of future distribution networks', in *Proceedings of IEEE International Symposium on Power Line Communications and Its Applications (ISPLC'07)*, pp. 499–504, March 2007.

[55] S. Bannister and P. Beckett, 'Enhancing power line communications in the smart grid using OFDMA', in *Proceedings of Power Engineering Conference (AUPEC'09)*, Australian Universities, pp. 1–5, September 2009.

[56] G. I. Tsiropoulos, A. M. Sarafi, and P. G. Cottis, 'Wireless-broadband over power lines networks: a promising broadband solution in rural areas', in *Proceedings of IEEE Bucharest PowerTech*, June–July 2009.

[57] O. A. Gonzalez, J. Urminsky, M. Calvo, and L. de Haro, 'Performance analysis of hybrid broadband access technologies using PLC and Wi-Fi', in *Proceedings of International Conference on Wireless Networks, Communications and Mobile Computing*, vol. 1, pp. 564–569, June 2005.

[58] S. Zhongwei, M. Yaning, S. Fengjie, and W. Yirong, 'Access control for distribution automation using Ethernet passive optical network', in *Proceedings of Power and Energy Engineering Conference (APPEEC)*, pp. 1–4, March 2010.

[59] J. Bertsch, C. Carnal, D. Karlson, J. McDaniel, and K. Vu, 'Wide-area protection and power system utilization', *Proceedings of the IEEE*, vol. 93, no. 5, pp. 997–1003, May 2005.

[60] M. Chenine, L. Nordstrom, and P. Johnson, 'Factors in assessing performance of wide area communication networks for distributed control of power systems', in *Proceedings of IEEE Lausanne PowerTech*, pp. 1682–1687, July 2007.

[61] A. G. Phadke and R. M. de Moraes, 'The wide world of wide-area measurement', *IEEE Power and Energy Magazine*, vol. 6, no. 5, pp. 52–65, September–October 2008.

[62] M. Hojo, K. Abe, Y. Mitani, H. Ukai, and O. Saeki, 'Real-time power system monitoring at demand sides by campus wide area measurement system', in *Proceedings of IEEE Conference on Industrial Electronics (IECON)*, pp. 3609–3614, November 2009.

[63] A. Bose, 'Smart transmission grid applications and their supporting infrastructure,' *IEEE Transactions on Smart Grid*, vol. 1, no. 1, pp. 11–19, April 2010.

[64] IEC, International Electrotechnical Commission, 'IEC 61970-301 energy management system application program interface part 301: common information model (CIM) base', IEC Reference number IEC 61970-301:2003(E).

[65] IEC, International Electrotechnical Commission, 'IEC 61850-7-4 Communication networks and systems in substations part 7-4: basic communication structure for substation and feeder equipment-compatible logical node classes and data classes', IEC, Reference number IEC 61850-7-4:2003(E).

[66] L. W.-Xia, L. Nian, F. Yong-Feng, Z. Li-Xin, and Z. Xin, 'Reliability analysis of wide area measurement system based on the centralized distributed model', in *Proceedings of IEEE/PES Power Systems Conference and Exposition (PSCE)*, pp. 1–6, March 2009.

[67] Y. Wang, W. Li, and J. Lu, 'Reliability analysis of wide-area measurement system', *IEEE Transactions on Power Delivery*, March 2010.

[68] D. Pendarakis, N. Shrivastava, Z. Liu, and R. Ambrosio, 'Information aggregation and optimized actuation in sensor networks: enabling smart electrical grids', in *Proceedings of IEEE International Conference on Computer Communications (INFOCOM)*, pp. 2386–2390, May 2007.

[69] Y. Yang, F. Lambert, and D. Divan, 'A survey on technologies for implementing sensor networks for power delivery systems', in *Proceedings of IEEE Power Engineering Society General Meeting*, June 2007.

[70] M. di Bisceglie, C. Galdi, A. Vaccaro, and D. Villacci, 'Cooperative sensor networks for voltage quality monitoring in smart grids,' in *Proceedings of IEEE Bucharest PowerTech*, June–July 2009.

[71] I. F. Akyildiz, W. Su, Y. Sankarasubramaniam, and E. Cayirci, 'Wireless sensor networks: a survey', *Computer Networks (Elsevier) Journal*, vol. 38, no. 4, pp. 393–422, March 2002.

[72] IEEE 802.15.4 Standard, 'Wireless medium access control (MAC) and physical layer (PHY) specifications for low-rate wireless personal area networks (LR-WPANs)', October 2003.

[73] V. C. Gungor, B. Lu, and G. P. Hancke, 'Opportunities and challenges of wireless sensor networks in smart grid', *IEEE Transactions on Industrial Electronics*, vol. 57, no. 10, pp. 3557–3564, September 2010.

[74] T. Roberts, 'Update on advanced metering for California's large utilities', *Energy Division California Public Utilities Commission's presentation in CEC Workshop on Advanced Metering Infrastructure*, May 2008.

[75] T. Itoh, Y. Zhang, M. Matsumoto, and R. Maeda, 'Wireless sensor network for power consumption reduction in information and communication systems', *Proceedings of IEEE Sensors*, pp. 572–575, October 2009.

[76] J. R. Pillion, 'Cyber security for PUC's', *Presented at the Mid-America Regulatory Conference*, June 2009 [online]. Available at: http://www.marc-conference.org/2009/presentations/.

[77] A. R. Metke and R .L. Ekl, 'Security technology for smart grid networks', *IEEE Transactions on Smart Grid*, vol. 1, no. 1, pp. 99–107, June 2010.

[78] The Smart Grid Interoperability Panel (SGIP) Cyber Security Working Group (CSWG), USA, 'Introduction to NISTIR 7628 guidelines for smart grid cyber security' [online]. Available at: http://www.nist.gov/smartgrid/, September 2010.

[79] P. McDaniel and S. McLaughlin, 'Security and privacy challenges in the smart grid', *IEEE Security & Privacy*, vol. 7, no. 3, pp. 75–77, May–June 2009.

[80] R. Shein, 'Security measures for advanced metering infrastructure components', in *Proceedings of Power and Energy Engineering Conference (APPEEC)*, pp. 1–3, March 2010.

[81] H. Li, L. Lai, and R. Qiu, 'Compressed meter reading for delay-sensitive and secure load report in smart grid', in *Proceedings of IEEE International Conference on Smart Grid Communications (SmartGridComm)*, 2010.

[82] G. N. Ericsson, 'Cyber security and power system communication-essential parts of a smart grid infrastructure', *IEEE Transactions on Power Delivery*, vol. 25, no. 3, pp. 1501–1507, July 2010.

[83] D. Wei, Y. Lu, M. Jafari, P. Skare, and K. Rohde, 'An integrated security system of protecting smart grid against cyber attacks', in *Proceedings of Innovative Smart Grid Technologies (ISGT)*, January 2010.

[84] International Electrotechnical Commission Technical Committee 57 (IEC TC 57) [online]. Available at: http://tc57.iec.ch/index-tc57.html, accessed: April 2011.

[85] The DNP User Group [online]. Available at: http://www.dnp.org/Modules/Library/, accessed: May 2011.

[86] T. Mander, L. Wang, R. Cheung, and F. Nabhani, 'Adapting the pretty good privacy security style to power system distributed network protocol', in *Proceedings of Large Engineering Systems Conference on Power Engineering*, pp. 79–83, July 2006.

[87] T. Mander, F. Nabhani, L. Wang, and R. Cheung, 'Open-access-compatibility security layer for enhanced protection data transmission', in *Proceedings of IEEE Power Engineering Society General Meeting*, June 2007.

[88] P. Zhang, O. Elkeelany, and L. McDaniel, 'An implementation of secured smart grid Ethernet communications using AES', in *Proceedings of IEEE Southeast Conferences (SoutheastCon)*, pp. 394–397, March 2010.

[89] Smart Grid News [online]. Available at: http://www.smartgridnews.com/, accessed: May 2010.

[90] A. Hamlyn, H. Cheung, T. Mander, L. Wang, C. Yang, and R. Cheung, 'Computer network security management and authentication of smart grids operations', in *Proceedings of IEEE Power and Energy Society General Meeting – Conversion and Delivery of Electrical Energy in the 21st Century*, July 2008.

[91] H. Cheung, T. Mander, A. Hamlyn, C. Yang, and R. Cheung, 'Network-integrated load-management collaborative computing for smart distribution system operations', in *Proceedings of IEEE Lausanne PowerTech*, pp. 1682–1687, July 2008.

[92] L. H. Tsoukalas and R. Gao, 'From smart grids to an energy Internet: assumptions, architectures and requirements', in *Proceedings of Third International Conference on Electric Utility Deregulation and Restructuring and Power Technologies (DRPT)*, pp. 94–98, April 2008.

[93] M. A. Azarm, R. Bari, M. Yue, and Z. Musicki, 'Electrical substation reliability evaluation with emphasis on evolving interdependence on communication infrastructure', in *Proceedings of International Conference on Probabilistic Methods Applied to Power Systems*, pp. 487–491, September 2004.

[94] K. Moslehi and R. Kumar, 'Smart grid – a reliability perspective', *Innovative Smart Grid Technologies (ISGT)*, pp. 1–8, January 2010.

[95] H. Li and R. C. Qiu, 'Need based communication for smart grid: when to inquire power price?', in *Proceedings of IEEE Global Communications Conference (GLOBECOM)*, 2010.

[96] H. Li and W. Zhang, 'QoS routing in smart grid', in *Proceedings of IEEE Global Communications Conference (GLOBECOM)*, 2010.

[97] Trilliant Inc. [online]. Available at: http://www.trilliantinc.com/, accessed: May 2011.

2 New models for networked control in smart grid

Anna Scaglione, Zhifang Wang, and Mahnoosh Alizadeh

2.1 Introduction

There is a growing communication and computation infrastructure in support of the transfer of electrical energy in both the high-voltage (HV) transmission network and the medium and low-voltage (MV/LV) distribution side. Rather than passively witnessing this trend, research efforts are ongoing to study systematically how information architectures can renew and advance power systems. They go under the umbrella of *smart grid*.

To advance this field, it is useful to understand why and in what form this information infrastructure has come about in the first place, and what challenges are intrinsic in the network design problem. Then, we can start questioning if new information networks can be a game changer in the energy sector, and can contribute, in more fundamental ways, to advance power-delivery systems. Can cheap information bits and computation flops make greener and cheaper joules flow in the system? That is *the* question. Some argue that a positive answer may amount to no less than ensuring prosperity for our species [1]. Clearly, the attention on *bits and flops* cannot replace other parallel investigations. But this research deserves some of the spotlight, along with carbon capture, nuclear fusion and other similarly motivated scientific quests centred around sustainable electrical energy systems.

The aim of this chapter is to envision what possible evolution of the power grid cyber-physical system can address the important issue of scaling up the generation capacity of the system, while relying increasingly on green energy and increasing the transmission efficiency. What bits and flops can help with is in turning upside down the notion that finding reliable generation sources is the key to mitigate future energy crises. Rather, we argue that intelligence and information can be used in real time to provide the service electricity is used for, without having a tight grip on the generation.

We start by describing (in Section 2.2) the functions of the information network that support power systems today, and how evolutionary advances in the cyber-infrastructure of the grid can improve these functions. In Sections 2.3 and 2.4 we discuss the ways in which information technology can determine a significant expansion of our reliance on green energy, at an advantageous economic cost.

2.2 Information in today's power system management operations

Initially, power transmission was divided between several providers, each relying on their own generation plants to serve their customers. It was soon recognized that interconnecting the systems could provide higher profitability, thanks to the access to a wider pool of resources and reserves. The electric system was therefore gradually transformed into the large interconnected grids we know today. This transformation introduced redundancy in case of equipment failure or unexpected demand fluctuations, but it is also increasingly required to pool together information to manage the system.

Since their inception in power systems, communication networks have been closely tied to two needs: (i) that of monitoring the safe operation of the grid and logistics of the power delivery, and (ii) that of gathering information needed to operate optimally the generation capacity and, later on, the energy market. In Sections 2.2.1 and 2.2.2 we describe how information is used to manage the network today.

To give an idea of the distance the information must travel, it suffices to think that the power grid in the USA, for example, is divided into three synchronous transmission systems (often referred to as transmission pools): Eastern and Western Interconnection, and Texas. With today's information network infrastructure, updates can take place in a matter of minutes but are not in real time. They are in most cases sufficient to exert global control of the power flow, through a constant correction of the operating point, to match closely generation and load. Fast feedback control in the power grid today is fragmented in local control functions, based on local real time sensor measurements. The existing controls are discussed in Section 2.2.4, after we give an overview of the mathematical models for power system controls in Section 2.2.3.

We conclude the discussion on today's system in Section 2.2.5, outlining the opportunities and challenges faced in advancing networked controls in general.

2.2.1 The management operations in today's power systems

Power outages have devastating economic consequences [2]. Recognizing the priority of maintaining system stability, there has been considerable cross-pollination between control theory and power systems theory over the years.

The stability of the power grid depends primarily on matching closely the load by adjusting the generation supply, under the congestion constraints of the grid. If the system is balanced and the generators are synchronized, the system is stable. Apart from the continuous re-dispatch of power, an automatic global control in the network is the regional frequency control, sometimes known as automatic generation control (AGC). Another one is the secondary regional voltage control that is, for example, performed in the European power grid. None of these controls occurs in real time (i.e., in seconds).

In this section we summarize the established framework to maintain secure operations in the grid through state information [3, 4]. The idea is to group states into four classes, shown in Figure 2.1(a), that define the type of actions applicable in that specific class of states. As shown, it is via the application of control actions that the system remains in or is restored to a secure normal state. The energy-management system awareness mechanism is formally the state-estimator module. In fact, the management of such control decisions is configured as shown in detail in Figure 2.1(b).

The more accurate and timely is the situation awareness, the more effective is the preventive action that can ensure the persistence of the normal operating state. As further shown in Figure 2.1(b), the ability to perform preventive action depends also critically on the security assessment and on the ability to forecast the load. The basic control mechanism exerted is the re-dispatch of power. In fact, load forecasting allows computing the optimal economic power dispatch, generally every 5 minutes to an hour, referred to as

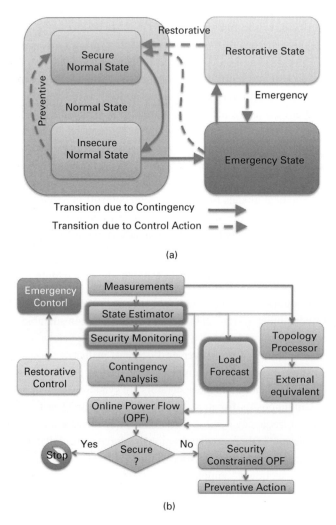

Figure 2.1 Models of the cyber-physical infrastructure: (a) power-system operating states, (b) energy-management system configuration.

the optimal power flow (OPF) [5] if the system is in a secure state. A security-constrained dispatch (SCOPF) [6] is used when the network is classified to be in the set of scenarios that are included in the insecure normal state. The control involves other physical control actions, including preventive, emergency or restorative procedures as well.

The key operation remains balancing demand and supply, however a variety of controls are present throughout the infrastructure, primarily switches but also other types of soft controls that can be remotely activated. As we have already mentioned, the information about the system state that operators and automatic controls act upon today is not in real time. The design of the information network follows the so-called supervisory control and data acquisition (SCADA) model, reviewed next.

2.2.2 Supervisory control and data acquisition (SCADA)

In addition to establishing the hybrid control framework discussed previously in Section 2.2.1, power-engineering research has bundled its essential cyber-infrastructure, for grid monitoring and control, under the SCADA model. Multiple SCADA systems are today deployed within control centres, at plants, and even at a substation. They are used for tasks such as switching on generators, controlling generator output, and switching in or out system elements for maintenance. Over the years, needs were identified, techniques were developed, device and sensors were invented, interfaces and standards were defined.

A SCADA system (see Figure 2.2) usually comprises the following components: a human–machine interface (HMI), a supervisory SCADA master server, a set of remote

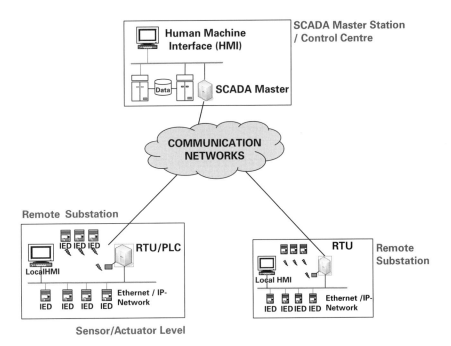

Figure 2.2 SCADA model.

terminal units (RTUs) and/or programmable logic controller, sets of intelligent electronic devices (IEDs) and, last but not least, the communication network infrastructure, connecting the supervisory master and the RTUs, and the RTUs and IEDs.

A SCADA HMI at a central control facility typically presents system-wide data on all generators and substations in the system every 2 to 10 seconds. In the transmission network, SCADA systems include a sensor network of IEDs, where sensor measurements are relayed to the RTUs and from the RTUs to the SCADA server, often using dedicated phone lines or Ethernet connections. The IEDs usually are various types of microprocessor-based controllers of power system equipment, such as circuit breakers, transformers, and capacitor banks.

The commendable activity in defining and producing cyber-physical equipment has, unfortunately, been quite insular. In SCADA, and more generally in hybrid systems control, the adoption of computer and networking hardware has been considered a system-engineering problem, and often overly simplistic assumptions about computer networks were made in the design integration process. Communication links and networks are treated as secure and perfect bit pipes, with some intrinsic delays that are physical features of the systems rather than results of the traffic conditions and the way the communications are managed.

If the grid is in a stable state and if synchronized measurements are only used for system visualization, then the communication latency of today's SCADA system is not critical, since it is sufficient to update operator displays every few seconds. However, if the grid experiences contingencies, security or oscillation monitoring, or prediction algorithms assessing whether the power system is moving into an unstable state, cannot lead to meaningful control actions at the pace information is delivered in SCADA systems today. The way failures are averted today is by setting wide margins, and by using redundancy that allows local controls to isolate individual parts, activating a redundant path on standby.

Today's SCADA model leaves a great deal of control to the human operator in the loop. This means that control actions based on global state information are relatively slow, which forces transmission equipment to be used inefficiently. As our electrical energy needs grow, the current trend of using the transmission resources inefficiently is not sustainable.

Recently, we have also become increasingly aware of the threats of computer attacks on SCADA units, causing significant disruptions to power systems [7]. That is why it is now of paramount importance to understand what are the right models for cyber-physical security, and how they differ from standard network security problems.

2.2.3 Basic models for power system controls

The dynamical characteristics of power grids may range widely from very fast to very slow. A large number of controllers have been developed over time to deal with different phenomena in order to enable stable operation and improve system efficiency. The evolution of power grid controls depends on several factors, such as the availability of new power electronics equipment and hardware, improved communications and computations devices, as well as advances in control theory. The so-called *fast* or *primary controls* (whose response time ranges from milliseconds to a few minutes) depress transient disturbances in the power grid, through protective relays and primary frequency or voltage controls. Other controls are much slower (from minutes to hours), and consist of regional or system-wide operational adjustments or planning; examples are the secondary voltage controls or the balancing of the slowly changing system load, by adjusting generation levels (the OPF or SOPF discussed in Section 2.2.1).

Some controllers are discrete, such as the protective relays, the on-line tap-changers (OLTC), and the switching of capacitor/reactor banks. Others are continuous, such as power-system stabilizers (PSS), and voltage controllers and generation control.

Existing power-grid controls are based on the following two models: steady-state and dynamic-state models.

1. *Steady-state model.* The power grid is said to be in a steady state if the system power is balanced and all the rotating generators rotate synchronously at the same frequency, $f_0 = 60$ or 50 Hz. The steady-state operation of a power grid can be captured by a set of power-flow equations [8]. The voltage on the mains is a narrow bandpass signal, whose power spectrum is centred around f_0. The complex phasor vectors of the bus voltages, $V = \text{V} \angle \theta_v$, and the current injections, $I = \text{I} \angle \theta_I$, are vectors listing the complex envelopes of the actual bandpass voltage vector $v(t)$ and injected current vector $i(t)$ respectively, around f_0. In the NB regime the power network dynamics are coupled by the algebraic equation

$$YV = I, \qquad (2.1)$$

where Y is the matrix of network admittances at frequency equal to f_0, which is defined not only by the connecting topology but also by its electrical parameters. And the network power flow equation is

$$S = g(V) = V \circ I^* = V \circ (YV)^*, \qquad (2.2)$$

where $\circ$ means element-wise multiplication and $(\cdot)^*$ indicates complex conjugation. $S = P + jQ$ is the vector of injected complex power with $P = \text{Re}(V \circ I^*) = \text{VI}\cos(\theta_V - \theta_I)$ and $Q = \text{Im}(V \circ I^*) = \text{VI}\sin(\theta_V - \theta_I)$, where P is the *real power* or *active power*, which is equal to the DC component of the instantaneous power $p(t) = v(t)i(t)$; whereas Q is the *reactive power* which corresponds to the $2f_0$ sinusoid component in $p(t)$ with zero average and magnitude Q.

A set of basic constraints needs to be satisfied for enforcing stability in the power grid: (a) the network power flow must be balanced; (b) the input power for generation or loads adjustment or power injects from other kinds of sources must lie within strict operational ranges; (c) voltage must take acceptable levels; (d) line thermal limits must be enforced, i.e., line current should keep its magnitude below a specified limit; (e) stability conditions must be satisfied, i.e., the Jacobian matrix $J(V) = [\frac{\partial P}{\partial \theta_V} \; \frac{\partial P}{\partial V}; \; \frac{\partial Q}{\partial \theta_V} \; \frac{\partial Q}{\partial V}]$ of the network power-flow equations must have negative real parts which keep a safe distance from zero.

Power system planning, steady-state analysis, and market operation are usually based on the steady-state model of (8.3) with various optimization objectives and/or additional operating constraints. Due to the non-linearity of the model and the large problem size related to a large interconnect grid, sometimes a linearized DC approximation is used instead. It is also typical practice to reduce the network size by taking an equivalency approximation for the outside areas except the concerned part of the grid, or to divide the whole system into a number of smaller-size subnetworks and solve the problem using an iterative algorithm.

2. *Dynamic model.* Any power imbalance in the system caused by disturbances, such as varying loads or random contingencies in the transmission network, tends to perturb the dynamic equilibrium of the generators. Relatively small changes, compared to the inertia of the total rotational mass of all the machines, can be continually corrected to keep the system close to synchronism. Large disturbances, like a short circuit on an important transmission line or the loss of a significant generator or load, however, may cause some generation machines to deviate from the rated frequency and become unstable [9]. These changes need to be isolated and compensated in time in order to protect devices, prevent disturbance propagation, and avoid cascading failures or even large interruption of power supply.

The dynamics of a rotating machine (inside a generator) can be represented in the simplest second-order differential equation as follows [8]:

$$M\dot{\omega} + D\dot{\delta} = P_m - P_e, \qquad (2.3)$$

where δ represents the deviation of the shaft rotational angle from synchronism, $\omega = \dot{\delta}$ is the deviation from synchronous speed $2\pi f_0$, M is the rotational inertia, and D the damping coefficient. The difference between the mechanical and electrical powers P_m (infused from outside energy conversion) and P_e (injected to the power grid network) is called the accelerating power, which tends to drive the generator shaft away from its steady-state synchronism. Equation (2.3) can be expanded to include higher-order terms capturing the dynamics of the machine's excitation system, the power-system stabilizer (PSS), etc. Thus, a generalized dynamic model for the whole grid can be written as a set of differential and algebraic equations (DAE):

$$\dot{X} = f(X, V, S; Y) \qquad (2.4)$$
$$S = g(V; Y). \qquad (2.5)$$

X is the vector of dynamic states which includes the generator shaft angle, rotation speed, terminal voltage, and/or other internal variables of the machine, exciter, and governor. Equation (2.4) indicates that the dynamic behaviour of X depends on $f(X, V, S; Y)$, a function of X, the network states V, and the electric power injections S from the generation and loads; and also related to the network conditions Y. It is clear that the number of differential equations scales by the total number of generators, while the number of algebraic equations scales by the total number of buses in a grid. Therefore, a large interconnected power system may typically be represented by several hundred differential and several thousand algebraic equations [9]. This large problem size brings significant challenges in the design of dynamic controls for the power grid. Hence it is not surprising that most fast controls in today's power grids are *near-sighted*, and are local controls based on approximations of the network model.

2.2.4 Existing power grid controls

Various controls have been developed over time to enable stable operations and improve system reliability and efficiency. Below we give a brief overview of existing power grid controls.

1. *Power-system protection* [10]. Power grids implement many types of protection technologies from wire fuses to microprocessor-based relays, to protect the system from damage by sensing and isolating short-circuit faults. A short-circuit fault is detected by a sudden decrease (or increase) in the voltage (or current) at the locale of the faulted equipment, which is thus identified and isolated by the circuit breakers. Power-system protection can be viewed as a fast method of control since it operates quickly, often in tens of milliseconds. Modern protective systems usually comprise five components: current/voltage transformers for measurement; protective relays for control; circuit breakers to open/close the circuits; batteries to provide power in case of disconnection; and communication channels to allow remote data analysis and remote circuit tripping.
2. *Frequency control* [9]. The power-system frequency is directly affected by the power balancing between loads and generation. The primary-level frequency control is fast and local at the generator. It is implemented on the governor, which senses the shaft rotation speed deviation and adjusts the mechanical input power accordingly. A secondary frequency control, also known as automatic generation control (AGC) or load frequency control (LFC), is implemented by the central controller to set the governor set points in order to ensure system-wide power balance and delivery efficiency. In a wider definition, market operations in a deregulated grid can be viewed as a system-wide frequency control, since the market task is balancing the generation and load requests based on the bids from both suppliers and buyers to achieve an optimal economic dispatch while keeping the whole system stable and efficient.
3. *Voltage control* [11]. Voltage control in an AC transmission network is closely related to the reactive power compensation. There are a number of ways to undertake voltage control in a grid: excitation adjustment of generators, tap changing of transformers (OLTC), manually or automatically switched shunt capacitors or reactors, etc. Power electronic control devices such as static var controllers (SVC) have been introduced more recently to provide more continuous and fast-acting reactive power, to handle dynamic voltage swings on high-voltage electricity transmission networks. Voltage control is mostly a local control, since reactive power cannot be transferred over a long distance. In some European countries, coordinated voltage controls have been implemented to control the reactive power compensation over an area, and some

research efforts and experiments have been made to introduce area voltage controls in the North American systems [9].

4. *Transmission power-flow control* [9]. Compared with the frequency and voltage controls, the flow control on transmission lines in a grid is much weaker. Although complete flow control of every transmission line in a system is neither feasible nor desirable, flow controls on some important lines have always been necessary in order to gain a better control of flow transfers and to avoid overload. This is especially true when it comes to power market operations, where the transactions between buyers and sellers need to be monitored and adjusted as scheduled. A more traditional type of flow control is through a phase-shifting transformer that allows the power flow to be controlled across itself, by direct manipulation of the phase angle, using taps. The control is local, discrete, and slow. Phase-shifting transformers have been used, especially on the Eastern Interconnection in North America. High-voltage DC (HVDC) lines [13] have been implemented in the transmission network for their advantages of lower long-distance transmission loss, enhanced system stability, and allowance for power transfer between unsynchronized systems. Flow over DC transmission lines is always controlled and the control is very fast. Flexible AC transmission systems (FACTS), a modern type of flow control of power electronics and other equipment, control one or more AC transmission system parameters, to adjust the power transfer capability of the network [12, 13]. Distributed FACTS devices are smaller in size and less expensive than traditional FACTS devices, and may be better candidates for wide-scale deployment [14–17].

As has been mentioned before, most fast controls are local, which means the control loops are confined within the same locale (substation). Unfortunately, most dynamic phenomena in power-systems have regional or sometimes even system-wide impact [9]. As a result, handling system-wide stability with local controllers is the constant struggle for designers of power-system controls. As we will discuss in the next section, the enhanced measuring functionality in smart grid will alleviate this dilemma. However, in order to achieve an effective design of fast global controls, one also needs to take into account several critical factors: (i) accurate system models; (ii) coherent, real-time measurement data available; (iii) adequate communication infrastructure; and (iv) effective networked control designs.

2.2.5 The intrinsic difficulties of networked control

Control under communication constraints has been an active area of investigation in the last 10 years. The theories developed in this context apply broadly to cyber-physical systems (see, e.g., [20, 21, 23, 24]). While the most recent literature gives a more positive and constructive outlook on the problem of networked control, modular and scalable solutions of networked control are still elusive in many cases. The most difficult class of problems in networked control arise when a separation of control and communications time scales is impossible. In these cases, networked control problems are hard to simplify and become intrinsically very complex, if not completely intractable, because of the lack of modularity between communication and control. It has long been known from the celebrated Witsenhausen's counterexample that, in these cases, the separation of estimation and controller design fails to hold even in the simplest settings [19]. Unlike transportation, water network, and other infrastructures that distribute commodities and are encountered in large-scale supply chains, the power grid delivers electricity as a commodity just as fast as communication signals do. Therefore, both the physical network dynamics and the cyber-system data spread at comparable speeds, exacerbating the difficulties of decoupling communications from control and management.

From a control perspective, it is then important to achieve a better understanding of *what* is sufficiently informative for stability, as well as *when* and *where* difficult decisions are to be made [25–43]. Part of the difficulty in the optimization of concurrent controllers is that each controller can infer information about unobservable events not only by pooling sensor information, but also by observing the other controllers' actions [35]. Due to the complex interdependencies that exist between *exploiting* and *exploring* information from the network, in some important cases the controllability of a discrete-event system is undecidable [44].

In light of this it is clear that, in principle, the optimal solution does not have a modular structure. In practice, cyber-infrastructures for managing physical systems can be designed, in spite of the challenges in producing optimal control strategies. In fact, the intrinsic theoretical challenges are in establishing an optimal control and communication strategy. A sensible trend prevailing today is what is labelled *co-design*: the control design starts from approximating the communication infrastructure as being packet switched, in most cases asynchronous, and it captures the effect of delays as packet losses [23]. The communication network design tries to leverage on the clear definition of the application objectives to control the quality of service. The concept of *co-design* is similar to that of *cross-layer* design, where information is shared among the application and network layer in order to improve performance.

However, optimality of protocols that emerge from these approaches cannot be claimed from either the control or the communication perspective.

2.3   Enhanced smart grid measuring functionalities

In Section 2.2.3 we have clarified that the state of the system, given the grid admittance matrix, depends on the voltage phasors at the load and generation buses. There cannot be an accurate measurement of phase if not relative to a common reference. Therefore, one of the key technologies that has changed significantly the outlook of sensing in the power grid is the wide availability of accurate global network synchronization, thanks to GPS receivers.

The protection and controls envisioned in smart grid will be highly enhanced by time-synchronized measurement technologies, which can be found at synchronous phase measurement units (PMUs) and the PMU function incorporated into a protective relay or other devices. *Synchrophasors* is a common name used to refer to these measurements.

The voltage or current measurements are taken with time stamps from global positioning system (GPS)-based clocks with microsecond time accuracy. And the reporting rate can be up to four times the 60 Hz mains frequency.

A decade ago, phasor measurements were taken only using stand-alone instruments, called PMUs, and only in the high-voltage (HV) transmission network. Today, such measurements may also become available from various intelligent electronic devices (IEDs) such as protective relays and fault recorders, and from the smart meters at customer households. That is, the coherent and real-time measuring functionality has become available at a much wider range of devices and even expanded to the distribution side as well. This trend dramatically lowers the cost of implementing synchrophasor-based control and protection strategies. In addition, synchronized measurements regarding the network topology, system frequency, line flows, and even devices settings, weather conditions, and predictions, etc. will also be provided by the enhanced measuring infrastructure of smart grid.

It can be expected that the enhanced measuring abilities in smart grid will bring fundamental improvements to existing power-system controls that will allow the transmission network to be used more efficiently.

However, the second challenge is to upgrade the data delivery infrastructure, to fully take advantage of the enhanced smart grid measuring functionalities to perform automatic networked control applications. In fact, parallel to the development of synchrophasors and the widespread deployment of synchronized measurement units, new wide-area measurement systems (WAMS) shall supplant SCADA systems. In the following sections we will discuss these issues in more detail.

2.3.1 State estimation

Power-system state estimation has always been an essential data-processing tool in the modern energy-management system (EMS), and has evolved in today's industry as a very important resource for the so-called locational marginal pricing (LMP) algorithms, used in the energy market to determine how to charge for congestion in the transmission network [18].

Since the 1970s [45, 46] the universally accepted problem formulation is to fuse available real-time PMU measurements and *denoise* the data through weighted non-linear least-squares SSE of $V(t) = (\text{V}(t), \angle V(t))$, i.e., voltage amplitude and phase at time t. In other words, the state is derived as the solution of a weighted least-squares problem fitting the measurements, say phasor $z_i(t) \in \mathbb{C}$ for sensor i, with the corresponding model equation, $\gamma_i(V(t))$. Denoting by R the measurements' noise covariance, the least-squares estimation solves for the argument $V(t)$ that gives $\min_{V(t)} (z(t) - \gamma(V(t)))^T R^{-1} (z(t) - \gamma(V(t)))$. Due to the non-synchronized measurements and model non-linearities, the traditional state estimator has limitations in computation time, solution errors, and convergence.

There are a number of papers in the power network literature that propose to distribute this computation: (i) [47] decomposes measurements based on geographical location; (ii) [48, 49] divide the process into two or more levels of aggregation, including simultaneous re-estimation of boundary states; (iii) [50] proposes a constrained optimization in which overlapping subsystems are formed with a common zero-injection boundary bus; (iv) virtual boundaries, overlapping subsystems, and the auxiliary problem principle are used in [51]; (v) multi-area estimation using synchronized phasor measurements in [52] and [53] uses updates at boundary buses; (vi) the use of the Dantzig–Wolfe decomposition algorithm [54] or of recursive quadratic programming and the dual method is considered in [55]. All these methods typically rely upon aggregation trees. The partial aggregation mitigates the problem of network bottlenecks but does so in a way that is rigid, without providing a natural way of reconfiguring the system if failure or attacks call for it. Also, these frameworks do not provide a clear indication of what is the impact of the communication network connectivity, how to trade off delay with accuracy for a given network fabric, and what is the impact of data loss and link failure. They do not address the issue of sharing only partial information without introducing approximations that reduce the accuracy of the resulting state estimate.

In smart grid, state estimation will increasingly rely on synchronized PMU measurements to achieve fast and accurate state calculation, which is critical for the quick response time requirements of wide-area protection and control loops. With more widely deployed PMU measuring functionalities across the network, states at more bus locations will be directly and more accurately obtained and the data can easily be synchronized according to the embedded time stamp, which will largely mitigate the time skew, convergence, and computation time issues of the traditional state estimator [76].

In the past, state estimation has been applied exclusively to the transmission network. However, as smart grid technology encompasses a variety of objectives, more attention is being paid to the operation of the distribution side. As will be discussed in Section 2.4, a large penetration of renewable energy sources (RES) and demand-side management (DSM), and demand response (DR) programs will be associated with an increasingly dynamic and complex distribution side of the grid.

The vast deployment of the advanced metering infrastructure (AMI) [56] has polarized considerable attention and is a first step in this direction. Synchronized sensor measurements may have a central role in the state estimation of this new and dynamic distribution grid. Unfortunately, current deployments of AMI have a rather inefficient topology: a data hub or concentrator service accrues measurement data from smart meters by polling the devices periodically, and then forwards the data to the utility database infrastructure. This model does not scale, and is subpar to what networking research can do today. The availability of scalable networking alternatives, as well as decentralized and fully automated processing, will allow the embedded intelligence in the system to be connected in a way that will support each of the physical devices with real-time feedback from its neighbouring devices [77, 78].

In the literature, there have been a variety of research efforts addressing the challenges to deploy state-estimation in distribution networks [79–84]. The extension of state estimation to the distribution network is not straightforward, since typically the medium-voltage/low-voltage (MV/LV) sections of the grid can be significantly different from the transmission network in their topology (mostly radial instead of meshed), network size ($10\sim100$ K vs. several hundred or thousand buses), degree of phase imbalance (large instead of negligibly small), and available measurement redundancy (modest compared with large redundancy).

2.3.2 Wide-area measurement system (WAMS) and GridStat

Deregulation, competition, and increase in complexity of today's power networks have exacerbated power-system stability issues. The specific problems faced today are system-wide disturbances, which are not ably covered by existing protection and networked control systems. As the load increases, sudden bulk power transfers make the system very vulnerable; even minor equipment failures can result in cascading relay-tripping events and eventually, blackouts.

To ensure system stability in a heavily loaded system, all or most installed components should remain in service and the correct actions must be taken quickly if the system has not recovered after a serious event. To cater to this requirement, the solution is to have real-time monitoring. Such a wide-area measurement system provides operators with real-time knowledge of various instability issues and events as they occur. This early warning system provides operators with much needed time for counteraction as well as choices for action. Eventually, such a system can provide leading operator guidance on the best course of action, as well as a base for automatic wide area control. However, conventional SCADA/EMS systems can control a limited portion of the grid and in a constrained fashion. They do not support dynamic control of a wide area power network [57].

Many utilities, government organizations, and a few manufacturers have been working on this technology and approach for the last decade, concluding that real-time monitoring with WAMS is an economically viable answer to ensure transmission reliability, as setting large margins and redundancy become too expensive to sustain. The implementation has, however, been hampered by a variety of reasons, the cost or the availability of standard solutions being the least significant of all. It was also determined that having an adaptable system is better than just strengthening the network support. To achieve this adaptability, real-time information is a necessity.

In the early 1990s the Bonneville Power Administration brought about a revolutionary design to monitor wide-area power networks based on synchronized measurements when expanding its smart grid research with prototype sensors that are capable of very rapid analysis of anomalies in electricity quality over very large geographic areas. The first operational WAMS became available in 2000 [58, 59]. The 14 August 2003 blackout in the North America Eastern Interconnection revealed the urgent need for better practices in wide-area information acquisition and stimulated the first implementations of WAMS [60].

WAMS utilize a back-bone synchrophasors network which consists of phasor measurement units (PMUs) dispersed throughout the transmission system, phasor data concentrators (PDCs) to collect the information, and/or a SCADA master system at the central control facility.

There are two kinds of WAMS network implementations: the first one is a stand-alone communication network and is primarily utilized for research and development efforts;

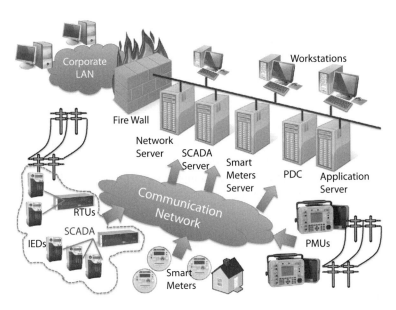

Figure 2.3 Integrated implementation architecture of WAMS and SCADA [61].

the second one is integrated into the control centre network (CCN) infrastructure and complements conventional SCADA/EMS systems, providing a wide-area system view [61] (see Figure 2.3).

Communication protocols serve as the basis for data transfer within the WAMS. The majority of the protocols used by a WAMS, however, do not include security mechanisms in their specifications. It is left to the implementation of the protocol to employ security mechanisms [61]. As WAMS advanced, new technologies were incorporated. Current WAMS include many transmission protocols such as serial, Modbus [62], OPC [63], IEEE 1334 [64], PDCStream [65, 66], and most recently IEEE C37.118 [64] to transmit data. Especially, IEEE standard 1344 is the first standardized protocol developed for synchronization and transmission of phasor measurements. This standard defines data formats for the transmission of phasor measurements and contains specifications for time synchronization and data conversion formats but does not address response time, measurement accuracy, or a process for calculating phasor measurements. The IEEE standard 1344 was superseded by IEEE C37.118 in 2005, which addresses several shortcomings identified in IEEE standard 1344. Unlike IEEE standard 1344, IEEE standard C37.118 accounts for phasor measurements from multiple PMUs. Additional fields have also been included to add needed functionality, as well as to conform to other standards and quantify the phasor measurement error. Data is usually streamed in this format over UDP/IP or across a serial link [60, 64]. WAMS in use today utilize several common communication technologies and models, as follows [60, 61]:

1. *Serial.* Serial communication is primarily used in SCADA networks and provided the original infrastructure for the WAMS. Some network owners still use serial communication in portions of their WAMS network. However, high-speed modems are

required to achieve the 30 samples/s or higher data rate of the WAMS communication protocols.
2. *Analogue microwave.* Analogue microwave radio systems have been used to transmit and receive information between two points dating from the 1950s. Analogue transmission uses a continuous signal to carry information from point to point. Analogue microwave is used by WAMS network owners to connect remote PMUs to a PDC when a wired connection is not feasible.
3. *Virtual private network (VPN).* VPN is a term used to describe various technologies that secure communication between two parties across a non-secure public network. Those network owners who employ VPNs in their WAMS use the IP Security (IPSEC) suite of protocols to establish the VPN. The IPSEC suite of protocols provides methods for authentication and encryption of WAMS data at the IP packet level. IPSEC also employs methods for cryptographic key establishment [67].
4. *Ethernet.* Ethernet connections have gained more popularity for local-area communications in WAMS.

As WAMS evolve, more fibre-optics and licensed digital microwave technologies will be incorporated into the WAMS communication infrastructure, to provide enhanced data delivery services.

During the past decade, various applications have been proposed and/or developed based on WAMS to enhance global system awareness, stability, and efficiency:

1. *System dynamic analysis.* Synchronized real-time data from WAMS enables global real-time dynamics display of phasors, power, etc.; enhances state estimation for SCADA/EMS applications; provides system-wide alarming and alert; and can be used to perform real-time data filtering and time/frequency-domain response analysis such as spectra analysis, machine model verification, and damping/oscillation analysis [68].
2. *Dynamic process record.* Generate system-wide dynamic disturbance recordings and log the system dynamic behaviour accordingly [68, 72].
3. *Transient stability prediction and control.* Includes dynamic transaction limits monitoring and dynamic performance monitoring [68].
4. *System-wide stability monitoring and control.* WAMS enables system-wide coordinated actions of protective relays. Especially the wide-area emergency protection and controls are designed to inhibit the fast propagation of cascading outages, enable intelligent system isolation, and avoid large-area or system-wide blackouts [68, 72].
5. *Low-frequency inter-system oscillation analysis and suppression.* Low-frequency ($< 2\,\text{Hz}$) oscillation is often observed in large interconnection of power systems and sometimes can be a serious stability threat if not suppressed [69]. This problem received considerable attention in the past. With synchronized measurements from WAMS, low-frequency oscillations between interconnected systems can be more accurately analysed (see, e.g., [70, 71]).
6. *Global feedback controls.* Fast system-wide frequency/voltage/flow controls will become available based on real-time WAMS data. One prominent example is given by FACTS control algorithms, based on synchronized PMU measurements. These

algorithms can potentially be implemented as an effective wide-area control aimed at mitigating subsynchronous oscillations. Global feedback controls bring challenges to current SCADA/WAMS systems as measurements must be inter-operable, consistent, and meet the real-time requirement of fast transient and voltage stability control [60, 68, 72].

However, to fully attain benefits of WAMS, the grid needs: (i) sufficient WAMS coverage for the whole interconnected grids; and (ii) an efficient design for data delivery services over this infrastructure. North America SynchroPhasor Initiative Network (NASPI or NASPInet) is an effort to develop an 'industrial grade', secure, standardized, distributed, and expandable data-communications infrastructure to support synchrophasor applications in North America. More than 200 PMUs are already installed and networked throughout North America and more are expected to be installed soon [73], see Figure 2.4.

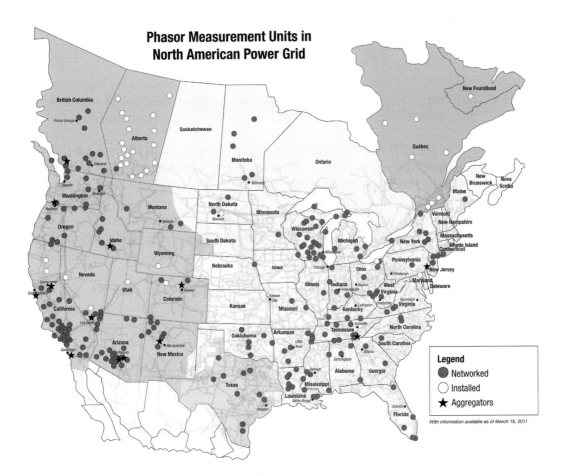

Figure 2.4   PMUs in North American power grid.

Power-system networked controls have a wide range of data-delivery requirements. In order to provide satisfactory communication service for various controls under different system conditions, it might be advisable to bypass the communication bottlenecks at the hierarchical SCADA architecture but design more efficient and adaptable data-delivery protocols directly on the WAMS network.

Middleware is necessary to streamline the design of control applications that can interact with WAMS. A notable example is *GridStat*, which is a publish–subscribe middleware architecture that has been proposed by a team of researchers led by EECS at Washington State University [74, 76] to provide an interoperable data delivery platform and address the communication QoS requirements including security for WAMS applications [74]. Specifically, key factors for these power applications that influence how the delivery system should be planned, implemented, and managed have been described, normalized, and then quantitatively compared. These factors include whether a person or computer is in the loop and application requirements such as latency, rate, criticality, quantity, and geographic scope.

For example, in the GridStat framework, the synchrophasors take the role of publishers which periodically announce status values, while the controllers choose to be subscribers, which periodically receive status values as specified. GridStat is implemented by simple and CORBA-compliant APIs for both publishers and subscribers, management/control infrastructure, etc. Subscribers have a transparent cache of latest status value. Networks of internal servers are managed to satisfy different communication QoS requirement for each data delivery. Therefore, data communications are optimized for semantics of status items instead of arbitrary event delivery, like generic publish–subscribe [74, 75].

2.4 Demand-side management and demand response: the key to distribute cheap and green electrons

We have mentioned that the transmission network is a bottleneck to deliver green electrons to consumers. While fast global optimal control can reduce the wide margins left in using transmission resources efficiently, this is insufficient to loosen the grip on the generation.

Currently, there are two very opposite views on how green generation can be integrated into the wholesale power grid. (i) *Generation reserves*: volatile and non-dispatchable plants using wind and solar energy should be backed by clean but controllable resources like hydro or natural gas units, that will start generating energy whenever the intermittent resources are not able to meet their scheduled generation requirements. This is the approach already followed today to handle the current portfolio of green energy. But the main problem with this solution is the limited availability of clean but dispatchable plants. The Department of Energy (DOE) reports that today 10% of all generation assets and 25% of distribution infrastructure are used less than 400 hours per year (5% of the time). To triple the penetration of renewables, these numbers will more than triple. (ii) *Demand-side management*: the volatility introduced by these intermittent resources on the generation side will be compensated by responsive and controllable loads, such as real-time pricing (RTP), or electrical appliances managed through direct load control (DLC) programs.

As we will see, DLC programs are mostly designed for emergency situations today, while RTP programs, although very appealing, face the very challenging problem of determining what these price signals should be.

Even if DSM is still a subject of intense debate and investigation, it is clear that the second approach is aimed at correcting the aberration of the current energy market and that the best pathway towards greener electrical energy systems is starting to control the load [89]. Namely, DSM addresses the key problem that the majority of the energy customers today are unaware of both the control as well as the market costs associated with turning on their electrical appliances at a given time. These costs exist, but they are socialized. The fact that customers are shielded from making aware economic decisions leads to a form of *tragedy of the commons* [90]. In the long term, this is the root cause of the lack of innovation that has characterized the power system infrastructure over the last 40 years. In this socialized model customers do not see the direct benefit of an increased electricity bill that is invested in research and development, because of the promise of granting to them and to future generations access to abundant, cheap, and green energy. In the short term, the market pressure to feed this mostly uncontrollable demand raises significantly the premium for employing green energy by forcing the large reserves that we mentioned above to be stockpiled.

To understand specifically why that is the case, some background on the way the energy market is operated today is necessary. Note that different pricing algorithms for DSM will be discussed in detail in Chapter 3 of this book.

2.4.1 The central electricity market

When we think of the global control of generation in the grid, we cannot separate it from the market where energy is sold. In this sense, this problem is well beyond any of the issues that are in the horizon of networked control theory.

Clearly, one of the advantages of resorting to electrical energy to power the world is that electricity can be produced in many ways and distributed rapidly, as long as the transmission lines can support the flow. Unfortunately, because the demand is not controlled, these are also the curses of this commodity market, since not all sources of electrical energy are equally ready to produce what is requested to satisfy this impatient and capricious marketplace, and the most volatile energy sources are also the ones which pollute the least. Currently, the electricity market consists of the following [85]:

1. *A central wholesale market.* In the USA, wholesale electricity markets are multi-unit uniform-price auctions, operated in most cases by an independent system operator (ISO). These auctions feature dual settlement systems: there is a day-ahead auction, run daily for each hour of the following day, as well as real-time auctions, run every few minutes during the day of the contract. In the day-ahead market, generators and retailers purchase and sell energy at financially binding day-ahead prices. Hourly bids to this market must be submitted at noon the day before the day of delivery. The

ISO then puts up a schedule of commitments for the purchase, sale, and ancillary services of energy, based on the participants' bids and constraints. *Of course, these transactions are just based on predictions of energy demand.* Real-time adjustments through additional balancing markets are necessary, as well as re-dispatch which we already discussed in Section 2.2.1 since, as we said before, keeping the power balanced is the most important factor in the operation of a power grid (mostly for frequency control). The important aspect to keep in mind for our work is that the market is designed such that deviating from the day-ahead bid in real time will result in higher costs for the participants due to deviation penalties (downward and upward balancing costs).

2. *Several retail markets* in which the retailers that purchased energy from the wholesale market on-sell it to end-use customers.
3. *A transmission services market* that determines the allocation and prices for transmission rights on power grid lines.

As has been mentioned, the wholesale electricity market is a dual-settlement auction for electricity run by an independent system operator (ISO): a day-ahead market and a spot (real-time) market. One of the most important benefits of having a dual-settlement system is price certainty. Setting day-ahead prices helps the retailers avoid volatility of real-time market prices. This will help the reliable function of the power grid and ensures the availability of services to end-use customers, whose demand is treated as being inelastic to price. Another reason for running a day-ahead market is the long start-up times of generation units.

1. *Day-ahead market.* The participants in the day-ahead market and the information they provide to this ISO are as follows:

    (a) Producers submit their generation capability and price bids for each hour of the next day. They will also provide ancillary service bids if applicable. Also, the ISO has information about each unit's start-up and shut-down costs, no-load costs, and ramping limits. The goal of generators is to maximize their profits. Since the wholesale market for electricity operates with a single market-clearing price, based on the marginal cost of the last unit of generation, generators usually provide the ISO with a low-enough price for electricity so that they are included in the generation schedule.

    (b) Retailers provide bids for each hour of the next day. Participating loads have two types: fixed and price-sensitive loads. Price-sensitive loads usually provide price caps for their bids.

    (c) Virtual participants (virtual supplies and loads) will also submit their bids to the day-ahead market. Virtual bidding allows participants with no real demand/supply capacity to participate in the day-ahead/real-time market.

Provided with the above information, the ISO schedules generation to meet specified (predicted) demand using economic dispatch strategies that take the transmission and generation limitations of the system into account. The economic dispatch solution identifies the generators that are called on for energy and ancillary services and informs them of their schedules. A simplified formulation is given here.

Assume that we have n_g generation units and n_l load buses. The cost of generating electricity for each unit is an arbitrary function $C_i(.)$ (There is a cost for the no-load state, in the event that a generator is on but not producing energy.) Also, each unit has its associated start-up and shut-down costs which depend on the state of the generating unit (hot, warm or cold state) and its type. The load forecasts are assumed to be constant for half-hour periods. Now, taking the costs due to start-up, shut-down, and generation of each utility into account, the economic dispatch problem for the next day can be formulated as an optimization problem (see, e.g., [86]):

$$\min_{G_{i,j}} \sum_{i=1}^{n_g} \sum_{j=1}^{48} C(G_{i,j}) + u_{i,48} SU_i + d_{i,48} SD_i, \quad (2.6)$$

where

$G_{i,j}$ = generation schedule of the ith unit at time j
$L_{i,j}$ = load bid at node i for time j
$C_i(.)$ = generation cost function of the ith unit
SU_i = cost of a single start-up for unit i (2.7)
SD_i = cost of a single shut-down for unit i
$u_{i,48}$ = total number of start-ups that occur over the day for unit i
$d_{i,48}$ = total number of shut-downs that occur over the day for unit i.

This optimization has multiple constraints, some of the most important ones being:

- $\sum_{i=1}^{n_g} G_{i,j} = \sum_{i=1}^{n_l} L_{i,j}$ load and generation balance.
- $0 \leq G_{i,j} \leq X_i$, generation of the ith unit limited by its capacity X_i.
- $|F_j| = D[G_{1,j} \ldots G_{n_g,j} L_{1,j} \ldots L_{n_l,j}]^T \leq F_{\max}$, element by element. (2.8)
- Required ancillary services.
- The system should not fail after 1 contingency.

An integer linear programming technique will then be used to simultaneously optimize energy and ancillary services in the above optimization problem. The out-coming schedules from this optimization are financially binding. The ISO also sets locational marginal prices for energy (using shadow price, the cost of serving the last MW at that particular load).

2. *Real-time market.* This market is designed to balance generation and demand every five minutes, since keeping energy balance is one of the most important constraints in the operation of power-delivery systems. An optimization problem similar to the above formulation will be solved to provide a least-cost dispatch for real-time resources while satisfying the line flow and ancillary services constraints. The participants in the day-ahead market and the information they provide to the ISO are as follows:

(a) Generation units provide adjusted price curves every 15 minutes to include real-time constraints and costs.
(b) Retailers provide bids for how much energy they want to buy or sell in the day-ahead market.
(c) Virtual bidders try to sell what they have purchased in the day-ahead market.

The ISO will then try to solve the following minimum-cost problem:

$$\min_{G_i} \sum_{i=1}^{n_g} C(G_i) \qquad (2.9)$$

subject to constraints similar to the day-ahead optimization.

In scenarios where the load exceeds its predicted value, the ISO takes bids from generators to increase their outputs, which leads to a market-clearing upward balancing price. On the other hand, if the load does not reach its predicted value, the ISO calls bids for generation decrease (downward balancing). The upward balancing price is the maximum of the day-ahead market price and the minimum balancing price, while the downward balancing price is the minimum of these two. Several papers have explored options on how to forecast these balancing prices [87].

3. *Retail market.* In most of the power systems in the world, inelasticity of demand is an accepted fact upon which the system operates. Almost all controls, monitors, and feedbacks required for a power system's safe operation are implemented in the generation and transmission section. As a result of this accepted principle, the retail price of electricity is usually constant and is determined using average pricing in a long horizon. In the retail markets for electricity, retailers that participate in the wholesale market on-sell energy to end-use customers in real time. The participants in this market therefore include:

(a) Retailers, who try to meet the demand as closely as possible to avoid contingencies and keep demand and supply balanced. To provide this energy, they forecast their consumers' load for the next day and participate in the day-ahead market. To balance the actual demand and supply, they also participate in the spot market and purchase energy at real-time prices. The goal of these participants is to maximize their profit from selling energy while avoiding any penalties from contingencies or blackouts.
(b) Consumers, who pay for their electricity use based on a predetermined price table, usually provided months before actual consumption. The goal of these participants is to maximize their utility from consuming electricity, i.e., if l_i is the consumption of a single consumer at time i, the consumer will try to maximize the following:

$$\max_{l_i} \sum_i [U(l_i, i) - C(l_i, i)], \qquad (2.10)$$

where $U(l_i, i)$ and $C(l_i, i)$ are respectively the utility and the cost of consuming l_i units of energy at time i.

As reported in [88], utilities have been reluctant to let distributed energy resources interconnect with the grid, citing safety and system stability concerns. In order to fit in the secure operation of the power grid, generators should be well able to forecast and control their output power for the next day and they should commit to generating the amount of electricity that they are scheduled to produce by the ISO. Failure to do so will result in system instability and price spikes. This is why fossil fuels, though finite and polluting, serve the largest part of our energy needs.

However, controlling the load can reverse this trend, since it is not the shear load of demand that skews the market this way, but rather the fact that the demand is inelastic, and cannot be held from drawing power in response to a variable generation profile. This is what DSM can change.

DSM systems have evolved over the past three decades through systematic activities of the power utilities, as well as government policies, designed to change the amount and timing of electricity consumption. Such DSM measures have been implemented for load management, for increasing energy efficiency, and for electrification (i.e., the strategic increase of electricity use) [91].

Initially, DSM programs mainly comprised reliability-driven load-management measures, used occasionally to manage emergency situations. Recently, more sophisticated and rapid forms of DSM have emerged, extending the level of consumer interaction for the services, through appropriate incentives [92]. One of the great promises of smart grid is to foster advanced forms of DSM that continuously control the load for potentially all consumers. The most notable idea emerging today is the inclusion of programs that enable direct price responsiveness, even of individual loads [91]. In a wider definition, DSM may also include items such as renewable energy systems, combined heat and power systems, independent power purchase, and all instruments that allow customer demand to be met with the highest efficiency [93].

The two most popular trends of demand management lie at opposite sides of the control spectrum. At one end are real-time pricing strategies, discussed in Section 2.4.2 and at the other end are direct load control (DLC) strategies, presented in Section 2.4.3.

2.4.2 Real-time pricing

One of the most serious contenders in the DSM research arena is RTP. From time to time, several operational issues such as lack of storage options, generator and transmission line failures, and the ignorance of customers about the real cost of the electricity they consume all contribute to cause spikes in the wholesale price of electricity. The purpose of price-based load control strategies is to transfer part of the risks of buying electricity through this market from the utility to the customers. Different techniques like time-of-use (TOU), critical peak pricing (CPP), or RTP all manage to do this to different extents.

The concept of real-time prices has been around for about three decades now [94]: instead of shielding the customer completely from fluctuations of energy costs in the spot market (as is done when using flat rates and time-of-use tariffs) in RTP, *price signals* delivered to the customers will provide incentives to modify their demand and alleviate the pressure on the grid, with the reward of lowering their bill. What is not obvious is how the utility should calculate and post price signals in order to attain a stable operating point for the system, balancing generation and demand.

Theoretically, the real-time price of electricity at each node in the power grid is the shadow price obtained from solving the optimization for the spot-market security-constrained generation dispatch [95, 96]. In a system with N generators and L load buses, each generating G_i or consuming D_i units of power, respectively, the optimization is

$$\min_{G_i} \sum_{i=1}^{N} C_i(G_i) \tag{2.11}$$

$$\text{s.t.} \sum_{i=1}^{N} G_i - \sum_{i=1}^{L} D_i - \text{loss} = 0$$

$$\left| H[G_1 \ldots G_N \; D_1 \ldots D_L]^T \right| \leq \bar{F}^{\max}$$

$$G_i^{\min} \leq G_i \leq G_i^{\max},$$

where $C_i(.)$ is the generation cost function and H is a matrix that relates power flow on transmission lines to nodal power inputs. The first constraint ensures power balance, the second one ensures that flow on each transmission line lies within its overload limit $\bar{F}^{\max}$, and the third one defines generation capacity limits. The Lagrangian of (2.11) is

$$\mathcal{L} = \sum_{i=1}^{N} C_i(G_i) - \theta(\sum_{i=1}^{N} G_i - \sum_{i=1}^{L} D_i - \text{loss})$$

$$- \bar{\mu}.(H[G_1 \ldots G_N \; D_1 \ldots D_L]^T - \bar{F}^{\max})$$

$$+ \nu_{\max}(G_i - G_i^{\max}) - \nu_{\min}(G_i - G_i^{\min}). \tag{2.12}$$

The locational marginal price (LMP) for load bus i is

$$\lambda_i = \frac{\partial \mathcal{L}}{\partial D_i}, \tag{2.13}$$

which represents the marginal cost of providing one additional unit of power at that bus. Simplifying assumptions or additional constraints can be added to the problem definition (2.11), such as adopting linear or quadratic generation costs and including security constraints like $(N-1)$ contingencies [97]. LMPs are the true costs of serving loads, which include costs of generation, grid losses, and congestion on transmission lines. Theoretically, one could calculate these prices before real-time operation and by releasing them the system should converge to its optimal operation point. This will only happen if:

- Perfect forecasts of demand value D_i's are available.
- The customers are shielded from this true cost and do not react to fluctuations of their associated LMP (as happens in most of the power grid today).

- Most importantly, all generators should be *price-taking rational agents*, so that posting the calculated LMPs will make them choose the dispatched value obtained from (2.11).

If, however, the customers are exposed to even limited information about these true costs, like all real-time pricing techniques aim to do, there will be an extra feedback loop added to the equation. That is the demand at the ith load bus, D_i, will be a function of λ_i and thus, (2.13) has to be modified to

$$\lambda_i = \frac{\partial \mathcal{L}(\lambda_1, \ldots, \lambda_i, \ldots, \lambda_L)}{\partial D_i}. \tag{2.14}$$

A diagram of the system is shown in Figure 2.5. Calculating the true cost of serving the demand if the customers react to real-time market fluctuations will require perfect knowledge of customer behaviour in reacting to price signals for every different load bus in the system at the time the LMPs are being calculated. [98] provides a dynamic model to analyse the dynamics of supply, demand, and clearing prices in a power grid with real-time retail pricing and information asymmetry. It shows that a power market accommodating common RTP techniques may possibly experience volatile prices and demand, or even lose its stability. Although proven with major simplifying assumptions, the important point made is that stability should be taken into account when designing a load control system. The authors conclude that more intricate models for demand in each area, knowledge about consumer behaviour in response to dynamic prices received, and a thorough understanding of the implications of different market mechanisms and system architectures are needed before real-time pricing techniques can be implemented in large scale.

Currently, power system operators have two major approaches to calculate LMPs for the real-time market. Ex-ante prices indicate the value of the LMPs before the true value of the demand is released, using predicted values of the stochastic variables like the demand and intermittent renewable resources. Ex-post values, on the other hand, are calculated after the load is served and with deterministic knowledge of all the values for demand and generation.

Since real-time price signals need to be delivered to customers beforehand to allow some planning time, they should be of the same nature as ex-ante LMPs. Also, it is

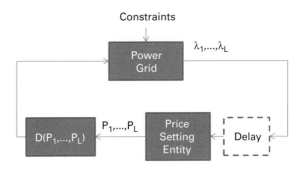

Figure 2.5 System model with RTP.

very unlikely that ex-post adjustments will be allowed to affect how customers are billed, since this would expose the public to unacceptable risks. The same approach just described is adopted by most RTP researchers in the literature. The real-time price sent to customers is derived either from a direct ex-ante analysis of wholesale market prices, or by adding some modifications to account for consumer satisfaction. To do so, either a term representing the benefit of customers from consuming electricity is added to the cost function in (2.11) or a cap on the variations of the price signal is enforced. For example, in [99], the authors maximize the social benefit, which includes known cost functions $C_i(.)$ representing the cost of production of energy and known benefit functions $B_i(.)$ representing the consumers:

$$\max \sum_{i=1}^{L} B_i(D_i) - \sum_{i=1}^{N} C_i(G_i). \qquad (2.15)$$

To calculate the price, the authors assume that both generation units and customers are price-taking agents with known limits on their consumption and generation values and declare the price at each bus as the marginal value of the objective function in this maximization problem (2.15) under various constraints. [100–102] follow a similar approach. The gap caused by the difference between ex-ante and ex-post prices should be compensated by the utility, similar to what is done today for the gap between flat or TOU rates and ex-post LMPs. Some of the problems that can arise from this approach are as follows:

- Perfect knowledge of the utility of the customers is assumed at every load bus (usually assumed to be simple analytic functions), which is unrealistic, at least in the current situation.
- If the price is set using an incorrect prediction of the behaviour of customers, prices that are posted may lead to system instability.
- Demand is only assumed to be dependent on the current price. This assumption is not valid for the demand for electricity since electricity is not delivered instantly in packets and appliances need time to finish their jobs.
- Generation owners may try to arbitrage the market.
- At least the major part of generation assets should have a fully deterministic and controllable nature in order to let the utility calculate valid prices in advance. This assumption may not hold with the addition of a considerable amount of intermittent resources to the grid.

To conclude this section, it is important to remark that in addition to concerns about market stability, any power imbalance resulting from volatile demand, if not compensated correctly and timely, may accumulate and cause the grid itself to experience oscillation or even lose stability. The disturbance in the generator angle, when neglecting the damping, can be approximated as $\Delta\delta(t) \approx \int M^{-1} \Delta G(t) dt$, where M is the system inertia matrix and $\Delta G(t)$ is the imbalance. When $\Delta\delta(t)$ goes beyond its stability margin, the grid will become unstable [8]. This is why the socio-economic feedback of the energy market, with all the uncertainty it carries, cannot be closed in real time and some type of direct control may prove necessary, blurring the boundary between RTP and direct load control (DLC).

2.4.3 Direct load control

Unlike RTP, direct load control, or *interruptible load* programs, have been widely and successfully practiced for over a decade. During peak load hours, utilities have the option to curtail the load due to certain appliances like air conditioners or water heaters belonging to participating customers for a predetermined duration of time (usually, 15–30 minutes). This is done by sending a curtailment *signal* to the target appliance from a central dispatch centre.

Ever since the 1980s, several researchers have worked on finding an optimum curtailment schedule that will maximize the benefits of the utility while avoiding unacceptable dissatisfaction of participating customers. Typically, an optimization problem is defined that minimizes a certain cost function of the load during a look-ahead horizon. Common constraints are maximum curtailment time and a minimum payback period between two successive curtailments, during which the appliance is allowed to function without interruption, to catch up on its duty. This type of control is effective only for certain appliances. Many researchers then use dynamic programming methods to solve this optimization problem [103, 104], while others use some form of simplification to turn it into a linear program, which is less computationally intensive. DLC solutions usually try to have a long enough look-ahead horizon in order to avoid rebound peaks due to payback periods, since the modified load $L'(t)$ at time t is given by

$$L'(t) = L(t) - \text{curtailed load from DLC} \\ + \text{payback load from previous curtailments}.$$

Note that DLC schedules are sometimes solved in coordination with the unit commitment problem (2.11) since they will affect the demand values D_i [105].

Another form of curtailment option that fits into this category is the Negawatts Program, where some market participants bid to use less electricity during peak times. Unlike DLC, these programs are market driven. Whether this idea is acceptable is still under debate, since the amount of energy reduction can only be measured based on the consumption history of the bidder and there are market design problems that arise due to this fact.

While interrupting certain loads can help alleviate the problems with high peak demand when facing generation shortage, most DLC solutions are designed merely for emergency situations. With the addition of a remarkable amount of unpredictable renewable resources like wind and solar energy to the grid, the frequency of these *emergency* situations will increase substantially and forced curtailment of the load, even if it is backed by customer participation, will no longer be a sufficient measure to match volatile and unpredictable generation with inelastic demand. Also, deciding how the customers should be paid for these interruptions will become increasingly unclear, bringing about similar complexity as RTP.

2.4.4 Possibilities and challenges at the edge of the network

A key innovation needed, which is not clearly described in any of the existing frameworks for load control, is that the load needs to be unbundled in individual specific requests of service, in order to understand what is flexible load and what is not. The vision parallels that of an *Internet of things*, which embodies a network of electrical appliances which can articulate and forward their energy requests to the retailer, and buy the energy service that makes sense to meet their actual needs and constraints, on the fly. More than a brand new communication infrastructure, what is needed is a service model that would allow paying for the network and computing resources that are necessary to support this service. The current emphasis on requirements is not well placed. The emphasis should be on developing new applications that speak the language of computers and networks and that manage large amounts of such transactions intelligently. This model is a departure from the control-based architecture, where the load is an amorphous conglomerate of requests, that the generation needs to catch up with. Google profits from businesses whose services are advertised to those who need them. It does not own exclusively the network resources that give the customer access to its services. Amazon profits by providing a network infrastructure to manage a variegated marketplace. There is space to create a marketplace of energy, but creative minds and policies have to converge to make this happen. Similarly, we need to develop applications and companies that would profit from matching in real time demand and supply of energy, and would be motivated to create the computational infrastructure to support their business. Queries to this service would come from such an Internet of things, i.e., smart loads whose request will be held until it is met, with the best effort possible. Just as in the search engine landscape, the best technology will prevail. Customers will have their devices select the energy search technology that provides the best service. And becoming the best service will require situation awareness, which will also naturally lead the providers of the service to pay for a faster and scalable communication and computation infrastructure.

Interestingly, this infrastructure would end up including some generation capability, quite naturally, to bypass the transmission bottleneck. In fact, the transmission, generation, and customer inconvenience costs can be mitigated if the provider can tap into local resources to make ends meet. Just as in the Internet caching data is a way to compensate for network congestion, this energy service will need to be able to leverage on local servers, which will be the microgrids of the future. This will stimulate a virtuous cycle that will allow producing energy locally, which will use very little transmission capacity and, with the right incentives, green energy, to feed the real needs of the customers rather than their assumed need. Generation capacity will proliferate close to where it is actually used. Hence, with the right mechanisms to place the micro-generation capacity in real time, decentralized generation will naturally complement the business of these energy search engines. This model will progressively meet an increasing portion of the needs of communities, without further burdening the grid with the task of relying on green generation capacity, and upgrading its transmission capacity to cope with increasing energy demand. Clearly this requires innovation both in the physical as well as in the cyber system. The main challenge in the physical system is to interface many asynchronous systems to the synchronous grid seamlessly. The cyber system should be able to accurately tune and predict the results of actions, much unlike the model we currently have. With a scalable model and a system of intelligent interfaces, it is natural to assume that these increasingly smart edges will rely sporadically on the traditional energy supply, and will also be able to trade among each other under the right model to exploit transmission resources. This is the only real positive scenario that can emerge from our increasing appetite for electrical energy, and the only scenario that can exhibit the impressive scalability of another pervasive infrastructure that has produced immense global benefits: the Internet.

2.5 Conclusion

The energy technology revolution is about making electrons clean and cheap to produce. It is unrealistic to think that green electrons can be just as reliable as the electrons produced with traditional fuel. We need to be able to tap into renewable resources in large quantities and the real bottleneck is neither the cost to produce solar panels nor windmills, nor their efficiency: it is their intermittence. What is clear is that, in many ways, the future of electrical energy technology is in being able to bear more volatility, both in the demand and the supply side. The transmission network needs to become more agile, aware, and able to respond in real time but this is not going to be what resolves the network problems. A centralized model is simply not reasonable, because of the issues of scalability it poses, which would imply tremendous costs. The periphery of the network needs to increase its capillary control over decentralized green energy sources, storage and, most importantly, it needs to be able to control directly or indirectly the demand, trying to realize an optimum microcosm of transactions locally, so that the grid can view a load that follows as closely as possible the profile it is committed to follow, based on a day-ahead market that reflects all the costs of the energy that is delivered, including a cost for CO_2 emissions, environmental impact, and social long-term cost of using up earth resources. In fact, the only way to deal with volatility is through a diversified form of resource allocation. The grid power should be viewed as one of the resources that the distribution cells should utilize, while trying to maximize its utilization of local green resources, relying also sparingly on local storage to maintain quality of service. Hence, to improve the integration of renewable resources, the market for these electrons needs to be a smarter and more agile place, working efficiently to serve its customers, not at any cost, but at the right cost for the specific service provided, given a profile of generation that is possible. What the system is not good at now is, in fact, seizing opportunities on the generation side. Of course, we can blame the fact that wind and solar power are abundant where transmission assets are not. But the deeper and truer dilemma lies in the scarce use that would be made of these resources, even if the transmission lines were in place. If demand and supply could be queried and matched in real time, and if pollution and collateral costs of fossil fuel were fairly priced, electrons from solar panels and windmills would find their customers.

References

[1] T. L. Friedman, 'Hot, flat and crowded: why we need a green revolution – and how it can renew America', *Farrar, Straus and Giroux, New York, 2008*, vol. 57, no. 1, pp. 232–241, January 2009.

[2] B. A. Carreras, D. E. Newman, I. Dobson, and A. B. Poole, 'Evidence for self-organized criticality in a time series of electric power system blackouts', *IEEE Transactions on Circuits and Systems I*, vol. 51, no. 9, pp. 1733–1740, September 2004.

[3] R. P. Schulz and W. W. Price, 'Classification and identification of power system emergencies', *IEEE Transactions on Power Apparatus and Systems*, vol. PAS-103, no. 12, pp. 3470–3479, December 1984.

[4] B. Stott, O. Alsac, and A. J. Monticelli, 'Security analysis and optimization', *Proceedings of the IEEE*, vol. 75, no. 12, pp. 1623–1644, December 1987.

[5] H. W. Dommel and W. F. Tinney, 'Optimal power flow solutions', *IEEE Transactions on Power Apparatus and Systems*, vol. PAS-87, no. 10, pp. 1866–1876, October 1968.

[6] A. Monticelli, M. V. F. Pereira, and S. Granville, 'Security-constrained optimal power flow with post-contingency corrective rescheduling', *IEEE Transactions on Power Systems*, vol. 2, no. 1, pp. 175–180, February 1987.

[7] K. Poulsen, 'Report: cyber attacks caused power outages in Brazil', WIRED Threat Level blog, Tech. Rep., November 2009.

[8] J. D. Glover and M. S. Sarma, *Power System Analysis and Design*. Brooks/Cole, 2001.

[9] A. Bose, 'Power system stability: new opportunities for control', in *Stability and Control of Dynamical Systems and Applications*, D. Liu and P.J. Antsaklis (eds). Birkhäuser (Boston), 2003 [online]. Available at: http://gridstat.eecs.wsu.edu/Bose-GridComms-Overview-Chapter.pdf.

[10] C. R. Mason, 'The art and science of protective relaying', General Electric [online]. Available at: http://www.geindustrial.com/pm/notes/artsci/artsci.pdf.

[11] M. Ilic-Spong, J. Christensen, and K. L. Eichorn, 'Secondary voltage control using pilot point information', *IEEE Transactions on Power Systems*, vol. 3, no. 2, pp. 660–668, May 1988.

[12] G. H. Narain and G. Laszlo, *Understanding FACTS: Concepts and Technology of Flexible AC Transmission Systems*. Wiley-IEEE Press, December 1999.

[13] V. K. Sood, *HVDC and FACTS Controllers: Applications of Static Converters in Power Systems*. Kluwer Academic Publishers, 2004.

[14] D. Divan and H. Johal, 'Distributed FACTS: a new concept for realizing grid power flow control', *IEEE Transactions on Power Electronics*, vol. 22, no. 6, pp. 2253–2260, November 2007.

[15] D. Divan and H. Johal, 'Design considerations for series-connected distributed FACTS converters', *IEEE Transactions on Industrial Applications*, vol. 43, no. 6, pp. 1609–1618, November 2007.

[16] K. Rogers and T. Overbye, 'Some applications of distributed flexible AC transmission system (D-FACTS) devices in power systems', in *40th North American Power Symposium (NAPS '08)*, Calgary, AB, 28–30 September 2008.

[17] A. Kechroud, J. Myrzik, and W. Kling, 'Taking the experience from flexible AC transmission systems to flexible AC distribution systems', in *42nd International Universities*

Power Engineering Conference (UPEC 2007), Brighton, UK, September 4–6 2007, pp. 687–692.

[18] A. Monticelli, *State Estimation in Electric Power Systems: A Generalized Approach*. Springer, 1999.

[19] H. Witsenhausen, 'A counter example in stochastic optimum control', *SIAM Journal on Control*, vol. 6, 1968.

[20] N. Elia and S. Mitter, 'Stabilization of linear systems with limited information', *IEEE Transactions on Automatic Control*, vol. 46, pp. 1384–1400, 2001.

[21] S. Tatikonda, 'Control under communication constraints', PhD Thesis, Department of Electrical Engineering and Computer Science, MIT, 2000.

[22] N. Elia and J. Eisembeis, 'Limitation of linear control over packet drop networks', *IEEE Transactions on Automatic Control*, vol. 5, pp. 5152–5157, December 2004.

[23] L. Schenato, B. Sinopoli, M. Franceschetti, K. Poolla, and S. Sastry, 'Foundations of control and estimation over lossy networks', *Proceedings of the IEEE*, vol. 95, no. 1, pp. 163–187, January 2007.

[24] D. Liberzon and J. Hespanha, 'Stabilization of nonlinear systems with limited information feedback', *IEEE Transactions on Industrial Applications*, vol. 50, no. 6, pp. 910–915, June 2005.

[25] G. Barrett and S. Lafortune, 'A novel framework for decentralized supervisory control with communication', in *Proceedings of IEEE Systems, Man, and Cybernetics Conference*, 1998.

[26] G. Barrett and S. Lafortune, 'On the synthesis of communicating controllers with decentralized information structures for discrete-event systems', in *Proceedings of IEEE Conference on Decision and Control*, 1998, pp. 3281–3286.

[27] G. Barrett and S. Lafortune, 'Some issues concerning decentralized supervisory control with communication', in *Proceedings of 38th IEEE Conference on Decision and Control*, 1999, pp. 2230–2236.

[28] G. Barrett and S. Lafortune, 'Decentralized supervisory control with communicating controllers', *IEEE Transactions on Automated Control*, vol. 45, pp. 1620–1638, 2000.

[29] R. Boel and J. van Schuppen, 'Decentralized failure diagnosis for discrete-event systems with constrained communication between diagnosers', in *Proceedings of Workshop on Discrete-Event Systems*, 2002.

[30] R. Debouk, 'Failure diagnosis of decentralized discrete-event systems', PhD Thesis, University of Michigan, Ann Arbor, 2000.

[31] R. Debouk, S. Lafortune, and D. Teneketzis, 'Coordinated decentralized protocols for failure diagnosis of discrete event systems', College of Engineering, University of Michigan, Ann Arbor, Technical Report CGR-97-17, 1998.

[32] R. Debouk, S. Lafortune, and D. Teneketzis, 'Coordinated decentralized protocols for failure diagnosis of discrete-event systems', *Discrete Event Dynamics Systems*, vol. 10, pp. 33–86, 2000.

[33] R. Radner, 'Allocation of a scarce resource under uncertainty: an example of a team', in *Decision and Organization*, C. McGuire and R. Radner (eds). North-Holland, 1972.

[34] S. Ricker and K. Rudie, 'Know means no: incorporating knowledge into decentralized discrete-event control', in *Proceedings of 1997 American Control Conference*, 1997.

[35] S. Ricker, 'Knowledge and communication in decentralized discrete-event control', PhD Thesis, Queen's University, Department of Computing and Information Science, August 1999.

[36] S. Ricker and G. Barrett, 'Decentralized supervisory control with single-bit communications', in *Proceedings of American Control Conference ACC'01*, 2001, pp. 965–966.

[37] S. Ricker and K. Rudie, 'Incorporating communication and knowledge into decentralized discrete-event systems', in *Proceedings of 38th IEEE Conference on Decision and Control*, 1999, pp. 1326–1332.

[38] S. Ricker and J. van Schuppen, 'Asynchronous communication in timed discrete event systems', in *Proceedings of American Control Conference ACC'01*, 2001, pp. 305–306.

[39] S. Ricker and J. van Schuppen, 'Decentralized failure diagnosis with asynchronous communication between supervisors', in *Proceedings of European Control Conference ECC'01*, 2001, pp. 1002–1006.

[40] D. Teneketzis and P. Varaiya, 'Consensus in distributed estimation with inconsistent beliefs', *Systems and Control Letters*, vol. 4, pp. 217–221, 1984.

[41] J. van Schuppen, 'Decentralized supervisory control with information structures', in *Proceedings of International Workshop on Discrete Event Systems WODES98*, 1998, pp. 36–41.

[42] J. van Schuppen, 'Chapter 1: Decentralized control with communication between controllers', in *Sixty Open Problems in the Mathematics of Systems and Control*, V. Blondel and A. Megretski (eds). Princeton University Press, 2003.

[43] K. Wong and J. van Schuppen, 'Decentralized supervisory control of discrete-event systems with communication', in *Proceedings of International Workshop on Discrete Event Systems WODES96*, 1996, pp. 284–289.

[44] S. Tripakis, 'Undecidable problems of decentralized observation and control', in *Proceedings of IEEE Conference on Decision and Control (CDC)*, Orlando, FL, 2001.

[45] F. Schweppe and D. Rom, 'Power system static-state estimation, Part I, II and III', *IEEE Transactions on Power Apparatus and Systems*, vol. PAS-89, no. 1, January 1970.

[46] R. Larson, W. Tinney, and J. Peschon, 'State estimation in power systems Part I: theory and feasibility', *IEEE Transactions on Power Apparatus and Systems*, vol. PAS-89, no. 3, pp. 345–352, March 1970.

[47] C. Brice and R. Cavin, 'Multiprocessor static state estimation', *IEEE Transactions on Power Apparatus and Systems*, vol. PAS-101, no. 2, pp. 302–308, February 1982.

[48] M. Kurzyn, 'Real-time state estimation for large-scale power systems', *IEEE Transactions on Power Apparatus and Systems*, vol. PAS-102, no. 7, pp. 2055–2063, July 1983.

[49] T. Van Cutsem, J. L. Horward, and M. R.-Pavella, 'A two-level static state estimator for electric power systems', *IEEE Power Engineering Review*, vol. PER-1, no. 8, pp. 34–35, August 1981.

[50] J. Carvalho and F. Barbosa, 'Parallel and distributed processing in state estimation of power system energy', in *Proceedings of 9th Mediterranean Electrotechnical Conference (MELECON'98)*, vol. 2, pp. 969–973, May 1998.

[51] D. Falcao, F. Wu, and L. Murphy, 'Parallel and distributed state estimation', *IEEE Transactions on Power Systems*, vol. 10, no. 2, pp. 724–730, May 1995.

[52] L. Zhao and A. Abur, 'Multi area state estimation using synchronized phasor measurements', *IEEE Transactions on Power Systems*, vol. 20, no. 2, pp. 611–617, May 2005.

[53] W. Jiang, V. Vittal, and G. Heydt, 'A distributed state estimator utilizing synchronized phasor measurements', *IEEE Transactions on Power Systems*, vol. 22, no. 2, pp. 563–571, May 2007.

[54] A. El-Keib, J. Nieplocha, H. Singh, and D. Maratukulam, 'A decomposed state estimation technique suitable for parallel processor implementation', *IEEE Transactions on Power Systems*, vol. 7, no. 3, pp. 1088–1097, August 1992.

[55] S.-Y. Lin, 'A distributed state estimator for electric power systems', *IEEE Transactions on Power Systems*, vol. 7, no. 2, pp. 551–557, May 1992.

[56] Advanced Metering Infrastructure (AMI), EPRI report, February 2007 [online]. Available at: http://www.ferc.gov/eventcalendar/Files/20070423091846-EPRI.

[57] C. Rehtanz and J. Bertsch, 'A new wide area protection system', *IEEE Porto Power Tech Conference*, vol. 4, pp. 186–191, Porto, September 2001.

[58] The Bonneville Power Administration, 'Breakthrough technology fights power outages', *BPA Journal*, March 2004 [online]. Available at: http://www.bpa.gov/corporate/pubs/journal/04jl/jl0304x.pdf.

[59] Pacific Northwest National Laboratory, 'GridWise history: how did GridWise start?', October 2007 [online]. Available at: http://gridwise.pnl.gov/foundations/history.stm.

[60] J. Y. Cai, Z. Huang, J. Hauer, and K. Martin, 'Current status and experience of WAMS implementation in North America', *IEEE/PES Transmission and Distribution Conference and Exhibition: Asia and Pacific*, pp. 1–7, Dalian, 2005.

[61] M. D. Hadley, J. B. McBride, T. W. Edgar, L. R. ONeil, and J. D. Johnson, 'Securing Wide Area Measurement Systems', *PNNL-17116 report*, June 2007 [online]. Available at: http://www.oe.energy.gov/DocumentsandMedia/Securing_WAMS.pdf.

[62] B. Drury, 'Control Techniques Drives and Controls Handbook (2nd edition)', 2009, Institution of Engineering and Technology [online]. Available at: http://knovel.com/web/portal/browse/display_EXT_KNOVEL_DISPLAY_bookid=2995&VerticalID=0.

[63] W. Mahnke and S. Leitner, 'OPC unified architecture – the future standard for communication and information modeling in automation', ABB Review, March 2009, pp. 56–61 [online]. Available at: http://www05.abb.com/global/scot/scot271.nsf/veritydisplay/75d70c47268d78bfc125762d00481f78/$file/56-61%203m903_eng72dpi.pdf.

[64] IEEE Digital Library [online]. Available at: http://ieeexplore.ieee.org/xpl/freeabs_all.jsp?arnumber=943067&isnumber=20419.

[65] J. Allen, 'TVA opens data collection software for industry use', *Tennessee Valley Authority*, October 2009 [online]. Available at: http://www.tva.gov/news/releases/octdec09/data_collection_software.htm.

[66] openPDC v1.0 Release [online]. Available at: http://openpdc.codeplex.com/releases/view/39621.

[67] Cisco Security Appliance Command Line Configuration Guide, Version 7.2 [online]. Available at: http://www.cisco.com/en/US/docs/security/asa/asa72/configuration/guide/conf_gd.html.

[68] D. Karlsson, M. Hemmingsson, and S. Lindahl, 'Wide area system monitoring and control – terminology, phenomena, and solution implementation strategies', *IEEE Power and Energy Magazine*, vol. 2, no. 5, pp. 68–76, September 2004.

[69] M. Klein, G. J. Rogers, and P. Kundur, 'A fundamental study of inter-area oscillations in power systems', *IEEE Transactions on Power Systems*, vol. 6, no. 3, pp. 914–921, August 1991.

[70] E. Grebe, J. Kabouris, S. Lopez Barba, W. Sattinger, and W. Winter, 'Low frequency oscillations in the interconnected system of continental Europe', *IEEE Power and Energy Society General Meeting*, pp. 1–7, 25–29 July 2010.

[71] K. Rahmani, G. B. Gharehpetian, and M. S. Naderi, 'Damping of low frequency oscillations in AC-DC interconnected power system using robust modulation controller', in *Proceedings of 18th Iranian Conference on Electrical Engineering (ICEE)*, pp. 801–805, 11–13 May 2010.

[72] M. Begovic, D. Novosel, D. Karlsson, C. Henville, and G. Michel, 'Wide-area protection and emergency control', *Proceedings of the IEEE*, vol. 93, no. 5, pp. 876–891, May 2005.

[73] Synchrophasor System Benefits Fact Sheet [online]. Available at: http://www.naspi.org/resources/2009_march/phasorfactsheet.pdf.

[74] C. H. Hauser, D. E. Bakken, I. Dionysiou, K. H. Gjermundrod, V. S. Irava, J. Helkey, and A. Bose, 'Security, trust, and QoS in next-generation control and communication for large power systems', *International Journal of Critical Infrastructures*, vol. 04, no. 1/2, pp. 3–16, 2008.

[75] S. F. Abelsen, E. S. Viddal, K. H. Gjermundrod, D. E. Bakken, and C. H. Hauser, 'Adaptive information flow mechanisms and management for power grid contingencies', Technical Report EECS-GS-012, School of Electrical Engineering and Computer Science, Washington State University, December 2007 [online]. Available at: http://gridstat.net/publications/TR-GS-009.pdf.

[76] D. Bakken, A. Bose, C. Hauser, D. Whitehead, and G. Zweigle, 'Smart generation and transmission with coherent, real-time data', *Proceedings of the IEEE*, vol. 99, no. 6, pp. 928–951, June 2011.

[77] S. Galli, A. Scaglione, and Z. Wang, 'Power line communications and the smart grid', in *IEEE International Conference on Smart Grid Communications (SmartGridComm)*, Gaithersburg, MD, October 4–6, 2010.

[78] S. Galli, A. Scaglione, and Z. Wang, 'For the grid and through the grid: the role of power line communications in the smart grid', *Proceedings of the IEEE*, vol. 99, no. 6, pp. 998–1027, June 2011.

[79] C. Lu, J. Teng, and W. Liu, 'Distribution system state estimation', *IEEE Transactions on Power Systems*, vol. 10, pp. 229–240, February 1995.

[80] K. Li, 'State estimation for power distribution system and measurement impacts', *IEEE Transactions on Power Systems*, vol. 11, pp. 911–916, May 1996.

[81] A. K. Ghosh, D. L. Lubkeman, M. J. Downey, and R. H. Jones, 'Distribution circuit state estimation using a probabilistic approach', *IEEE Transactions on Power Systems*, vol. 12, pp. 45–51, February 1997.

[82] Y. Deng, Y. He, and B. Zhang, 'A branch-estimation-based state estimation method for radial distribution systems', *IEEE Transactions on Power Delivery*, vol. 17, no. 4, pp. 1057–1062, October 2002.

[83] H. Wang and N. N. Schulz, 'A revised branch current-based distribution state estimation placement impact', *IEEE Transactions on Power Systems*, vol. 19, pp. 207–213, February 2004.

[84] J. J. M. E. Baran and T. McDermott, 'Including voltage measurements in branch current estimation for distribution systems', in *IEEE PES General Meeting*, 2009.

[85] S. Stoft, *Power System Economics: Designing Markets for Electricity*. Wiley-IEEE Press, 2002.

[86] A. Mazer, *Electric Power Planning for Regulated and Deregulated Markets*. John Wiley and Sons, Inc., 2006.

[87] M. Olsson and L. Soder, 'Estimating real-time balancing prices in wind power systems', in *Proceedings of IEEE/PES Power Systems Conference and Exposition (PSCE'09)*, pp. 1–9, March 2009.

[88] D. E. King, 'Electric power micro-grids: opportunities and challenges for an emerging distributed energy architecture', PhD Thesis, 2006.

[89] 'A National Assessment of Demand Response Potential', Federal Energy Regulatory Commission (FERC), Technical Report, June 2009.

[90] G. Hardin, 'The Tragedy of the Commons', *Science*, vol. 162, no. 3859, pp. 1243–1248, 1968.

[91] 'Primer on demand-side management with an emphasis on price-responsive programs', Charles River Associates, Tech. Rep., 2005 [online]. Available at: http://siteresources.worldbank.org/INTENERGY/Resources/Primeron-Demand-SideManagement.pdf.

[92] J. Eto, C. Goldman, G. Heffner, B. Kirby, J. Kueck, M. Kintner-Meyer, J. Dagle, T. Mount, W. Schultze, R. Thomas, and R. Zimmerman, 'Innovative developments in load as a reliability resource', in *IEEE Power Engineering Society Winter Meeting*, vol. 2, pp. 1002–1004, 2002.

[93] 'Demand side management', Ministry of Power Government of India, Tech. Rep., 2011 [online]. Available at: http://www.powermin.nic.in/distribution/demand_side_management.htm.

[94] M. Caramanis, R. Bohn, and F. Schweppe, 'Optimal spot pricing: practice and theory', *IEEE Transactions on Power Apparatus and Systems*, vol. PAS-101, no. 9, pp. 3234–3245, September 1982.

[95] W. Hogan, E. Read, and B. Ring, 'Using mathematical programming for electricity spot pricing', *International Transactions in Operational Research*, vol. 3, no. 3–4, pp. 209–221, 1996 [online]. Available at: http://dx.doi.org/10.1111/j.1475-3995.1996.tb00048.x.

[96] E. Litvinov, 'Design and operation of the locational marginal prices-based electricity markets', *Generation, Transmission Distribution, IET*, vol. 4, no. 2, pp. 315–323, February 2010.

[97] M. Huneault and F. D. Galiana, 'A survey of the optimal power flow literature', *IEEE Transactions on Power Systems*, vol. 6, pp. 762–770, May 1991.

[98] M. Roozbehani, M. A. Dahleh, and S. K. Mitter, 'Volatility of power grids under real-time pricing', arXiv.org, Quantitative Finance Papers 1106.1401, June 2011 [online]. Available at: http://ideas.repec.org/p/arx/papers/1106.1401.html.

[99] J. Y. Choi, S.-H. Rim, and J.-K. Park, 'Optimal real time pricing of real and reactive powers', *IEEE Transactions on Power Systems*, vol. 13, no. 4, pp. 1226–1231, November 1998.

[100] M. Roozbehani, M. Dahleh, and S. Mitter, 'Dynamic pricing and stabilization of supply and demand in modern electric power grids', in *Proceedings of First IEEE International Conference on Smart Grid Communications (SmartGridComm)*, pp. 543–548, October 2010.

[101] R. Sioshansi, 'Evaluating the impacts of real-time pricing on the cost and value of wind generation', *IEEE Transactions on Power Systems*, vol. 25, no. 2, pp. 741–748, May 2010.

[102] P. Samadi, A. Mohsenian-Rad, R. Schober, V. Wong, and J. Jatskevich, 'Optimal real-time pricing algorithm based on utility maximization for smart grid', in *Proceedings of First IEEE International Conference on Smart Grid Communications (SmartGridComm)*, pp. 415–420, October 2010.

[103] A. Cohen and C. Wang, 'An optimization method for load management scheduling', *IEEE Transactions on Power Systems*, vol. 3, no. 2, pp. 612–618, May 1988.

[104] K.-Y. Huang and Y.-C. Huang, 'Integrating direct load control with interruptible load management to provide instantaneous reserves for ancillary services', *IEEE Transactions on Power Systems*, vol. 19, no. 3, pp. 1626–1634, August 2004.

[105] Y.-Y. Hsu and C.-C. Su, 'Dispatch of direct load control using dynamic programming', *IEEE Transactions on Power Systems*, vol. 6, no. 3, pp. 1056–1061, August 1991.

3 Demand-side management for smart grid: opportunities and challenges

Pedram Samadi, Hamed Mohsenian-Rad, Vincent W. S. Wong, and Robert Schober

3.1 Introduction

Demand-side management (DSM) is one of the key components of the future smart grid to enable more efficient and reliable grid operation [1]. To achieve a high level of reliability and robustness in power systems, the grid is usually designed for peak demand rather than for average demand. This usually results in an under-utilized system. To remedy this problem, different programs have been proposed to shape the daily energy consumption pattern of the users in order to reduce the peak-to-average ratio in load demand and use the available generating capacity more efficiently, avoiding the installation of new generation and transmission infrastructures. However, the increasing expectations of the customers both in quantity and quality [2], emerging new types of demand such as plug-in hybrid electric vehicles (PHEVs), which can potentially double the average household energy consumption [3], the limited energy resources, and the lengthy and expensive process of exploiting new resources give rise to the need for developing some more advanced methods for DSM.

Since electricity cannot be stored economically, wholesale prices (i.e., prices set by competing generators to regional electricity retailers) vary drastically between the low-demand times of day and the high-demand periods. However, these changes are usually hidden from retail users. That is, end users are usually charged with some average price. To alleviate this problem, various time-differentiated pricing methods have been proposed in the literature. Some examples include day-ahead pricing, time-of-use pricing, critical-peak-load pricing, and adaptive pricing [4–7].

By equipping users with two-way communication capabilities in smart grid systems and by adopting real-time pricing (RTP) methods, it is possible to reflect the fluctuations of wholesale prices to retail prices. Considering the environmental issues in the current power system, there is a high incentive to adopt renewable sources of energy. However, the random nature of some of these resources (e.g., wind and solar energy) may bring even more uncertainty into the generation side. This has further increased the need to develop new DSM methods.

A wide range of DSM techniques such as voluntary load management programs [8–10], direct load control (DLC) [11], and smart pricing [12–21] have been proposed. In

DLC programs, based on an agreement between the utility company and the customers, the utility company can remotely control the operation and energy consumption of certain appliances in a household. As an alternative to DLC, smart pricing is an effective tool to *encourage* users to consume electricity wisely and more efficiently. In general, DSM programs using smart pricing aim at one or both of the following design objectives: *reducing consumption* and *shifting consumption*. The level of success in achieving these goals for different pricing methods depends on different factors such as the amount of information being provided to each user, the effectiveness of mapping the wholesale prices to the retail prices, and the knowledge and abilities of users to respond to price information. Another important factor is the effectiveness of the home automation systems. For example, it is important whether the decisions about the schedule and the amount of power consumption are made automatically or manually. Some examples showing the disadvantages and limitations of manual control can be found in [22].

In this chapter, we review different pricing algorithms for DSM purposes aiming to achieve different objectives. The system model is introduced in Section 3.2. Energy consumption scheduling and power control are two main categories of different DSM programs. Different objectives and algorithms proposed for energy consumption scheduling are discussed in Section 3.3. To generalize the objectives introduced in Section 3.3, the concept of utility function from microeconomics adopted to model different user objectives is discussed in Section 3.4. Conclusions are drawn in Section 3.5.

3.2 System model

In this section, we provide a mathematical representation of the power system and the energy cost [23]. A system with multiple users and one energy provider is considered. It is assumed that each user is equipped with a smart meter that has energy consumption control or scheduling capabilities. Smart meters are not only connected to the power grid but also to the energy provider through a local area network (LAN). The block diagram of such a power distribution system is shown in Figure 3.1. Throughout this chapter, let $\mathcal{N}$ denote the set of users, where $N \triangleq |\mathcal{N}|$ represents the number of users. The intended time of operation is divided into K equal-length time slots, where $K \triangleq |\mathcal{K}|$, and $\mathcal{K}$ is

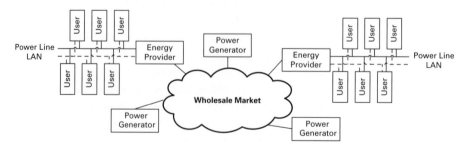

Figure 3.1 Illustration of the regional energy providers, several users, and multiple power generators as parts of the general wholesale energy market in smart grid.

the set of all time slots. This division can be based on the behaviour of users and their demand pattern: *on-peak*, *mid-peak*, and *off-peak* time slots, or it can simply represent the 24 hours of a day. For each user $n \in \mathcal{N}$, let x_n^k denote the power consumption of user n in time slot k. Based on this definition, the total power consumption of all users in each time slot $k \in \mathcal{K}$ can be calculated as

$$L_k = \sum_{n \in \mathcal{N}} x_n^k. \tag{3.1}$$

The energy cost model is as follows. Let $C_k(L_k)$ denote the cost of providing L_k units of energy offered by the energy provider at each time slot $k \in \mathcal{K}$. We make the following assumptions.

ASSUMPTION 1. The cost functions are *increasing* in the total provided energy.

ASSUMPTION 2. The cost functions are *strictly convex*.

ASSUMPTION 3. There exists a differentiable, convex, non-decreasing function $p_k(q)$ over $q \geq 0$ for each $k \in \mathcal{K}$, with $p_k(0) \geq 0$ and $p_k(q) \to \infty$ as $q \to \infty$, such that for each $q \geq 0$ we have

$$C_k(q) = \int_0^q p_k(z) dz. \tag{3.2}$$

Note that *quadratic functions* are among the practical examples satisfying assumptions 1-3. In this chapter, we consider quadratic cost functions [24, 25]:

$$C_k(L_k) = c_1^k L_k^2 + c_2^k L_k + c_3^k, \tag{3.3}$$

where $c_1^k > 0$, $c_2^k \geq 0$, and $c_3^k \geq 0$ are predetermined parameters.

3.3 Energy-consumption scheduling model

Several incentive-based energy-consumption scheduling algorithms have been proposed recently for smart grid [22, 24, 26, 27]. They usually do not aim to change the amount of energy consumption, but instead to shift it to off-peak hours systematically. Minimizing the energy cost or reducing the *peak-to-average ratio* (PAR) are among the most common objectives considered for scheduling algorithms.

3.3.1 Residential load-scheduling model

For each user $n \in \mathcal{N}$, let $\mathcal{A}_n$ denote the set of appliances of user n such as washer, dryer, refrigerator, dishwasher, air conditioner, PHEV, etc. For each appliance $a \in \mathcal{A}_n$, $\mathbf{x}_{n,a} \triangleq (x_{n,a}^1, \ldots, x_{n,a}^K)$ is defined as an energy-consumption scheduling vector, where $x_{n,a}^k$ denotes the energy consumption scheduled for appliance a at time slot k. The task of the *energy-consumption scheduler* (ECS), to be embedded in smart meter as the key home energy management function, is to properly select vector $\mathbf{x}_{n,a}$ for each appliance

$a \in \mathcal{A}_n$. However, the feasible set of choices for the energy-consumption scheduling is determined by the set of constraints which are based on each user's energy needs. These constraints include the required energy for each appliance $a \in \mathcal{A}_n$, $E_{n,a}$, and also the feasible scheduling interval for each appliance $a \in \mathcal{A}_n$, which will be determined by two parameters, starting time $\alpha_{n,a} \in \mathcal{K}$ and ending time $\beta_{n,a} \in \mathcal{K}$. Thus, for proper operation of the appliance, we need to have

$$\sum_{k=\alpha_{n,a}}^{\beta_{n,a}} x_{n,a}^k = E_{n,a}, \qquad (3.4)$$

and

$$x_{n,a}^k = 0, \quad \forall\, k \in \mathcal{K} \backslash \mathcal{K}_{n,a}, \qquad (3.5)$$

where $\mathcal{K}_{n,a} \triangleq \{\alpha_{n,a}, \ldots, \beta_{n,a}\}$. Note that we always have $\alpha_{n,a} \leq \beta_{n,a}$. For each appliance $a \in \mathcal{A}_n$, the minimum and maximum power levels are defined as $\gamma_{n,a}^{\min}$ and $\gamma_{n,a}^{\max}$, respectively. Therefore, we have

$$\gamma_{n,a}^{\min} \leq x_{n,a}^k \leq \gamma_{n,a}^{\max}. \qquad (3.6)$$

In summary, the feasible set of power consumption for each user n is defined as

$$\mathcal{X}_n = \Bigg\{ \mathbf{x}_n \,\Bigg|\, \sum_{k=\alpha_{n,a}}^{\beta_{n,a}} x_{n,a}^k = E_{n,a},\; x_{n,a}^k = 0,\quad \forall\, k \in \mathcal{K} \backslash \mathcal{K}_{n,a}, \\ \gamma_{n,a}^{\min} \leq x_{n,a}^k \leq \gamma_{n,a}^{\max},\quad \forall\, k \in \mathcal{K}_{n,a},\; \forall\, a \in \mathcal{A}_n \Bigg\}. \qquad (3.7)$$

We also define $\mathcal{X} \triangleq \mathcal{X}_1 \times \ldots \times \mathcal{X}_N$ as the feasible set of all users.

3.3.2 Energy-consumption scheduling problem formulation

The energy-consumption scheduling model is mainly based on two assumptions [22]. First, at the beginning of each operation cycle, each user knows the list of all appliances to run and their corresponding feasible working intervals. Second, the designed energy consumption scheduler aims to shift the energy consumption to achieve certain design objectives. In the following, we will discus some of these objectives. The energy-scheduling model is appropriate in illustrating the time coupling of different time-slots, i.e., how users may shift their demand from high price time slots to low-price time slots. However, in reality, the needed energy of each user may change during the operation period based on different real-time situations, or users may change their total energy consumption in response to different price values.

PAR minimization
The PAR in aggregate load demand can be formulated as a function of energy consumption scheduling vectors $\mathbf{x}_1, \cdots, \mathbf{x}_N$. Given complete information about the users'

appliances and their energy consumption needs, an optimal energy consumption schedule aimed at to minimizing the PAR can be characterized as the solution to the following optimization problem [22]:

$$\underset{\mathbf{x} \in \mathcal{X}}{\text{minimize}} \quad \frac{K \max_{k \in \mathcal{K}} \left(\sum_{n \in \mathcal{N}} \sum_{a \in \mathcal{A}_n} x_{n,a}^k \right)}{\sum_{n \in \mathcal{N}} \sum_{a \in \mathcal{A}_n} E_{n,a}}, \tag{3.8}$$

where $\mathbf{x} = (\mathbf{x}_1, \ldots, \mathbf{x}_N)$. Next, we note that since K and $\sum_{n \in \mathcal{N}} \sum_{a \in \mathcal{A}_n} E_{n,a}$ are fixed and independent of optimization variable $\mathbf{x}$, problem (3.8) can be simplified and written as

$$\underset{\mathbf{x} \in \mathcal{X}}{\text{minimize}} \quad \max_{k \in \mathcal{K}} \left(\sum_{n \in \mathcal{N}} \sum_{a \in \mathcal{A}_n} x_{n,a}^k \right). \tag{3.9}$$

However, due to the max term in the objective function of (3.9), it is still difficult to solve problem (3.9). Therefore, by introducing the new auxiliary variable Γ, we reformulate problem (3.9) to the following equivalent linear programming problem:

$$\begin{aligned} \underset{\Gamma, \mathbf{x} \in \mathcal{X}}{\text{minimize}} \quad & \Gamma \\ \text{subject to} \quad & \Gamma \geq \sum_{n \in \mathcal{N}} \sum_{a \in \mathcal{A}_n} x_{n,a}^k, \quad \forall\, k \in \mathcal{K}. \end{aligned} \tag{3.10}$$

Energy cost minimization

Mohsenian-Rad *et al.* [22] also considered minimizing the total energy costs in the system as an alternative design objective, which can be formulated as

$$\underset{\mathbf{x} \in \mathcal{X}}{\text{minimize}} \quad \sum_{k \in \mathcal{K}} C_k \left(\sum_{n \in \mathcal{N}} \sum_{a \in \mathcal{A}_n} x_{n,a}^k \right). \tag{3.11}$$

Unlike problem (3.10), problem (3.11) always has a unique solution which can be found using convex optimization techniques such as interior-point method (IPM) [28].

Joint energy payment and waiting time minimization

Users are mainly interested in two issues [24]. First, each user wishes to minimize its electricity payment. Second, it is desirable for each user to complete different tasks as soon as possible. However, these two objectives can be conflicting in many scenarios. As shown in [24], the waiting cost can be modelled as

$$\sum_{k=1}^{K} \sum_{a \in \mathcal{A}_n} \rho_{n,a}^k x_{n,a}^k, \tag{3.12}$$

where $\rho_{n,a}^k$ is the waiting parameter assigned to each appliance $a \in \mathcal{A}_n$ in time slot $k \in \mathcal{K}$. Clearly, $\rho_{n,a}^k = 0$ for all $k < \alpha_{n,a}$ and $k > \beta_{n,a}$ as the concept of waiting may only be defined within the valid scheduling interval $[\alpha_{n,a}, \beta_{n,a}]$. It is also assumed that

$$\rho_{n,a}^{\alpha_{n,a}} \leq \cdots \leq \rho_{n,a}^{\beta_{n,a}}, \quad \forall a \in \mathcal{A}_n. \tag{3.13}$$

The model used in [24] to determine the waiting parameter is

$$\rho_{n,a}^k = \frac{(\delta_{n,a})^{k-\alpha_{n,a}}}{E_a}, \quad \forall a \in \mathcal{A}_n, \ k \in [\alpha_{n,a}, \beta_{n,a}], \tag{3.14}$$

where $\delta_{n,a} \geq 1$ is a control parameter. The higher the value of $\delta_{n,a}$, the higher will be the cost of waiting to finish the operation of the appliance.

The objective of the optimization problem in [24] is defined as the weighted sum of the total electricity payment and the total waiting cost across all appliances. Thus, the optimal energy consumption scheduling in this case is formulated as the solution of the following optimization problem:

$$\begin{aligned}\underset{\mathbf{x}_n \in \mathcal{X}_n}{\text{minimize}} \quad & \sum_{k \in \mathcal{K}} \mu_k \left(\sum_{a \in \mathcal{A}_n} x_{n,a}^k \right) \left(\sum_{a \in \mathcal{A}_n} x_{n,a}^k \right) \\ & + \lambda_{\text{wait}} \sum_{k \in \mathcal{K}} \sum_{a \in \mathcal{A}_n} \frac{(\delta_{n,a})^{k-\alpha_{n,a}}}{E_{n,a}},\end{aligned} \tag{3.15}$$

where $\mu_k \left(\sum_{a \in \mathcal{A}_n} x_{n,a}^k \right)$ is the price charged to each user n at time slot k and parameter λ_{wait} is used to control the importance of the waiting cost terms in the objective function of the optimization problem (3.15).

Maintaining system stability with minimum curtailment

As opposed to economic-based programs which are designed to minimize price spikes during high-demand periods, stability-based programs are mainly concerned with having a balance between demand and generation. In this context, interruptible loads (ILs) are consumers who agree to be interrupted, as required and within constraints. In return, this interruption will be compensated by paying reduced tariffs [29]. In [29], the scheduling problem is defined as a constrained multi-objective optimization problem, i.e., the algorithm should derive a curtailment schedule that has the smallest total payment to the ILs and at the same time has the minimum number of interruptions.

To minimize the number of interruptions, a penalty is assigned to each interruption. To avoid further interruption of the same user, a penalty, C_{int}, is incurred whenever a previously interrupted load is interrupted again, and it is doubled at each subsequent interruption. However, since keeping the system stable is the most important objective, insufficient curtailment should be avoided. In this regard, a penalty P_{UC} is incurred in each time slot for insufficient curtailment demand. A penalty P_V is also considered when an IL constraint is violated which is set to be much smaller than P_{UC}. Thus, the optimal energy consumption can be characterized as the solution of the following optimization problem [29]:

$$\begin{aligned}\underset{Sch_n^k\in\{0,1\},\forall n\in\mathcal{N}, k\in\mathcal{K}}{\text{minimize}} \quad & \sum_{k\in\mathcal{K}}\sum_{n\in\mathcal{N}} Sch_n^k P_n \lambda_n \\ & + \left(C_{\text{int}} 2^0 FI_2 + C_{\text{int}} 2^1 FI_3 + \cdots + C_{\text{int}} 2^{\beta-2} FI_\beta\right) \\ & + P_V V + P_{UC} UC,\end{aligned} \quad (3.16)$$

where Sch_n^k denotes the status of the nth IL during the kth time slot, i.e., Sch_n^k is equal to 1 if the nth IL is curtailed during the kth time slot, otherwise, it is equal to 0. P_n represents the capacity of the nth IL, and λ_n denotes the curtailment rate in which it is payed to user n. FI_i is the number of ILs interrupted i times, β is the maximum number of interruptions incurred by any IL, V represents the number of violations of IL constraints, and UC denotes the number of time slots with insufficient curtailments.

3.3.3 Energy-consumption scheduling algorithm

The various objective functions defined in the previous section can be solved in a centralized fashion using efficient optimization algorithms. However, since users are essentially independent decision makers and their behaviour is not controlled by the grid, there are also several advantages in solving these optimization problems autonomously in a distributed fashion [22, 26].

Considering the cost minimization model in [22], it has been shown that an elaborately calculated energy bill for each user can provide incentives for each user to adopt the optimal energy-consumption schedule obtained by its smart meter. Assuming that users are charged proportionaly to their daily energy consumption, it can be shown that for each user $n \in \mathcal{N}$, we have [22]

$$t_n = \Omega_n \sum_{k\in\mathcal{K}} C_k\left(\sum_{m\in\mathcal{N}}\sum_{a\in\mathcal{A}_m} x_{m,a}^k\right), \quad (3.17)$$

where

$$\Omega_n \triangleq \frac{\kappa \sum_{a\in\mathcal{A}_n} E_{n,a}}{\sum_{m\in\mathcal{N}}\sum_{a\in\mathcal{A}_m} E_{m,a}}. \quad (3.18)$$

Here, κ is defined as

$$\kappa \triangleq \frac{\sum_{n\in\mathcal{N}} t_n}{\sum_{k\in\mathcal{K}} C_k\left(\sum_{n\in\mathcal{N}}\sum_{a\in\mathcal{A}_n} x_{n,a}^k\right)} \geq 0. \quad (3.19)$$

In [22], Mohsenian-Rad et al. model the behaviour of users using techniques from game theory. They showed that the following game is defined among the users.

- Players: Registered users in set $\mathcal{N}$.
- Strategies: Each user $n \in \mathcal{N}$ selects its energy consumption schedule $x_n \in \mathcal{X}_n$ to maximize its payoff.
- Payoffs: Negative of bill amount, i.e., $-t_n$ for each user $n \in \mathcal{N}$ as in (3.17).

The Nash equilibrium of the above game always exists, is unique, and is the optimal solution of the cost-minimization problem in (3.11).

3.3.4 Performance evaluation

Among different algorithms proposed for scheduling of energy consumption, in this section, we mainly focus on the cost minimization model introduced in (3.11). It is assumed that there are $N = 10$ users. For the purpose of study, each user is selected to have between 10 and 20 appliances with non-shiftable operation, i.e., with strict energy consumption scheduling constraints. Such appliances may include refrigerator/freezer (daily usage: 1.32 kWh), electric stove (daily usage: 1.89 kWh for self-cleaning and 2.01 kWh for regular), lighting (daily usage for 10 standard bulbs: 1.00 kWh), heating (daily usage: 7.1 kWh) [24]. Moreover, each user is selected to also have between 10 and 20 appliances with shiftable operation, i.e., with soft energy-consumption scheduling constraints. Recall that the smart meter with ECS capability may schedule only the appliances with soft energy-consumption scheduling constraints. Such appliances may include dishwasher (daily usage: 1.44 kWh), washing machine (daily usage: 1.49 kWh for energy-star, 1.94 kWh for regular), clothes dryer (daily usage: 2.50 kWh), and PHEV (daily usage: 9.9 kWh) [24]. In our simulation model, we assume that each user has a randomly selected combination of the considered shiftable and non-shiftable loads to be used at different times of the day by taking into account that the load demand is higher in the evening and lower during the night. The energy cost function is assumed to be quadratic as in (3.3). For simplicity, we assume that $c_2^k = 0$ and $c_3^k = 0$, $\forall k \in \mathcal{K}$. We also have $c_1^k = 0.3$ cents at daytime hours, i.e., from 8:00 am to 12:00 pm and $c_1^k = 0.2$ cents during the night, i.e., from 12:00 pm to 8:00 am the day after. The power system is assumed budget-balanced, i.e., $\kappa = 1$ and the aggregate payment of the users is equal to the energy cost incurred to the energy provider.

Simulation results for the total scheduled energy consumption and the energy cost for the energy cost minimization scenario are shown in Figures 3.2 and 3.3 without and with the deployment of the ECS function in the smart meters, respectively. For the case without ECS deployment, each appliance for each user is assumed to start operation right at the beginning of the time interval and at its typical power level. For the case with ECS deployment, the timing and the power level for the operation of each household appliance are determined as a solution to the problem (3.11). By comparing the results in Figures 3.2 and 3.3, we can see that when the ECS functions are not used, the PAR is 2.1 and the energy cost is \$44.77. However, when the ECS feature is enabled, the PAR reduces to 1.8 (i.e., 17% less) and the energy cost reduces to \$37.90 (i.e., 18% less). In

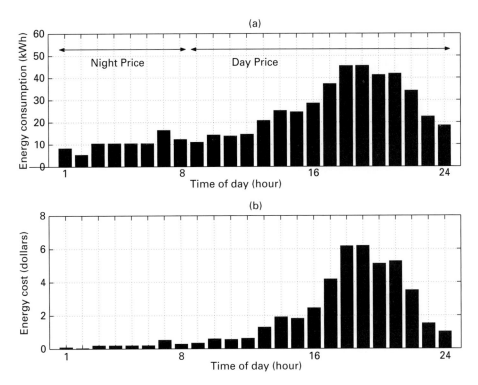

Figure 3.2 Scheduled energy consumption and corresponding cost when ECS units are not used. In this case, PAR is 2.1 and the total daily cost is $44.77.

fact, in the latter case, the load is more evenly distributed across the different hours of the day. Note that each user consumes the same amount of energy in the two cases but schedules its consumption more efficiently when the ECS units are used.

3.4 Energy-consumption control model using utility functions

3.4.1 User preference and utility function

In general, consumers may have different objectives than minimizing the waiting time. To generalize the concept of minimizing joint energy cost and waiting cost, Li *et al.* [26] adopted the concept of utility function from microeconomics to model different objectives that each user may consider for its different appliances [30]. In this regard, an appliance $a \in \mathcal{A}_n$ is characterized by two parameters. First, a utility function $U_{n,a}(\mathbf{x}_{n,a})$ that quantifies the utility (a quantitative measure of satisfaction of the appliance's operation) that user n obtains as a function of its power consumption vector. Second, a set of linear inequalities $A_{n,a}\mathbf{x}_{n,a} \leq \boldsymbol{\eta}_{n,a}$ for each appliance's energy consumption schedule $\mathbf{x}_{n,a}$. The details of defining matrix $A_{n,a}$ and vector $\boldsymbol{\eta}_{n,a}$ for each appliance can be found in [26].

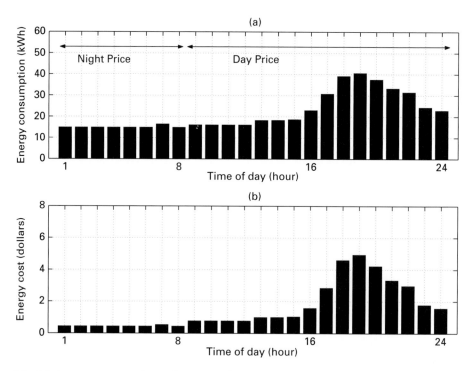

Figure 3.3 Scheduled energy consumption and corresponding cost when ECS units are deployed. In this case, PAR is 1.8 and the total daily cost is $37.90.

Instead of defining a utility function for each appliance, sometimes it is more convenient for the purpose of analysis to model the general behaviour of each user by adopting the concept of utility functions. That is, the level of satisfaction of each user is modelled as a function of its total power consumption in each time slot. This approach is more appropriate for industrial users which can work in different operation modes. In this regard, we represent the utility function of each user n in each time slot k by $U_n^k(x_n^k) \triangleq U(x_n^k, \omega_n^k)$, where x_n^k is the power consumption level of user n and ω_n^k is a parameter which may vary among users and also at different times of day representing the value of electricity for each user in that particular time of day. In this model, we define $\tilde{\mathcal{X}}_n^k \triangleq [l_n^k, L_n^k]$ as the feasible set of power consumption of user n in time slot k, where l_n^k is the minimum and L_n^k is the maximum power requirement of user n in time slot k. $\tilde{\mathcal{X}}^k$ represents the feasible set of power consumption of all users in time slot k. Regardless of whether the utility functions are defined for each appliance or for each user, it is commonly assumed that the utility functions fulfil the following properties:

PROPERTY 1. Utility functions are *non-decreasing*. That is, users are always interested to consume more power if possible until they reach their maximum consumption level.

PROPERTY 2. Utility functions are concave.

While the class of utility functions that fulfil Properties 1 and 2 is very large, it is convenient to have a linear marginal benefit [8, 9].

PROPERTY 3. Considering the utility functions defined for each user, we should be able to rank the customers based on their utilities. We assume that for a fixed consumption level x, a larger ω gives a larger $U_n^k(x_n^k)$.

PROPERTY 4. No power consumption brings no benefit.

Samadi *et al.* have considered *quadratic utility* functions corresponding to *linearly decreasing marginal benefit* for each user [10]. Quadratic utility functions satisfy the above four properties and are defined as

$$U(x,\omega) = \begin{cases} \omega x - \frac{\delta}{2}x^2, & \text{if } 0 \leq x \leq \frac{\omega}{\delta}, \\ \frac{\omega^2}{2\delta}, & \text{if } x \geq \frac{\omega}{\delta}, \end{cases} \qquad (3.20)$$

where δ is a predetermined parameter. Sample utility functions from this class are shown in Figure 3.4. As illustrated in Figure 3.4, the point where the utility function of the user gets saturated and does not change represents the maximum power requirements of the user.

3.4.2 Energy consumption-control problem formulation

In this section, we consider the problem of power control as well as energy-consumption scheduling. Broad categories of load shaping objectives include *peak clipping*, *valley*

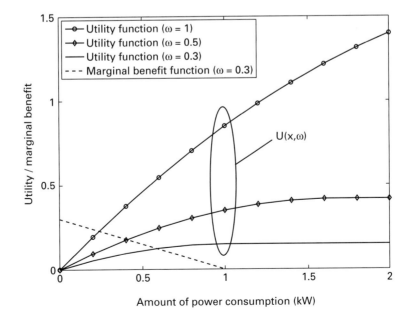

Figure 3.4 Sample utility functions and marginal benefit for power users.

filling, and *flexible load shaping* [31]. However, by equipping users with two-way communication capabilities, it becomes possible to consider new objectives for power system optimization.

From a social fairness point of view, it is desirable to utilize the power generated by the energy provider in such a way that the sum of the utility functions of *all* users is maximized and the cost imposed on the energy provider is minimized.

Case I: Power-control formulation

If centralized control is feasible and the grid operator can collect all information about the users' utility functions, an efficient power-control formulation can be characterized as the solution of the following optimization problem:

$$\underset{\mathbf{x}^k \in \tilde{\mathcal{X}}^k,\, k \in \mathcal{K}}{\text{maximize}} \sum_{k \in \mathcal{K}} \sum_{n \in \mathcal{N}} U_n^k(x_n^k) - C_k\left(\sum_{n \in \mathcal{N}} x_n^k\right), \quad (3.21)$$

where $\mathbf{x}^k$ is the vector of power consumption of all users in time slot k, $U_n^k(x_n^k)$ is as in (3.20), and C_k is defined in (3.3). We solve optimization problem (3.21) separately for each time slot. Thus, we can suppress the dependency of all variables on time slot k, and at each time slot k, problem (3.21) reduces to

$$\underset{\mathbf{x} \in \tilde{\mathcal{X}}}{\text{maximize}} \sum_{n \in \mathcal{N}} U_n(x_n) - C\left(\sum_{n \in \mathcal{N}} x_n\right). \quad (3.22)$$

The objective function in problem (3.22) is the sum of all utility functions *minus* the total energy cost in the system. The power-control model is appropriate in showing the real-time responses of users to different real-time situations. However, this model is limited to time-separable demands, i.e., this model extends each time slot to cover those demands which should be finished by the end of that time slot, and they do not affect the demands in other time slots.

Case II: Energy-consumption scheduling formulation

In addition to appliances, Li *et al.* [26] assumed that each user may benefit from a battery which provides further flexibility in power-consumption optimization. The cost of operating the battery is modelled by a cost function $D_n(r_n)$, where r_n is defined as $r_n \triangleq \{r_n^k \mid k \in \mathcal{K}\}$, and r_n^k denotes the power charged to ($r_n^k > 0$) or the power discharged from ($r_n^k < 0$) the battery at time slot k. Therefore, in this case, the DSM optimization problem to be solved by the energy provider can be formulated as

$$\underset{\mathbf{x},\mathbf{r}}{\text{maximize}} \sum_{n \in \mathcal{N}} \left(\sum_{a \in \mathcal{A}_n} U_{n,a}(\mathbf{x}_{n,a}) - D_n(r_n)\right)$$

$$- \sum_{k \in \mathcal{K}} C\left(\sum_{n \in \mathcal{N}} Q_n^k\right) \quad (3.23)$$

$$\text{subject to} \quad A_{n,a} x_{n,a} \leq \eta_{n,a} \qquad (3.24)$$

$$0 \leq Q_n^k \leq Q_n^{\max}, \quad \forall k \in \mathcal{K}, n \in \mathcal{N} \qquad (3.25)$$

$$r_n \in \mathcal{R}_n, \quad \forall n \in \mathcal{N}, \qquad (3.26)$$

where $\mathcal{R}_n$ is defined as, for all k, the vectors $r_n \in \mathcal{R}_n$ if and only if

$$0 \leq b_n^k \leq B_n, \quad b_n^K \geq \gamma_n B_n \qquad (3.27)$$

$$r_n^{\min} \leq r_n^k \leq r_n^{\max}, \qquad (3.28)$$

$b_n^k \triangleq \sum_{\tau=1}^{k} r_n^\tau + b_n^0$, B_n is the battery capacity, b_n^0 is the initial charge of the battery, $\gamma_n \in (0,1]$, and Q_n^k is defined as $Q_n^k \triangleq \sum_{a \in \mathcal{A}_n} x_{n,a}^k + r_n^k$. The inequality in (3.24) models different sets of power-consumption constraints on different appliances. The lower-bound inequality in (3.25) indicates the fact that user n's battery cannot provide more power than the total amount consumed by all of user n's appliances. The upper-bound inequality in (3.25) limits the maximum power drawn by each user.

3.4.3 Equilibrium among users

In general, users may have different approaches in responding to the price values set by the energy provider. This can lead to different equilibria among users. Considering the problem formulation in (3.22), one can analyse *competitive equilibrium* and *Nash equilibrium*. In competitive equilibrium, each user acts as a *price taker*. That is, it does not consider the effect of its actions on the price. However, in Nash equilibrium, users are assumed to be price anticipators, i.e., they consider the effect of their actions on the price set by the energy provider.

Marginal cost pricing with price-taking users

If users are price takers, i.e., they do not consider the effect of their actions on the price, then we need to analyse the *competitive equilibrium* among the users.

Case I: Power-control algorithm

Given a price $\lambda > 0$, the *payoff* function for each user n is obtained as

$$P_n(x_n) = U_n(x_n) - \lambda x_n, \qquad (3.29)$$

where the first term represents the utility of user n as a function of its power consumption, and the second term represents its payment to the energy provider. We call a pair $(\mathbf{x}, \lambda)$, where $\mathbf{x} \triangleq \{x_n \mid n \in \mathcal{N}\}$, a *competitive equilibrium* if users maximize their own payoff function defined in (3.29) for a given price λ, i.e.,

$$P_n(x_n) \geq P_n(\bar{x}_n), \quad \bar{x}_n \in \tilde{\mathcal{X}}_n, \, n \in \mathcal{N}. \qquad (3.30)$$

By applying the dual decomposition method to the primal problem in (3.22), the objective function of the *dual optimization problem* can be written as

$$\mathcal{D}(\lambda) = \underset{\mathbf{x}\in\tilde{\mathcal{X}},\, q\in\mathcal{Q}}{\text{maximize}}\ \mathcal{L}(\mathbf{x},q,\lambda)$$

$$= \sum_{n\in\mathcal{N}} B_n(\lambda) + S(\lambda), \quad (3.31)$$

where λ is the Lagrange multiplier, $\mathcal{L}(\mathbf{x},q,\lambda)$ is the Lagrangian for the problem (3.22),

$$B_n(\lambda) = \underset{x_n\in\tilde{\mathcal{X}}_n}{\text{maximize}}\ U_n(x_n) - \lambda x_n, \quad (3.32)$$

and

$$S(\lambda) = \underset{q\in\mathcal{Q}}{\text{maximize}}\ \lambda q - C(q), \quad (3.33)$$

where q is an auxiliary variable, $\mathcal{Q} \triangleq [Q_{\min}, Q_{\max}]$, $Q_{\min} = \sum_{n\in\mathcal{N}} l_n$, and $Q_{\max} = \sum_{n\in\mathcal{N}} L_n$. The dual problem is

$$\underset{\lambda\geq 0}{\text{minimize}}\ \mathcal{D}(\lambda). \quad (3.34)$$

One can show that *strong duality* holds, and we can solve the dual problem (3.34) iteratively instead of the primal problem (3.22). In this case, we can obtain the solution of the dual problem λ^*, and in each iteration, each user and the energy provider can simply solve their own local optimization problems determined by (3.32) and (3.33) to obtain x_n^* and q^*, respectively. For the tth iteration of the algorithm, the price is updated as

$$\lambda_{t+1} = \left[\lambda_t - \gamma \frac{\partial \mathcal{D}(\lambda_t)}{\partial \lambda}\right]^+$$
$$= \left[\lambda_t + \gamma \left(\sum_{i\in\mathcal{N}} x_i^*(\lambda_t) - q^*(\lambda_t)\right)\right]^+, \quad (3.35)$$

where γ is the step size, $[\cdot]^+$ denotes projection onto positive numbers, $x_i^*(\lambda_t)$ is the local optimizer of (3.32), and $q^*(\lambda_t)$ is the local optimizer of (3.33) for a *given* λ_t. The interaction between the energy provider and the subscribers is depicted in Figure 3.5.

Case II: Energy-consumption scheduling algorithm

Given the price λ^k, the local energy-consumption scheduling problem of each price-taker user is defined as [26]

$$\underset{x_n, r_n}{\text{maximize}}\ \sum_{a\in\mathcal{A}_n} U_{n,a}(x_{n,a}) - D_n(r_n) - \sum_{k\in\mathcal{K}} \lambda^k Q_n^k \quad (3.36)$$
$$\text{subject to}\ (3.24) - (3.26).$$

It has been shown that setting the prices equal to the marginal cost of the energy provider is indeed optimal, i.e., the solution to local problem (3.36) also solves the system-wide problem (3.23). In fact, one can check that the Karush–Kuhn–Tucker (KKT)

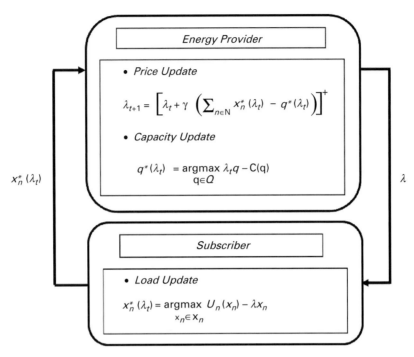

Figure 3.5 Illustration of the operation of the proposed algorithm and the interactions between the Energy Provider and Subscribers in the system.

conditions of problem (3.36) are identical to the KKT conditions of problem (3.23). Based on the gradient-projection algorithm and by using marginal cost pricing, users and energy provider will be able to iteratively solve problem (3.23). In this regard, at the tth iteration we have [26]:

- The energy provider sets the price to the marginal cost based on the total demands from all users

$$\lambda_t^k = C'\left(\sum_{n \in \mathcal{N}} Q_{n,t}^k\right). \tag{3.37}$$

- Each user updates its demand $x_{n,t}^k$ as well as charging schedule $r_{n,t}^k$ after receiving the updated λ_t^k, according to

$$\begin{aligned}
\hat{x}_{n,a}^{k,t+1} &= x_{n,a}^{k,t} + \gamma \left(\frac{\partial U_{n,a}(x_{n,a}^{k,t})}{\partial x_{n,a}^{k,t}} - \lambda_t^k \right) \\
\hat{r}_{n,t+1}^k &= r_{n,t}^k - \gamma \left(\frac{\partial D_n(r_{n,t}^k)}{\partial r_{n,t}^k} + \lambda_t^k \right) \\
(x_{n,t+1}^k, r_{n,t+1}^k) &= [\hat{x}_{n,t+1}^k, \hat{r}_{n,t+1}^k]^{S_n},
\end{aligned} \tag{3.38}$$

where $\gamma > 0$ is a constant step size, and $[\cdot]^{S_n}$ denotes projection onto set S_n specified by constraints (3.24)–(3.26).

Price-anticipating users

In this subsection, we consider the problem of power-level selection in (3.22) for price-anticipating users. If users are price anticipators, i.e., they consider the effect of their actions on the price, then we need to analyse the Nash equilibrium of the *game* which is played among multiple users who compete for the available power provided by the energy provider. In this game-theoretic model [32], the strategies of the users represent their power consumption level. We consider the following pricing scheme for resource allocation. Given $\mathbf{x} = (x_1, \ldots, x_N)$, the energy provider sets a single price $\mu(\mathbf{x}) = p(\sum_{n \in \mathcal{N}} x_n)$. User n pays $x_n \mu(\mathbf{x})$ as bill payment to the energy provider. We use the notation $\mathbf{x}_{-n}$ to denote the vector of all consumption powers chosen by users *other than* user n, i.e., $\mathbf{x}_{-n} = (x_1, \ldots, x_{n-1}, x_{n+1}, \ldots, x_N)$. Then, given $\mathbf{x}_{-n}$, the payoff of each user n is obtained as

$$Q_n(x_n; \mathbf{x}_{-n}) = U_n(x_n) - x_n p\left(\sum_{m \in \mathcal{N}} x_m\right). \tag{3.39}$$

The payoff function Q_n is similar to the payoff function P_n defined for price-taking users in (3.29). The key difference is that while the payoff function P_n takes the price λ as a fixed parameter, price-anticipating users realize that the price is set according to $p(\sum_{m \in \mathcal{N}} x_m)$, and adjust their payoffs accordingly.

From (3.39), each user's payoff depends on its power consumption and the power consumptions of other users. Hence, the following game is played among users.

- Players: Registered users in set $\mathcal{N}$.
- Strategies: Each user $n \in \mathcal{N}$ selects its energy-consumption level $x_n \in \hat{\mathcal{X}}_n$ to maximize its payoff.
- Payoffs: $Q_n(x_n; \mathbf{x}_{-n})$ for each user $n \in \mathcal{N}$ as in (3.39).

A *Nash equilibrium* of the game defined by $(Q_1, \ldots, Q_N)$ is a vector $\mathbf{x}$ such that for all $n \in \mathcal{N}$, we have

$$Q_n(x_n; \mathbf{x}_{-n}) \geq Q_n(\bar{x}_n; \mathbf{x}_{-n}) \quad \bar{x}_n \in \tilde{\mathcal{X}}_n. \tag{3.40}$$

It can be shown that a Nash equilibrium exists for this game [32]. In general, the Nash equilibrium of a resource allocation game may not be optimal [32, 33]. That is, the energy consumption profile obtained at the Nash equilibrium in a distributed pricing scenario may not necessarily be the same as the optimal solution of the optimization problem in (3.22). Next, we investigate how the price values can be set carefully by the utility company such that the system performance becomes optimal at the aforementioned Nash equilibrium.

3.4.4 The Vickrey–Clarke–Groves (VCG) approach

In the previous section, we considered a mechanism which uses only a single price for all users to solve the problem of power-level selection in (3.22). Despite its simplicity and ease of implementation, the introduced mechanism suffers from a loss in efficiency if users are price anticipators, and are able to evaluate the effect of their actions on the price function. This is particularly the case for regional utilities with relatively small numbers of users. As mentioned before, the main obstacle in solving problem (3.22) in a centralized fashion is the lack of information about the utility functions of the users. However, if we remove the restriction that the mechanism only chooses a single price, it is possible to elicit the utility information of the users. One of the best-known approaches to convince users to declare their utility functions is the Vickrey–Clarke–Groves (VCG) mechanism [34].

In the VCG class of mechanisms, each user is asked to specify its utility function, which in case of the utility functions in (3.20) reduces to revealing a single utility parameter ω_n and the required range of power consumptions. For each user n, we use $\hat{U}_n$ to denote the declared utility function and $\hat{\mathbf{U}} = (\hat{U}_1, \ldots, \hat{U}_N)$ to denote the vector of declared utility functions. If user n consumes x_n units of power, but has to pay t_n, then the payoff function of user n is

$$U_n(x_n) - t_n.$$

For a given vector of declared utility functions $\hat{\mathbf{U}}$, the VCG mechanism chooses the energy-consumption allocation $\mathbf{x}(\hat{\mathbf{U}})$ as an optimal solution to problem (3.22) and calculates optimal energy-consumption vectors as

$$\mathbf{x}(\hat{\mathbf{U}}) = \arg\max_{x \in \tilde{\mathcal{X}}} \left\{ \sum_{n \in \mathcal{N}} \hat{U}_n(x_n) - C\left(\sum_{n \in \mathcal{N}} x_n\right) \right\}, \quad (3.41)$$

and the payments are structured such that

$$t_n(\hat{\mathbf{U}}) = -\left(\sum_{m \in \mathcal{N}_{-n}} \hat{U}_m(x_m) - C\left(\sum_{m \in \mathcal{N}} x_m\right) \right) + h_n(\hat{\mathbf{U}}_{-n}), \quad (3.42)$$

where $\mathcal{N}_{-n}$ is the set of all users except user n, and h_n is an arbitrary function of the declared utility functions other than n, denoted by $\hat{\mathbf{U}}_{-n}$. We also note that the definition of the payments in (3.42) is a natural way to align user objectives with the objective of a social planner, in this scenario, the grid operator. Here, we will use the most popular choice for h_n, which is referred to as Clarke tax [34],

$$h_n(\hat{\mathbf{U}}_{-n}) = \sum_{m \in \mathcal{N}_{-n}} \hat{U}_m(x_m(\hat{\mathbf{U}}_{-n})) - C\left(\sum_{m \in \mathcal{N}_{-n}} x_m(\hat{\mathbf{U}}_{-n}) \right), \quad (3.43)$$

where $\mathbf{x}(\hat{\mathbf{U}}_{-n})$ is the VCG allocation choice introduced in (3.41) but with user n excluded from the system. Therefore, the payment of user n is

$$t_n(\hat{\mathbf{U}}) = -\left(\sum_{m \in \mathcal{N}_{-n}} \hat{U}_m(x_m(\hat{\mathbf{U}})) - C\left(\sum_{m \in \mathcal{N}} x_m(\hat{\mathbf{U}})\right)\right)$$
$$+ \left(\sum_{m \in \mathcal{N}_{-n}} \hat{U}_m(x_m(\hat{\mathbf{U}}_{-n})) - C\left(\sum_{m \in \mathcal{N}_{-n}} x_m(\hat{\mathbf{U}}_{-n})\right)\right). \quad (3.44)$$

The payment of user n is the difference in the social welfare of other users with and without the presence of user n. It has been shown that the VCG-based mechanism defined in (3.41) and (3.42) is *truthful* and *efficient*, i.e., declaring $\hat{U}_n = U_n$ is a dominant strategy for each user n, and following this strategy results in an efficient power allocation across the grid [35].

3.4.5 Performance evaluation of power-level selection algorithms

In this section, simulation results of different power-control algorithms are presented. It is assumed that all users have concave quadratic utility functions as described in (3.20). The parameter δ of the utility function is chosen to be 0.5. The parameters of the cost function in (3.3) are set to $c_1 = 0.01$, $c_2 = 0$, and $c_3 = 0$.

Performance gains from real-time interaction with users

Given the two-way communication capabilities of smart meters, the VCG-based mechanism reflects the cost of energy to the demand side and improves the performance of the power grid. To compare the performance of the VCG-based mechanism, a *peak-load pricing* (PLP) method is also considered in which the price value for each time slot is calculated based on the average parameter ω of the users to maximize the payoff of the energy provider, which is its revenue minus total energy cost. For the PLP method, it is assumed that the energy provider has some prior information about the distribution of parameter ω in each time slot.

We assume that there are $N = 50$ users. We consider a 24-hour period consisting of three time slots of mid-peak, on-peak, and off-peak hours, respectively. Parameter ω of each user is selected from the sets $\{1.5, 2, 3, 5\}$, $\{2, 3, 5, 8\}$, and $\{1, 1.5, 2, 2.5\}$ representing mid-peak, on-peak, and off-peak hours, respectively. However, random events are modelled via a small perturbation in the ω value of each user. Parameter c_1 of the cost function is set equal to 0.1, 0.2, and 0.05 for mid-peak, on-peak, and off-peak hours, respectively. As illustrated in Figure 3.6, the VCG-based mechanism in [35] improves the performance of the system by reducing the PAR from 2.5852 to 1.2231.

Impact of reflecting the generating cost

The VCG-based mechanism is used to maximize the social welfare. Maximizing the aggregate utility of all users while minimizing the cost imposed on the energy provider is beneficial for both users and energy provider. The capability to reflect the fluctuations of the wholesale price into the customer side is one of the main advantages of the VCG-based mechanism. This aspect becomes more important, especially in situations where the cost imposed on the energy provider is high. To have a baseline scheme to compare

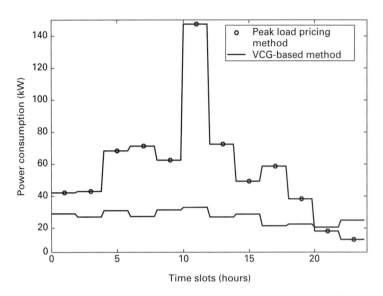

Figure 3.6 Average power consumption for the VCG-based method and a peak-load pricing method.

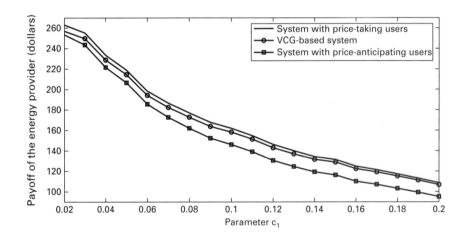

Figure 3.7 Average payoff of the energy provider for the VCG-based system, the system with price-anticipating users, and the system with price-taking users.

with, a system which has price-anticipating users and employs marginal cost pricing is considered. It has been shown that in a system with price-taking users, marginal cost pricing not only maximizes the social welfare, but also maximizes the payoff of the energy provider [23]. As an upper bound on the payoff of the energy provider, we also consider a system which has price-taking users and employs marginal cost pricing. It is assumed that there are 50 users, and parameter ω of each user is selected from set $\{2, 3, 5, 8\}$.

Focusing on one time slot, the average payoff of the energy provider and the average total power consumption of the VCG-based system, the system with price-anticipating

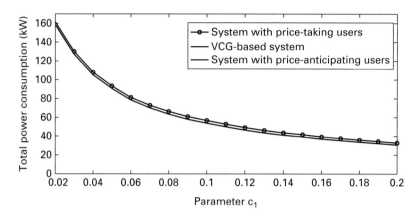

Figure 3.8 Average power-consumption of all users for the VCG-based system, the system with price-anticipating users, and the system with price-taking users.

users, and the system with price-taking users are presented in Figures 3.7 and 3.8 as functions of parameter c_1 of the cost function, respectively [35]. As illustrated in Figure 3.7, since the VCG payment (3.44) is structured to consider the cost imposed on the energy provider, the payoff of the energy provider is higher compared to that in the system with price-anticipating users. As illustrated in Figure 3.8, the VCG-based system and the price-taking system are both efficient systems with the same power allocation. Hence, they have the same total power consumption and lead to the same system efficiency.

3.5 Conclusion

In this chapter, we have reviewed different algorithms for demand-side management (DSM) in smart grid. We have distinguished two different groups of algorithms. The first group consists of algorithms designed to shift the energy consumption of users, and the second group consists of those algorithms designed to change the energy-consumption level of users. We have addressed different design objectives considered for different DSM programs including energy-cost minimization, PAR reduction, and joint cost and waiting time minimization. Furthermore, we have explained that users may have other objectives than minimizing waiting time or minimizing the electricity payment. The concept of utility functions from microeconomics has been adopted to model different objectives of users. Next, different methods for both categories of energy-consumption scheduling and power-control algorithms have been studied in which they all use the concept of utility functions to model different objectives of users. We have studied the concept of social welfare maximization and also different methods proposed to achieve the maximum social welfare for both categories of algorithms. Simulation results have been presented to compare the performance of different methods.

Different models discussed in this chapter can be extended in several directions. For example, a system with multiple energy providers could be considered, and the effect of malicious users could be explored. Load uncertainties and generation uncertainties due to the penetration of renewable energy resources are other aspects that could also be studied.

References

[1] G. M. Masters, *Renewable and Efficient Electric Power Systems*. John Wiley & Sons, Inc., 2004.

[2] L. H. Tsoukalas and R. Gao, 'From smart grids to an energy internet: assumptions, architectures, and requirements', in *Proceedings of 3rd International Conference on Electric Utility Deregulation and Restructuring and Power Technologies*, Nanjing, China, April 2008.

[3] A. Ipakchi and F. Albuyeh, 'Grid of the future', *IEEE Power Energy Magazine*, vol. 7, no. 2, pp. 52–62, March.–April 2009.

[4] P. Luh, Y. Ho, and R. Muralidharan, 'Load adaptive pricing: an emerging tool for electric utilities', *IEEE Transactions on Automatic Control*, vol. 27, no. 2, pp. 320–329, April 1982.

[5] Y. Tang, H. Song, F. Hu, and Y. Zou, 'Investigation on TOU pricing principles', in *Proceedings of IEEE PES Transmission and Distribution Conference Exhibition: Asia and Pacific*, Dalian, China, August 2005.

[6] M. Crew, C. Fernando, and P. Kleindorfer, 'The theory of peak-load pricing: a survey', *Journal of Regulatory Economics*, vol. 8, no. 3, pp. 215–248, November 1995.

[7] S. Zeng, J. Li, and Y. Ren, 'Research of time-of-use electricity pricing models in China: a survey', in *Proceedings of IEEE International on Conference on Industrial Engineering and Engineering Management*, Singapore, December 2008.

[8] M. Fahrioglu, M. Fern, and F. Alvarado, 'Designing cost effective demand management contracts using game theory', in *Proceedings of IEEE Power Engineering Society 1999 Winter Meeting*, New York, January 1999.

[9] M. Fahrioglu and F. Alvarado, 'Using utility information to calibrate customer demand management behavior models', *IEEE Transactions on Power Systems*, vol. 16, no. 2, pp. 317–322, May 2001.

[10] R. Faranda, A. Pievatolo, and E. Tironi, 'Load shedding: a new proposal', *IEEE Transactions on Power Systems*, vol. 22, no. 4, pp. 2086–2093, November 2007.

[11] N. Ruiz, I. Cobelo, and J. Oyarzabal, 'A direct load control model for virtual power plant management', *IEEE Transactions on Power Systems*, vol. 24, no. 2, pp. 959–966, May 2009.

[12] L. Exarchakos, M. Leach, and G. Exarchakos, 'Modelling electricity storage systems management under the influence of demand-side management programs', *International Journal of Energy Research*, vol. 33, no. 1, pp. 62–76, January 2009.

[13] S. Vandael, N. Boucke, T. Holvoet, and G. Deconinck, 'Decentralized demand side management of plug-in hybrid vehicles in a smart grid', in *Proceedings of First International Workshop on Agent Technologies for Energy Systems*, Toronto, Canada, May 2010.

[14] V. Bakker, M. G. C. Bosman, A. Molderink, J. L. Hurink, and G. J. M. Smit, 'Demand side load management using a three step optimization methodology', in *Proceedings of IEEE International Conference on Smart Grid Communications*, Gaithersburg, MD, October 2010.

[15] K. Spees and L. Lave, 'Impacts of responsive load in PJM: load shifting and real time pricing', *The Energy Journal*, vol. 29, no. 2, pp. 101–122, February 2008.

[16] A. Kowli, M. Negrete-Pincetic, and G. Gross, 'A successful implementation with the smart grid: demand response resources', in *Proceedings of IEEE Power and Energy Society General Meeting*, Minneapolis, MN, July 2010.

[17] M. Ann-Piette, G. Ghatikar, S. Kiliccote, D. Watson, E. Koch, and D. Hennage, 'Design and operation of an open, interoperable automated demand response infrastructure for commercial buildings', *Journal of Computing and Information Science in Engineering*, vol. 9, pp. 1–9, June 2009.

[18] G. Xiong, C. Chen, S. Kishore, and A. Yener, 'Smart (in-home) power scheduling for demand response on the smart grid', in *Proceedings of IEEE PES Innovative Smart Grid Technologies Conference*, Anaheim, CA, January 2011.

[19] Z. A. Vale, H. Morais, and H. Khodr, 'Intelligent multi-player smart grid management considering distributed energy resources and demand response', in *Proceedings of IEEE Power and Energy Society General Meeting*, Minneapolis, MN, July 2010.

[20] K. Hamilton and N. Gulhar, 'Taking demand response to the next level', *IEEE Power and Energy Magazine*, vol. 8, no. 3, pp. 60–65, May 2010.

[21] B. Daryanian, R. Bohn, and R. Tabors, 'Optimal demand-side response to electricity spot prices for storage-type customers', *IEEE Transactions on Power Systems*, vol. 4, no. 3, pp. 897–903, August 1989.

[22] A. H. Mohsenian-Rad, V. W. S. Wong, J. Jatskevich, R. Schober, and A. Leon-Garcia, 'Autonomous demand-side management based on game-theoretic energy consumption scheduling for the future smart grid', *IEEE Transactions on Smart Grid*, vol. 1, no. 3, pp. 320–331, December 2010.

[23] P. Samadi, A. H. Mohsenian-Rad, R. Schober, V. W. S. Wong, and J. Jatskevich, 'Optimal real-time pricing algorithm based on utility maximization for smart grid', in *Proceedings of IEEE International Conference on Smart Grid Communications (SmartGridComm)*, Gaithersburg, MD, October 2010.

[24] A. H. Mohsenian-Rad and A. Leon-Garcia, 'Optimal residential load control with price prediction in real-time electricity pricing environments', *IEEE Transactions on Smart Grid*, vol. 1, no. 2, pp. 120–133, September 2010.

[25] A. Wood and B. Wollenberg, *Power Generation, Operation, and Control*. Wiley-Interscience, 1996.

[26] N. Li, L. Chen, and S. H. Low, 'Optimal demand response based on utility maximization in power networks', *IEEE Power Engineering Society General Meeting*, July 2011.

[27] P. Samadi, R. Schober, and V. W. S. Wong, 'Optimal energy consumption scheduling using mechanism design for the future smart grid', in *Proceedings of IEEE International Conference on Smart Grid Communications (SmartGridComm)*, Brussels, October 2011.

[28] S. Boyd and L. Vandenberghe, *Convex Optimization*. Cambridge University Press, 2004.

[29] M. A. A. Pedrasa, T. D. Spooner, and I. F. MacGill, 'Scheduling of demand side resources using binary particle swarm optimization', *IEEE Transactions on Power Systems*, vol. 24, no. 3, pp. 1173–1181, August 2009.

[30] A. Mas-Colell, M. D. Whinston, and J. R. Green, *Microeconomic Theory*, 1st edn. Oxford University Press, 1995.

[31] C. Gellings, 'The concept of demand-side management for electric utilities', *Proceedings of the IEEE*, vol. 73, no. 10, pp. 1468–1470, October 1985.

[32] R. Johari, S. Mannor, and J. Tsitsiklis, 'Efficiency loss in a network resource allocation game: the case of elastic supply', *IEEE Transactions on Automatic Control*, vol. 50, no. 11, pp. 1712–1724, November 2005.

[33] R. Johari and J. Tsitsiklis, 'A scalable network resource allocation mechanism with bounded efficiency loss', *IEEE Journal on Selected Areas in Communications*, vol. 24, no. 5, pp. 992–999, May 2006.

[34] Y. Shoham and K. Leyton-Brown, *Multiagent Systems: Algorithmic, Game-Theoretic, and Logical Foundations*. Cambridge University Press, 2008.

[35] P. Samadi, A. H. Mohsenian-Rad, R. Schober, and V. W. S. Wong, 'Advanced demand side management for the future smart grid using mechanism design', submitted to *IEEE Transactions on Smart Grid*, 2011.

4 Vehicle-to-grid systems: ancillary services and communications

Chenye Wu, Amir-Hamed Mohsenian-Rad, and Jianwei Huang

4.1 Introduction

Recent studies have shown that about 70% of the total oil extracted worldwide is consumed in the transportation sector [1]. With rising oil prices, the USA and many other countries have set long-term plans to electrify their transportation system and manufacture electric vehicles (EVs) to reduce their oil consumption. It is foreseen that by 2013, approximately 700,000 grid-enabled electric vehicles will be on the road in the USA. The expected trend in the automotive market share for EVs is shown in Figure 4.1 [2]. A large number of EVs can not only help to reduce the amount of oil and gas consumption, but also provide great opportunities for the power grid, as the batteries of millions of EVs can be used to boost *distributed electricity storage*. Depending on the type and class, the battery storage capacity for an existing EV varies from 1.8 kW [3] to 17 kW [4, 5]. Note that, currently, the only major electricity storage unit in most power grids are the pumped storage systems [6].

In general, EVs have the capability to work in two main modes of operation: *stand-alone mode* and *grid-connected mode* [7]. These two modes and their transition cycles are shown in Figure 4.2. In the stand-alone mode, the storage capacity of EVs is used as a back-up energy source at the time of electricity shortage or blackout. In addition, it helps to smooth down possible fluctuations in local renewable generation units, such as rooftop solar panels and wind turbines [8–11]. In the grid-connected mode, the EV storage units can be synchronized with the grid to participate in *demand-side management* programs [12, 13] or to provide reserve power capacity and other *ancillary services* [14–17] in a distributed vehicle-to-grid (V2G) infrastructure. Our focus in this chapter is on EVs' grid-connected operation mode in V2G systems.

In order to be successful, the V2G systems require a reliable and secure communications and networking infrastructure, which enables *two-way message exchanges* among EVs and the grid operation, control, and monitoring centres [18]. The type of message exchanges and the communications technologies and architectures needed depend mainly on the ancillary services provided and the centralized and distributed management strategies to be implemented. In this chapter, we will overview such services and a variety of existing communications technologies that facilitate efficient and practical V2G systems in future smart grid systems. The rest of this chapter is organized as follows. In Section 4.2, we overview different types of ancillary services that can be offered in future V2G systems. In Section 4.3, we compare two different V2G system architectures

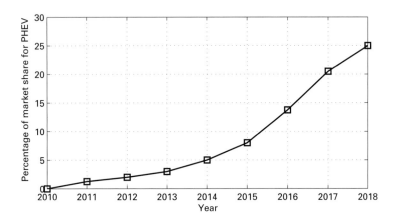

Figure 4.1 Expected increase in market share for electric vehicles in the USA.

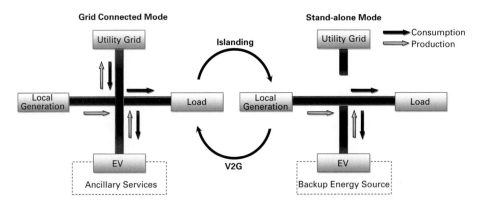

Figure 4.2 Operating modes in future home energy systems in the presence of EVs.

to be implemented, by introducing EV aggregators and their different roles in V2G systems. Different communications and networking technologies to support V2G systems are discussed in Section 4.4. Research challenges and open problems are discussed in Section 4.5. The chapter is concluded in Section 4.6.

4.2 Ancillary services in V2G systems

As has been mentioned before, a key benefit of a V2G power system is to facilitate and encourage EV participation in offering various ancillary services to the power grid through adequate communications. To start, in this section, we will overview different services that can potentially be offered in a V2G power system.

- *Reserve power supply.* A large-scale V2G system can help maintain the balance between supply and demand in power grid by injecting power. For example, by simultaneously discharging their batteries, thousands of EVs will be able to provide the additional power required by a medium-sized factory at a certain time period, acting similarly to a so-called *spinning reserve* power generation source in the existing power distribution systems [19]. While the supply capacity for each individual EV is small, a synchronized aggregated capacity can be both noticeable and manageable, as we explain in Section 4.3.

- *Peak shaving*. A group of EVs can also participate in peak shaving by coordinating charging and discharging of their batteries. Note that, in general, the operation cost of a grid depends highly on the *peak-to-average* ratio (PAR) in aggregate load demand [20]. For example, as shown in Figure 4.3, there is usually at least one major peak in a daily residential load demand profile, e.g., in the afternoon. To assure reliable service, the grid generation capacity should essentially match these peak demands. Therefore, a high PAR can significantly increase the generation cost, as the grid will be highly under-utilized most of time. By charging the EV batteries at off-peak hours and discharging them at peak hours, the EVs can help reduce the PAR significantly. Peak shaving participation can be coordinated by implementing various demand-side management (DSM) programs [12, 22–27]. The impact of EVs on DSM programs will be significant, as the charging load of EVs is expected to *double* the average residential load in the near future [14]. Interestingly, such a major load is *controllable*, as EV charging can potentially be scheduled using advanced *energy-consumption scheduling* (ECS) features in smart meters [12, 26].

- *Renewable energy integration*. Due to the stochastic and intermittent nature of solar and wind-power generation, their large-scale integration into the current power grid requires large-capacity storage systems [28, 29]. Take wind power as an instance, its stochastic nature is due to the changes in wind speed, since other on-site condition changes are relatively slow [30]. For example, a recent measurement in Crosby County in Lubbock, Texas showed that the wind speed in this region can fluctuate between 2 and 12 m/s within a few hours [29]. While a centralized control using a massive battery bank is very expensive, and thus may not always be practical, a distributed V2G power storage system can be implemented to solve this problem.

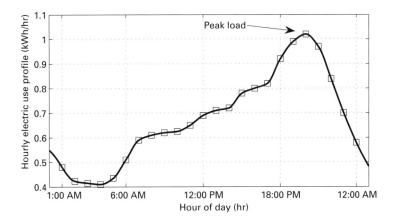

Figure 4.3 Hourly average residential load profile in Southern California [21].

To achieve a better wind-power penetration, He *et al.* proposed a multiple time-scale pricing model in [31]. They considered a power grid with two types of energy source: conventional and wind-energy. The conventional energy is drawn from two sources: base-load generation and peaking generation, with generation cost c_1 and c_2 per unit, respectively. Peaking generation is typically from fast-start generators (e.g., gas turbines), with a higher generation cost, and thus $c_2 > c_1$. Due to the start-up time and ramp rate of generators, the base-load generators are scheduled day-ahead for each T_1 slot of the next day, and the generation cost c_1 contains the start-up cost and other operating costs. In real-time scheduling of each T_2 slot, peaking generation and wind generation are used, as needed, to clear the balance between demand and the base-load generation. Clearly, with a distributed V2G storage system in place, the EVs can be charged at off-peak hours as a base load and discharged at peak hours to act as an extra power source to help the peaking generator, and thus reduce the cost of power generation.

- *Regulation.* A V2G system can help to regulate frequency and voltage in a power grid. In the USA, the grid frequency needs to be maintained very close to its nominal frequency of 60 Hz. Any deviation from this requires action by the grid operator [32, 33]. If the frequency is too high, then there is too much power being generated in relation to load. Therefore, the load must be increased or the generation must be reduced to keep the system in balance. Currently, such regulation is achieved mainly by reducing power generation via turning on/off fast-responding generators, which is very costly. Alternatively, EVs can help by charging their batteries and increasing the load demand. On the other hand, if the frequency is too low, then there is too much load in the system and the generation must be increased or the load reduced. This can be done by terminating charging or starting discharging a number of EVs connected to the grid. Such adjustment is called frequency regulation [32–34], as shown in Figure 4.4. It is usually

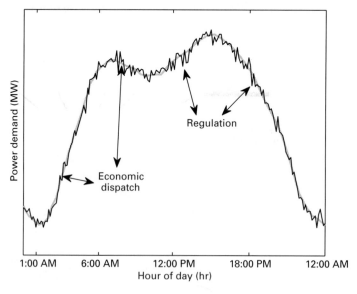

Figure 4.4 Frequency regulation by adjusting active power supply and load demand.

performed frequently, e.g., once every few seconds [28, 35]. Note that the focus in frequency regulation is to remove *small mismatches* between supply and demand. Major load following is achieved separately via *economic dispatch* of major generators [36]. A large group of EVs can also help the power grid to regulate voltage. In the power system operation, voltage profile is strongly correlated with *transmission and distribution system losses*. The voltage drops from the normal voltage (110 V in the United States and Canada) when the load and consequently transmission losses increase [37]. This can cause damage to the grid equipment and user appliances. Voltage drops can be compensated by adjusting (injecting or consuming) *reactive power* across the power grid. Interestingly, given the right power electronics devices in place, EVs can help by changing their reactive and active power load, without any major impact on their battery life [38]. This makes reactive power compensation a very promising ancillary service in future V2G systems [39, 40].

4.3 V2G system architectures

Various ancillary services that we listed in Section 4.2 can be provided and managed using either *direct* or *indirect* V2G system architectures [5], as illustrated in Figures 4.5 and 4.6, respectively. In a direct architecture, there exists a *direct line of communication* between the grid system operator and the vehicle, so that each vehicle can be treated as a deterministic resource to be commanded by the grid system operator. Under this paradigm, each vehicle is allowed to individually bid and perform services while it is at

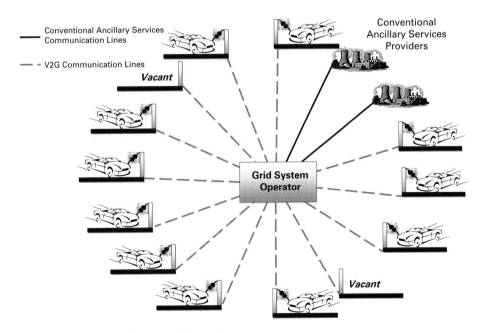

Figure 4.5 Direct V2G system architecture without using aggregators.

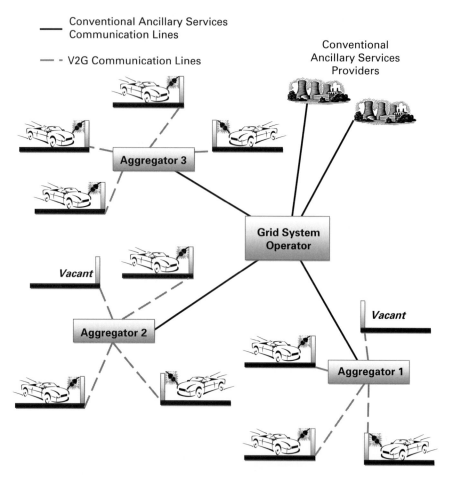

Figure 4.6 Indirect V2G system architecture involving several aggregators.

the charging station. When the vehicle leaves the charging station, the contracted payment for the previous full hours is made and the ancillary service contract is ended until the next time when the vehicle is parked and available again. The direct and thus deterministic architecture is conceptually simple, but it has recognized problems in terms of *near-term feasibility* and *long-term scalability* [5]. The challenges with direct architecture are twofold. First, the amount of signalling and control task overhead imposed on the grid operator is significant and overwhelming, as the operator needs to interact directly with a larger number of individual EVs [35]. As these millions of vehicles engage and disengage from the grid, the grid system operator must constantly update the contract status, connection status, power availability, state-of-charge, and driver requirements to contract the power it can command from the vehicle [5]. Second, the geographically distributed nature of vehicles and their limited individual storage capacity is incompatible with the existing contracting frameworks with minimum 1 MW threshold for many ancillary services' hourly contracts [19].

The alternative indirect V2G system architecture involves several aggregators as shown in Figure 4.6. In this regard, each aggregator aggregates the ancillary services provided by individual EVs to make a single controllable power resource. This architecture is indirect as aggregators are intermediate between the vehicles and the grid operator. The aggregator receives ancillary service requests from the grid system operator and issues *charging or discharging commands* to contracted vehicles that are both available and willing to perform the required services. Alternatively, the aggregator may interact with its corresponding vehicles through *smart pricing*, where the prices are set according to the grid's service requests [33]. Given an estimate of the EV participation, the aggregator can then *bid* to perform ancillary services for the power grid at any time, while the individual vehicles can engage and disengage from the aggregator as they arrive at and leave charging stations. The individual EVs are then compensated according to, for example the number of minutes that they have participated in offering ancillary services. As such, this aggregative architecture attempts to address the key problems with the direct architecture that we mentioned earlier. In fact, the larger scale of the aggregated V2G power resources commanded by the aggregator, the more improved reliability of aggregated V2G resources connected in parallel; this allows the grid operator to treat the aggregator just like a *conventional ancillary services provider*. It means that the aggregator can utilize the same communication infrastructure for contracting and command signals that conventional ancillary services providers use, which eliminates the concern of additional communications workload placed on the grid operator [5].

4.3.1 Aggregation scenarios

The key elements in an indirect V2G system architecture are aggregators. They act as an interface between the grid and several EVs. In general, an aggregator may take one of the following three roles in an indirect V2G system:

- it may represent the grid operator;
- it may represent a group of EVs;
- it may act as an independent dealer.

The first case is a common scenario in the current literature [32, 41–43]. By representing the grid operator, the aggregator tries to coordinate the ancillary service to best serve the grid. This includes maximizing EV participation in providing ancillary services while minimizing the cost of obtaining such services. In addition, the aggregator tries to keep the quality of supply within the set limits. This means that the voltage has to be kept within upper and lower limits at all points of the distribution network, and the power flowing through the transformers, cables, overhead lines, and other network components must not exceed their limits. The grid operator also aims to minimize energy losses [44].

Alternatively, an aggregator may represent a *coalition* of EVs to maximize their profit when offering ancillary services. Such an aggregator can enter the ancillary service market and negotiate with the grid operators on behalf of its EVs in order to receive the best offer [45]. In order to be successful, the aggregator should assure *efficiency* and *fairness* among the participating EVs.

Finally, an aggregator can be an *independent entity*, representing neither the grid nor the EVs, e.g., as in [19, 28, 46]. In this scenario, an aggregator acts as a coordinator and dealer, trying to maximize its own profit. A number of parties might want to serve as such aggregators: an automobile manufacturer or automotive service organization, who are increasingly using on-vehicle telematics to deliver information services between repairs; a battery manufacturer/distributor, who could offer battery replacement discounts in exchange for sharing part of the profit in providing ancillary services; a cell phone network provider, who might provide the communications functions and whose business expertise focuses on automated tracking and billing of many small transactions distributed over space and time cell phone networking, similar to the V2G in terms of communications, control, value per transaction, and billing [28].

4.3.2 Charging scenarios

Depending on how EVs and aggregators interact, there can be different charging scenarios. In [47], the main charging control methods are classified as follows:

- opportunity charging;
- price-signal charging;
- load-signal charging;
- renewable energy-signal charging.

The opportunity charging scenario assumes that electric vehicles charge their batteries at fixed rates and efficiency as soon as they are parked and continue charging until the battery pack is fully charged. Clearly, no communication takes place and no V2G ancillary services are offered by these vehicles. The price-signal-based charging presumes a one- or two-way communication network available and is based on the *real-time*, *day-ahead*, or *time-of-use price* of electricity tariffs [12]. In this scheme, the EVs passively listen to the aggregator's broadcast of the pricing information for both active and reactive power. Once the price of offering electricity is high enough, the EVs may consider injecting (selling) electricity back to the grid. With the load-signal charging, the EVs may receive direct commands on the charging or discharging rates or their allowed ranges. Similar to the price-signal-based charging scenario, availability of a communication network is presumed with the load-signal charging. Last but not least, renewable-energy-signal charging is based on the premise that the EVs can be charged *exclusively* using renewable energy, with EVs acting as an energy sink. During windy or sunny conditions, the EV fleet absorbs bulk power generated by wind and solar farms. During low-renewable-power conditions, the vehicles charge at a slower rate [47]. This scenario can be implemented not only in a V2G system, but also in a stand-alone islanding scenario, as shown in Figure 4.2, where the EVs use the excessive renewable local power generation for charging their batteries at off-peak hours. In that case, the charging command signals may come directly from the local renewable power generator.

4.4 V2G systems communications

Considering the indirect architecture in Figure 4.6, communications in a V2G system may include message exchanges between the grid operator and the aggregators and between each aggregator and its corresponding group of EVs. The former can be done using the existing communication infrastructure for contracting and command signals that conventional ancillary service providers use, mostly based on fibre-optic and broadband communications [5]. However, the latter may involve a variety of communication technologies as we explain next.

4.4.1 Power-line communications and HomePlug

Broadband communication over power line, also known as power-line communication (PLC), is a technology that utilizes existing power line conductors for data transmission [48]. High-frequency data signals are superimposed on top of the distribution voltage. Typically, transformers prevent the PLC signals propagating, thus making it difficult to use such communications over high-voltage lines. However, PLC can be transmitted well across medium-voltage lines, providing a desirable last-mile service that can then be tied to the nearest wide-area communications network. Recent implementation of the PLC technology mainly limits the communication data transmission over residential-side power lines only leading up to neighbourhoods. This has successfully helped to reduce the damaging antennae effect from medium-voltage power lines [5]. With a PLC infrastructure in place, the command and price signals can be sent by aggregators towards the residential PLC receivers, such as HomePlug devices, which have been widely deployed recently [49]. The PLC technologies have also particularly found several applications in home energy management, including handling message exchanges between EVs and aggregators [50].

4.4.2 Wireless personal-area networking and ZigBee

ZigBee refers to a combination of high-level wireless communication technologies and protocols using small, inexpensive, and low-power digital radios based on the IEEE 802.15.4-2006 wireless personal area networking standard [51]. ZigBee operates in both the 2.4 GHz and 900 GHz frequency bands, which enjoys the flexibility of choosing the most proper frequency band in noisy radio environments [5]. ZigBee is mainly designed for sensing and automation applications, including home-energy management. The ZigBee transmission rates of 20–250 kbps are fast enough to transmit the updated data, e.g., once every second that is required for frequency and voltage regulation services. The range for the ZigBee communications can be as large as 400 m, making it adequate to reach every EV in a large parking lot with only a small number of transceivers [52]. Moreover, the use of wireless technology has offered the flexibility to allow adding more devices to the network without modifying its structure. Using two-bytes local addressing, ZigBee can accommodate up to 65,000 devices on a single network [53]. Experience to date indicates that ZigBee is reliable for home appliances and shows remarkable performance [54], making it a good candidate for automated demand–response applications [55].

While ZigBee is a general technology with a variety of applications, there are also recently proposed customized low-power wireless communications technologies, specifically for smart grid applications. One example is presented in [56], where the authors tailor wireless personal-area networking protocol for load-management problems. In this regard, they introduce *power update*, *power request*, and *power command* message frames to carry information such as frequency and voltage-regulation commands, pricing information, service and usage deadlines, power-scheduling information, power-usage duration, power-curtailment information, charging and discharging rates, and number of appliances [56].

4.4.3 Z-Wave

Z-Wave is a proprietary wireless communications protocol designed for automation and energy management in residential and light commercial environments, involving lighting, security, heating, ventilating, air conditioning (HVAC), and electric vehicles [5]. The Z-Wave technology is optimized for reliable and low-latency communication of small data packets for low-data-rate communications. Z-Wave devices can automatically set up an ad hoc mesh network which allows high implementation flexibility. Z-Wave operates in the sub-gigahertz frequency range, around 900 MHz. This means that it competes with some cordless telephones and other consumer electronics devices, but avoids interference with IEEE 802.11 and other systems that operate on the crowded 2.4 GHz band [57].

4.4.4 Cellular networks

The cellular network is a widely available long-range wireless data transmission infrastructure with high coverage, making it a good option for highly mobile devices such as EVs [5]. With the cellular connectivity, EVs can inform aggregators about their trip schedules (e.g., when and where they will be parked and connected to the grid), in advance. Moreover, as pointed out in Section 4.3.1, cellular network providers can serve as aggregators, focusing on automated tracking and billing of many small distributed transactions. Cellular network-based V2G systems can also benefit from the existing cell phone applications for EV management, e.g., those provided by General Motors and OnStar for Chevrolet Volt on iPhone, Blackberry, and Droid devices. Such applications currently have the capability to monitor state of charging and to adjust charging (and possibly discharging) schedules, e.g., based on electricity pricing information [58]. These features can be coordinated by the aggregator via commands sent from the cellular towers, as illustrated in Figure 4.7. This will require minimum changes in the existing communications infrastructure [59].

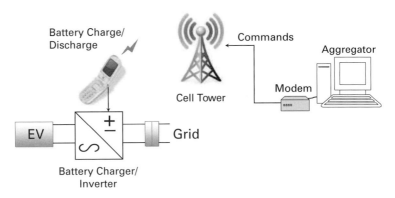

Figure 4.7   Interactions among an Aggregator and an EV over cellular network.

4.4.5 Interference management and cognitive radio

Since some of the existing V2G communications technologies and home area networking devices utilize the same frequency bands, wireless interference and congestion can be a major issue in a network populous area. For example, the 2.4 GHz band for ZigBee can interfere with IEEE 802.11 b/g/n Wi-Fi and Bluetooth technologies. The higher frequency band for ZigBee and the frequency band for Z-Wave also interfere with each other and with cordless phone services [5, 60]. Therefore, the coexistence strategies (such as dynamic and distributed channel allocations) between various technologies need to be carefully designed and implemented. In addition, cognitive radio techniques can be used for spectrum sharing across different V2G and home-area networking technologies [61].

4.5 Challenges and open research problems

In order to fully benefit from the new opportunities that EVs can offer to smart grid, there are two types of challenges that need to be addressed: (i) fulfilling the *communications needs* to facilitate efficient interactions among EVs, aggregators, and utilities by either adjusting the existing technologies (such as those mentioned in Section 4.4) or developing new communications technologies; (ii) using the available V2G communications infrastructure to *coordinate* the interactions among EVs to efficiently offer the various ancillary services mentioned in Section 4.2. Next, we explain these two types of challenges in detail.

4.5.1 Fulfilling communications needs

The requirements for V2G communications can be identified with respect to five key factors: *bandwidth*, *latency*, *reliability*, *security*, and *mobility*.

The United States Department of Energy has recently estimated that the bandwidth required for V2G communications can be up to 100 kbps per EV [62]. Furthermore, it is estimated that the latency in V2G communications needs to be as low as two seconds, in particular when EVs offer ancillary services. Although there is some ongoing related research, e.g., in the United States National Institute of Standards and Technology [63], there is still no comprehensive study in the existing literature to investigate the capabilities of various communications technologies in fulfilling the bandwidth and latency requirements mentioned above. As the number of EVs increases in a neighbourhood, achieving these requirements can become even more challenging due to the need for higher bandwidth. Some of the tools and techniques that have recently been proposed to support higher bandwidth and lower latency in V2G and smart grid communications systems include cognitive radio and spectrum sharing [61, 64], MIMO communications [65], and using multiple orthogonal frequency channels [66]. Moreover, reliability in V2G communications systems should be moderately high, i.e., in the 99%–99.99% range [62]. Reliable communications can be achieved by using reliable transport layer protocols, such as TCP, or through enhanced error detection, error correction, and source coding at lower layers [63].

Security is another key challenge in V2G communications. In fact, although the two-way communication capabilities and distributed intelligence in V2G systems can improve efficiency and offer new opportunities such as users' participation in the ancillary service market, these new features can also create new vulnerabilities in the power infrastructures if they are not accompanied with proper security enforcements. Some of the security concerns in V2G systems are as follows. First, to assure user's privacy, the charging status and the EVs' locations should not be disclosed to any unauthorized third party. Second, it is critical to avoid unauthorized discharging of vehicles' batteries by potential intruders. Third, given the fact that EVs are one of the most common types of *controllable load*, V2G systems need to be strictly protected against the recently introduced *distributed Internet-based load-altering attacks* [67]. This latter case refers to a scenario where a hacker uses a software-intruding agent to remotely access the EVs through the V2G communications infrastructure and to *simultaneously trigger the charging phase* for a large number of EVs to cause a major *load spike* in the power grid. If the system is not protected, such an attack can degrade power quality, damage utility and consumer equipment, and even cause a blackout.

In addition to certain levels of bandwidth, latency, reliability, and security, V2G systems present additional requirements due to mobility, which are not needed in most other smart grid communications applications. Because most EVs will likely charge at a variety of locations, including their home premises, office parking lots, and other public or private locations during long-distance travel, it will be important to maintain compatibility of communications technologies in V2G systems [62]. On the other hand, although most existing V2G applications focus on vehicle and grid interactions only when the vehicles are parked and are connected to the grid, a vehicle-to-vehicle or a vehicle-to-roadside communications infrastructure [68, 69] may find interesting new applications in the future smart grid, for example by tracking EVs' movements and charging levels in order to forecast where and when the new charging load of each EV will be connected to the power grid.

4.5.2 Coordinating charging and discharging

To be effective, the operation of a large group of EVs needs to be coordinated when they offer ancillary services. In general, such coordination can be done in either a *centralized* or a *decentralized* fashion as we will explain next.

In a centralized control scenario, a utility or an aggregator can remotely control users' EVs by sending appropriate command signals that enforce charging and discharging of batteries when needed. Some of the recent studies on central optimization of EVs' operations are provided in [70–72]. There are various challenges that need to be addressed in this line of research. First, they need to make sure that all EVs maintain some minimum charging level at all times. Note that participation in ancillary services becomes undesirable for users if it makes them unable to use their vehicles when they need to, e.g., in case of an emergency. Second, given the flexibility needed to be offered to users, the coordination of EVs requires to take into account the *randomness* in availability of EVs, as some EVs may leave the V2G system and some new EVs may join it at any time. This important aspect is particularly under-explored and needs to be investigated further. Third, a major concern for users is the depreciation of their EV batteries due to participation in ancillary services. Therefore, the number of charging and discharging cycles imposed on each user must be minimized.

Centralized coordination schemes may not always be scalable. Moreover, users can be reluctant to relinquish full control of their EVs to utilities and aggregators [73]. Therefore, a decentralized control approach can be more desirable in V2G systems. It can not only provide more scalable solutions with reduced control overheads on utilities and aggregators, but also allow users to maintain full control of the operation of their own electric vehicles, thus further encouraging users to participate in V2G systems and various ancillary service programs. The main challenge in decentralized coordination of EVs' charging and discharging is to implement elaborate pricing rules that can leverage optimal EVs' operations without direct involvement of utilities and aggregators. As recently shown in [33, 74], *game theory* and *mechanism design* have promising applications in this line of research, where we can design good pricing strategies by understanding users' rational reactions to various pricing and incentive mechanisms.

4.6 Conclusion

We have discussed the advantages of vehicle-to-grid systems in terms of providing various ancillary services such as reserve power supply, peak shaving, integration of renewable energy sources, and frequency and voltage regulation. We have also introduced two general vehicle-to-grid system architectures, namely, direct and indirect, where the latter is more scalable and involves aggregators. Such aggregators can represent the grid operator, the EVs, or independent dealers. Furthermore, we have summarized several vehicle-to-grid system communications technologies such as power-line communications, HomePlug, ZigBee, Z-Wave, cellular, and cognitive radio. The advantages and applications of each technology have been discussed. Finally, a wide range of challenges and open research problems have been discussed with respect to not only V2G communications requirements such as bandwidth, latency, reliability, security, and mobility but also centralized and decentralized coordination of EVs' charging and discharging in order to assure effective participation of users in large-scale offerings of various ancillary services.

References

[1] Electrification Coalition, *Electrification roadmap: revolutionizing transportation and achieving energy security*, November 2009.
[2] S. W. Hadley, 'Impact of plug-in hybrid vehicles on the electric grid', Oak Ridge National Laboratory, October 2006.
[3] T. H. Bradley and A. A. Frank, 'Design, demonstrations and sustainability impact assessments for plug-in hybrid electric vehicles', *Renewable and Sustainable Energy Reviews*, vol. 13, no. 1, pp. 115–128, January 2009.
[4] T. B. Gage, 'Development and evaluation of a plug-in HEV with vehicle-to-grid power flow', AC Propulsion Inc., ICAT 01-2, 2003.
[5] Y. Tang, H. Song, F. Hu, and Y. Zou, 'The effect of communication architecture on the availability, reliability, and economics of plug-in hybrid electric vehicle-to-grid ancillary services', *Journal of Power Sources*, vol. 195, no. 5, pp. 1500–1509, 2010.
[6] Energy Information Administration, 'The theory of peak-load pricing: a survey', Inventory of Electric Utility Power Plants in the United States 2000, US DOE, Washington, DC. DOE/EIA-0095, 2002.
[7] I. Cvetkovic, T. Thacker, D. Dong, G. Francis, V. Podosinov, D. Boroyevich, F. Wang, R. Burgos, G. Skutt, and J. Lesko, 'Future home uninterruptible renewable energy system with vehicle-to-grid technology', in *Proceedings of IEEE Energy Conversion Congress and Exposition*, September 2009.
[8] F. Giraud and Z. M. Salameh, 'Steady-state performance of a grid-connected rooftop hybrid wind-photovoltaic power system with battery storage', in *Proceedings of IEEE Power Engineering Society Winter Meeting*, Columbus, OH, January 2001.
[9] N. Mithraratne, 'Roof-top wind turbines for micro-generation in urban houses in New Zealand', *Energy and Buildings*, vol. 41, no. 10, pp. 1013–1018, October 2009.
[10] Y. Gurkaynak, Z. Li, and A. Khaligh, 'A novel grid-tied, solar powered residential home with plug-in hybrid electric vehicle (PHEV) loads', in *Proceedings of IEEE Vehicle Power and Propulsion Conference (VPPC'09)*, Dearborn, MI, September 2009.
[11] K. Sedghisigarchi, 'Residential solar systems: technology, net-metering, and financial payback', in *Proceedings of IEEE Electrical Power and Energy Conference (EPEC'09)*, Montreal, QC, October 2009.
[12] H. Mohsenian-Rad, V. W. S. Wong, J. Jatskevich, R. Schober, and A. Leon-Garcia, 'Autonomous demand-side management based on game-theoretic energy consumption scheduling for the future smart grid', *IEEE Transactions on Smart Grid*, vol. 1, no. 3, pp. 320–331, December 2010.
[13] M. D. Galus and G. Andersson, 'Demand management of grid connected plug-in hybrid electric vehicles (PHEV)', in *Proceedings of IEEE Energy 2030 Conference*, November 2008.

[14] A. Ipakchi and F. Albuyeh, 'Grid of the future', *IEEE Power and Energy Magazine*, vol. 7, no. 2, pp. 52–62, March 2009.

[15] A. Vojdani, 'Smart integration', *IEEE Power and Energy Magazine*, vol. 6, no. 6, pp. 71–79, November 2008.

[16] U.S. Department of Energy, 'The smart grid: an introduction', 2008.

[17] S. M. Amin and B. F. Wollenberg, 'Toward a smart grid: power delivery for the 21st century', *IEEE Power and Energy Magazine*, vol. 3, no. 5, pp. 34–41, September 2005.

[18] G. Xiong, C. Chen, S. Kishore, and A. Yener, 'Communication requirements for risk-limiting dispatch in smart grid', in *Proceedings of IEEE International Conference on Communications (ICC) Workshops*, Cape Town, South Africa, May 2010.

[19] W. Kempton and J. Tomic, 'Vehicle-to-grid power fundamentals: calculating capacity and net revenue', *Journal of Power Sources*, vol. 144, no. 1, pp. 268–279, 2005.

[20] E. Lakervi and E. J. Holmes, *Electricity distribution network design*, Peter Peregrinus Ltd, 1998.

[21] NAHB Research Center Inc., 'Review of residential electrical energy use data', July 2001.

[22] H. Mohsenian-Rad and A. Leon-Garcia, 'Optimal residential load control with price prediction in real-time electricity pricing environments', *IEEE Transactions on Smart Grid*, vol. 1, no. 2, pp. 120–133, September 2010.

[23] F. Saffre and R. Gedge, 'Demand-side management for the smart grid', in *Proceedings of IEEE/IFIP Network Operations and Management Symposium (NOMS) Workshops*, April 2010.

[24] C. W. Gellings and J. H. Chamberlin, *Demand-side Management: Concepts and Methods*. PennWell Books, 1993.

[25] A. L. Conejo, J. M. Morales, and L. Baringo, 'Real-time demand response model', *IEEE Transactions on Smart Grid*, vol. 1, no. 3, pp. 236–242, December 2010.

[26] S. Caron and G. Kesidis, 'Incentive-based energy consumption scheduling algorithms for the smart grid', *Proceedings of IEEE International Conference on Smart Grid Communications (SmartGridComm)*, Gaithersburg, MD, October 2010.

[27] V. Bakker, M. G. C. Bosman, A. Molderink, J. L. Hurink, and G. J. M. Smit, 'Demand side load management using a three step optimization methodology', in *Proceedings of IEEE International Conference on Smart Grid Communications (SmartGridComm)*, October 2010.

[28] W. Kempton and J. Tomic, 'Vehicle-to-grid power implementation: from stabilizing the grid to supporting large-scale renewable energy', *Journal of Power Sources*, vol. 144, no. 1, pp. 280–294, August 2009.

[29] C. Wu, H. Mohsenian-Rad, and J. Huang, 'Wind power integration with user participation: a game theoretic approach', in *Proceedings of IEEE Conference on Innovative Smart Grid Technologies*, Washington, DC, January 2012.

[30] M. Lange, 'On the uncertainty of wind power predictions – analysis of the forecast accuracy and statistical distribution of errors', *ASME Journal of Solar Energy Engineering*, vol. 127, no. 2, pp. 177–184, 2005.

[31] M. He, S. Murugesan, and J. Zhang, 'Multiple timescale dispatch and scheduling for stochastic reliability in smart grids with wind generation integration', in *Proceedings of IEEE INFOCOM*, Shanghai, China, April 2011.

[32] S. Han, S. Han, and K. Sezaki, 'Development of an optimal vehicle-to-grid aggregator for frequency regulation', *IEEE Transactions on Smart Grid*, vol. 1, no. 1, pp. 65–72, June 2010.

[33] C. Wu, H. Mohsenian-Rad, and J. Huang, 'Vehicle-to-aggregator interaction game', *IEEE Transactions on Smart Grid*, vol. 3, no. 1, March 2012.

[34] W. Kempton, V. Udo, K. Huber, K. Komara, S. Letendre, S. Baker, D. Brunner, and N. Pearre, 'A test of vehicle-to-grid (V2G) for energy storage and frequency regulation in the PJM system', November 2008 [online]. Available at: http://www.magicconsortium.org/_Media/test-v2g-in-pjm-jan09.pdf.

[35] B. J. Kirby, 'Frequency regulation basics and trends', Technical Report, Oak Ridge National Laboratory, 2004.

[36] A. J. Wood and B. F. Wollenberg, *Power Generation, Operation, and Control*. Wiley-Interscience, 1996.

[37] M. Singh, I. Kar, and P. Kumar, 'Influence of EV on grid power quality and optimizing the charging schedule to mitigate voltage imbalance and reduce power loss', in *Proceedings of International Power Electronics and Motion Control Conference*, September 2010.

[38] M. C. Kisacikoglu, B. Ozpineci, and L. M. Tolbert, 'Examination of a PHEV bidirectional charger system for V2G reactive power compensation', in *Proceedings of IEEE Applied Power Electronics Conference and Exposition (APEC)*, February 2010.

[39] J. Zhong and K. Bhattacharya, 'Toward a competitive market for reactive power', *IEEE Transactions on Power Systems*, vol. 17, no. 4, pp. 1206–1215, November 2002.

[40] P. Frias, T. Gomez, and D. Soler, 'A reactive power capacity market using annual auctions', *IEEE Transactions on Power Systems*, vol. 23, no. 3, pp. 1458–1468, August 2008.

[41] S. Han, S. Jang, K. Sezaki, and S. Han, 'Quantitative modeling of an energy constraint regarding V2G aggregator for frequency regulation', in *Proceedings of International Conference on Environment and Electrical Engineering (EEEIC)*, May 2010.

[42] S. Jang, S. Han, S. H. Han, and K. Sezaki, 'Optimal decision on contract size for V2G aggregator regarding frequency regulation', in *Proceedings of International Conference on Optimization of Electrical and Electronic Equipment*, May 2010.

[43] E. Sortomme and M. A. El-Sharkawi, 'Optimal charging strategies for unidirectional vehicle-to-grid', *IEEE Transactions on Smart Grid*, vol. 2, no. 1, pp. 131–138, March 2011.

[44] N. Belonogova, T. Kaipia, J. Lassila, and J. Partanen, 'Demand response: conflict between distribution system operator and retailer', in *Proceedings of 21st International Conference on Electricity Distribution*, Frankfurt, June 2011.

[45] S. Kamboj, K. S. Decker, K. Trnka, N. Pearre, C. Kern, and W. Kempton, 'Exploring the formation of electric vehicle coalitions for vehicle-to-grid power regulation', in *Proceedings of AAMAS Workshop on Agent Technologies for Energy Systems (ATES 2010)*, October 2010.

[46] A. Brooks and T. Gage, 'Integration of electric drive vehicles with the electric power grid – a new value stream', in *Proceedings of International Electric Vehicle Symposium and Exhibition*, Berlin, October 2001.

[47] T. Markel, M. Kuss, and P. Denholm, 'Communication and control of electric drive vehicles supporting renewables', in *Proceedings of IEEE Vehicle Power and Propulsion Conference*, September 2009.

[48] H. C. Ferreira, L. Lampe, J. Newbury, and T. G. Swart, *Power Line Communications: Theory and Applications for Narrowband and Broadband Communications over Power Lines*. John Wiley & Sons, 2010.

[49] K. H. Afkhamie, S. Katar, L. Yonge, and R. Newman, 'An overview of the upcoming HomePlug AV standard', in *Proceedings of IEEE International Symposium on Power Line Communications and Its Applications*, Vancouver, April 2005.

[50] H. Farhangi, 'The path of the smart grid', *IEEE Power and Energy Magazine*, vol. 8, no. 1, pp. 18–28, January 2010.

[51] P. Barontib, P. Pillaia, V. W. C. Chooka, S. Chessab, A. Gottab, and Y. Fun Hua, 'Wireless sensor networks: a survey on the state of the art and the 802.15.4 and ZigBee standards', *Computer Communications*, vol. 30, no. 7, pp. 1655–1695, 2007.

[52] C. Guille and G. Gross, 'A conceptual framework for the vehicle-to-grid (V2G) implementation', *Energy Policy*, vol. 37, no. 11, pp. 4379–4390, 2009.

[53] M. Galeev, 'Home networking with ZigBee', *Electrical Engineering Times*, April 2004.

[54] C. Guille and G. Gross, 'Design of a conceptual framework for the V2G implementation', in *Proceedings of IEEE Energy 2030*, Atlanta, GA, November 2008.

[55] M. LeMay, R. Nelli, G. Gross, and C. Gunter, 'An integrated architecture for demand response communications and control', in *Proceedings of IEEE Hawaii International Conference on System Science*, Waikoloa, Big Island, HI, January 2008.

[56] G. Xiong, C. Chen, S. Kishore, and A. Yener, 'Smart (in-home) power scheduling for demand response on the smart grid', in *Proceedings of IEEE Innovative Smart Grid Technologies (ISGT)*, January 2011.

[57] C. Gomez and J. Paradells, 'Wireless home automation networks: a survey of architectures and technologies', *IEEE Communications Magazine*, pp. 92–101, January 2010.

[58] General Motors, 'Chevy Volt iPhone, Blackberry, and Droid Apps Unveiled'. Available at http://gm-volt.com/2010/01/06/chevy-volt-iphone-blackberry-and-droid-apps-unveiled/, accessed January 2010.

[59] B. Kramer, S. Chakraborty, and B. Kroposki, 'A review of plug-in vehicles and vehicle-to-grid capability', in *Proceedings of 34th Annual Conference of the IEEE Industrial Electronics Society*, Orlando, FL, November 2008.

[60] M. Zeghdoud, C. Pascal, and M. Terre, 'Impact of clear channel assessment mode on the performance of ZigBee operating in a Wi-Fi environment', in *Proceedings of 1st IEEE Workshop on Operator-Assisted (Wireless Mesh) Community Networks*, Berlin, September 2006.

[61] R. Ranganathan, R. C. Qiu, Z. Hu, S. Hou, M. P. Revilla, G. Zheng, Z. Chen, and N. Guo, 'Cognitive radio for smart grid: theory, algorithms, and security', *International Journal of Digital Multimedia Broadcasting*, 2011, to appear.

[62] Department of Energy, 'Communications requirements of smart grid technologies'. Available at http://www.gc.energy.gov/documents/Smart_Grid_Communications_Requirements_Report_10-05-2010.pdf, accessed October 2010.

[63] M. Souryal, C. Gentile, D. Griffith, D. Cypher, and N. Golmie, 'A methodology to evaluate wireless technologies for the smart grid', *Proceedings of IEEE International Conference on Smart Grid Communications (SmartGridComm)*, Gaithersburg, MD, October 2010.

[64] A. Ghassemi, S. Bavarian, and L. Lampe, 'Cognitive radio for smart grid communications', *Proceedings of IEEE International Conference on Smart Grid Communications (SmartGridComm)*, Gaithersburg, MD, October 2010.

[65] P. P. Parikh, M. G. Kanabar, and T. S, Sidhu, 'Opportunities and challenges of wireless communication technologies for smart grid applications', in *Proceedings of IEEE Power and Energy Society General Meeting*, Minneapolis, MN, July 2010.

[66] Bonneville Power Authority, 'Comments – request for information on smart grid communications requirements'. Available at: http://www.gc.energy.gov/documents/Bonneville Power_Comments\CommsReqs.pdf, accessed 2010.

[67] H. Mohsenian-Rad and A. Leon-Garcia, 'Distributed Internet-based load altering attacks against smart power grids', *IEEE Transactions on Smart Grid*, vol. 2, no. 3, September 2011.

[68] X. Yang, L. Liu, N. H. Vaidya, and F. Zhao, 'A vehicle-to-vehicle communication protocol for cooperative collision warning', in *Proceedings of ICST International Conference on Mobile and Ubiquitous Systems: Computing, Networking and Services (MOBIQUITOUS)*, Boston, MA, August 2004.

[69] S. Biswas, R. Tatchikou, and F. Dion, 'Vehicle-to-vehicle wireless communication protocols for enhancing highway traffic safety', *IEEE Communications Magazine*, vol. 44, no. 1, pp. 74–82, January 2006.

[70] S. Han, S. Han, and K. Sezaki, 'Development of an optimal vehicle-to-grid aggregator for frequency regulation', *IEEE Transactions on Smart Grid*, vol. 1, no. 1, pp. 62–72, June 2010.

[71] A. Y. Saber and G. K. Venayagamoorthy, 'Optimization of vehicle-to-grid scheduling in constrained parking lots', in *Proceedings of IEEE Power and Energy Society General Meeting*, Calgary, AB, July 2009.

[72] C. Hutson, G. K. Venayagamoorthy, and K. A. Corzine, 'Intelligent scheduling of hybrid and electric vehicle storage capacity in a parking lot for profit maximization in grid power transactions', in *Proceedings of IEEE Energy 2030 Conference*, Atlanta, GA, November 2008.

[73] P. Hoffert, 'Automated homes and offices on the infoway', CulTech Collaborative Research Centre, York University, Toronto, Canada, Technical Report, November 1994.

[74] P. Samadi, H. Mohsenian-Rad, R. Schober, and V. Wong, 'Advanced demand side management for the future smart grid using mechanism design', submitted to *IEEE Transactions on Smart Grid*, March 2011.

Part II

Physical data communications, access, detection, and estimation techniques for smart grid

5 Communications and access technologies for smart grid

Sara Bavarian and Lutz Lampe

5.1 Introduction

Availability of reliable and real-time information is essential for the integration of intermittent renewable energy resources and improving the efficiency and performance of the aging electrical power grid. Hence, an integrated high-performance, pervasive, and secure communications infrastructure is one of the key foundations of smart grid evolution. Much of the recent standardization efforts, such as that led by the US National Institute of Standards and Technology (NIST) [1], and the IEEE P2030 [2], has focused on defining high-level, technology-neutral architecture and reference models for smart grid communications networks. An abstract architecture offers a framework of logical connections between different system domains and high-level requirements to be followed by specific solutions. While such a conceptual architectural model is imperative for ensuring interoperability, it is not mapped directly to specific solutions, nor does it address detailed implementation issues.

This chapter is focused on physical communications and access techniques that support current and upcoming smart grid applications. We discuss in detail a variety of communications media and technologies and how they can be applied in smart grid communications networks. The rest of this section provides some background information on our discussion. Section 5.1.1 begins with a look at the history of utility communications networks. Such knowledge is important in understanding the existing utility communications infrastructure and necessary improvements needed *en route* to smart grid. The key objectives in establishing smart grid communications networks are discussed in Section 5.1.2, followed by data classification and requirements in smart grid in Section 5.1.3.

The remainder of this chapter is organized in four main sections. Section 5.2 examines a variety of wired and wireless communications solutions in utility communications networks. Sections 5.3 and 5.4 review a number of relevant power-line communications (PLC) and wireless standards, respectively. A number of applied networking options including IP-based networking and a discussion on public vs. private infrastructure are considered in Section 5.5.

5.1.1 Legacy grid communications

Utility communications networks have evolved over the years by adopting the technologies developed for general voice and data communications with a delay of about 5–15 years [3]. The main reason for this lag has been to make sure that the technologies are well matured to be safely applied in utilities' acute functions. Understanding legacy utility communications systems and their limits is important in outlining the areas that need to be upgraded to accommodate envisioned smart grid applications. Additionally, the path towards realizing smart grid is of evolutionary nature and legacy systems are (generally expected) to be supported by the new systems.

In the 1950s, utilities started phasing out manual control of major substations by application of supervisory control and data acquisition (SCADA) systems. Further advances in automation occurred in the 1980s by the introduction of intelligent electronic devices (IEDs) that incorporated processing power and transceivers to communicate with external resources. Smart devices such as remote telemetry units (RTUs) have been applied to communicate the status of the electric grid, and traditional grid components such as transformers, capacitors, and batteries have been equipped with monitoring devices that were connected to a local annunciator system. Table 5.1 provides a summary of the evolution of utility communications systems [3]. Although there are many exceptions to these general categories, this information is useful in illustrating the trends towards standardized, integrated, and high-performance solutions in utility networks.

Development of utility communications technologies has mostly been motivated by the introduction of new policies as well as economical concerns. Deregulation policies of the 1990s, for example, required utilities to establish interconnections with neighbouring utilities and other organizations. The need for asset management and downsizing the protection and SCADA departments are examples of economical concerns that necessitated increased networking of grid components. Customer demand for increased services has also led to integration of some automation tools with corporate information technology functions. In order to withstand harsh environments and extremely noisy channels, the communications solutions used so far in utility networks have been rather simple and highly robust. Most had to be low cost because the business case for automation projects was unproven. Communications links had low bandwidth and computing power and wireless technologies were originally rarely applied.

5.1.2 Smart grid objectives

Although utility communications networks have been evolving over the years, they have to be expanded and upgraded substantially to support desired smart grid features. This section outlines some of the key principles in smart grid communications networks including ubiquity, interoperability, standardization, scalability, manageability, upgradability, backward compatibility, and security.

Smart grid communications networks are expected to have ubiquitous reach over the grid. Current utility networks generally connect offices, control centres, and major

Table 5.1. Overview on the history of utility communications systems, adapted from [3]

Phase	Years	Characteristics	Architecture	Media	Standards
Proprietary systems	Up to 1985	• Single vendor • Basic data collection	• Hierarchical tree • Isolated substations	• RS232 and RS485 • Dial-up • Trunked radio • Power-line carrier • Less than 1200 bps	• Modbus • SEL • WISP • Conitel 2020
Early standards	1985–1995	• Multi-vendor systems • Protocol conversion	• Hierarchical tree • Redundant links	• Leased lines • Packet radio • 9600 to 19,200 bps	• DNP3 Serial • IEC 60870 • TASE 2
Area networks	1995–2000	• Substation LANs • Merging protection and SCADA networks	• Peer-to-peer communications in substation • Joining substations via WAN	• Ethernet • Spread spectrum radio • Frame relay • Megabit data rates	• TCP-IP • FTP • Telnet • HTTP • DNP3 • WAN/LAN • UCA.2
Business integration	2000 to now	• Merging automation and business networks • Corporate IT departments • Asset management	• Linking of utility WANs to corporate networks • Extension of network to customer premises • Use of Internet	• Digital cellular • IP radios • Wireless Ethernet • Gigabit backbones	• TCP-IP • IEC 61850 • XML

transmission substations. The distribution domain, including substations, relays, and other components, has limited connectivity access. This lack of communications capability is a major barrier for distribution automation (DA) and advanced distribution automation (ADA). Traditional DA refers to automated control of basic distribution circuit-switching functions. ADA, however, involves automation of all the equipment and functions in the distribution system facilitating the exchange of both electrical energy and information [3].

Lack of visibility in the distribution domain has been tolerable so far because of the central architecture of the existing grid, where power generally flows in one direction from the major generation facilities to the consumers. A smart grid architecture, however, is expected to be distributed and much more sophisticated. The grid of the future is going to include large numbers of smaller, mostly renewable distributed energy resources (DERs). Small-scale microcosms (generally referred to as microgrids) are anticipated to form within the grid. Distributed control centres will manage the bi-directional flow of electricity to and within these microgrids [4]. Maintaining the stability of a power grid in such complex environments requires reliable access to timely information and an ability to remotely control grid components through widespread deployment of sensors and actuators.

Moreover, real-time links to customers are currently missing in most utility communications networks. These are, however, necessary for many smart grid applications such as advanced metering infrastructure (AMI), real-time pricing, and demand response (DR). AMI involves installation of smart meters and establishing two-way communication links between consumers and utilities. AMI enables the utilities to remotely connect/disconnect power, read usage information, detect a service outage, and unauthorized use of electricity. As electricity demand changes throughout the day, utilities currently generate or purchase power to meet the estimated demand. Much of utilities' (costly) resources are reserved to handle the peak load. DR and demand-side management (DSM) are attributes envisioned in smart grid that seek to shape and balance electrical loads to improve grid efficiency. Load management can either be achieved actively through command and control signals and by shedding or reducing non-critical loads or passively through dynamic pricing.

Many utilities around the world have started major AMI projects or are testing technologies through pilot projects. For example, the main Italian energy provider, Enel, has successfully installed PLC-based smart meters for more than 30 million customers. With the proceeding deployment of these assets, there are growing concerns that without interoperable standards, these investments can become prematurely obsolete. Interoperability in the smart grid refers to the ability of the systems, networks, devices, and applications to communicate and exchange meaningful information, although they may be using different communications protocols or hardware and software products from different providers. Interoperability is considered to be the key to success in evolving smart grid initiatives.

Several national and international standards development organizations including NIST are coordinating the development of a framework for protocols and standards of interoperability in the smart grid system. The NIST framework and roadmap for smart grid interoperability standards [1] provides a conceptual reference model and identifies the standardization gaps in realizing an interoperable smart grid framework. NIST has recognized 25 existing standards for smart grid communications and 50 additional standards for further investigation. It also identifies 15 areas that urgently need new or revised standards, and offers a set of priority action plans (PAPs) to establish these critical standards.

The International Electrotechnical Commission (IEC), through its Strategic Group 3, also published a smart grid roadmap in 2010 that is similar to the one developed by NIST. In 2009, IEEE launched the IEEE P2030 *Draft Guide for Smart Grid Interoperability of Energy Technology and Information Technology Operation with the Electric Power System, and End-Use Applications and Loads* to provide an open standard for integration of information and communications technology in the electrical power grid. IEEE P2030 is currently in balloting phase and provides a base for upcoming smart grid applications, guidelines in defining smart grid interoperability, a knowledge-based architectural design addressing terminology, and characteristics.

High-level smart grid architectural guidelines are technology-neutral and open to all stakeholders and they define a set of principles to be followed in implementing different aspects of smart grid. Scalability is one of these measures that recommends application of technologies with no inherent limitations on their size so there is no barrier in expanding the networks throughout the power system. An example for lack of scalability is the address exhaustion problem in the original version of ZigBee standard (see Section 5.4.1). Another example is the application IP-based networking in smart grid, which is discussed later in Section 5.5.3.

Design for manageability is another important objective in smart grid communications networks. Networks have to be remotely manageable, have to have the ability to configure themselves, identify faults, isolate them (fault-tolerant), and to take the necessary steps to repair themselves (self-healing). Section 5.2.2 discusses the self-healing quality of wireless mesh networks that makes them desirable in smart grid applications. Backup links are also desired, particularly in critical applications such as teleprotection in order to ensure safety and prevent major blackouts. Grid network components are expected to function for decades so they should have the ability to be remotely and securely upgraded. Software defined radio (SDR) devices that can accommodate modification of radio functions through software updates are, hence, desirable in smart grid applications for the ease of upgradability. Many applications are expected to evolve, once smart grid communications infrastructure is in place, and the networks have to be flexible enough to serve these upcoming needs.

Smart grid networks have to be predominantly compatible with legacy systems to ensure a smooth transition. Utilities own massive amounts of legacy communications and networking assets that are not as powerful as the new smart grid compatible devices. Millions of meters with automatic meter reading (AMR) capability are currently in service and are examples of outdated legacy assets. AMR meters have limited capability with one-way flow of information from the meter to the utility and do not provide enhanced features of AMI systems such as remote connect/disconnect. It is going to take years, maybe decades to upgrade the legacy infrastructure, however, there are innovative methods to integrate them into smart grid via gateways. In the case of AMR meters, the data can be published on the Internet via bridging solutions to enable real-time usage monitoring for the customers.

Security and privacy of smart grid data communications, particularly through public communications networks or the Internet is a critical issue, and it is going to be discussed in detail later in the book. Utilities' main mission and expertise is to provide reliable electrical energy to consumers, so ease of installation and operation is an important factor in smart grid communications. In addition, the nature of different applications sets distinct requirements for traffic in terms of bandwidth and latency. Smart grid networks have to be designed for reliable communication of different classes of data with diverse requirements. The next section covers the principles of data classification in smart grid communications.

5.1.3 Data classification

Classifying data characteristics and requirements is an important step in determining the suitability of communications interfaces in order to ensure that data is transferred securely and effectively. Data characteristics are determined by how the data is created as well as how it is used in the application. For example, meter data can be processed daily or even monthly for billing applications, but needs to be processed within minutes or hours for DR purposes. Smart grid applications include a variety of purposes such as protection, control, or monitoring.

Many different (usually independent) factors have to be considered in classifying data. Some information is intended to be used locally (within customers' premises or a microgrid) and some have a far reach (national or regional). Rate of occurrence and volume (bytes, kB, MB, or GB) determine the average capacity needed for the communications link. Time-sensitivity is an important factor in selecting communications technology and involves both latency (channel access and propagation delay) and processing time. Some information is broadcast all over the grid (broadcast), some targets many defined nodes (multicast), and some is transmitted from one device to the other (singlecast).

Some information, such as the data gathered by phasor measurement units (PMUs), needs to be synchronized all over the grid. PMU data is time-stamped by using the signals from the global positioning system (GPS). Data security is another important factor that reflects protection level (high, medium, and low) against unauthorized access to ensure confidentiality, integrity, and assurance. Information has to be categorized and treated according to importance (priority level) and different data have diverse reliability requirements in terms of availability and level of assurance (LOA). It is critical that some information is available within a desired time frame, while other information can be retransmitted until it is received. LOA refers to the certainty of timely arrival of accurate data. When applications metrics are available, LOA is a quantitative measure, else it is a qualitative specification (high, medium, or low).

As smart grid applications are evolving, the above-mentioned factors have to be assessed by users for each data type. A well-defined process for categorizing communications requirements for a smart grid application is needed to facilitate the design of a cohesive smart grid communications system. The IEEE P2030 lists a set of smart grid evaluation criteria (SGEC) to assess smart grid communications use cases in terms of protocols and technologies [2]. The SGEC considers the three aspects of LOA, minimum latency, and impact on operation, and classifies data into three tier classes:

- Tier 1: Critical

    - Priority level 1
    - LOA: High
    - Safe operation, control of grid
    - Loss of life or injury, and damage to assets
    - Latency: very low/low (relaying), medium–high (distribution)

Table 5.2. Examples of smart grid application tier classifications, adapted from [2]

Application	Related standard	Tier
AMI and smart grid end-to-end security	AMI–SEC system security requirements	2
Revenue metering	ANSI C12.19/MC1219	2
Building automation	BACnet ANSI ASHRAE 135-2008/ISO 16484-5	2
Substation and feeder device automation	DNP3	2
Inter-control centre communications	IEC 60870-6/TASE.2	1
Substation automation and protection	IEC 61850	1
Energy-management system interface	IEC 61968/61970	1
PMU communications	IEEE Std C37.118	1
Security for IEDs	IEEE Std 1686-2007	1
Cyber security standards for the bulk power system	NERC CIP 002-009	1
HAN device communication and information model	ZigBee/HomePlug smart energy profile	2, 3

- Tier 2: Important
  - Priority level 2
  - LOA: Medium
  - Control with limited impact
  - Damage to assets
  - Latency: medium/high
- Tier 3: Informative
  - Priority level 3
  - LOA: Low
  - Informative
  - No damage to assets
  - Latency: high–very high

Table 5.2 shows examples of smart grid applications, related references, and their tier classifications [2].

5.2 Communications media

The future smart grid communications infrastructure is expected to include a hybrid mix of technologies. One reason for this variety is the assets inherited from the existing utility networks. Some utilities have a fibre-optics backbone between their main offices and major substations. Diverse standardized or proprietary PLC and licensed or unlicensed radio links have been installed, usually for application-specific purposes. Within offices and substations local-area network (LAN) technologies such as Ethernet on coaxial cables or twisted pairs, and/or wireless LAN (WLAN) options such as Wi-Fi are commonly used. Internet access is available in corporate IT infrastructure and some utility LANs. Microwave links, digital subscriber line (DSL), satellite and cellular links have been used to connect regional/remote stations. Also, as utility communications networks evolve to accommodate the smart grid vision, an increased variety of communications solutions is expected to be deployed to meet diverse requirements of upcoming smart grid applications. This section covers the pros and cons of most relevant wired and wireless physical communications links. Virtual communications links through the Internet are discussed later in Section 5.5.3.

5.2.1 Wired solutions

Power-line communications

PLC refers to data transmission over existing power lines. It is an old concept, with the first patents dating back to the early 1900s. Utilities have long been using low-capacity PLC technologies for remote control of grid components as an alternative to constructing a dedicated communications infrastructure. In recent years, the interest in PLC has grown due to deregulation of the electricity market, advances in PLC technology, and its potential application in smart grid communications. Detailed discussions on characteristics of specific PLC standardized technologies are provided later in Section 5.3. The general issues relating to the reuse of power grid as a communications network are as follows.

- *Infrastructure*. The main motivation for PLC is that the wiring infrastructure already exists, i.e., the reuse of infrastructure.
- *Coverage*. Power lines provide a pervasive media for communications over the grid.
- *Cost*. Among the wired communications technologies, PLC is the only option with costs comparable to wireless because its wiring infrastructure is already installed. Another advantage of using the existing wiring is that links can be rapidly and cost-effectively installed even in remote, rural areas.
- *Noise and attenuation*. Power line is a noisy, harsh communications medium. Signals are distorted by coloured background noise, and suffer from periodic and aperiodic impulsive noise. Power-line cables are usually unshielded, suffer from electromagnetic interference (EMI), and can be a source of interference for other wired and wireless telecommunications systems. Depending on the frequency range, signal attenuation is considerable. Furthermore, bridging is necessary at the transformers (except at very low frequencies) [5, 6].
- *Channel modelling*. Power line channels are extremely hard to model. Grid topology and structure, including wiring and grounding practices, are highly variable in different areas even within a country. PLC channels are frequency-selective and vary with time. This time-variability can be either abrupt, when devices are plugged in or out or switched on or off, or periodically changing according to main frequencies (50 or 60 Hz) [7].
- *Standardization*. Absence of a unified standard was one of the key problems impeding PLC market success. Many standardization efforts have been going on in parallel in recent years. Now, PLC might be facing the opposite problem of non-interoperable

standards dividing the industry. Hence, standardized coexistence mechanisms are essential for success in PLC technologies [5].
- *Capacity*. Achievable capacity of communications over noisy power lines that suffer from high interference used to be a concern. Today, high-speed power line devices for in-home communication with data rates up to 200 Mbps are commercially available. It has to be noted, however, that power line is a shared media and the average capacity per user will be lower than the total capacity.
- *Reliability and open circuit problem*. Connection is lost when power lines are damaged, e.g., in a blackout. Grid components such as switches, reclosers, and sectionalizers on the other side of an open circuit may not be accessible via power-line technology. So a backup communications link to PLC is necessary in critical applications.
- *Security*. As power line is a shared medium, there are also concerns regarding security of PLC technologies. Signals must be encrypted properly to avoid unauthorized access to confidential data.

Other wired communications technologies

Various types of wiring have long been providing voice and data connections in utilities' and customers' premises networks and are expected to continue to function in smart grid communications networks. Most commonly available are coaxial cable, phone lines, and other twisted-pair copper wires such as Cat-5 and Cat-6.

Coaxial cable is a shielded electrical cable and consists of an inner conductor that is surrounded by an inner flexible tubular insulating layer and an outer tubular conducting shield. A key advantage of coaxial cables is the excellent protection against external EMI. They can be bent and moderately twisted, and strapped to conductive supports without inducing unwanted currents in them. Coaxial cables have been used for carrying radio frequency (RF) signals in applications such as feedlines from antennas to transceivers and distributing television cable signals. Application of coax in data communications networks is mostly abandoned in favour of twisted-pair cables.

Twisted-pair cabling is a type of wiring that cancels out EMI by twisting two conductors and it is the primary wiring for telephone usage and computer network connections due to cost-effectiveness and high flexibility of the cables. Although landline subscriptions have been decreasing with the rise of cellular communications, legacy landlines are widely used to provide data connections using DSL technologies. While voice is transmitted on lower frequencies ($f \leq 5$ kHz), DSL uses higher frequencies ($f \geq 25$ kHz), and it can easily be filtered out enabling simultaneous use. Many residential data connections use asymmetric DSL (ADSL), which is a form of DSL technology with faster download speeds (slower upload), and higher data rates can be achieved through the application of improved DSL technologies such as very-high-bit-rate DSL (VDSL or VHDSL).

Many businesses, utility offices, and substations have wired LANs within their premises using Cat-5 or Cat-6 cables and leased lines (T1/E1 carriers) to provide voice/data connectivity between remote offices or to the service providers. A major issue with wired communications is that the market is divided between many non-interoperable technologies. For example, for in-home communications, there are at least three power line, two phone line, and two coax communications technologies which is very inconvenient for consumers, service providers, and consumer electronics companies [8]. In order to solve this issue, as we discuss later in Section 5.3.1, ITU-T recommendation G.9960/61 (formerly known as G.hn) provides most of the physical (PHY) and medium access control (MAC) layer of a unified home networking solution.

Optical-fibre communications

Optical-fibre communication systems have been around for decades and have largely replaced traditional electrical wire communications in long-distance, high-demand applications and in high-speed core communications networks. Optical-fibre links are part of the existing utility networks and are expected to grow extensively to accommodate demanding data requirements in smart grid. In the following, some of the deciding factors for the application of optical links in utility communications networks are discussed.

- *Capacity*. Optical-fibre channels have an extremely high capacity. Current commercial optical-fibre transmission systems offer bit rates of up to 10 Gbps using single-wavelength transmission, and 40 Gbps to 1600 Gbps using wavelength division multiplexing (WDM) [9]. Channel access and queuing delays are also minimal in fibre links as a result of high capacity.
- *Attenuation*. Optical-fibre communication systems require repeaters about every 100–1000 km to relay the signal along the fibre, ensuring that the signal does not become too distorted or weak. For comparison, typical T1 or coaxial communication systems require repeaters every 2 km. This low attenuation quality leads to overall cost-effectiveness in long-distance links.
- *EMI*. Fibre can be installed in environments with high EMI, such as high/medium-voltage substations, alongside utility lines, power lines, and railroad tracks. Furthermore, and in contrast to electrical transmission lines, there is no crosstalk between fibre links even when run alongside each other for long distances.
- *Safety and security*. Optical links are not electromagnetically radiating, and unauthorized physical access is difficult without disrupting the signal, which is important in high-security applications They do not cause any sparks, which is safe in flammable or explosive gas environments. Non-metallic all-dielectric cables are also ideal for areas of high lightning-strike incidence.
- *Cost*. Although fibre-optics has so many technical advantages, it has been slow to reach consumers directly because of the higher costs of installation, fibre material, and transceivers. With decreasing prices, fibre links have become more affordable over the years. Moreover, the enormous bandwidth of fibre can be shared between many applications such as broadband access of residential customers, making the installation of fibre backbones more cost-effective. Still, infrastructure development within cities is relatively difficult and time-consuming. A recent example is the smart grid city project in Boulder, Colorado that is one of the most widely publicized experiments in the USA, bringing smart grid systems to an entire city. The project went over budget mainly because of the cost over-runs in fibre optics installation and caused a rate hike, customer dissatisfaction, and negative press attention for smart grid projects in general.

5.2.2 Wireless solutions

Various wireless technologies are available in existing utility communications networks [3]. Wireless links allow for quick installation and save the cost of new wiring infrastructure. Unlike PLC, wireless links can continue to function (with battery power) when the electrical connection is lost. Wireless networks are more flexible than wired networks as new nodes can easily be installed within the coverage area.

Availability of radio spectrum for smart grid applications is a source of concern. In many countries, most of the spectrum except for the unlicensed industrial, scientific, and medical (ISM) band is dedicated to licensed users. When available, utilities may acquire spectrum for their communications traffic and deploy customized communications products. For example, 30 MHz of spectrum in the 1.8 GHz band was recently allocated to electric utilities in Canada [10].

In less critical or local networking applications where interference can be tolerated, utilities may choose to apply technologies such as Wi-Fi or ZigBee that use the unlicensed ISM band. In other cases utilities may start partnerships with wireless service providers or a spectrum licensee, for sharing resources such as spectrum and/or network equipments. Section 5.5.2 covers the ongoing debate on public or private networking infrastructure, and Section 5.4.2 discusses the option of adopting cognitive radio technology to reuse TV white band in smart grid communications.

There are many measures such as required coverage, capacity, latency, and security that have to be considered in designing an efficient wireless communications network. The first step in this way is to choose the topology according to the system requirements. Once the proper network structure is identified, specific standard or proprietary technologies can be selected to deploy new networks or to bridge existing systems. The rest of this section discusses the pros and cons of different wireless network topologies. Section 5.4 is devoted to the most relevant standardized wireless technologies in smart grid communications.

Point-to-point/point-to-multipoint

Point-to-point (P2P) communications refers to a dedicated wireless link between two (usually) fixed points. High-speed, bi-directional P2P links are widely applied in wireless backhaul applications to replace or complement wired solutions (such as fibre or leased lines). In some cases a number of P2P links are connected in a daisy chain topology, as illustrated in Figure 5.1, to provide redundancy in the system. Most P2P connections require a clear line-of-sight (LOS) between the two nodes and use either free-space optical (FSO) communications or a variety of standardized or proprietary microwave wireless technologies in licensed or unlicensed bands.

FSO is an optical communications technology that takes advantage of rays of light propagating in free space to establish a data link between two points. Terrestrial links usually cover a range of 2 to 3 km and by using efficient modulation schemes, data rates of a few hundred megabits per second can be achieved. The advantages of using FSO links include ease of installation, licence-free operation (relative to wireless option), high capacity, immunity to EMI, full duplex operation, secure connection (beams are

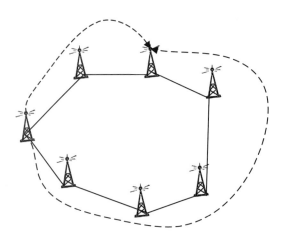

Figure 5.1 Route redundancy in P2P daisy chain architecture.

narrow and highly directional). The main disadvantage of using FSO is that it is easily affected by weather conditions, background light, shadowing, or any other factor that could stop, disperse, or attenuate a ray of light.

Microwave radios mostly cover the frequencies of 1–30 GHz and are widely used for P2P communications. With wavelengths of 1–30 cm, microwave links allow for conveniently sized, narrow-beam antennas that can be pointed directly to the receiving terminal. These highly directional antennas offer better antenna gains (better power efficiency) and cause less interference for co-channel equipment. A variety of solutions available in licensed and unlicensed bands typically offer data rates of tens to a few hundred megabits per second. Unlicensed 60 GHz and registered 80 GHz millimetre devices are also available that can provide gigabit wireless links.

A disadvantage in these links is that high-frequency microwave signals are highly attenuated by an obstructed path. P2P microwave links are mostly designed for LOS propagation and rain fade can be a problem in high frequencies ($f > 11$ GHz). Non-LOS (NLOS) microwave technologies are also available that apply orthogonal frequency-division multiplexing (OFDM) and multiple-input, multiple-output (MIMO) techniques to mitigate the negative effect of multipath fading. They also apply adaptive code modulation (ACM) to automatically adapt the data rate to the channel quality. Depending on the channel characteristics, these NLOS solutions can provide data rates of up to a few hundred megabits per second.

In point-to-multipoint (P2MP) communications a central antenna or an antenna array called the base station unit (BSU) or access point (AP) communicates to several receiving antennas called subscriber units (SUs) or customer premise equipments (CPEs). The wireless P2MP backhaul system is an easy to deploy, economical solution for connecting multiple remote sites to a network. P2MP systems typically offer data rates of a few hundred megabits per second over a few kilometres range. Many different proprietary technologies are commercially available and standards-based technology of worldwide inter-operability for microwave access (WiMAX) backhaul is gaining interest as it allows interoperability between software and hardware solutions from multiple vendors (see Section 5.4.2 for more detail).

Cellular networks

Cellular networks provide wireless coverage for a (usually large) number of (fixed or mobile) users over an extensive geographical area that is divided into multiple cells. Each cell has a fixed main transceiver called the base station (BS) that communicates with and controls the users within the cell. The basic premise behind the design of cellular systems is that the signal from the BSs and users is mostly confined to the cell occupied by the transceivers. The key advantage of cellular systems is the efficient use of spectrum as the radio channels are reused in other cells. As wireless signals attenuate sharply with distances, intercell interference is mitigated by separating co-channel cells spatially in the case of narrowband systems or by spectral spreading in the case of code-division multiple access (CDMA).

Cellular communications is a fast-paced industry that has grown exponentially in the last few decades. Cellular systems are now widely deployed globally. A detailed account of the evolution of cellular networks can be found in many wireless communications references, such as [11]. Utilities have long been using public or private cellular networks to connect their assets, mobile workforce, or to backhaul metering data particularly in urban environments. Cellular communications standards, particularly the latest Fourth-generation (4G) WiMAX and the Third-generation partnership project long-term evolution (3GPP-LTE) are expected to deliver wide-area coverage in many smart grid communications networks. These standard-based solutions offer interoperability between different vendors and competitive equipment pricing. The characteristics of WiMAX and 3GPP-LTE are discussed in more detail in Section 5.4.2, and Section 5.5.2 includes the arguments for and against using existing public cellular systems vs. establishing private networks.

Wireless mesh networks

Unlike the infrastructure-based design of cellular networks, where central BSs connect and control the users in the networks, a mesh network refers to a communications system where a number of nodes self-configure their connection without the help of an established infrastructure. Figure 5.2 illustrates a basic view of mesh networking topology, where the nodes collaborate in routing the traffic. That is, each node not only sends and receives its own data but also serves as relay for other communications signals. A wide variety of routing or flooding techniques are applied to ensure efficient and reliable delivery of traffic across mesh networks. Although mostly used in wireless scenarios, the mesh networking concept is sometimes applied in wired networks and software interaction.

Wireless mesh networks (WMNs) provide dynamic and cost-effective connectivity over a certain geographic area. They are typically quite reliable because (as seen in Figure 5.2) there is often more than one path between a data source and its destination. Therefore, when one (few) connections fail, traffic can often be rerouted and reach the destination. Lack of wired infrastructure in WMN significantly reduces the initial

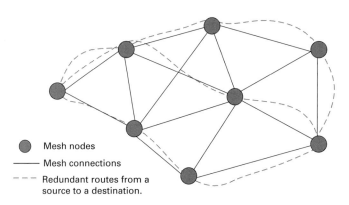

Figure 5.2 Basic view of a wireless mesh network.

installation cost and ongoing maintenance expenses. The reliability, robustness, and self-healing qualities of mesh topology make it an attractive option in smart grid applications. WMNs can be implemented using a variety of proprietary and standardized wireless technologies including the IEEE 802.11s (draft amendment for mesh networking in Wi-Fi networks, see Section 5.4.1) or the IEEE 802.15.4 (ZigBee) which is discussed in more detail in Section 5.4.1.

Satellite communications

Satellite technology offers extensive coverage and has long been used for voice and data communications, particularly in remote areas where terrestrial communications systems are not available. Some utilities already use satellite communications for remote control and monitoring in rural substations, and customized very small aperture terminals (VSATs) for substation monitoring are commercially available [9]. GPS, which provides microsecond accuracy in time synchronization, is also widely deployed in transmission substations for time-stamping PMUs signals in wide-area monitoring (WAM) applications. Satellite links also provide a safe backup system for utility communications networks. Critical data can be rerouted through satellite links in case the terrestrial connections fail.

Satellite connections can easily be set up by installing a transceiver at the desired location (self-installation is even possible in some cases) without any need for cabling or relay systems. In situations where no other communications infrastructure exists, satellite communication is a cost-effective solution. Satellites are classified into the three categories of geostationary earth orbit (GEO) satellites, middle earth orbit (MEO), and low earth orbit (LEO) satellites according to the orbit altitude above the earth's surface. Different types of satellite systems have a wide range of different features and technical limitations, which can greatly affect their usefulness and performance in specific applications.

GEOs orbit above the equator in the direction of the earth's rotation, at an altitude of 35,786 km. They are attractive as they appear stationary with respect to a fixed point on the rotating earth, allowing a fixed antenna to maintain a link with the satellite. GEO signals reach most of the globe except for polar regions. The one-way latency (from ground antenna to satellite or vice versa) is about a quarter of a second, which presents problems for time-sensitive applications such as voice communication or WAM in substations. Data delay in GEO satellites is substantially higher than terrestrial communications links. Some networking protocols such as TCP that are developed for terrestrial communication links do not function properly in these high-delay links. Virtual private network (VPN) connections also suffer significantly and have to be modified to function in such a high-latency environment.

LEOs are usually located 160–2,000 km above the earth's surface. LEOs can provide worldwide coverage for many communication applications such as remote sensing, and they need less powerful amplifiers for communication with the earth. It takes less energy to launch an LEO satellite, but because LEO orbits are not geostationary, a network of satellites is required to provide continuous coverage. MEOs such as GPS satellites are located in the region between LEO and GEO satellites, and are commonly used for navigation applications. LEO satellites are more desirable in time-sensitive applications as their round-trip delay is comparable to that of terrestrial networks. For example, the current LEO constellations of Globalstar and Iridium satellites have delays of less than 40 ms round-trip, and the planned COMMStellation (scheduled for launch in 2015) will orbit the earth at 1000 km with a latency of approximately 7 ms.

Satellite links require a clear line-of-sight that can be challenging to set up, particularly in areas where features can change with growing foliage. Channel characteristics are variable and highly dependent on weather conditions. The throughput of satellite links (particularly higher frequencies) is deeply affected by moisture and various forms of precipitation. Special techniques such as large rain margins (larger dishes to compensate for fading effect), adaptive uplink power control, and reduced bit rates during precipitation are applied to handle rain fade.

Limited capacity and cost of satellite links should also be taken into account while evaluating their application in electric system automation. In general, satellite links have lower data rates compared to terrestrial communications technologies. Globalstar and Iridium satellites, for example, offer a throughput of 64 kbit/s per channel. Satellite capacity is expected to improve in future satellite constellations (such as Eutelsat's KA-SAT and COMMStellation with overall throughput of 900 Mbps and 1.2 Gbps, respectively). Satellite communication can be a cost-effective option for connecting remote assets where any other communication infrastructure is not available. The higher initial cost of satellite modems (depending on the provider, it can vary from hundreds to thousands of dollars), and ongoing high subscription fees (two to three times more than DSL links for residential customers) are some of the disadvantages of satellite communications. However, the prices could become more competitive in the future.

5.3 Power-line communication standards

As mentioned in Section 5.2.1, PLC has been with us for as long as twisted-pair wired and wireless communications. However, different from wireless communications and communications over dedicated wires (i.e., DSL), it was not widely used for high-data-rate consumer applications and therefore received relatively little attention not only in the public but also in industry and research communities and standardization. PLC was used mainly for low-data-rate communication for utility telemetry and control and home automation applications. This changed in the 1990s with the rising interest in PLC for high-speed access and in-home communications. The arrival of smart grid has now led to a second wave of research and standardization activities in PLC. One of the main advantages of PLC for smart grid is that communication is 'through the grid' [5], and thus the communications medium is in place and (largely) under the control of power utilities.

Following [5], PLC technologies considered for a smart grid communications infrastructure can be classified into four categories: ultra narrowband (UNB), low data-rate (LDR), narrowband (NB), high data-rate (HD) NB, and broadband (BB) technologies. UNB PLC uses frequencies below 3 kHz. It includes ripple control systems, which establish a unidirectional communication with low data rates for load control [12], and the Turtle and two-way automatic communications systems (TWACS) for AMR and DR [5]. While UNB PLC provides only very low data rates for individual links (on the order of bps), its advantage is a high connectivity since network nodes can be reached over large distances (on the order of 100 km). LDR NB PLC systems operate in the frequency band from about 3 kHz to 500 kHz, depending on the country of deployment. These systems typically use single-carrier communication signals with phase-shift or frequency-shift keying modulation, possibly with frequency hopping, and provide data rates in the kbps range. There are a number of LDR NB PLC specifications which have also been ratified by standards developing organizations (SDOs), including ISO/IEC 14908-3 (LonWorks), ISO/IEC 14543-3-5 (KNX), CEA-600.31 (CEBus), and IEC 61334-5 [5, 13].

The above-mentioned second wave of PLC technology refers to recent and still ongoing developments and standardization of HDR NB PLC. But also BB PLC, originally developed for Internet access and multimedia communication, is being reconsidered for smart grid applications. In accordance with the chronology of these developments, we start with a discussion of BB PLC followed by NB PLC.

5.3.1 Broadband power-line communications

BB PLC systems use frequency bands starting at about 2 MHz. The high signal attenuation of power lines at those frequencies limits the range of PLC links, with absolute numbers strongly dependent on the cable type, grid topology, and loading conditions. It is therefore considered as an option mainly for the home networking segment of smart grid.

A number of existing BB PLC specifications have recently been consolidated into two SDO ratified standards, namely IEEE 1901 and ITU-T G.9960/61, also known as ITU-T G.hn. Another SDO BB PLC standard is TIA-1113, which is based on the HomePlug 1.0 specification [13].

IEEE 1901

The IEEE 1901 standard was ratified in September 2010 as a standard for high-speed BB PLC with transmission rates above 100 Mbps at the physical layer. The standard addresses both in-home communication networks (<100 m between devices) and last-mile access connections (<1500 m to the premise). It includes two different PHY layers, one based on windowed OFDM modulation and another based on Wavelet-OFDM modulation [7]. Each PHY is optional, and implementers of the specification may, but are not required to include both. The OFDM PHY is derived from and is backward compatible with HomePlug AV technology by the HomePlug Powerline Alliance. Similarly, wavelet-OFDM is based on and backward compatible with the HD-PLC specifications of the HD-PLC Alliance led by Panasonic [5]. Both PHY layers use the frequency band from about 2 MHz to about 30 MHz with an optional extension to about 50 MHz. The maximal PHY layer transmission rates are about 420 Mbps using the optional frequency band.

The IEEE 1901 standard also defines a common MAC layer that can manage the two different PHY layers through intermediate layers referred to as physical layer convergence protocols. Medium access is accomplished by TDMA and carrier-sense multiple access with collision avoidance (CSMA/CA). The coexistence between the non-interoperable PHY layers is managed through the mandatory inter-system protocol (ISP). The ISP also handles the coexistence of IEEE 1901 compatible access and in-home systems and ITU-T G.hn systems (see Section 5.3.3 on coexistence). In-home systems share the medium through time-division multiplexing synchronous to the mains cycle, while coexisting access and in-home systems may also apply frequency-division duplexing with a frequency split point at 10 MHz or 14 MHz.

ITU-T G.9960/61

ITU-T G.hn was launched in 2006 in order to develop a high-speed (rates up to 1 Gbps) unified standard for home area networking that can operate over all types of legacy in-home wiring including phone lines, power lines, coax and Cat 5 cables [8]. Recommendations G.9960 (ratified in October 2009) and G.9961 (ratified in June 2010) define the PHY layer and the data link layer (DLL) of ITU-T G.hn respectively. ITU-T G.9960/61 includes two profiles for PLC: 50 MHz (2–50 MHz) and 100 MHz (2–100 MHz). It targets BB networking and entertainment in residential houses and public places.

The G.9960/G.9961 devices are designed to be flexible and capable of operating over different types of media. The standard is based on OFDM and uses different sets of parameters and frequency ranges to (somewhat) optimize the signalling scheme and to address channel characteristics of different media. A G.9960 network may include up to 16 separate domains, which may be established over any type of in-home wiring. In each domain, a designated domain master coordinates the communication between all the nodes (up to 250 nodes in a domain). Devices located in different domains can communicate via an inter-domain bridge (IDB). A global master (GM) manages resources, priorities, and operational characteristics between domains of a G.9960 network. G.9960 domains can also be bridged to alien (wired or wireless non-G.9960) domains.

The G.9960/61 also defines a low-complexity profile (LCP) for less demanding applications. LCP operates in the 2–25 MHz range over all in-home wired media (power line, coax, phone line), and it is interoperable with the other G.9960 profiles (50 MHz, 100 MHz). LCP offers reduced bit rate (5–20 Mbps), low complexity, and power consumption, while maintaining most of the desirable characteristics of full G.9960/61 including low latency, full mesh networking, and advanced security. LCP targets typical smart grid applications such as AMI, smart appliances, in-home energy management devices, and charging plug-in electric vehicles (PEVs)/plug-in hybrid electric vehicles (PHEVs).

HomePlug Green PHY

The HomePlug Powerline Alliance published the HomePlug Green PHY (HPGP) specification in June 2010. It may be considered a replacement of the LDR NB PLC HomePlug Command and Control specification as the specification for smart grid applications in the HomePlug standards portfolio. HPGP is essentially a scaled down and amended version of the HomePlug AV (HPAV) specification, to which the IEEE 1901 standard is backward compatible. As such it uses the 2 MHz to 30 MHz frequency band also used by HPAV and IEEE 1901 BB PLC. But to better meet the requirements of low cost and low power consumption for HAN smart grid devices, HPGP uses only the basic set of coding and modulation schemes provided in HPAV. In particular, only robust transmission modes, so-called ROBO modes, with peak data rates between 4 Mbps and 10 Mbps are enabled and adaptive bit loading is disabled.

A new feature that has been added compared to HPAV is the power save mode. Each HPGP device can operate in a synchronized periodic awake/sleep cycle. Through message exchange and periodic beacon signals, the network coordinator schedules to power save periods of all nodes such that network connectivity is maintained. The power save mode is an effective means to reduce the energy-per-bit metric of devices supporting applications with average data-rate requirements much below the megabits per second peak data rates offered by HPGP. According to [14], using the power save mode, power consumption can be reduced by up to 85% and 97% depending on latency requirements. Other features are a so-called distributed bandwidth control, which limits the aggregate use ('time on wire') by HPGP devices to 7% in the presence of HPAV devices, and an optional transmit power control to further lower power consumption in case of links with low signal attenuation.

5.3.2 Narrowband power-line communications

Different from HPGP, which uses a wide frequency band with possibly low duty cycle to achieve a moderate average data rate of a few hundred kilobits per second, HDR NB PLC transmits in a much narrower band of less than 500 kHz. This significantly simplifies the analogue-to-digital conversion and digital signal processing requirements in PLC modems. HDR NB PLC systems also operate at below 500 kHz compared to above 2 MHz for BB PLC, which leads to much reduced signal attenuation, cf. [15] and [5, Table II], and unwanted radiation. Therefore, larger distances can be bridged, making HDR NB PLC an option for HAN *and* neighbourhood area network (NAN) segments of smart grid. However, while signal attenuation through skin effect and signal reflection due to impedance mismatches are much less pronounced at these lower frequencies, low-impedance loads can be challenging for HDR NB PLC, cf. [15]. Furthermore, in areas with only few households served by one distribution transformer, the typically high attenuation inflicted by the transformer is a possible disadvantage for NB PLC compared to UNB signals, which pass easily through transformers [5].

HDR NB PLC systems stemming from the second wave of PLC developments are multicarrier based. In particular, OFDM has been adopted as the common modulation format. This development is likely in part due to the experience gained with OFDM in BB PLC, and not unlike trends in wireless communications. In Europe, they operate in the so-called CENELEC bands established in 1992 in the standard EN 50065 [16]. The CENELEC bands are from 3–95 kHz (A), 95–125 kHz (B), 125–140 kHz (C), and 140–148.5 kHz (D), of which the CENELEC A band is reserved for power utilities and thus particularly interesting for smart grid applications. Members of the PLC community are lobbying for an extension of the usable band up to 500 kHz, which is already usable for PLC in North America (USA, Canada) and Asia (Japan, China).

In the following, we take a brief look at the prominent HDR NB PLC specifications and recently launched standardization activities.

Industry and project specifications

The two perhaps most known HDR NB PLC specifications are the PoweRline Intelligent Metering Evolution (PRIME) and G3-PLC [17–19]. Both specifications are available online. They are intended to connect meters to concentrator points, typically located at transformer substations.

PRIME [17] describes a PHY, MAC, and convergence layer for HDR NB PLC operating in the CENELEC A band. The OFDM physical layer data rates are between 21.4 kbps and 128.6 kbps. The MAC layer considers subnetworks defined by nodes at the transformer substation and meters, with the substation node taking on the role of a master in a tree topology. Medium access is via CSMA/CA and time-division multiple access (TDMA), and repeater functionality is included to reach all meters.

The G3-PLC specification [18] shows similar features. It also operates in the CENELEC A band and provides a data rate of up to 33.4 kbps. Different from PRIME, the MAC layer is based on IEEE 802.15.4 and an IPv6 over low-power wireless personal-area networks (6LoWPANs) adaptation layer is included for transmission of IPv6 packets over PLC. Recently, a supplement of G3-PLC (also referred to as G3-FCC) was released that operates between about 145 kHz and 478 kHz and provides data rates of up to 208 kbps.

Field trial results reported in [18, 20] for G3-PLC transmission over medium-voltage (MV) and low-voltage (LV) distribution lines in European and US distribution networks demonstrate that reliable communication across MV–LV transformers is possible. This allows us to place concentrators on the MV side and thus connect more meters to one concentrator, which makes PLC an economically more viable option especially in rural areas and most of the USA, where a transformer serves only few homes. The use of OFDM over a preferably wide band is advantageous here, as it allows us to dynamically select sub-bands with favourable transmission conditions [18]. Field trials for PRIME technology reported in [21] indicate that a large number of nodes could reliably be connected with two levels of repeating.

Another HDR NB PLC-based system for connecting utility servers and equipment at customers' premises over the MV and LV grid segments is the real-time energy management over power lines and Internet (REMPLI) system [22]. It also uses an OFDM physical layer and features an interesting repeater mechanism based on single-frequency networking. This is advantageous for fast delivery of messages, in particular broadcast and multicast messages, over large distances making autonomous use of network redundancies and avoiding network congestion [23].

Standardization activities

In early 2010, almost simultaneously, IEEE and ITU-T launched the working groups IEEE P1901.2 and ITU-T G.hnem (G.9955 and G.9956), respectively, to develop standards based on HDR NB PLC technology. Both groups consider essentially the same target features, which are low-complexity PLC technology for communication over LV and MV grid segments, including communications through MV–LV transformers, and over DC power lines with data rates of up to 500 kbps and 1 Mbps, respectively. While the former is to support two-way communication for distribution automation, AMI, DSM, as well as home energy management, the latter is mainly to enable communication between PEVs/ PHEVs and charging stations. With standardization still ongoing, it can be expected that both standards will include modes such that G3-PLC and PRIME devices will be standard compliant [24].

5.3.3 PLC coexistence

PLC coexistence typically concerns two aspects: firstly, coexistence with other communication media, mainly wireless systems, but also VDSL, and secondly, coexistence among PLC systems. The former is a problem of unwanted radiation from PLC signals and ingress of wireless signal and thus mainly restricted to BB PLC. Approaches to deal with this include static and dynamic notching and (cooperative) sensing, rendering PLC devices a cognitive 'radio'. In the context of PLC for smart grid, coexistence of PLC systems is the more relevant aspect. Until recently, coexistence has received relatively little attenuation, which is due to the low 'density' of deployed PLC systems. But coexistence needs to be addressed in the future due to the increasing proliferation of PLC. The underlying problem is that power-line networks represent a shared medium in which signals propagate without well-defined boundaries, which is not unlike the situation in wireless communications.

A coexistence protocol for BB PLC has recently been adopted as part of the IEEE 1901 standard and as ITU-T G.9972 standard. This so-called inter-system protocol (ISP) enables three in-home and one access BB PLC system to coexist, using frequency and time division multiplexing. In the NB PLC domain, the CENELEC standard EN 50065 [16] prescribes a CSMA/CA mechanism using a beacon signal at 132.5 kHz, but only for the CENELEC C band. Recognizing the need for coexistence, NIST PAP 15 seeks to harmonize PLC standards. PAP 15 has been successful to achieve harmonization between IEEE 1901 and ITU G.hn devices using ITU-T G.9972, and it has also considered an extension of the CSMA/CA mechanism from CENELEC to the entire 500 kHz band used by NB PLC [25]. The current HDR NB PLC standardization efforts, IEEE P1901.2 and ITU-G.hnem, include coexistence working groups. These are concerned with coexistence to legacy NB (single-carrier) as well as mechanisms to enable coexistence among each other [24]. Although in the initial phase of smart grid applications, only one HDR NB PLC system will likely be installed in a certain part of the grid, and thus coexistence will not be an issue, coexistence mechanisms for non-interoperable PLC standards will become instrumental for the unimpeded proliferation of PLC solutions for smart grid.

5.4 Wireless standards

While some proprietary wireless protocols such as Z-Wave, MiWi, INSTEON, or Wavenis offer a number of advantages including ease of optimization for a specific application and faster development (as there is no need for a certification process), standardized solutions such as ZigBee or WiMAX are generally favoured because they are designed for general purposes, their specifications are (publicly, or through membership) available. They are generally more cost-effective to deploy as multiple vendors offer competing, interoperable products, and they are easier to maintain in the long run. This section elaborates on some of the relevant wireless networking standards in smart grid communications.

5.4.1 Short-range solutions

Short-range wireless connections are widely used in setting up indoor home/building area networks as well as in connecting sensors or smart meters to local (neighbourhood) data aggregators in outdoor environments, usually through setting up a wireless mesh network. This section covers ZigBee and Wi-Fi as they are currently the most prevalent standardized solutions for short-range wireless networking in smart grid applications.

ZigBee

ZigBee is a protocol developed by an industry interest group called the ZigBee Alliance for low-data-rate and short-range (10–100 m) wireless networking applications. The PHY and MAC layers of ZigBee are based on the IEEE 802.15.4 standard. The IEEE 802.15.4 operates in the licence-free ISM radio bands 868 MHz (Europe), 915 MHz (USA and Australia), and 2.4 GHz (worldwide) with data rates of 20 kbps, 40 kbps, and 250 kbps, respectively. ZigBee has extensive hardware products from multiple vendors primarily in the 2.4 GHz band.

Three types of devices are defined in ZigBee. The ZigBee coordinator is the most sophisticated node in a ZigBee network. It acts as the bridge to other networks, controls the authentication process and security keys, or is the root node in tree topology. ZigBee routers can both run application functions and act as a router by passing on the data they receive from other devices. The ZigBee end-device is a simple node with limited functionality that is in sleep mode most of the time in order to maintain long battery life. An end-device does not have relaying capability and can only communicate with routers or the coordinator.

ZigBee supports addressing and routing for both tree and mesh networking topologies. The original version of ZigBee (ratified in 2004) did not support IP and suffered from scalability issues. The 2004 ZigBee standard used a tree structure for addressing and could run out of addresses in networks connecting only a few hundred nodes. The ZigBee PRO standard (ratified in 2007) offers a more scalable addressing scheme based on the random assignment of addresses. The devices are identified with 16-bit addresses in ZigBee PRO, and collision is uncommon even in networks supporting a few thousand devices. An address conflict resolution mechanism is also provisioned in the network stack. In addition, ZigBee PRO offers an efficient solution for aggregating data through many-to-one routing, i.e., communication between several devices and a central controller or sink node [26, 27].

ZigBee networks operate in unlicensed ISM bands and are prone to interference from numerous wireless devices working in these popular bands. The 802.15.4 standard utilizes several mechanisms such as dynamic channel selection (DCS), CSMA/CA, acknowledged transmission and retry (ATR) to enable coexistence. In DCS, the physical layer has the ability to measure the present interference on a particular channel. Then it is possible to build mechanisms against interference based on selecting the least interfered channel. The CSMA technique listens to see if a channel is busy before transmitting and ATR ensures successful reception of data by an acknowledged frame delivery. However, the fast-varying interference in time, frequency, and space particularly from 802.11 wireless LANs can critically affect the performance of ZigBee-based networks.

ZigBee is expected to be widely applied in wireless sensor networks (WSNs) (see Section 5.5.4) and in smart metering applications. For example, massive AMI projects for deployment of 30 million ZigBee-equipped smart meters are currently underway in North America [27]. In 2009, ZigBee announced integration of IP into its specification portfolio. ZigBee networks can then readily be connected to the Internet with a router, and there would be no need for bridging or address mapping hardware. IP-based networking in smart grid is discussed in more detail in Section 5.5.3, and Section 5.5.1 elaborates on the MAC/PHY agnostic smart energy profile 2.0 (SEP 2.0) that enables technology-neutral, plug-and-play interoperability between a variety of wired and wireless standards and technologies.

Wi-Fi

The term Wi-Fi is a trademark of the Wi-Fi alliance and refers to a family of protocols based on the IEEE 802.11 standards that are widely applied in WLANs. There are approximately over 2 billion Wi-Fi-certified devices worldwide and the industry alliance has close to 300 members. The technology is well matured with proven management systems, native IP-network compatibility encryption, authentication and end-to-end network security. Wi-Fi connections and hot spots provide wireless Internet connection in many utility facilities as well as both residential and commercial customers' premises.

It is only natural to assume that a smart grid communications network can take advantage of this widespread technology, particularly in cost-effective deployment of customers' premises' networks. The drawback of Wi-Fi devices, however, is that IEEE 802.11 is optimized for fast data rates and connection powerful computing devices but not for large numbers of energy-efficient low-power devices. Power consumption of a typical Wi-Fi device (more than 700 mW) is more than 7 times that of a ZigBee chip (around 100 mW) [28].

There are, however, ongoing extensive efforts in the industry to develop embedded Wi-Fi technologies that are optimized for low power consumption and compatible with existing Wi-Fi transceivers. Low-power Wi-Fi devices have to be integrated on a single chip in order to shorten connections, and reduce the waste of power. They have to consume extremely low power in standby mode and be able to quickly change state from active to idle and vice versa. According to [29], currently available embedded Wi-Fi products employing these power-saving techniques would be capable of functioning for more than a decade with a single battery charge. They benefit from the widespread IEEE 802.11 infrastructure, and they can also take advantage of the standards' evolution in areas such as security (802.11i), meshing (802.11s), and quality-of-service (802.11e). Embedded Wi-Fi, hence, is expected to be an attractive technology for short-range wireless links in smart grid communications networks.

5.4.2 Long-range solutions

Long-range wireless solutions are expected in many smart grid applications such as wireless backhauling, connecting mobile workforce, or establishing direct links to endpoints such as smart meters. Cellular standardized technologies are typically applied not only when utilities subscribe to public access networks, but they are also preferred in deployment of private networks because of availability and cost-effectiveness of the equipment. Some of the existing equipments and pilot projects apply general packet radio service (GPRS) or 3G technologies. However, it is expected that most utilities will be applying 4G standards in their smart grid initiatives [30]. WiMAX and 3GPP-LTE are the two contenders for 4G cellular networks. While most cellular carriers have chosen 3GPP-LTE, WiMAX seems to be gaining momentum in smart grid networks.

WiMAX

WiMAX is based on the IEEE 802.16 series of standards, and WiMAX products are promoted and certified by WiMAX Forum. Hence, WiMAX products and services from different vendors are expected to be interoperable, which is an important concern for electric utilities. WiMAX provides flexible broadband links of up to 140 Mbps (20 MHz band and 2×2 MIMO), features low latency (10–50 ms), and supports both fixed (IEEE 802.16d) and mobile (IEEE 802.16e) connections. Depending on the carrier frequency, density of users, and environment, WiMAX can realize long-range connections (e.g., more than 20 km of coverage for a 1.8 GHz link in a suburban area [31]).

WiMAX takes advantage of modern wireless communications techniques such as OFDM to combat delay spread, MIMO systems for increased bandwidth efficiency, and adaptive modulation and iterative coding for robustness. Orthogonal frequency-division multiple access (OFDMA) is the access method for both uplink and downlink connections. Managed use of different sub-carriers in neighbouring sectors and cells limits co-channel interference (fractional frequency reuse). WiMAX also has the inherent

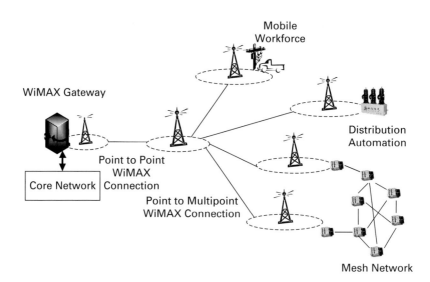

Figure 5.3 WiMAX backhaul in smart grid communications networks.

feature of supporting different levels of quality-of-service (QoS), allowing the system operator to prioritize time-sensitive traffic.

Many electric utilities have (or are in the process of) adopting WiMAX solutions in their upcoming smart grid communications networks. WiMAX can provide P2P or P2MP wireless backhaul for a variety of different smart grid applications [9, 32], as illustrated in Figure 5.3. Unlike most traditional microwave solutions, WiMAX links can operate in NLOS situations due to obstacles in the environment. In some networks, WiMAX technology is applied to provide broadband connection to the end-user (e.g., a smart meter). The direct approach is currently more expensive but with the expected decline in the price of WiMAX devices, it will become more competitive particularly in situations where low latency is desired.

NIST PAP 2 *Guidelines for Assessing Wireless Standards for Smart Grid Applications* provides some guidelines for network planning of WiMAX backhaul systems in AMI applications [31]. More work is needed to analyse the system requirements when other smart grid applications (such as DR, distribution automation, etc.) are considered in network planning. With support from numerous companies such as GE, Silver Spring Networks, and Motorola, WiMAX-based AMI networks are currently under development and expected to grow in many areas of the USA (such as San Diego, Michigan, and Texas) and parts of Australia [30].

3GPP LTE

Although WiMAX has had an early lead in smart grid wireless backhaul applications, LTE is expected to erode this head start. LTE is designed for backward-compatibility with previous generations of 3GPP standards (3G UMTS), and LTE base stations can use existing 3G towers to reduce the cost of network upgrade. Many major cellular carriers are currently in the process of adopting LTE, and heavy competition will likely drive low hardware prices and broadband network coverage. The carriers have also been marketing hard and reducing their subscription fees in order to attract utility customers.

The technical frameworks of LTE and WiMAX standards are very similar. They both take advantage of advanced communications techniques such as MIMO, allow for scalable channel bandwidth between 1.25 MHz and 20 MHz, and apply OFDMA for downlink access. One notable difference between LTE and WiMAX is that LTE applies single-carrier frequency-division multiple access (SC-FDMA) instead of OFDMA for uplink access. SC-FDMA reduces the peak-to-average power ratio (PAPR) compared to OFDMA by adding an extra level of processing in the frequency domain and the lower PAPR relieves the requirement for highly linear (usually more expensive) power amplifiers.

The NIST PAP 2 document elaborates on coverage and capacity analysis in planning LTE-based AMI networks. It provides examples of estimating the maximum coverage radius of a cell, based on a channel propagation model and parameters of the LTE deployment. The number of customers in each area (cell) determines the required system capacity. For example, typical meter density in urban areas is 2000 meters/km^2 compared to 800 and 10 meters/km^2 in suburban and rural areas, respectively. PAP 2 also includes a detailed example of estimating the overall capacity of a sector, based on the geographic density of smart meters [31].

IEEE 802.22

Limited availability (and high cost of acquiring) radio spectrum is a key disadvantage of deploying wireless backhaul in smart grid applications. A promising solution for scarcity of spectrum is the cognitive radio (CR) concept, which refers to opportunistic usage of frequency bands that are not densely occupied by licensed primary users. In the USA, the Federal Communications Commission (FCC) has allowed cognitive users to exploit unoccupied digital television (DTV) spectrum or TV white space. TV channels are highly desirable in long-range coverage and difficult terrain for their ability to provide NLOS coverage and penetrate foliage.

The IEEE 802.22 is a standardized CR solution (currently in balloting phase) for a wireless regional area network (WRAN). This standard specifies PHY and MAC layer P2MP WRANs including fixed base stations connecting with fixed and mobile terminals operating in the VHF/UHF TV broadcast bands between 54 MHz and 862 MHz. Application of IEEE 802.22 in smart grid WANs has recently been proposed [33]. Two different scenarios are considered: the use of CR as a primary broadband access system in sparsely populated rural areas, and CR as a smart secondary radio in urban centres and suburban areas to provide a backup connection and/or for opportunistic transfer of non-urgent data.

Wireless connectivity based on CR provides an alternative, cost-effective broadband solution to utilities, particularly in rural areas. Spectrum Bridge in partnership with Google has reported a successful trial deployment of a TV whiteband network for Plumas-Sierra Rural Electric Cooperative [34]. The base stations in this trial are connected to a national database to check the availability of spectrum in different geographical areas.

This network delivers broadband connections to a remote community and real-time link for monitoring and control of remote substations and switchgear. CR networking based on a central server and a few optimization methods for remote management of cognitive functions are discussed in [35].

5.5 Networking solutions

The selection of physical communication media and technologies only covers part of the engineering process in designing smart grid communications networks. In practice, hybrid networking solutions may have to be devised to implement a pervasive, reliable, and cost-effective communications infrastructure. Many applied networking issues such as the choice between public and private networks, and application of IP protocols need to be considered in the context of smart grid communications. This section discusses some of these issues and relevant networking paradigms such as WSNs and machine-to-machine (M2M) communications.

5.5.1 Hybrid solutions

Existing utility communications networks are a hybrid mix of different communications technologies (see Section 5.1.1). This trend is likely to continue as these networks are evolving *en route* to smart grid, not only to maintain backward compatibility but also to provide cost-effective, omnipresent connectivity in a wide variety of environments throughout grid. No single communications link can provide 100% reliability, so backup connections are required in critical applications such as teleprotection. One way to reduce the chance of failure is to provision multiple routes in network design. Another way to increase the reliability is to use different communications technologies, e.g., installing wireless or satellite receivers in addition to wired links in transmission substations.

The SEP 2.0 protocol that is proposed by the ZigBee Alliance and the HomePlug Forum is an example of a hybrid solution for smart grid communications, and is one of the main standards identified in the NIST roadmap [1]. ZigBee-based networks can face difficulties with coverage, particularly in larger homes or apartment buildings with concrete structures requiring wired access points in different areas, or floors. They can also become less robust to node failure when there is less redundancy in routes. PLC-based HomePlug solutions, however, can suffer from excessive noise in the power lines and are not suitable for nodes that are not connected to the grid (such as an outdoor temperature sensor). SEP 2.0 develops an integrated system architecture and application profile interface for home area networking that is supported both by ZigBee and HomePlug devices [36, 37]. Because of the technology-agnostic design of SEP 2.0 protocol, other wired and wireless technologies are also able to support it. It was recently announced that the HomePlug Alliance, Wi-Fi Alliance, HomeGrid Forum, and ZigBee Alliance have agreed to create a consortium for SEP 2.0 interoperability [38].

5.5.2 Public vs. private networks

There is extensive debate about the application of public and private networks in smart grid communications. While it is challenging to finance, deploy, maintain, and run sophisticated private communications networks, particularly for smaller utilities, the public option requires low capital investment and may result in declining operating expense as a result of competition between commercial service providers. The communications service provider will be responsible for guaranteed performance, security, and upgrading the infrastructure. Furthermore, involving multiple public providers can improve fault-tolerance or extend coverage.

On the other hand, many utilities, particularly the larger ones, prefer to build and manage their own networks. A private network can provide access to all their customers, not just the ones in population centres, and cover remote areas where their facilities and infrastructure are located. Some utilities are concerned about issues such as data priority, lack of control, security, reliability, availability, cumulative operational costs, survivability in harsh environment and weather conditions. Utilities are also concerned that their devices will not be able to keep up with the fast rate of technology change in public networks and become prematurely obsolete. These issues have to be considered on a case-by-case basis, and smart grid communications infrastructure will most likely include both public and private networks.

5.5.3 Internet and IP-based networking

Availability of broadband Internet connections is growing globally, and in some countries such as Finland, service providers are mandated by law to provide reasonably priced broadband connections to every permanent residence and office. Hence, the existing infrastructure for Internet access provides an alternative cost-effective high-speed communication core network for smart grid communications. In countries where broadband Internet is less pervasive, broadband access and smart grid initiatives can be combined either to take advantage of government incentives [39] or to give utilities other sources of income by providing data services or by charging the Internet service providers for using their infrastructure.

As is emphasized by the NIST smart grid roadmap [1], future utility networks are expected to be predominantly based on IP. IP is a mature technology that is prevalent in both public and private networks. IP-based networking, hence, facilitates development of interoperable smart grid applications by different vendors. Various applications can seamlessly share the links and information (without the need for proxies and bridges) independent of underlying physical communications technologies. IP is a scalable technology as any smart device, sensor, and actuator that is added to the network is assigned a new IP-address. The dynamic routing capabilities of IP technology increase the reliability of communications in case of failure.

With large numbers (trillions) of potential Internet-ready (IP-enabled) IEDs to be applied in remote sensing, monitoring, control, automation, and video surveillance applications across the grid, smart grid is considered to be a major driver in realizing the so-called *Internet of things* or *Internet of embedded devices*. Internet of things envisions integration of universally IP-enabled smart devices, and it is believed to be the next frontier of Internet evolution [40].

When the public Internet is used in realizing smart grid networks, QoS measures are a source of concern. Internet traffic is generally delivered through best effort and usually is not capable of guaranteeing strict requirements of many smart grid applications (such as reliability measures and delay). In order to address QoS concerns, utilities that need to connect a large number of substations and remote control centres can apply multiprotocol label-switching (MPLS) technology to establish VPN architectures and reduce the communication cost significantly compared to private communication links [9]. Although Internet-based VPN technology has the ability to prioritize data and provide a high-performance networking solution through the public Internet, there are limits in using a shared infrastructure, particularly with respect to transferring critical, time-sensitive information.

There are also concerns about the limited availability of IPv4 addresses. With the substantial increase in the numbers of IP-enabled IEDs in the electrical grid, large numbers of IP addresses are needed to identify these devices. One way to solve this issue is to design alternative addressing scheme combined protocols enabling translation/mapping into IP addresses [1]. Another option (which is encouraged by NIST) is to apply the IPv6 protocol in smart grid devices. IPv6 has been developed specifically to enhance IP networks, and solve the issue with IPv4 address exhaustion.

Traditional Internet protocols, however, might be too challenging to implement in low-cost, low-power, and low-bandwidth embedded devices. IPv6, for example, involves complex IP Security (IPSec) authentication, relies on web services, and long frame-lengths that require substantial bandwidth. These requirements have limited the use of IP to devices with powerful processing power. Standard IP suites also fail to address the unique features of wireless networks such as the difference between a packet loss due to congestion or link quality, power conservation through limited duty cycle, enabling multicast, and multihop mesh networking.

6LoWPAN is a set of recommendations by the Internet Engineering Task Force (IETF) that enables efficient use of IPv6 for low-power, low-cost wireless embedded devices [40]. 6LoWPAN fragments IPv6 long packets (minimum 1280 bytes) to fit into short IEEE 802.15.4 frame size (127 bytes). It also compresses the 40-byte IPv6 header to only two bytes and offers mechanisms for IPv6 address auto-configuration and neighbour discovery for low-power mesh networks [27].

Mesh under and route over are two IP-based routing paradigms in low-power wireless mesh networks. In the mesh under scheme, the whole mesh network is a single IP link and routing is performed under IP using the IEEE 802.15.4 protocol. In route over, every node has an IP address, and routing happens in the IP layer. A standardized protocol for route over in IPv6, called IPv6 routing protocol for low-power and lossy networks (RPL), is currently being developed by the IETF routing over low-power and lossy networks (ROLL) working group [27].

As we have discussed, smart grid is a complex network of networks in different environments. For ubiquitous use of IP, a variety of suitable protocols need to be identified to address specific applications for network control, management, and security. NIST's PAP 01 is dedicated to developing guidelines for the use of an IP protocol suite in the smart grid and to perform the necessary analysis for cyber security and desired performance characteristics. Detailed objectives of this mission and its progress can be followed in [41].

5.5.4 Wireless sensor networks

WSNs were first developed for military applications in the mid-1990s. With the advances in capabilities, and decreasing cost and size of sensor nodes, WSNs have since become popular in a variety of control and monitoring applications including automation and monitoring in the electrical power grid [9]. A WSN is generally formed by a large number of low-cost, low-power ad hoc wireless sensors and actuators (nodes). Traffic generally flows from the nodes all over the network to a special sink node that monitors the overall network and communicates with external networks and processing centres.

WSN nodes are usually small in size, are densely scattered, and the networks have the ability to extract localized information over a large area. The nodes are usually battery-powered and apply short-range communications technologies to save energy (extend the battery life). Messages can travel multiple hops cooperatively from node to node following the mesh networking topology (as discussed in Section 5.2.2). Due to data redundancy, and availability of multiple routing paths, WSNs can robustly handle node failures (self-configurability and fault-tolerance). Other advantages of WSNs include ease of deployment, cost-effectiveness, flexibility, and aggregated intelligence through parallel processing. WSN technology, hence, is a promising networking solution for realizing low-cost embedded electric utility monitoring, equipment fault diagnostics, or meter reading infrastructures in smart grid.

Reliability of WSNs in critical smart grid applications, however, is a source of concern. The harsh conditions of an electric system environment can affect the performance of WSN nodes with high levels of noise, interference, corrosion, humidity, vibrations, dirt and dust, or other conditions. Throughput and latency of WSN links is variable and makes it hard to meet QoS requirements. Also, the nodes have limited processing power, memory, and (usually) energy (battery-powered). Many of the WSN protocols are designed to increase the battery life [42]. WSNs have mostly been optimized for a particular application and typically apply proprietary communications technologies. These networks are traditionally operating in isolation, and the individual nodes are not IP-compatible. There is, however, a clear trend towards applying standardized IP-enabled solutions such as ZigBee, low-power Wi-Fi, and 6LoWPAN in WSNs [40].

There are three options for providing power for WSN nodes in smart grid applications. The nodes that are connected to indoor or outdoor electrical wires can be powered directly from the grid. The second option is that nodes are connected to the grid but have a backup source of power like a battery. This backup system is essential in critical applications to avoid system failure in case of a blackout. Finally, some of the nodes such as thermal sensors or pressure sensors on high-voltage lines may not be connected directly to the grid. In this case the nodes either have to be battery-powered or use harvesting techniques to use ambient sources of energy such as solar, thermal, vibration, or electromagnetic fields. In all these cases, the nodes must consume very low power to keep the overall power consumption low and extend the battery life (in battery-powered nodes).

5.5.5 Machine-to-machine communications

M2M or machine-type communications (MTC) refers to data communication between devices without human interaction. The early roots of M2M communications go back to the development of SCADA systems in the 1980s [43]. This area has recently gained a lot of attention in industry, research, and standardization bodies such as 3GPP LTE. The main objective is to foster more efficient use of network resources by opening public networks for a much wider audience of smart devices. M2M applications are evolving in many different sectors such as health care, public safety, vehicular, manufacturing, utilities and consumer products.

Smart grid is a key area where M2M technologies are expected to grow, providing the utilities with low-cost connections to grid assets such as circuit breakers, transformers, and other substation equipment. Smart grid automation and control applications such as home energy management, AMI, and DA require widespread and automatically controlled infrastructure of communicating devices. Hence, smart grid presents one of the best potential growth opportunities in M2M communications. Here, we present an overview of M2M communications and discuss some of the enabling technologies as well as open issues pertinent to smart grid applications.

M2M creates a ubiquitous environment of networked interconnections commonly referred to as the Internet of embedded devices or the Internet of things. There are three main factors enabling the widespread adoption of M2M technologies. Pervasive (mainly wireless cellular) public networks provide the infrastructure for data communication and access to the Internet. Advanced software components process the data, make decisions autonomously, and generate control signals. Low-cost, high-performance, and easy to deploy sensors and actuators that can collect data and carry out control commands. Public wireless networks have been designed for traditional human-to-human (H2H) connections with high requirements for mobility, user experience, and QoS. The novelty of M2M communications lies in providing low-cost, low-effort, reliable, and scalable interconnectivity solutions to a large number of users with small and infrequent traffic transmission per user.

An M2M network generally consists of five main components: module, gateway, server, area network, and core network. Figure 5.4 depicts a basic overview of an M2M network. An M2M module is a transceiver that is usually embedded within a smart device or sensor. The M2M gateway links M2M modules with the core network. The M2M area network provides connectivity between M2M modules and M2M gateways. M2M servers process data from M2M modules and provide information for the application layer. The M2M core network connects M2M gateways with M2M servers. An AMI

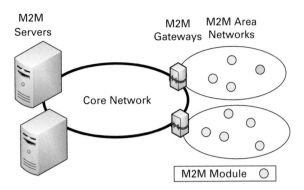

Figure 5.4 Overview of M2M communications.

system can readily be mapped into the M2M communications model. Smart meters are M2M modules and are connected to local base stations or concentrators (M2M gateway) through NANs (M2M area network). AMI data is then aggregated through a backhaul network (M2M core network) to be processed by utility meter data-management (MDM) servers (M2M server). AMI data is then available to be shared for different applications such as billing and DR.

Advanced wireless technologies are very important for easy installation of M2M area and core networks, eliminating the need for wiring infrastructure. Nowadays, cellular networks are accessible extensively around the world and are able to connect large numbers of modules either directly or through gateways. Imminent deployment of fourth-generation systems such as WiMAX (IEEE 802.16) and 3GPP LTE is going to increase the capacity and performance of cellular networks. M2M area networks can be established by short and mid-range wireless technologies. Wi-Fi (IEEE 802.11) is an attractive solution for M2M area networking because of scalability and widespread installation (particularly within buildings and residences). Wireless mesh networking based on Zig-Bee (IEEE 802.15.4) is another popular solution offering low-cost, low-power devices, and self-healing network architecture. Other existing solutions include Bluetooth (IEEE 802.15.1), ultra-wide band (UWB) (IEEE 802.15.3a), infrared data association (IrDA), and a variety of proprietary options [44].

In [28], the authors compare four existing standardized technologies including Zig-Bee, Wi-Fi, Bluetooth, and UWB for AMI and HAN applications. They conclude that ZigBee is the preferred option because of its low power, ease of network configuration, scalability, and adequate range. This preference, however, is not universal, and the best technology for each application has to be determined according to its particular requirements. Another reason why there is no clear winning technology for M2M applications is because of the modified low-power versions of many standards, which are designed to efficiently handle infrequent traffic of M2M communications. Embedded Wi-Fi, as we have discussed, is an energy-efficient version of Wi-Fi that can take advantage of the widespread Wi-Fi infrastructure and integrated IP-based networking [29]. While conventional Wi-Fi is optimized for fast response and high data rates, embedded Wi-Fi is optimized for power efficiency, particularly in idle mode. Bluetooth v4.0 (or Blutooth low-energy) is another example of a previously used standard that is modified for use in sensors and actuators [45]. Fast connection set-up time (a few milliseconds), sleep mode, and shorter message are some of the strategies for increasing the power efficiency in Bluetooth v4.0. Also, out of 37 available channels, three control channels and nine data channels are located between wireless LAN channels in the 2.4 GHz band to avoid interference with Wi-Fi devices.

There are many technical challenges that need to be addressed towards mass adoption of M2M communications in critical applications such as smart grid. Standardization of integrated M2M architecture can facilitate widespread deployment of M2M systems. M2M communications technologies or their modified versions (such as UWB, Bluetooth, WiMAX) need to be identified and specific comparisons have to be made for different use cases. Traffic patterns of M2M networks have to be studied and applied in optimizing network protocols and architectures.

Data characterization is also essential for guaranteeing the required QoS for M2M applications. M2M networks are expected to support large numbers of devices and in order to increase data throughput, innovative algorithms can be applied for data concentration or modifying the messaging schemes. In [28], for example, the authors recommend that HAN devices should be programmed to only communicate with the controller when there is a change in their power requirement, and have shown that this technique allows for seamless support of many more devices within a home network. Developing smart methods for improving the efficiency and reducing the cost of M2M through optimal grid topology measures (such as the ratio of modules to gateways) is another interesting area of research. In [44], the authors propose a clustering algorithm based on dynamic programming.

Many M2M smart grid networks are expected to share the ISM bands with other communications devices. Studying the effect of interference in reliability and meeting the required QoS in these networks is of interest. There is also potential in developing smart interference avoidance and spectrum management techniques for M2M networks. As previously discussed, coexistence and interoperability of different technologies are also vital for the successful realization of smart grid communications networks.

A comprehensive treatment on M2M communications will be provided in Chapter 6 of this book.

5.6 Conclusion

An integrated communications infrastructure is a fundamental requirement of smart grid. Smart grid communications networks are evolving from the existing utility communications networks and expanding throughout the electrical grid including utility offices, generation, transmission, distribution domains, and into customers' premises. These networks, hence, are expected to involve a hybrid mix of different standards and technologies to provide connectivity solutions in different environments and to accommodate widely variable data requirements. Considering the long lifespan required of utility communications equipment (at least a decade), connectivity solutions have to be studied carefully and evaluated for different smart grid applications. This chapter has provided an overview of communications and access technologies, and the pros and cons of their application in smart grid communications. As we have discussed, there is a clear trend in the industry towards open, standardized network architecture and communications solutions in order to ensure interoperability. In recent years, there have been extensive national and international efforts such as the one led by the NIST, to identify communications technologies suitable for smart grid applications and offer solutions to fill the standardization gaps. To provide a reference for interested professionals, we have reviewed most relevant PLC and wireless communications standards and their characteristics. In addition, a number of applied networking schemes including public vs. private infrastructure, IP networking, sensor networks, and M2M communications have been discussed.

References

[1] United States, Department of Commerce, National Institute of Standards and Technology, Office of the National Coordinator for Smart Grid Interoperability, NIST Special Publication 1108, NIST Framework and Roadmap for Smart Grid Interoperability Standards, Release 1.0, January 2010.

[2] IEEE P2030, Draft Guide for Smart Grid Interoperability of Energy Technology and Information Technology Operation with the Electric Power System (EPS), and End-Use Applications and Loads, Draft. 4.1, February 2011.

[3] F. Goodman et al., Technical and system requirements for advanced distribution automation, Electrical Power Research Institute (EPRI) Technical Report 1010915, June 2004.

[4] H. Farhangi, 'The path of the smart grid', *IEEE Power Energy Magazine*, vol. 8, no. 1, pp. 18–28, January/February 2010.

[5] S. Galli, A. Scaglione, and Z. Wang, 'For the grid and through the grid: the role of power line communications in the smart grid', *Proceedings of the IEEE*, vol. 99, no. 6, pp. 998–1027, 2011.

[6] S. Galli, A. Scaglione, and K. Dostert, 'Broadband is power: Internet access through the power line network', *IEEE Communications Magazine*, vol. 41, no. 5, pp. 82–83, May 2003.

[7] S. Galli and O. Logvinov, 'Recent developments in the standardization of power line communications within the IEEE', *IEEE Communications Magazine*, vol. 46, no. 7, pp. 64–71, July 2008.

[8] V. Oksman and S. Galli, 'G.hn: the new ITU-T home networking standard', *IEEE Communications Magazine*, vol. 47, no. 10, pp. 138–145, October 2009.

[9] V. C. Gungor and F. C. Lambert, 'A survey on communication networks for electric system automation', in *Computer Networks Journal (Elsevier)*, vol. 50, pp. 877–897, May 2006.

[10] K. C. Budka, J. G. Deshpande, T. L. Doumi, M. Madden, and T. Mew, 'Communication network architecture and design principles for smart grids', *Bell Labs Technical Journal, Special Issue: Green Information and Communications Technology (ICT) for Eco-Sustainability*, vol. 15, no. 2, pp. 205–227, September 2010.

[11] A. Goldsmith, *Wireless Communications*. Cambridge University Press, 2005.

[12] H. C. Ferreira, H. Grove, O. Hooijen, and A. J. H. Vinck, 'Power line communication', in *Encyclopedia of Electrical and Electronics Engineering* (ed. J. Webster), John Wiley & Sons Ltd, 1999, pp. 706–716.

[13] H. Latchman, 'PLC standardization by industrial alliances', in *Power Line Communications: Theory and Applications for Narrowband and Broadband Communications over Power Lines* (eds H. C. Ferreira, L. Lampe, J. Newbury, and T. G. Swart), John Wiley & Sons Ltd, 2010.

[14] HomePlug Powerline Alliance, HomePlug Green PHY whitepaper [online]. Available at: http://www.homeplug.org/tech/whitepapers/HomePlug_Green_PHY_whitepaper_100614.pdf

[15] P. Amirshahi, F. Cañete, K. Dostert, S. Galli, M. Katayama, and M. Kavehrad, 'Channel characterization', in *Power Line Communications: Theory and Applications for Narrowband and Broadband Communications over Power Lines* (eds H. C. Ferreira, L. Lampe, J. Newbury, and T. G. Swart), John Wiley & Sons Ltd, 2010.

[16] European Committee for Electrotechnical Standardization (CENELEC), 'Signalling on low voltage electrical installations in the frequency range 3 kHz to 148.5 kHz', Standard EN 50065-1, 1991.

[17] I. Berganza, A. Sendin, and J. Arriola, 'PRIME: powerline intelligent metering evolution', *IET-CIRED Seminar: SmartGrids for Distribution*, pp. 1–3, 2008.

[18] K. Razazian, M. Umari, A. Kamalizad, V. Loginov, and M. Navid, 'G3-PLC specification for powerline communication: overview, system simulation and field trial results', in *Proceedings of IEEE International Symposium on Power Line Communications and Its Applications (ISPLC)*, pp. 313–318, Rio de Janeiro, Brazil, March–April 2010.

[19] I. B. Valmala, G. Bumiller, and A. S. Escalona, 'PLC smart grid systems', in *Power Line Communications: Theory and Applications for Narrowband and Broadband Communications over Power Lines* (eds H. C. Ferreira, L. Lampe, J. Newbury, and T. G. Swart), John Wiley & Sons Ltd, 2010.

[20] K. Razazian, A. Kamalizad, M. Umari, Q. Qu, V. Loginov, and M. Navid, 'G3-PLC field trials in U.S. distribution grid: initial results and requirements', in *Proceedings of IEEE International Symposium on Power Line Communications and Its Applications (ISPLC)*, pp. 153–158, Udine, Italy, April 2011.

[21] A. Arzuaga, I. Berganza, A. Sendin, M. Sharma, and B. Varadarajan, 'PRIME interoperability tests and results from field', in *Proceedings of IEEE International Conference on Smart Grid Communications (SmartGridComm)*, pp. 126–130, Gaithersburg, MD, USA, October 2010.

[22] A. Treytl, T. Sauter, and G. Bumiller, 'Real-time energy management over power-lines and Internet', in *Proceedings of International Symposium on Power Line Communications and Its Applications (ISPLC)*, Zaragossa, Spain, March–April 2004.

[23] G. Bumiller, L. Lampe, and H. Hrasnica, 'Power line communications for large-scale control and automation systems', *IEEE Communications Magazine*, vol. 48, no. 4, pp. 106–113, April 2010.

[24] Panel on standardization, *IEEE International Symposium on Power Line Communications and Its Applications (ISPLC)*, Udine, Italy, April 2011 [online]. Available at: http://www.ieee-isplc.org/2011/programme.html#panels

[25] NIST Priority Action Plan (PAP) 15: Harmonize power line carrier standards for appliance communications in the home, 'Coexistence of narrow band power line communication technologies in the unlicensed FCC band [Online]. Available: http://collaborate.nist.gov/twiki-sggrid/pub/SmartGrid/PAP15PLCForLowBitRates/PL_coexistence_paper_rev3.doc

[26] A. Wheeler, 'Commercial applications of wireless sensor networks using ZigBee', *IEEE Communications Magazine*, vol. 45, no. 4, pp. 70–77, April 2007.

[27] C. Gomez and J. Paradells, 'Home automation networks: a survey of architectures and technologies', *IEEE Communications Magazine*, vol. 48, no. 6, pp. 92–101, June 2010.

[28] Z. M. Fadlullah, M. M. Fouda, N. Kato, A. Takeuchi, N. Iwasaki, and Y. Nozaki, 'Toward intelligent machine-to-machine communications in smart grid', *IEEE Communications Magazine*, vol. 49, no. 4, pp. 60–65, April 2011.

[29] D. M. Dobkin and B. Aboussouan, 'Low power Wi-Fi (IEEE 802.11) for IP smart objects [online]. Available at: http://www.gainspan.com/docs2/Low_Power_Wi-Fi_for_Smart_IP_Objects_WP_cmp.pdf

[30] M. Anderson, 'WiMax for smart grids', *IEEE Spectrum*, vol. 47, no. 7, pp. 14, July 2010.

[31] NIST Priority Action Plan 2, Guidelines for Assessing Wireless Standards for Smart Grid Applications version 1, 2011 [online]. Available at: http://collaborate.nist.gov/twiki-sggrid/pub/SmartGrid/PAP02Objective3/NIST_PAP2_Guidelines_for_Assessing_Wireless_Standards_for_Smart_Grid_Applications_1.0.pdf

[32] D. M. Laverty, D. J. Morrow, R. Best, and P. A. Crossley, 'Telecommunications for smart grid: backhaul solutions for the distribution network', *IEEE Power and Energy Society General Meeting*, pp. 1–6, 25–29 July 2010.

[33] A. Ghassemi, S. Bavarian, and L. Lampe, 'Cognitive radio for smart grid communications', in *Proceedings of IEEE International Conference on Smart Grid Communications (SmartGridComm)*, pp. 297–302, 4–6 October 2010.

[34] Spectrum Bridge, The future is now: nation's first smart grid TV white space network trial [online]. Available at: http://www.spectrumbridge.com/WhiteSpacesSolutions/success-stories/plumas.aspx

[35] Q. D. Vo, J.-P. Choi, H. M. Chang, and W. C. Lee, 'Green perspective cognitive radio-based M2M communications for smart meters', in *Proceedings of 2010 International Conference on Information and Communication Technology Convergence (ICTC)*, pp. 382–383, 17–19 November 2010.

[36] ZigBee+HomePlug Joint Working Group Smart Energy Profile Market Requirements, 2009 [online]. Available at: http://www.zigbee.org/imwp/download.asp?ContentID=16081

[37] ZigBee Smart Energy Profile 2.0 Technical Requirements Document, March 2010 [online]. Available at: http://www.zigbee.org/Standards/Downloads.aspx#821

[38] R. Merritt, 'Test body formed for SEP 2 spec', *EE Times*, August 2nd 2011 [online]. available at: http://www.eetimes.com/electronics-news/4218436/Test-body-formed-for-SEP-2-spec

[39] P. Swire, 'Smart grid, smart broadband, smart infrastructure: melding federal stimulus programs to ensure more bang for the buck', Center for American Progress, April 2009 [online]. Available at: http://www.americanprogress.org/issues/2009/04/pdf/smart_infrastructure.pdf

[40] Z. Shelby and C. Bormann, *6LoWPAN: The Wireless of Embedded Internet*. John Wiley & Sons Ltd, 2009.

[41] SGIP NIST Collaboration Site, PAP01. http://collaborate.nist.gov/twiki-sggrid/bin/view/SmartGrid/PAP01InternetProfile

[42] V. C. Gungor, B. Lu, and G. P. Hancke, 'Opportunities and challenges of wireless sensor networks in smart grid', *IEEE Transactions on Industrial Electronics*, vol. 57, no. 10, pp. 3557–3564, October 2010.

[43] S. Krishnamurthy, J. Falco, and K. Kent, 'Guide to Supervisory Control and Data Acquisition (SCADA) and Industrial Control Systems Security', NIST Technical Report, September 2006 [online]. Available at: http://www.cyber.st.dhs.gov/docs/NIST%20Guide%20to%20Supervisory%20and%20 Data%20Acquisition-SCADA%20and%20Industrial%20Control%20Systems%20Security %20(2007).pdf

[44] D. Niyato, L. Xiao, and P. Wang, 'Machine-to-machine communications for home energy management system in smart grid', *IEEE Communications Magazine*, vol. 49, no. 4, pp. 53–59, April 2011.

[45] Bluetooth Special Interest Group (SIG). http://www.bluetooth.com/Pages/Low-Energy.aspx

6 Machine-to-machine communications in smart grid

Jesus Alonso-Zarate, Javier Matamoros, David Gregoratti, and Mischa Dohler

6.1 Introduction

This chapter reviews the emerging paradigm of machine-to-machine (M2M) communications in the context of smart grids. Commencing here with an introduction to the topic at hand, we then introduce in subsequent sections available M2M communications technologies as well as the applicability of said technologies. We then dwell in greater detail on M2M architectural standards bodies, such as ETSI M2M and 3GPP MTC. We finally position the use of M2M in smart grids and identify open challenges for a symbiotic development of both technologies.

A machine-to-machine network is defined to be a network formed by devices that communicate with each other without (or with very little) human intervention in order to accomplish some specific task(s). The prime driver for this networking paradigm is the ability of a large number of devices/machines to execute tasks in an autonomous (and often distributed) manner which is beyond the ability of humans. From a technical point of view, it requires the system to be scalable, power-efficient, autonomous, intelligent; among many other properties, some of which are discussed below. Indeed, as highlighted throughout this book as well as below in Section 6.5, a huge number of points in the power grid need to be constantly monitored and controlled to ensure smart operation of the system.

Although the above design aims have been conceptually the core to various prior networking design efforts, the idea of M2M is currently receiving great attention from both academia and industry. Having a world of interconnected and automated devices facilitates the development of unprecedented applications and breaks with the current human-driven and human-oriented (and thus saturated) markets.

Predictions for growth in the M2M domain are thus optimistic. However, the degree of optimism varies greatly: the Wireless World Research Forum envisions a total of 7 trillion (7,000,000,000,000) devices by 2017 [1], other market research such as [2] suggests 'only' billions of M2M devices. While the difference in orders of magnitude is partially due to the fact that there is no widely accepted definition of M2M, it is fairly irrelevant since M2M technology is happening as we speak and will surely be a ubiquitous part of the communication landscape of the 21st century.

As suggested by its name, machine-to-machine comprises three core components:

1. *Machine*. On one end, there is a device (e.g., electricity meter) which is monitored by means of a sensor (in 'uplink's) or a device (e.g., a switch) which is instructed to actuate (in 'downlink').
2. *'To'*. This is the networking part which facilitates seamless and autonomous end-to-end connectivity between machines.
3. *Machine*. On the other end of the communication link, there is a device (e.g., a computer) which extracts, processes and displays the information gathered from the monitoring devices in order to make decisions and, possibly, send instructions to the actuators to perform some task(s).

While sensor and control software are an important constituent in an M2M system, of prime interest in the context of smart grid is the communication part. In this chapter, we will thus focus on the communication and networking part of M2M in the context of smart grids.

The design and growth of M2M system has been driven by the following high-level requirements:

- *Number of nodes*. Possibly the strongest differentiator with regard to current systems is the need and ability to support a huge amount of devices, which stretches well beyond the currently used cellular or Wi-Fi users. For example, the use of an automated meter in each household or an automated phasor monitor in critical points of the grid, quickly yields a huge number of nodes to be supported by the system.
- *Dispersive applications*: These devices are being driven or driving a very wide spectrum of applications, some of them being critical in delay, others critical in security, etc. For example, automated meters may only need to report their non-critical readings every few minutes, whereas phasor monitoring needs to be done several times per second and data delivery ought not to fail.
- *Affordable cost*. None of the above can be accomplished if cost is not kept to a minimum. It implies that M2M technology must be more affordable at the same or better technical characteristics than any alternative approach.

All of these have a strong percussion on the technical design requirements, which lead to the prime novelty of M2M system design and are summarized as follows:

- *Autonomous operation*. One of the most important technical design drivers is the need to facilitate a truly autonomous operation. This is due to the fact that the large number of devices cannot be serviced by humans as is traditionally done in human-centric networks. Autonomy here includes auto-configuration, auto-optimization, auto-healing, scalability, etc. For example, with tens of thousands of meters installed in an urban district, a utility company has no means of attending to each of these meters individually – the only viable way forward is to guarantee autonomous operation.
- *Mobility and remote operation*. Driven by many M2M applications, the system must support some degree of mobility and must be usable in remote regions. A current network which supports this requirement is the cellular network; it is hence no surprise

that cellular operators see an enormous opportunity, at the caveat of requiring a major technical overhaul of their cellular networks. For example, many places in the grid which need to be monitored are in remote locations with little networking connectivity, apart from the typically available cellular network.

- *Comparably low rate.* The very large majority of M2M applications requires fairly low communication rates, rarely exceeding hundreds of kilobits per second (kbps). This contrasts very much with current trends in human-centric communication networks, where capacity densities of 1 Gbps/km^2 [3] are needed. For instance, automated metering and associated demand–response systems rarely exceed rates of 100 kbps per link; however, note that the large number of devices yield aggregated rates which may be well beyond human-centric networks.
- *Critical delay requirements.* Some applications are very critical in delay, i.e., the M2M readings need to be delivered with a hard-delay constraint which is often significantly more stringent than current delay-sensitive human-centric networks. For instance, phasor readings in the grid are considered critical and need to be reported within tens of milliseconds (ms) to be able to react meaningfully to any impeding outage.
- *Highly energy efficient.* Due to the unattended field deployment, power supply is not always guaranteed and, if it is not, batteries cannot be replaced on a regular basis. A major design driver is hence the energy efficiency of the communication system. For instance, while electric meters can naturally be powered by the underlying power grid, gas and water meters may not enjoy this luxury and hence need a highly efficient M2M system.
- *Highly secure.* Again, due to unattended and remote field deployments, the devices as well as the communication links must be highly secure. For instance, to avoid false outage reports from the phasor monitors in the grid, security is one of the major technical design drivers.

Without depicting the novelty along all of the above dimensions, Figure 6.1 illustrates the positioning of M2M systems w.r.t. current communication systems along range and

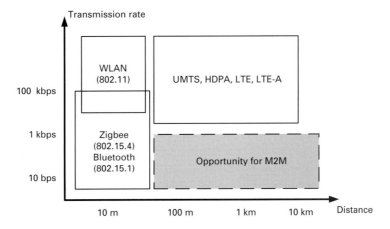

Figure 6.1 The low-rate and long-range communication opportunity for M2M.

rate requirements. Notably, a short-range high rate is provided by IEEE 802.11-like wireless local-area networks (WLANs), which have not been designed to operate over large distances nor to be very energy efficient. A short-range low rate is provided by IEEE 802.15.4-like wireless sensor networks (WSNs), which have not been designed to operate over large distances. A long-range high rate is provided by cellular systems, which have not been designed to be very energy efficient. Finally, a long-range low-rate is provided by M2M networks, which ought to fulfil the above technical requirements to be able to viably support M2M applications.

However, there is no optimized technology capable of offering a cost and energy-efficient technological solution to date to meet the above requirements. This remains one of the most important challenges and is briefly treated in the subsequent section.

6.2 M2M communications technologies

6.2.1 Wired vs. wireless

The different networking technologies essentially boil down to wired and wireless ones, as well as hybrids thereof. In the context of smart meters, this has been illustrated in Figure 6.2, and is known to trade the following:

- *Wired M2M*. A wired solution physically connects sensors and actuators to a network/Internet-enabled gateway by means of dedicated cables, power-line communication, Ethernet cables, etc. It provides an extremely reliable communication channel, typically at very high transmission rates, low delays, and extremely high security. Information cannot be overheard by a third party unless cables are physically intercepted. However, although wired solutions are cheap to maintain, their deployment might be an important entry barrier for some M2M applications. The main drawback of wired solutions is the lack of mobility and scalability, since the addition of new devices to an M2M network requires the cabling of the new devices. Traditional SCADA systems, the early ancestor of M2M, are almost exclusively based on cabled systems. In the context of smart grids, critical decision points will continue being cabled due to security, reliability, and delay reasons.
- *Wireless capillary M2M*. The wireless capillary embodiment is a very cost-effective solution in that radios are fairly cheap and no subscription is needed. On the downside, multihop operation is often needed to reach the gateway which introduces delay and reduces effective bandwidth, among others. In addition, gateways need to be installed at fairly specific locations and mobility is not supported, which harms flexibility. In the context of smart grids, capillary networks are in use for smart metering and are likely to be used for the monitoring of non-critical parts of the grid.
- *Wireless cellular M2M*. The cellular embodiment is a very convenient solution since it guarantees virtually complete coverage, mobility, roaming, hand-overs, subscription management, etc. On the downside, as of today the technology is not ready yet since power consumption, delays, and other technological key issues remain a concern. In addition, a subscription has to be paid which makes the solution inherently more

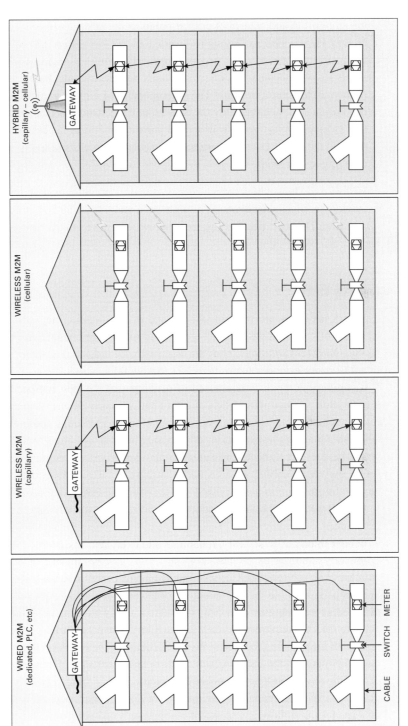

Figure 6.2 Wired, wireless, and hybrid solutions for M2M communications.

expensive. In the context of smart grids, cellular M2M networks may prove vital for the monitoring and control of critical points in the grid.
- *Hybrid M2M solutions.* Trying to leverage on the strength of all technologies whilst mitigating their weaknesses, a hybrid approach is common in real-world M2M deployments today. This being so, many standards bodies have adapted the hybrid approach in their architectural designs. The heterogeneity of a hybrid solution, however, may become an issue and therefore the definition of the standards that set the rules to drive this heterogeneous scenario is a must for the long-term success of M2M applications. In the context of smart grids, hybrid solutions are in use today mainly for smart metering applications.

History will show which technological embodiment will prevail, but with the cellular technology sufficiently advanced it might be a real contender for global M2M market dominance. This will become apparent below where we contrast mainly the capillary and cellular M2M solutions.

6.2.2 Capillary M2M

Capillary networks are medium/short-range communication networks that allow the coverage of cellular networks to be extended to provide connectivity to the machines forming the M2M solutions. They are essentially the industrial adaptation of the concept of wireless sensor and actuator networks (SANETs).

A capillary network is composed of a number of devices equipped with a sensor, a radio chip, a micro-controller, and an energy supply. These devices are low cost, low complexity, low size, and very energy efficient. Although these are requirements desirable for any communications device, they become essential for M2M applications, where devices are expected to run for many years without human intervention. The (often multihop) communication between these devices is essential in order to convey the information to a centralized server (or to a gateway that provides connectivity to the Internet to reach the server), which can process the data and make appropriate decisions.

Driven by the commercial opportunities but the lack of suitable standards, several proprietary solutions have emerged in past years, such as those commercialized by the companies Dust Networks, Coronis, Worldsensing, CrossBow, Sensinode, among many, many others. To avoid the fragmentation of the market and to enable the interoperability of different technologies, standard bodies are actively working to define standards adequate to meet the requirements of M2M capillary networks. Among the different bodies, it is worth highlighting the role of the IEEE (defining the PHY and the MAC layers) and the IETF (defining the network and transport layer protocols). They face the challenges of needing to support a huge number of devices, most of them duty-cycled; the presence of strong fading and interference in bands of choice; short communication ranges and thus need for multihop; and stringent requirements on energy efficiency.

Without being exhaustive, the favourite standards at PHY and MAC layers trying to address the above are as follows:

- *IEEE 802.15.4 (ZigBee) and .15.4e/g/k*. The obvious technology is the ZigBee-like products which rely on the PHY and MAC of the IEEE 802.15.4-2006 standard [4]. The design has explicitly considered energy efficiency over short communication ranges. However, it suffers from many industrial short-comings, such as many serious security vulnerabilities, no ability to duty-cycle relaying nodes, no support for routing, inability to handle interference, and fading in an efficient manner. For this reason, the .15.4e working group has created a standard which overcomes these shortcomings and is likely to find its applicability in the M2M arena in upcoming years. In addition, the .15.4g, dealing with smart utility networks (SUN), and .15.4k, dealing with low-power critical infrastructure monitoring, are viable M2M extensions and certainly of use in smart grid deployments.
- *IEEE 802.11 (Wi-Fi)*. While ZigBee is commonly mentioned in the context of smart grid, the emergence of low-power Wi-Fi [5] should not be underestimated. From a technology point of view, it is Wi-Fi-compliant but with a significantly thinned protocol stack which allows the saving of energy and thus lifetime extensions. From a deployment point of view, it enjoys significant benefits over ZigBee networks in that coverage is largely guaranteed through already available networks and larger communication distances are generally supported, among others. Therefore, once these type of networks are able to be fully duty-cycled, they will be serious competitors to ZigBee-like networks.

If IP (Internet Protocol) and/or multihop communication is required to ensure end-to-end connectivity and coverage, then IETF standardization developments are pertinent. Notably, for IPv6 connectivity and multihop routing through the capillary M2M network, the following working groups are of interest:

- *IETF 6LoWPAN*. This working group [6] has enabled IPv6 connectivity to embedded nodes (mainly using IEEE 802.15.4) and thus seamless end-to-end connectivity without the need for proprietary gateways and network address translation (NAT). It is achieved by means of IPv6 header compression and packet fragmentation/defragmentation. The protocol also caters for neighbourhood discovery and has also lately aimed for a tailored routing protocol. In the light of the smart grid, it essentially allows a utility to manage its monitoring and control M2M network the same way as it has done with its IT infrastructure.
- *IETF ROLL*. This working group [7] has enabled routing through embedded M2M networks suffering from unreliable channels as well as resource constraints. A further challenge successfully addressed in this working group is that the traffic flow is mainly converge-cast in that information typically needs to be routed from many sensors to a single or few gateways. The designed routing protocol is highly adaptive in that if a link suffers an outage, the network automatically reconfigures to find the next best path. In the context of the smart grid, it means that various meters or control sensors can be networked without compromising end-to-end connectivity.
- *IETF CoRE*. This working group [8] provides a framework for resource-oriented applications intended to run on constrained IP networks. It essentially facilitates applications to run end-to-end from an Internet client to the embedded M2M node. It is

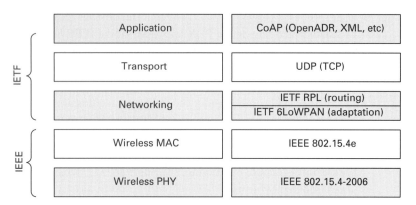

Figure 6.3 Most likely industrial protocol stack for M2M capillary multihop networks.

a needed step to run feed-based middleware platforms, such as provided by [9], and started to be used by smart grid utility companies.

Based on the above, a likely M2M capillary multihop protocol stack is shown in Figure 6.3 which is likely to prevail for some time to come. Note that IPv6 means more than just the basic networking capabilities: it is a great asset in guaranteeing global reachability, true scalability, reliable security and, since IP-enabled networks have been successfully engineered and deployed for decades now, the same engineering skills can be used for maintaining and troubleshooting these types of emerging networks. This is an enormous advantage over proprietary solutions and allows utilities to capitalize on a similar workforce as for their IT networks.

Note, however, that lately a trend is emerging facilitating communication in a single hop for M2M while consuming little power. The result is new standards and systems, such as ANR+ [10], DASH7 [11], and the Wavenis Open Standards Alliance [12] (which has recently liaised with the European Smart Metering Industry Group (ESMIG)) as well as cellular M2M systems to be discussed subsequently.

6.2.3 Cellular M2M

There are a few major disadvantages that embedded M2M suffer, such as lack of global coverage, global interference governance, support for roaming and handovers, etc. While any issues related to mobility are not a major issue in smart grid, the uncontrolled interference and global coverage have and will be preventing a major roll-out of M2M solutions for smart grid applications. A technology which facilitates interference control and enables global geographical coverage has successfully been designed and used for decades: cellular systems. It is no surprise that M2M has found solid support within the cellular community, with many standards bodies aiming for a viable technology solution.

Historically, global system for mobile communications (GSM) broke into the market in 1990, providing a cellular network infrastructure designed for the transmission of voice calls. This technology showed an extraordinary market penetration and today provides connectivity worldwide. With the emergence of the Internet by the mid-1990s, the transmission of data and the sense of ubiquitous connectivity has become almost a commodity for humans, who demand more and more data transmission capabilities and some degree of guaranteed quality-of-service (QoS). Unfortunately, GSM was designed for the transmission of voice, and its capability to provide efficient data transmission was very limited.

The response of the standards bodies towards this emerging need was the continuous evolution of the cellular technology. General packet radio service (GPRS) was the first evolution of GSM to provide increased data transmission capabilities, raising the concept of the 2.5 generation of mobile communications (2.5G). Then came the third generation networks (3G), with technologies such as enhanced data rates for GSM evolution (EDGE), universal mobile telecommunications systems (UMTS), CDMA2000, and WiMAX. High-speed downlink packet access (HSDPA) is often referred to as 3.5G, offering increased data transmission capabilities still. Even though the data transmission capabilities with 3G and 3.5G are considerably higher than the capability of GSM, the increasing demand for multimedia contents and interactive applications such as social networking still feeds the need for increased data rates and higher performance. This has been the motivation for the fourth generation (4G) of cellular mobile communications, with envisioned transmission rates of up to 1 Gbps in the downlink. The evolution of cellular networks has been driven by the need for higher transmission rates and increased QoS to meet the increasing demands of humans to use multimedia and interactive applications.

However, M2M networks pose a completely different scenario. Although the range of M2M applications is typically large with requirements substantially varying for different applications, it is clear that machines have entirely different requirements from humans. For instance, in most applications, devices will need to transmit a few data bits; in some cases, this transmission will be very delay-tolerant, but in some other cases, this little information will be critical and must be transmitted with strict timing requirements. Cellular networks have not been designed for this purpose, and it is necessary to redesign the architecture and access networks of cellular systems to meet the requirements of the impending growing market of M2M, one of which is the smart grid market.

The motivation to make this effort is twofold; first of all, there is an enormous potential market that attracts mobile operators and manufacturers. The cellular market is reaching saturation in developed countries, and thus equipping devices with communication capabilities is very attractive. From a more technological point of view, cellular networks offer almost ubiquitous coverage and users are already familiar with this kind of communication system, thus making it appealing for inclusion in day-to-day lives. Therefore, although cellular networks have played in the past an indirect role (providing a pipe for data transmission), the role of network providers is to become a more active player in the M2M arena. Pioneering companies here were the Swedish company Maingate, who commenced as early as 1998 with cellular M2M solutions, followed by giants Ericsson and Nokia.

Today, cellular M2M standardization is seen as the only facilitator for a truly global growth of this technology. While discussed in greater detail in the subsequent section, the main challenges that the cellular community currently faces in order to accommodate M2M for smart grids within the current cellular communication architecture are:

- *Reduce complexity*. To enable cheap, long-lasting M2M devices in the field, it is necessary to allow simple detection of radio frequency (RF) signals in the downlink. For the very same reason, transmission in the uplink ought ideally to have constant envelope in order to be able to use simple and cheap amplifiers for transmission. Current cellular systems constitute a total overkill for such M2M systems, making it a very inefficient choice.
- *Emphasis on uplink rate*. Although cellular systems have been designed with major focus on the downlink, M2M applications are likely to show a different behaviour from humans by having stronger requirements for the uplink than for the downlink. Indeed, the downlink traffic associated with M2M is expected to be low (e.g., instructions to devices), and thus may be integrated in the downlink control channel without requiring dedicated resources for M2M traffic. M2M devices in the smart grid, however, may report an important alert in a correlated manner, yielding sudden spikes of uplink data which need to be catered for.
- *Provision of hard-delay*. End-to-end communication delays are generally tolerable but there are some smart grid applications, mostly related to critical monitoring and control, which require a tight hard delay that is beyond current cellular designs. Another issue is the ability to connect quickly once powered on because most M2M devices are likely to be duty-cycled to save power; however, cellular networks yield connection delays which are well beyond the tolerable.
- *Architecture and coexistence*. To accommodate smart grid requirements with many nodes monitoring wide areas in the grid or providing metering data in an area, the cellular networking architecture may need to be redesigned to be capable of handling this high number of nodes, managing devices into groups, etc. Furthermore, given that a heterogeneous technology roll-out in the context of smart grids seems very likely in the medium-term time frame, with ZigBee-type and wireless local-area networks playing a role as important as cellular networks, coexistence issues need to be clarified. This does not primarily pertain to interference but rather to data flow management.
- *Billing*. Finally, M2M introduces a new market where new business models have to be defined. It seems clear that existing models for charging humans do not fit well the nature of M2M networks, which show a very low average revenue per unit (ARPU). Hence, billing issues (mainly in the triangle of users, grid utility, and cellular service provider) need to be resolved.

These are very stringent architectural requirements, but key to the success of global M2M uptake.

6.3 M2M applications

The application range of M2M communications is vast, with current markets pertaining to telemetry but currently expanding into smart grids, smart metering, smart cities, etc.

There is no doubt today that M2M communications will be a key technology in the deployment of smart grids. Since the release of the National Broadband Plan in March 2010 [13], smart grids have received a lot of attention from both academia and industry. Parts of this report outline how ICT can improve the efficiency of the electric grid, reduce dependency on natural resources, and improve the distribution of the electricity. The use of automated communication devices monitoring the state of the grid and the instantaneous use of the resources is essential to attain the objective of achieving a truly smart grid. Indeed, the core part of the remainder of this chapter will be devoted to the applicability of M2M communications to smart grids.

Core to the front-end of smart grids is the use of smart meters, which not only pertains to electrical but to utility meters (gas, water, etc.) in general. A smart and automated reading of meters is beneficial for all involved parties: consumers can accurately track their consumption and thus learn to be more efficient; on the other side, utilities, besides avoiding the manual reading and reporting of meters, can better dimension the production of resources, thus leading to cost savings.

Extending the deployment of sensors and actuators out of the utility framework, and applying it to public spaces has led to the concept of smart cities. Automated parking search, public light control, container-level monitoring, notification in case of incident, warning in case of vandalism and security threat, are some example applications that could make cities more efficient by enabling communications between devices.

Extending the above vision to a more localized domain has given birth to the concept of smart building. Populating buildings with sensors measuring temperature, humidity, presence, quality of air, etc. can help improve well-being and even the productivity of business.

Another example of application for M2M is in the automotive sector, where M2M communications enable remote monitoring, in-vehicle diagnosis, or car-to-car communication to improve safety on the road. Indeed, the European Union has launched the eCall initiative [14] that targets the installation of SIM cards in each vehicle for automatic notification in case of accident, and to enable more sophisticated vehicle-based applications.

The use of automated communications among machines clearly increases the systems' efficiency. The ones described above are just some of the examples of the wide variety of applications that can benefit from M2M. A long road is still ahead in defining not only the technology towards making M2M a reality, but also in identifying and creating new application domains.

6.4 M2M architectural standards bodies

Given the emerging importance of cellular M2M systems, several standard bodies and associations around the world are actively defining suitable standards for M2M cellular mobile communications. The definition of standards is essential for the long-term development of M2M technology because there is a clear market demand to have standards-compliant solutions that can interoperate with each other, regardless of the subscribed operator or equipment manufacturer. The list of interested bodies is long, but there are clearly two key players: the European Telecommunications Standard Institute (ETSI), original promoter of GSM, and the 3rd Generation Partnership Project (3GPP), a global initiative promoting mobile standards such as UMTS (and subsequent evolutions) of LTE and LTE-Advanced.

6.4.1 ETSI M2M

In 2009, the ETSI created a dedicated technical committee (TC) to develop a standardized architecture for M2M communications. ETSI aims to shift the existing proprietary vertical applications using dedicated devices, to a framework where different applications share a common infrastructure and network elements and can thus interact with each other.

ETSI is working towards the definition of six technical reports. Some of them have already been published as of Q2 2011, with the remaining ones planned to be released in upcoming months. One of these documents is devoted to M2M definitions, while the other five are separated into the main envisioned applications: smart metering, eHealth, connected consumer, automotive, and city automation. The output of these reports will be used to create two technical specification documents, one devoted to define the M2M service requirements, and another one devoted to define the M2M functional architecture.

The vision for the M2M architecture is illustrated in Figure 6.4 [15]. The core network is largely based on existing specifications of the respective cellular systems, with the addition of support for M2M functionalities. The true visionary contribution, however, comes from the architectural design in the device domain. Here, the support and coexistence of different short-range capillary M2M technologies is considered. For the first time, a cellular system has to be designed considering the cooperation of different technologies, such as IEEE 802.15.4, IEEE 802.15.1 or IEEE 802.11, just to mention a few.

In this architecture, two options are considered for the interconnection of the device domain with the network domain:

- *Hybrid capillary and cellular M2M*. Devices might create a short-range M2M network with a gateway providing connectivity to the core network (which is the hybrid deployment architecture alluded to above in this chapter).
- *Purely cellular M2M*. Devices connect to the core network through a cellular interface, each of the devices equipped with a cellular network interface, SIM card, etc.

This second option has triggered an active discussion about the subscriber identity module (SIM) card. If devices are to be equipped with direct cellular connectivity, the concept of the SIM card has to be revisited for several reasons:

- *Harsh operating conditions*. Some industrial applications may have to work in harsh environments, where SIM cards have to resist severe temperature or humidity conditions. Therefore, a more robust design of SIM card is a must.

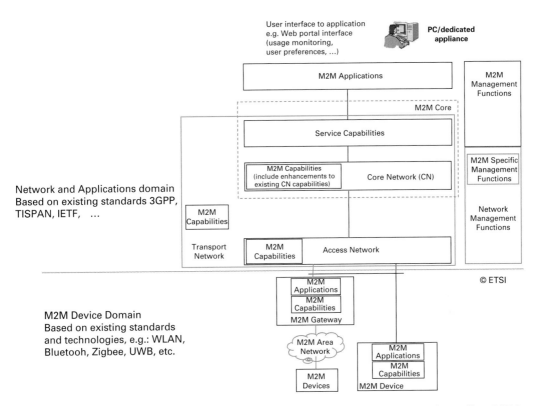

Figure 6.4 ETSI's M2M architecture, where the true novelty lies in the explicit support for capillary M2M devices.

- *SIM card size.* The current size of SIM cards may be too big for small sensors or actuators. Smaller SIM cards, if inserted or soldered, have hence been designed to meet these industrial demands.
- *Soft vs. hard SIM.* Today, the SIM card is bound to the operator. In M2M applications potentially involving several thousands of devices, it is not viable to replace all the SIM cards in all the devices due to, for example, a change in the operator providing the service. This issue has opened the discussion on the suitability of software-based SIM identification rather than the traditional card approach.

As of early 2011, ETSI M2M has also commenced to explicitly consider smart grid issues in the design process. How to take this forward has been discussed in great detail in the first ETSI workshop on smart grid standardization at the ETSI Headquarters, Sophia Antipolis, France, 13 April 2011. It has been recognized that developing and deploying the smart grid requires close collaboration between the power industry and the ICT industry, using the best experience from both. Smart grids will contain a mix of legacy and new technologies integrated into a common architecture, enabling new services and applications such as integration of renewable energy sources, widespread adoption of electric vehicles, and better energy consumption management.

The European Commission (EC) has been establishing a Smart Grids Task Force and issuing a mandate to the three European standards organizations (ETSI, CEN, and CENELEC) to develop standards for smart grid. The three standards bodies are responding by working in partnership to coordinate their activities and take account of all industry needs.

ETSI M2M has henceforth commenced working on a possible M2M solution tailored to smart grids. Notably, the OCG Remote Consensus under ETSI coordination of Mandate M/490 for smart grids has been initiated. To this end, it has been proposed that ETSI TC M2M is appointed lead for the ETSI coordination of Mandate M/490 following principles outlined in the document ETSI/OCG(10)0026 available on the ETSI portal. Following the approach for Mandate M/441 it is expected, but not limited to, that the following ETSI TBs will become supporting/affected TBs of Mandate M/490: ATTM, ERM, PLT, SCP, TISPAN, and MSG. Negotiation about each player's involvement is still (Q2 2011) under way. ETSI M2M, through document M2M(11)0345, is also monitoring the ITU-T Smart Grid Focus Group, facilitating interoperability in the long run.

Another recent activity of ETSI M2M is the study of the impact that smart grids have on M2M platforms. It considers the application of the M2M architecture for the needs of smart grids. Based on the high-level framework developed by the ETSI Board (energy plane, control and connectivity plane, service plane), it will consider the applicability of the M2M platform to smart grids and derive a standard gap analysis: e.g., applicability of M2M APIs for smart grid applications, mechanisms to manage energy for end-users, etc. It is also foreseen to develop an annex containing the most relevant use cases. This initial M2M smart grid framework will be used as a basis for deriving initial high-level requirements, developing recommendations for future work within ETSI TC M2M, and highlighting dependencies and relationships with other work within ETSI TCs and other standards organizations including the ones covering the energy plane: NIST, CEN/Cenelec, and IEC in particular. Some alignment work has already commenced through document TR102.935.

6.4.2 3GPP MTC

In 2009, the 3GPP started to work actively on the definition of the requirements and means to integrate M2M capabilities in the existing 3GPP networking architecture. Two main documents had been released by the 3GPP back then: (i) TS22.368, defining the service requirements for machine-type communications (MTC), and (ii) TR23888, defining the system improvements for M2M. Therein, 3GPP has defined different communication architec tures, considering the case where many devices communicate with one or more servers, independently of the operator behind the application, and the case where devices have to communicate among them.

Traditionally, cellular systems have been designed with voice and later data calls in mind. For example, UMTS defines four classes of traffic, namely: conversational class, streaming class, interactive class, and background class. Each of these classes has its specific requirements, and the network architecture is able to handle them differently.

However, the inclusion of M2M functionalities in the architecture is a real challenge as the range of applications is very wide with the variety of requirements.

In order to cope with this heterogeneity of requirements, the 3GPP has defined a number of features (i.e., particular characteristics associated with certain applications), for which the network needs to be optimized. These features are as follows:

- *Low mobility.* This feature is suitable for MTC devices that do not move, move infrequently, or move only within a certain region. The lack of mobility allows for the simplification and reduction of the frequency of mobility management procedures. This is a feature which is of particular interest to smart grid applications, since most monitoring and control devices will be placed in a stationary environment.
- *Time controlled.* This is suitable for those applications that transmit and receive data during predefined time intervals and can thus avoid unnecessary signalling outside these time intervals. The network operator may allow such MTC applications to send/receive data and signalling outside these defined time intervals but charge differently for such traffic. This is also a feature which is of particular interest to those applications in the smart grid which do not require alert capabilities, such as smart metering.
- *Time tolerant.* This feature is suitable for MTC devices that can delay their data transfer. The purpose of this functionality is to allow the network operator to prevent MTC devices that are time tolerant from accessing the network (e.g., in case of radio access network overload). Again, this is a feature of interest to applications which require no critical transfer of data, such as metering.
- *Packet switched (PS) only.* This is intended to provide PS-only subscriptions with or without assigning an MSISDN (i.e., a telephone number). Remote MTC device triggering will be supported with or without assigning an MSISDN. Remote MTC device configuration will still be supported for subscriptions without an MSISDN. If used in large quantities, this is a feature of definite interest to M2M smart grid deployments.
- *Mobile originated only.* This feature is suitable for applications where it is possible to reduce the frequency of mobility management procedures per MTC device; the network shall provide a mechanism for the network operator to dynamically instruct the MTC devices to perform mobility management procedures only at the time of the mobile originated communications. While this is a feature which is of interest to many mobile M2M applications, it has little relevance to smart grids today. There is possibly an application in the context of electric vehicles, as will be discussed below.
- *Small data transmissions.* This is suitable for applications that have to transmit very little information and can thus ensure minimum signalling overhead and use of network resources. This feature is likely of interest to most smart grid applications since data volumes per entity are typically very low.
- *Infrequent mobile terminated.* This feature is suitable for devices that mainly utilize mobile originated communications, and thus enable the reduction of mobility control information. Again, due to low mobility, application of this feature is not of priority in smart grids.

- *MTC monitoring.* This is suitable for applications that require monitoring the state of all the devices and possible events that occur in the network. This is a feature vital to all critical M2M applications, as the utility needs to be sure that deployed devices are operational.
- *Priority alarm message (PAM).* This feature is suitable for devices that issue a priority alarm in the event of theft, vandalism, or other needs for immediate attention. This feature is of great use in applications which require attention but are not too critical (since very critical applications are likely to use other systems until delays have been addressed in cellular systems); an example is the detection of a leak which requires some valves or switches to be closed.
- *Secure connection.* This is suitable for MTC devices that require a secure connection between the MTC device and MTC server (even in the case of roaming). This is a must-have feature in all smart grid applications. Security is often underestimated but vital to the success of M2M in smart grids.
- *Location-specific trigger.* This is suitable for applications where MTC devices are known to be in a particular area and thus triggering can be done by using the location information. This feature could be of interest to applications pertaining to electrical vehicles.
- *Network-provided destination for uplink.* This feature is suitable for MTC applications that require all data from an MTC device to be directed to a network-provided destination IP address. This feature, with impact on core network, is of use if meters, etc. are reporting to the same and only utility platform.
- *Infrequent transmission.* This is suitable for applications with long periods between two subsequent data transmissions. This feature is less likely to be used in the context of smart grids.
- *Group-based policing and addressing.* This feature is suitable for applications where devices can be managed in groups for tasks. In the context of smart grids with M2M bulk deployments, this is of great use since meters and similar monitoring equipment can be handled on a group basis.

More features are likely to emerge over the design process. According to the 3GPP ongoing work, the system architecture will be designed to handle all these features to which M2M applications will be able to subscribe or unsubscribe depending on their needs. To facilitate the above features, the 3GPP needs to address the following challenges:

- *Addressing.* The current phone number allocation may not be suitable for M2M applications, which may need to make use of IPv6 addresses. Therefore, cellular networks should include mechanisms to connect IMSI-based with IP-based devices. This will become critical in smart grids if metering is to become reality through cellular M2M technologies.
- *Device triggering.* In the majority of current M2M deployments (e.g., based on SMS), devices can only be triggered when they are online and have a connection established. However, for the sake of energy efficiency and to reduce control overheads, it is

necessary to develop mechanisms to trigger a device regardless of whether it has an established connection or not.
- *Charging and billing.* The business model for M2M applications is not yet clear for mobile operators. The traditional way of charging does not seem to be the right approach, as some MTC applications will have a huge number of devices with the capability to transmit, but with just a few of them transmitting very little amounts of information. Therefore, although the infrastructure has to support a great number of potentially connected devices, the charge for these devices has to be affordable to the end-users which, in the context of smart grids, are the utilities.
- *Security.* MTC should offer the same security as regular voice and data traffic. Information conveyed by M2M applications might be protected in terms of confidentiality, integrity, and authentication (CIA), and must ensure privacy and include trust mechanisms. Provisioning of security is paramount to the success of cellular M2M applications.

These are currently, as of Q2 2011, actively being discussed within 3GPP standards bodies. Apart from ETSI M2M and 3GPP, there are other bodies currently dealing with the definition of communication architectures in smart grids, such as GSMA, WiMAX, WFA, OMA, TIA, CCSA NITS, and ITU; however, these often take a more holistic view with less emphasis on M2M communications and are thus not considered further. The interested reader is referred to [16–18].

6.5 M2M application in smart grid

In spite of local specificities, power grids mainly consist of four basic components: generation, transmission network, distribution network, and customers. Briefly, high-voltage transportation lines bring electricity from generation plants to distribution substations. Here, electricity is 'step-down' converted to medium and low voltage to be delivered to customers [19].

The smart grid concept refers to the integration of the power grid with communication and information technologies aimed at increasing performance and reliability. The term 'smart' comes from the fact that the power grid has to be an automated system that benefits from self-controlling and self-healing capabilities in case of failure. It is necessary to stress that this is completely aligned with what M2M communications are conceived for, i.e., communications between machines without human intervention.

We now discuss the application of M2M technologies to specific smart grid needs. We first dwell on the communication and service architecture which is suitable to the smart grid, before discussing more application-related matters in subsequent sections.

6.5.1 M2M architecture

As shown in Figure 6.5, the main constituents of a successful M2M system are the use of capillary and cellular M2M devices, feeding data into the operator's core network, which

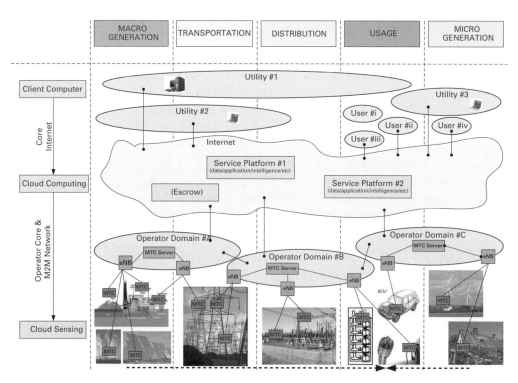

Figure 6.5   Holistic M2M architecture in the context of smart grids, stretching from generation to usage.

then gets delivered to the Internet where the utility's service platforms are placed. With the exception of some operators, it is deemed unlikely that the M2M service platforms are placed in the operator's core network; only M2M device control, billing, and similar functionalities will remain in the realm of operators. This is important for the scalability of smart grid services and third-party service provisioning.

As mentioned above, a grid is composed of energy generation, either at macro scale (left) or micro scale (right), the transportation and distribution networks, and the end-user consumption. M2M monitoring and control devices are typically rolled out in the field in the various parts of the grid, and report to the associated cellular base stations (eNB). These deliver the information to the control centres within the operator domain, the role of which is to control the M2M node, subscription, etc. The data is then typically delivered into a service platform placed or connected via the Internet. For sensitive data, such as metering data from private properties, the data may pass through an escrow-type system to decouple technical from private data. Finally, the data is displayed, used, analysed by the end-users, which are the various utilities, end-users, and other possible stake-holders.

In the light of the above M2M architecture, we shall subsequently discuss its applicability to various constituents of the smart grid with emphasis on transmission and distribution networks as well as end-user appliances.

6.5.2 Transmission and distribution networks

We describe and summarize here the most important communication requirements for smart grid applications at the transmission and distribution networks.

Wide-area situational awareness

One of the main challenges of the smart grid initiative is to increase the reliability of the power grid. This is in part motivated by the high costs derived from large blackouts. Roughly speaking, a large blackout occurs as a result of a cascading series of failures in the grid. An example of a large blackout is the one experienced in North America in August 2003, which caused losses on the order of billions of dollars [20]. Clearly, these situations occur due to the high interconnection and interdependence of the elements of the grid and the lack of wide-area situational awareness in today's power grid.

Wide-area situational awareness refers to the monitoring of the grid across large geographical areas aimed at obtaining a detailed and accurate picture of the overall grid performance. To that end, the so-called synchrophasors or phasor measurement units (PMUs) provide voltage and current phasor measurements of the power line (i.e., magnitude and angle information). These measurements are time-tagged with an accurate global reference clock (by means of global positioning system (GPS)). By exploiting time synchronization, measurements taken by different synchrophasors can provide a snapshot of the state of the grid for each time instant. For more information about the requirements in terms of accuracy of the measurements and data communication formats, the interested reader is referred to the IEEE standard C37.118-2005 [21].

Currently, several initiatives have started for the massive deployment of synchrophasors in the transport network [22]. Based on this, the NASPI working group is developing a global, secure, and standardized communications architecture for wide-area situational awareness called NASPInet [23]. As shown in Figure 6.6, the NASInet architecture is composed of several components: PMUs, phasor data concentrators (PDCs), and phasor gateways (PGWs). Within a utility level, PMUs monitor current and voltage phasors. These measurements are transmitted to the PDC, whose main function is to time-align synchrophasor data and provide a wide measurement set. The PDC output data stream is used for regional application purposes or sent out to the PGW for wide-area monitoring.

Regarding phasor applications, the NASPI working group has identified four classes (A, B, C, and D) based on the latency and data rate requirements [25]. The most critical applications are within Class A: automatic arming of remedial action schemes, out-of-step protection and transient stability; with latencies lower than 100 ms and data rates of 30 samples/s. In a broader sense, latency requirements for real-time monitoring and control have been identified in [26] to be in the range 20–200 ms, whereas data rate requirements for synchrophasors are expected to be 600–1500 kbps. The selected communication technology will depend on its availability and deployment costs. Among the possible candidates are [23, 26]: optical communications, public/private internet, microwave communications and, in some cases, broadband power-line communications. The use of cellular M2M technologies, as shown in Figure 6.5, is possible under these

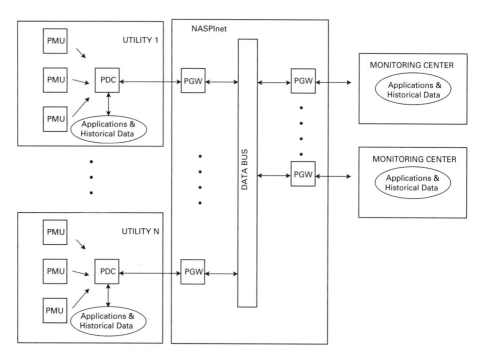

Figure 6.6 NASPInet architecture [23].

delay constraints since LTE is expected to meet the required latencies; this has been quantified in [24].

Distributed energy resources (DER)

Today's power grid is designed to operate in a centralized fashion. Namely, the vast majority of the power delivered to the customers is generated in a few very large capacity stations. This approach has several disadvantages. First, energy is generated far from the loads, thus entailing significant transportation losses. Second, the dependency of the whole grid on a few generation plants causes major reliability issues.

To cope with these limitations, distributed generation (DG) has attracted considerable attention. To be precise, distributed generation consists of generating electricity in smaller energy plants at, or close to, the loads as depicted on the right-hand side of Figure 6.5. Distributed generation units (based on renewable sources or fossil fuels), along with distributed energy storage systems, constitute a microgrid, which consists of a semi-autonomous system connected to the main utility grid [27].

This new architecture, as alluded to in Figure 6.5, will need an upgrade of the transportation and distribution network. With that aim, the IEEE group has started to develop the IEEE standard P1547.4 [28], which deals with the design, operation, and interconnection of distributed energy resources (DER) systems. M2M, in one form or another, will play a central role in this development.

As commented above, DER and microgrids include distributed energy generation units and energy storage devices. Energy storage has been identified as one of the cornerstones of the future power grid [29]. Future developments will reduce costs in energy storage systems, which will entail an affordable massive and distributed deployment. The increasing number of energy devices will bring several benefits to the grid: higher stability and power quality, and higher storage reserves in case of generator failure. In the current power grid, energy storage supports the main grid with a number of applications and ancillary services [30]. Among the most important applications and ancillary services are: renewable integration, voltage support, peak shaving, spinning reserves, frequency regulation, etc. Clearly, to implement these applications, storage systems will exchange information and control signals with the utility and distributed generation units.

Microgrids are envisaged to provide a number of benefits. Among these benefits are: reliability in power delivery (e.g., by islanding), efficiency and sustainability by increasing the penetration of renewable sources, scalability and investment deferral, and the provision of ancillary services. From this list, it is necessary to stress the islanding capability [31–33]. Islanding is one of the highlighted features of microgrids, which refers to the ability to disconnect some loads from the main grid and energize them via distributed energy resources. Intended islanding will be executed in those situations where the main grid cannot support the aggregated demand and/or an outage power event occurs. To detect such situations, low latency two-way communications between the utility and the distributed generation units is a must [33]. Besides, in order to implement the broader range of applications and ancillary services that a microgrid can potentially offer, an efficient coordination of the different DG units and storage systems is mandatory. To that end, one may adopt a centralized control, in which a central entity is in charge of coordinating the elements of the microgrid, or a decentralized approach, in which each DER element behaves with the maximum autonomy but in an intelligent manner [31].

The communications requirements for DER systems were reported in [26]. In particular, data-rate requirements range from 9.6–56 kbps, whereas the lowest latency requirement is 20 ms for switching protection devices during faults.

Distribution automation

Distributed automation is defined as the set of automatic actions for an efficient management of the distribution grid [34, 35]. These automated actions encompass: bus voltage and current control, reactive power compensation, fault detection, fault location and isolation, and equipment monitoring. Obviously, these functionalities require continuous monitoring. The monitoring of the distribution grid is carried out by the so-called intelligent electronic devices (IEDs) [36]. IEDs report the sensed information to the distribution management system (e.g., supervisory control and data acquisition (SCADA) system) which, based on this information, selects the appropriate control actions to be executed by each IED. In summary, an IED is a communication device that usually operates as a sensor and actuator.

The benefits of implementing distributed automation are broad [34, 35]. Distributed automation will reduce costs associated with power outages, due to a quicker restoration of the system. Indirect costs will also be reduced, since manual operation and maintenance will be performed automatically. Besides the financial benefits, distributed automation will enhance the quality of the power delivered to customers. For instance, based on the IEDs information, failures in the grid will be localized and isolated, without affecting other sectors of the distribution network. This fact will substantially diminish the number and duration of power outages.

According to [26], distributed automation requires stringent latency constraints below 1 s, whereas data-rate requirements are 9.6–100 kbps. Given these constraints, cellular M2M seems a viable solution where the respective features need to be enabled.

6.5.3 End-user appliances

This section gives a brief description of the most peripheral parts of the energy grid, namely customers' appliances and their interface. The subject is vast and we limit the exposition to communications requirements. Opposite to transmission and distribution, customers' applications are, generally, non-critical, meaning that they can tolerate probabilities of failure in the range of 0.01–1%. Besides, they do not require back-up energy sources since, during blackouts, they have no reason to be active.

Advanced metering infrastructure

In classic electric grids, metering has mainly been considered as the simple reading of customers' consumptions. To that purpose, some communications protocols for automatic reading have been developed with the only aim of simplifying customers' lives and avoiding operators' displacement. These protocols require very little bandwidth for one-way information transmissions from customer premises to the utility control centre.

Future smart grids, however, are envisioned to be much more interactive systems and they will require two-way communications infrastructures (see, e.g., demand response below). The first step in this direction is to deploy the so-called advanced metering infrastructure (AMI). In a wide sense, AMI is intended as the interconnection of smart meters, smart appliances, control devices, access points, and the communications network connecting one to another and to the utility processing centre. Note that most applications will still be related to informing the users about their energy consumption and, thus, bandwidth, reliability, and latency required by each device will not be critical. Difficulties will come, nevertheless, from scalability issues. Indeed, it is not difficult to imagine that the total number of devices will grow rapidly as we will be connecting entire buildings of flats and offices. Some proposals [37, 38] foresee two different types of traffic:

- *Real-time traffic*. Limited to home-area networks and dedicated to managing energy consumption devices and appliances.
- *Non-real-time traffic*. Composed of aggregated data to be backhauled to some central processing unit.

The need for differentiating the type of traffic, together with the scalability issue, make Internet Protocol (IP) a valuable candidate for the network core (see, e.g., [37–39]). Being the basis of the modern Internet, IP has proven to work well in large, scalable networks since it benefits from an important number of valuable features. We refer to

[40] for a detailed analysis of the applicability of IP to AMI and two IP-based standard proposals.

Note also that employing well-known and working solutions (as is the case with IP) will reduce deployment times and costs. Regarding the latter variable, it is worth recalling that an economic approach is important in terms of both initial investment and following management, and it has to be evaluated in relative terms with respect to the overall benefits that smart grid AMI will bring.

Note further that if privacy issues arise, the metering data may need to pass escrow-like entities, as shown in Figure 6.5. Furthermore, given the comparably loose delay requirements, cellular M2M is a technology enabler. The major issue is coverage, since meters deep down in the basement also need to be reached; this has been a major deployment block to using purely cellular technologies in the AMI context.

Somehow related to the last point, there also exists an ongoing discussion about whether the advanced metering infrastructure should be public rather than a patchwork of private solutions. It is to be expected that utilities will push for private networks, trying to recycle old existing infrastructure and claiming the integrity of private information. Conversely, public solutions are probably more suitable for guaranteeing interoperability and faster renovation cycles. It is true, though, that some efforts will be needed to preserve customers' privacy.

Demand response

One of the most anticipated features of future smart grids is their capability to reduce peak loads, thus obtaining a flatter (and possibly lower) consumption profile over time. Indeed, utilities spend considerable sums to maintain spare generators that are activated a few times per year to cope with rare demand peaks [41, 42]. Apart from building more efficient electrical appliances, some procedures have been proposed to reduce users' consumption (automatically) at high load occurrences. All together, these procedures are generally gathered under the wide concept of demand response (DR).[1]

Looking more into the details of DR strategies, the most trivial solution consists of selling energy at a variable price that increases with the system load. Decisions are thus left to the customer, who is aware of the price curve and decides when to connect his/her loads according to a personal cost/comfort tradeoff (a basic example are the day/night tariffs, quite diffuse in different countries). The opposite solution, which does not extend any decision capability to users, is direct load control (DLC) [43–45]. According to this DR approach, utilities have direct control over customers' appliances and can shut them down to avoid load rises above safety level. Both approaches present pros and cons: DLC is probably the most efficient of the two, but requires customers to renounce privacy and liberty to manage their appliances to their liking. Hence, intermediate solutions are sought, where intelligent devices can automatically shift their load over time. Inputs to these machines will be utility messages about the grid state, on the one hand, and customers' constraints, on the other hand [46, 47].

---

[1] Differences between DR and demand-side management (DSM) are, if any, minimal.

The first DR approach does not require great communications infrastructures. Periodically, utilities only need to broadcast the time/price look-up tables (the duration of each time interval will probably depend on utility policies and/or customer classes) that they are going to use to charge customers and no feedback is required. Similarly, DLC implementations only require one-way communications from the utility control centre to switch on or off customers' loads. Instead, things will be different if the last DR approach with intelligent devices were to take place. Different works (e.g., [46–48]) envision a real-time pricing mechanism that requires a continuous interchange of load forecasts and resulting price figures between the two parties.

Independently of which DR strategy will take place, bandwidth requirements should be quite low, on the order of 100 kbps [49, 50] and thus can be met by the cellular M2M architecture shown in Figure 6.5. The question is still open of how to manage the aggregate traffic. Latency requires particular attention. Even though customers see DR simply as a way to reduce their expenses, utilities may rely on it to manage overload situations. These are critical emergencies that need to be tackled in real time.

Plug-in hybrid electric vehicles

Plug-in hybrid electric vehicles (PHEVs) have been gaining a lot of public interest in recent years, since they promise a noticeable reduction of greenhouse gas and noise emissions. Evidence of this trend is the number of subsidy and regulation programmes that are promoted by public administrations to push PHEV research activities within car manufacturers (see, e.g., [51] for the California case).

Focusing on the impact on the electrical grid, a wide deployment of PHEV fleets will bring both disadvantages and benefits. Indeed, the energy needed to charge a large number of vehicles can easily overload the grid if no control is put in place. However, parking time is typically longer than charging time, leaving quite some margin for a safe load distribution. Even better, batteries offer an important distributed storage capability, which can be exploited by utilities to implement the ancillary services described in Section 6.5.2 (the so-called vehicle to-grid (V2G) power [52–54] or energy 'roaming').

Note that a discharging storage device can easily be modelled as a negative load. Thus, charge management of PHEV batteries may be included in the DR procedures discussed above and does not bring new particular requirements in terms of bandwidth, latency, or reliability [49]. On the other hand, there exists a characteristic feature of PHEVs that needs special attention: mobility. We should expect that PHEV users will need to charge their vehicles at different locations (e.g., home garage, office car park), possibly served by different utilities. It is then important to guarantee interoperability among charging technologies and, also, within the supporting communications infrastructure. For example, let us consider the billing issue. When recharging their batteries away from their home premises, PHEV users have to be charged for the energy they draw. While a direct solution would be a credit-card-based payment system (which needs its communications infrastructure), other experts in the field [37] propose to develop *electric usage roaming*. Like cellular network operators, utilities will exchange information about the consumption of out-of-place customers, who will be charged in a single bill issued by their reference utility. In this way, PHEV users do not need to worry about each single energy refill. However, new security issues come in: customers need to be identified while, possibly, they do not want to leave track of their movements.

Cellular M2M, in the embodiment of Figure 6.5, is an enabler for managing a mobile system of electrical vehicles. This is because it allows for support of mobility, handovers, roaming, and many other features required when exchanging information in the presence of mobility. Indeed, the main growth of cellular M2M is today (Q2 2011) attributed to telemetric applications from the car industry.

6.6 Conclusion

ICT plays a central role in facilitating a more efficient and thus smarter grid. At the centre of this are the emerging M2M technologies which have been the focus of this chapter. Machine-to-machine is indeed the solution to many of the problems currently occurring in the grid, most of them boiling down to being able to monitor variables of interest in often remote and geographically widely spread locations in the grid. The sheer number of nodes to be deployed also enjoys the ability of M2M networks to auto-bootstrap, configure, heal, and thus minimize the intervention of humans.

While conceptually and from an architecture point of view well developed, many challenges lie ahead. Arguably the biggest challenge is to meet all technical requirements if the M2M system were to be supported by a complete cellular architecture. This is because the current cellular network has been designed and optimized to satisfy the needs of human-related communications.

To meet the requirements of often small and embedded machines, cellular communications need to decrease their complexity since the devices are expected to be very low cost and low power with limited processing capabilities. The management of a huge number of devices is another challenge, and the allocation of resources has to be modified to avoid the total overkill that the current resource allocation policy constitutes for M2M applications often only need little traffic. In turn, short-range capillary networks have to reduce delays and increase coverage, and especially, need to improve their security, as M2M applications will be used mainly to transmit confidential information that may compromise the privacy of the users with regard to personal data.

The real challenge, though, is to optimize the operation of these systems together in an efficient manner, so that the needs of M2M applications in smart grids and other application domains can be met, without hampering the performance of human communications. New standards have to be and are currently being defined to run a heterogeneous technology landscape where different systems coexist and cooperate in order to improve performance. Users, such as smart grid utilities, need solutions, not technology, and the achievement of this seamless operation is the main challenge for M2M applications. An international initiative worth mentioning to address these challenges is the ICT research project EXALTED, composed of academics and industries, and running from September 2010 to February 2013 [55], which aims at defining an end-to-end LTE-M system suitable for M2M communications.

References

[1] http://www.wireless-world-research.org/
[2] S. Lucero and S. Carlaw, 'Cellular M2M connectivity service providers – the market opportunity for MNOs, MMOs, and MNVOs,' *ABI Research Report*, 2009.
[3] http://www.ict-bungee.eu/
[4] 'Low-rate wireless personal area network medium access control (MAC) and physical layer (PHY) specifications', IEEE Standard 802.15.4.
[5] 'Wireless LAN medium access control (MAC) and physical layer (PHY) specifications', IEEE Standard 802.11-2007.
[6] http://datatracker.ietf.org/wg/6lowpan/charter/
[7] http://datatracker.ietf.org/wg/roll/charter/
[8] http://datatracker.ietf.org/wg/core/charter/
[9] http://www.pachube.com
[10] http://www.anr-connect.org/
[11] http://www.dash7.org/
[12] http://www.wavenis-osa.org/
[13] 'The national broadband plan – connecting America', http://www.broadband.gov
[14] http://ec.europa.eu/information_society/activities/esafety/ecall
[15] 'Technical specification machine-to-machine communications (M2M); functional architecture', Draft ETSI TS 102 690 V0.11.2, April 2011.
[16] W. Geng, S. Talwar, K. Johnsson, N. Himayat, and K. D. Johnson, 'M2M: from mobile to embedded internet', *IEEE Communications Magazine*, April 2011.
[17] S.-Y. Lien, K.-C. Chen, and Y. Lin, 'Toward ubiquitous massive accesses in 3GPP machine-to-machine communications', *IEEE Communications Magazine*, April 2011.
[18] Z. M. Fadlullah, M. M. Fouda, N. Kato, A. Takeuchi, N. Iwasaki, and Y. Nozaki, 'Toward intelligent machine-to-machine communications in smart grid', *IEEE Communications Magazine*, April 2011.
[19] G. L. Johnson, S. Rahman, and R. A. Messenger, *Electric Power Generation: Non-Conventional Methods*. CRC Press, 2000, ch. 1.
[20] 'Technical analysis of the August 14, 2003, blackout: what happened, why, and what did we learn?' North American Electric Reliability Council Report, 13 July, 2004 [online]. Available at: http://www.nerc.com/docs/docs/blackout/NERC_Final_Blackout_Report_07_13_04.pdf
[21] 'IEEE standard for synchrophasors for power systems', IEEE Std C37.118-2005 (Revision of IEEE Std 1344-1995), pp. 0_1–57, 2006.
[22] North American Synchrophasor Initiative [online]. Available at: http://www.naspi.org/
[23] R. Bobba, E. Heine, H. Khurana, and T. Yardley, 'Exploring a tiered architecture for NASPInet', in *Innovative Smart Grid Technologies (ISGT)*, January 2010, pp. 1–8.
[24] 'LS on LTE latency analysis. Response to: Release: 8 Work Item: LTE-L23', *3GPP TSG-RAN WG2 Meeting* #58, Tdoc R2-072193, Kobe, Japan, 7–11 May 2007, RAN WG2.
[25] 'Phasor application classification', North American Synchrophasor Initiative, Report, 7, August, 2007 [online]. Available at: http://www.naspi.org/resources/dnmtt/phasorapplicationclassification_20080807.xls
[26] 'Communications requirements of smart grid technologies', Department of Energy, United States of America, Report, October 2010 [online]. Available at: http://www.gc.energy.gov/documents/Smart_Grid_Communications_Requirements_Report_10-05-2010.pdf

[27] A. Vojdani, 'Smart integration', *IEEE Power Energy Magazine*, vol. 6, no. 6, pp. 71–79, 2008.

[28] 'IEEE draft guide for design, operation, and integration of distributed resource island systems with electric power systems', IEEE P1547.4/D11, March 2011, pp. 1–55, 2011.

[29] '"Grid 2030" a national vision for electricity's second 100 years', USA Department of Energy Report, July 2003 [online]. Available at: http://www.oe.energy.gov/DocumentsandMedia/Elec_Vision_2-9-4.pdf

[30] R. Walawalkar and J. Apt, 'Market analysis of emerging electric energy storage systems', National Energy Technology Laboratory, Technical Report, 2008 [online]. Available at: http://www.netl.doe.gov/energy-analyses/pubs/Final%20Report-Market%20Analysis%20of%20Emerging%20Electric%20Energy%20Sto.pdf

[31] F. Katiraei, R. Iravani, N. Hatziargyriou, and A. Dimeas, 'Microgrids management', *IEEE Power Energy Magazine*, vol. 6, no. 3, pp. 54–65, May/June 2008.

[32] I. Balaguer, Q. Lei, S. Yang, U. Supatti, and F. Z. Peng, 'Control for grid-connected and intentional islanding operations of distributed power generation', *IEEE Transactions on Industrial Electronics*, vol. 58, no. 1, pp. 147–157, January 2011.

[33] V. K. Sood, D. Fischer, J. M. Eklund, and T. Brown, 'Developing a communication infrastructure for the smart grid', in *IEEE Electrical Power Energy Conference (EPEC)*, October 2009, pp. 1–7.

[34] C. Smallwood and J. Wennermark, 'Benefits of distribution automation', *IEEE Industrial Applications Magazine*, vol. 16, no. 1, pp. 65–73, January/February. 2010.

[35] R.-L. Chen and S. Sabir, 'The benefits of implementing distribution automation and system monitoring in the open electricity market', in *Canadian Conference on Electrical and Computer Engineering*, vol. 2, 2001, pp. 825–830.

[36] J. D. McDonald, 'Substation automation. IED integration and availability of information', *IEEE Power Energy Magazine*, vol. 1, no. 2, pp. 22–31, March/April 2003.

[37] 'Comments – requests for information on smart grid communications requirements', Alcatel-Lucent, RFI, 12 July 2010 [online]. Available at: http://www.gc.energy.gov/1662.htm

[38] 'Comments – requests for information on smart grid communications requirements', Honeywell, RFI (undated) 2010 [online]. Available at: http://www.gc.energy.gov/1662.htm

[39] 'Comments – Requests for information on smart grid communications requirements', Florida Power & Light Company, RFI, 12 July 2010 [online]. Available at: http://www.gc.energy.gov/1662.htm

[40] J. Wang and V. C. M. Leung, 'A survey of technical requirements and consumer application standards for IP-based smart grid AMI network', in *International Conference on Information Networking (ICOIN)*, Kuala Lumpur, Malaysia, 26–28 January 2011.

[41] 'Assessment of demand response and advanced metering', FERC Federal Energy Regulatory Commission, US Staff Report, February 2011 [online]. Available at: http://www.ferc.gov/legal/staff-reports.asp

[42] 'European technology platform smartGrids: strategic research agenda for Europe's electricity networks of the future', Directorate-General for Research Cooperation Energy Report, 2007 [online]. Available at: http://www.smartgrids.eu/

[43] A. Molina, A. Gabaldon, J. A. Fuentes, and C. Alvarez, 'Implementation and assessment of physically based electrical load models: application to direct load control residential programs', *IEE Proceedings on Generation, Transmission & Distribution*, vol. 150, no. 1, pp. 61–66, 2003.

[44] D. Bargiotas and J. D. Birdwell, 'Residential air conditioner dynamic model for direct load control', *IEE Proceedings on Generation, Transmission & Distribution*, vol. 3, no. 4, pp. 2119–2126, 1998.

[45] H. Lee and C. Wilkins, 'A practical approach to appliance load control analysis: a water heater case study', *IEEE Transactions on Power Application Systems*, vol. 102, no. 4, April 1983.

[46] A.-H. Mohsenian-Rad and A. Leon-Garcia, 'Optimal residential load control with price prediction in real-time electricity pricing environments', *IEEE Transactions on Smart Grid*, vol. 1, no. 2, pp. 120–133, September 2010.

[47] A.-H. Mohsenian-Rad, V. W. S. Wong, J. Jatskevich, R. Schober, and A. Leon-Garcia, 'Autonomous demand-side management based on game-theoretic energy consumption scheduling for the future smart grid', *IEEE Transactions on Smart Grid*, vol. 1, no. 3, pp. 320–331, 2010.

[48] C. Ibars, M. Navarro and L. Giupponi, 'Distributed demand management in smart grid with a congestion game', in *Proceedings of IEEE SmartGridComm 2010*, Gaithersburg, MD, USA, 4–6 October 2010.

[49] 'Comments – Requests for information on smart grid communications requirements', Utilities Telecom Council, RFI, 12 July 2010 [online]. Available at: http://www.gc.energy.gov/1662.htm

[50] 'Comments – requests for information on smart grid communications requirements', Avista Corporation, RFI, 2 July 2010 [online]. Available at: http://www.gc.energy.gov/1662.htm

[51] P. Fairley, 'California to rule on fate of EVs [news]', *IEEE Spectrum*, vol. 44, no. 11 (NA), pp. 10–12, November 2007.

[52] W. Kempton and J. Tomić, 'Vehicle-to-grid power fundamentals: calculating capacity and net revenue', *Journal of Power Sources*, vol. 144, no. 1, pp. 268–279, June 2005.

[53] W. Kempton and J. Tomić, 'Vehicle-to-grid power implementation: from stabilizing the grid to supporting large-scale renewable energy', *Journal of Power Sources*, vol. 144, no. 1, pp. 280–294, June 2005.

[54] J. Tomić and W. Kempton, 'Using fleets of electric-drives vehicles for grid support', *Journal of Power Sources*, vol. 168, no. 2, pp. 459–468, June 2007.

[55] *EXALTED, Project Full Title: EXpAnding LTE for Devices*, Contract Number: INFSO-ICT-258512, Call Identifier: FP7-ICT-2009-5, Duration: September 2010–February 2013 (30 months).

7 Bad-data detection in smart grid: a distributed approach

Le Xie, Dae-Hyun Choi, Soummya Kar, and H. Vincent Poor

7.1 Introduction

This chapter is motivated by the fact that wide-area monitoring, control and protection (WAMPAC) are becoming increasingly important in the vision for future smart grid operations [1]. Technological advances in sensing, communication, and computation could enable smart grid operations with improved situational awareness. This improved situational awareness could lead to more reliable and economical integration of renewable energy resources, as well as to the prevention of potential blackouts [1].

Given the need for improved situational awareness in large interconnected power systems, a key research challenge is to develop fast and robust state-estimation techniques for wide-area monitoring. State estimation converts redundant measurements into reliable estimates of the state of an interconnected electric power system [2]. For wide-area state estimation, which involves multiple system operators or utilities, it is more desirable to develop distributed approaches to obtaining the system-wide states through limited information exchange among the system operators [3, 4]. In our recent work [5], a fully distributed and fast state-estimation method is proposed with provable convergence with centralized state-estimation results.

One essential function of a state estimator is to detect, identify, and eliminate measurement errors if possible. Such functions in power system operations are defined as 'bad-data processing' [6]. The main objective of this chapter is to review the bad-data processing techniques and propose a fully distributed bad-data detection algorithm for wide-area state estimation. In particular, the focus of this chapter is to (i) formulate the bad-data processing problem in a fully distributed manner; and (ii) design an information-exchange scheme among different control centres for *provable* distributed bad-data detection performance. In summary, the main contribution of this chapter is twofold:

- Distributed/hierarchical state estimation and bad-data processing methods are reviewed.
- A fully distributed approach for bad-data detection is proposed. This proposed algorithm is shown to have the same convergence of chi-square statistics as in the same centralized bad-data detection. The information exchange and communication requirements are also discussed.

The rest of this chapter is organized as follows. In Section 7.2 the formulation of distributed multi-area state estimation and bad-data processing is presented. The state-of-the-art bad-data detection techniques are briefly reviewed. In Section 7.3 a fully distributed bad-data detection algorithm, along with the necessary information exchange and communication structure, is proposed. An illustrative case study based on an IEEE 14-bus system is presented in Section 7.4. Concluding remarks are presented in Section 7.5.

7.2 Distributed state estimation and bad-data processing: state-of-the-art

7.2.1 Wide-area state-estimation model

An interconnected multi-area power system is assumed to be partitioned into a total of N regions, each region n corresponding to a geographically non-overlapping control area. Each control area is allowed, if necessary, to exchange information with its neighbouring areas. The measurement model for control area n is formulated as follows:

$$\mathbf{z}_n = h_n(\mathbf{x}) + \mathbf{e}_n, \tag{7.1}$$

where $\mathbf{z}_n$ is the measurements vector in control area n (real power, reactive power, voltage magnitudes of buses, etc.), $\mathbf{x}$ is the state vector of the *entire* interconnected power system, $h_n(\mathbf{x})$ is a non-linear measurement function for control area n, and $\mathbf{e}_n$ is a measurement-error vector with zero mean in area n. The measurement error vector is assumed to be Gaussian with zero mean. Based on the static power-flow equations, if a measurement is located inside area n (not on the boundary), then it is only a function of state variables $\mathbf{x_n}$ corresponding to area n, which is a subset of $\mathbf{x}$.

In this chapter, we study the DC power-flow state-estimation problem (i.e., voltage magnitudes at all buses are assumed to be 1.0 per unit, and voltage phase angles at all buses are assumed to be small). The non-linear measurement model for wide-area state estimation (7.1) can be linearized to

$$\mathbf{z}_n = H_n \theta + \mathbf{e}_n, \tag{7.2}$$

where H_n corresponds to the measurement Jacobian matrix for control area n, and the state vector θ represents the vector of voltage phase angle at all buses.

Given the model above, the state-estimation function obtains the optimal estimate of θ by minimizing the weighted least squares of the measurement error:

$$\text{minimize } J(\theta) = \mathbf{r}^T R^{-1} \mathbf{r} \tag{7.3}$$

$$\text{s.t. } \mathbf{r} = \mathbf{z} - H\theta, \tag{7.4}$$

where $R = \text{cov}([\mathbf{e}_1 \cdots \mathbf{e}_N]^T) = \text{diag}(R_1, \ldots, R_N) R_n$, and denotes the covariance of the noise vector $\mathbf{e}_n$ for area n. Defining $\bar{\mathbf{z}}_n = R_n^{-1/2} \mathbf{z}_n$ and $\overline{H}_n = R_n^{-1/2} H_n$ for each control area n, the centralized weighted least squares estimate of θ is given by

$$\hat{\theta} = \left(\overline{H}^T \overline{H}\right)^{-1} \overline{H}^T \overline{z} = \overline{G}^{-1} \overline{H}^T \overline{z}, \qquad (7.5)$$

where $\overline{H}^T = [\overline{H}_1^T \overline{H}_2^T \cdots \overline{H}_N^T]$ and $\overline{G} = \overline{H}^T \overline{H}$.

In [5], it is shown that fully distributed state estimation can be performed by an iterative algorithm as shown below:

$$\hat{\mathbf{x}}_n(i+1) = \hat{\mathbf{x}}_n(i) - a \bigg[b \sum_{l \in \Omega_n} (\hat{\mathbf{x}}_n(i) - \hat{\mathbf{x}}_l(i))$$

$$- \overline{H}_n^T \left(\overline{\mathbf{z}}_n - \overline{H}_n \hat{\mathbf{x}}_n(i) \right) \bigg]. \qquad (7.6)$$

Initially, each control area n has access to only its local measurement $\overline{\mathbf{z}}_n$ and measurement Jacobian $\overline{H}_n$. Then, by exchanging its own state $\hat{\mathbf{x}}_n(i)$ at time $i+1$ with other control areas, each control area can obtain an estimate of the state of the *entire interconnected system*. This distributed iterative algorithm works as long as the inter-area communication graph is connected and the whole interconnection is globally observable. In this chapter, we propose an algorithm for detecting bad data in such a distributed manner.

7.2.2 Bad-data processing in state estimation

In power system state estimation, due to the existence of large measurement bias, drifts or wrong connections, there is a need to detect and identify such 'bad data' [6]. Bad-data processing typically consists of two procedures: detection and identification [6]. Bad-data detection determines whether the measurement set contains bad data. Then, bad-data identification is subsequently performed to find which measurements contain bad data. The procedure is briefly reviewed as follows.

Chi-square test for bad-data detection
Consider the estimation objective function

$$J(\hat{\theta}) = \mathbf{r}^T R^{-1} \mathbf{r}, \qquad (7.7)$$

where $\mathbf{r} = \mathbf{z} - H\hat{\theta}$ is defined as the estimated residual vector. Since the measurement errors are normally distributed, the estimated objective function $J(\hat{\theta})$ obeys a chi-square distribution with $m - n$ degrees of freedom, i.e., $J(\hat{\theta}) \sim \chi^2_{m-n}$. m and n represent the number of measurements and state variables, respectively. Therefore, bad data will be detected if

$$J(\hat{\theta}) \geq \chi^2_{(m-n), p}, \qquad (7.8)$$

where p is the detection confidence probability.

Largest normalized residual test for bad-data identification

Once bad data is detected, normalized residuals of all the measurements are used for identifying bad data. The measurement residual vector $\mathbf{r}$ can be represented as

$$\mathbf{r} = \mathbf{z} - H\theta = S\mathbf{e}, \tag{7.9}$$

where the residual sensitivity matrix S represents the relationship between the measurement residuals and the measurement errors:

$$S = I - HGH^T R^{-1}, \tag{7.10}$$

where we define the gain matrix $G = H^T R^{-1} H$. Therefore, the normalized residual vector can be represented as

$$\mathbf{r}^N = \frac{|\mathbf{r}|}{\sqrt{\operatorname{diag}(SR)}}. \tag{7.11}$$

If the measurement corresponding to the largest normalized residual is greater than a prespecified identification threshold, that measurement is considered to be bad data and is eliminated for another round of state estimation. Note that bad-data processing requires the system-wide knowledge of the matrix S. In the next section, a distributed bad-data processing algorithm without a requirement of system-wide information is proposed.

7.2.3 Related work

Over the past three decades a large body of literature has been accumulated in the area of state estimation and bad-data processing for multi-area power interconnections. In [7–8], a *star-like* hierarchical state estimation method was proposed. Two-level state estimation for a multi-area power system has been studied in [9–13], driven by the capability and need to conduct WAMPAC. The state estimation results obtained at the local control area level are coordinated at a higher level via synchronized phasor measurements. Most recently, a multi-level state estimator (feeder, substation, transmission system organization, and regional levels) was described for the purpose of monitoring large-scale interconnected power systems [14]. However, as the number and sampling rate of measurements increase, the hierarchical state estimation approach may suffer from a communication bottleneck and computational complexity issues inherent in the architecture with a single coordination centre.

A parallel, distributed state-estimation algorithm was first proposed in [15]. Enabled by the naturally decoupled characteristic of weighted least squares (WLS) estimation, the state-estimation problem is shown to be decomposable into each area's local estimator with a coupling constraint optimization technique to ensure convergence of the boundary buses' estimates. The numerical study illustrates that the distributed algorithm not only speeds up the computational time, but also yields acceptable accuracy. However, local observability of each control area is always required in the aforementioned algorithms. In other words, all the local control areas need to have enough measurement redundancy in order to compute the locally decoupled weighted least-squares estimate (excluding the boundary bus measurements). This assumption may not always hold due to (i) the increasing vulnerability of measurements subject to potential bad/malicious data, and (ii) the emergence of smaller control areas such as micro-grids. A fully distributed algorithm for static state estimation is proposed in [5, 16] without requirement of local observability.

As an important component of state estimation, bad-data processing, along with other functions of state estimation, is conventionally performed at one control centre. The first attempt for fully distributed bad-data processing was presented in [17], in which the concept of error residual spread area was introduced. For any given measurement, its measurement error spreads only to measurement residuals within the error residual spread area. The entire interconnected power system is decomposable into several non-overlapping error residual spread areas. In other words, no measurement error in one error residual spread area contaminates measurement residuals in the other error residual spread areas. In this setting of error residual spread area, a reduced model for distributed bad-data processing was proposed in [18]. For an error residual spread area m, the local measurement Jacobian matrix is denoted by H_m. It is proven in [18] that the residual sensitivity matrix S ordered based on error residual spread areas exhibits block diagonal structure. Namely, the measurement error in error residual spread area m does not contaminate any measurement outside area m. Moreover, the block matrix S_m is expressed as

$$S_m = I_m - H_m[G_m]^{-1} H_m^T R_m^{-1}. \qquad (7.12)$$

Using (7.12), the normalized residual vector in the mth error residual spread area is defined as

$$\mathbf{r_m^N} = \frac{|\mathbf{r_m}|}{\sqrt{\mathrm{diag}(S_m R_m)}}, \qquad (7.13)$$

where $\mathbf{r_m^N}$ is the normalized residual vector at the mth error residual spread area. $[G_m]^{-1}$ is the inverse of the the sub-block gain matrix corresponding to the mth error residual spread area. Note that the computation of G_m requires knowledge from only the mth error residual spread area. Therefore, independent bad-data processing is performed for each error residual spread area by the existing bad-data detection and identification techniques. A fully distributed bad-data detection and identification scheme with a distributed state estimator was proposed based on error residual area decomposition in [19]. A heuristic measurement design algorithm for information exchange between local areas was proposed in [20]. By exchanging local measurements or estimates with neighbouring local areas, each local area could achieve improved performance of bad-data processing.

However, the aforementioned work of distributed bad-data processing is primarily based on *error residual spread decomposition*. For a large multi-control-area power system, the decomposition of administrative control areas is typically not overlapped with the error residual spread decomposition. The research challenge is to design the information exchange scheme among administrative control areas so that *through communication, a distributed bad-data detection and identification scheme becomes possible*. A heuristic approach for distributed bad-data detection and identification is proposed in [21], in which each administrative control area performs its own chi-square test and largest normalized residual test. Only the largest normalized residual for each control area is exchanged with a selected set of control areas, all of which belong to the same error residual spread area. However, it is still an open question to find a distributed bad-data processing algorithm that can be shown analytically to converge to the centralized algorithm's performance. In the next section, a fully distributed bad-data detection algorithm based on administrative control area will be presented. It will be shown that this proposed algorithm yields the same convergence of chi-square statistics as in centralized bad-data detection.

7.3 Fully distributed bad-data detection

7.3.1 Preliminaries

In k-dimensional Euclidean space $\mathbb{R}^k$, the $k \times k$ identity matrix is denoted by I_k, while $\mathbf{1}_k$ and $\mathbf{0}_k$ represent the column vectors with all ones and all zeros in $\mathbb{R}^k$, respectively. The operator $\|\cdot\|$ applied to a vector and matrices corresponds to the standard Euclidean 2-norm and the induced 2-norm, respectively. The induced 2-norm is equivalent to the matrix spectral radius for symmetric matrices.

In this chapter, we assume that all the random variables are defined in a common measurable space, $(\Omega, \mathcal{F})$. In addition, all inequalities involving random variables are to be considered a.s. (almost surely), see [22].

In the spectral graph theory literature, for an *undirected* graph $G = (V, E)$, $V = [1 \cdots N]$ is the set of nodes or vertices with $|V| = N$, and E is the set of edges with $|E| = M$, where $|\cdot|$ denotes the cardinality. The unordered pair (n, l) belongs to the set E if nodes n and l are connected to each other through an edge. We consider only simple graphs which contain no self-loops and multiple edges. A graph is connected if there exists a path[1] between each pair of nodes. The neighbourhood of node n is defined as

$$\Omega_n = \{l \in V \mid (n, l) \in E\}. \tag{7.14}$$

Node n has degree $d_n = |\Omega_n|$, the number of neighbouring edges of node n. The structure of the graph can be expressed by the symmetric $N \times N$ adjacency matrix $A = [A_{nl}]$, in which element $A_{nl} = 1$, if $(n, l) \in E$, $A_{nl} = 0$, otherwise. Assuming that the degree matrix is the diagonal matrix $D = \text{diag}(d_1 \ldots d_N)$, the graph Laplacian matrix, L, is

$$L = D - A. \tag{7.15}$$

Due to a positive semidefinite property of the Laplacian matrix, its eigenvalues can be ordered as

$$0 = \lambda_1(L) \leq \lambda_2(L) \leq \cdots \leq \lambda_N(L). \tag{7.16}$$

The smallest eigenvalue $\lambda_1(L)$ is always equal to zero, with $\left(1/\sqrt{N}\right)\mathbf{1}_N$ being the corresponding normalized eigenvector. The multiplicity of the zero eigenvalue equals the

---

[1] A path between nodes n and l of length m is a sequence $(n = i_0, i_1, \ldots, i_m = l)$ of vertices such that $(i_k, i_{k+1}) \in E \; \forall \; 0 \leq k \leq m - 1$.

number of connected components of the network; for a connected graph, $\lambda_2(L) > 0$. This second eigenvalue is the algebraic connectivity or the Fiedler value of the network; see [23–25] for a detailed treatment of graphs and their spectral theory. For the computation of vectors, Kronecker products will be involved in most of the matrix manipulations. For example, the Kronecker product of the $N \times N$ matrix L and I_M will be an $NM \times NM$ matrix, denoted by $L \otimes I_M$.

7.3.2 Proposed algorithm for distributed bad-data detection

Recall the multi-area power system DC state-estimation model:

$$\mathbf{z}_n = H_n \mathbf{x} + \mathbf{e}_n. \tag{7.17}$$

Under nominal operating conditions, the observation vector $\mathbf{z}_n$ is supposed to have independent Gaussian components, the noise variance at each sensor m being σ_m^2. In the framework of hypothesis testing, the above idealized operating condition corresponds to the null hypothesis H_0.

However, due to the existence of large measurement bias, drifts or wrong connections, there is a need to detect such 'bad data' [6]. The problem of bad-data detection may then be modelled as a binary (composite) hypothesis-testing problem, in which one tests the alternative H_1 (presence of bad data) against the null H_0.

In the following we develop a distributed version of the so-called '$J(\widehat{\mathbf{x}})$ test' for bad-data detection (see Appendix B of [26] for details). To this end, we recall some basic concepts from centralized power system state estimation.

The optimum solution $\widehat{\mathbf{x}}$ in (7.5) is obtained under the following assumption: *(E.1) Global observability*: The matrix G

$$G = \sum_{n=1}^{N} H_n^T H_n \tag{7.18}$$

is full-rank.

REMARK 7.1. *Under assumptions **(E.0)**, the weighted Gramian*

$$\overline{G} = \sum_{n=1}^{N} H_n^T R_n^{-1} H_n \tag{7.19}$$

is also full-rank.

Under H_0, the weighted residual

$$J(\widehat{\mathbf{x}}) = (\mathbf{z} - H\widehat{\mathbf{x}})^T R^{-1} (\mathbf{z} - H\widehat{\mathbf{x}}) = \sum_{n=1}^{N} (\mathbf{z}_n - H_n\widehat{\mathbf{x}})^T R_n^{-1} (\mathbf{z}_n - H_n\widehat{\mathbf{x}}) \tag{7.20}$$

has a χ^2 distribution with $K = M - \bar{N}$ degrees of freedom, where M and $\bar{N}$ denote the total number of measurements and the system state (global) dimension, respectively (see Appendix A of [26] for details). The central-limit theorem suggests that for large K ($K \geq 30$ in practice [26]), the standardized random variables

$$\zeta_1 = \frac{J(\widehat{\mathbf{x}}) - K}{\sqrt{2K}} \quad \text{and} \quad \zeta_2 = \sqrt{2J(\widehat{\mathbf{x}})} - \sqrt{2K} \qquad (7.21)$$

converge in distribution to the standard normal (with unit variance).

The $J(\widehat{\mathbf{x}})$ test then takes the following form:

$$\begin{aligned}\text{Accept } \mathsf{H}_0, &\quad \text{if } \zeta_1 < \gamma \text{ (or } \zeta_2 < \gamma) \\ \text{Reject } \mathsf{H}_0, &\quad \text{otherwise.}\end{aligned} \qquad (7.22)$$

The threshold γ in the above test is chosen so as to guarantee a desired level of false alarm probability P_f (see Appendix B of [26] for details).

From the above, we note that the quantity $J(\widehat{\mathbf{x}})$ is the key statistic that needs to be computed to perform the desired tests. In the following we provide a distributed scheme $\mathcal{DBD}$ for computing $J(\widehat{\mathbf{x}})$, such that each substation is eventually able to perform the desired (centralized) test.

Starting from some initial deterministic estimate of $(1/N)J(\widehat{\mathbf{x}})$ (N denotes the number of substations), denoted by $\bar{\mathbf{s}}_n(0)$, each node generates (by a distributed iterative algorithm) a sequence of estimates, $\{\bar{\mathbf{s}}_n(i)\}_{i \geq 0}$. The estimate $\bar{\mathbf{s}}_n(i+1)$ of $(1/N)J(\widehat{\mathbf{x}})$ at the nth node at time $i+1$ is a function of: (i) its previous estimate; (ii) the communicated estimates at time i from its neighbouring nodes; and (iii) the local observation $\mathbf{z}_n$.

Algorithm $\mathcal{DU}$: Based on the current state $\bar{\mathbf{s}}_n(i)$, the exchanged data $\{\bar{\mathbf{s}}_l(i)\}_{l \in \Omega_n}$, and the observation $\mathbf{z}_n$, we update the estimate at the nth node by the following distributed iterative algorithm, termed the $\mathcal{DU}$ algorithm:

$$\begin{aligned}\bar{\mathbf{s}}_n(i+1) = \bar{\mathbf{s}}_n(i) &- \left[a \sum_{l \in \Omega_n} (\bar{\mathbf{s}}_n(i) - \bar{\mathbf{s}}_l(i))\right] \\ &- \delta(i) \left[\left\|R_n^{-1/2}(\mathbf{z}_n - H_n \widetilde{\mathbf{x}}_n(i))\right\|^2 - \bar{\mathbf{s}}_n(i)\right].\end{aligned} \qquad (7.23)$$

The auxiliary state sequence $\{\widetilde{\mathbf{x}}_n(i)\}$ at each node n is also generated by a distributed scheme:

$$\begin{aligned}\widetilde{\mathbf{x}}_n(i+1) = \widetilde{\mathbf{x}}_n(i) &- \left[b \sum_{l \in \Omega_n} (\widetilde{\mathbf{x}}_n(i) - \widetilde{\mathbf{x}}_l(i)) - \alpha(i) \overline{H}_n^T (\mathbf{z}_n - \overline{H}_n \widetilde{\mathbf{x}}_n(i))\right].\end{aligned} \qquad (7.24)$$

The algorithm in (7.24) is called the $\mathcal{M-CSE}$ algorithm. In (7.23) and (7.24), a and b are positive constants and $\{\alpha(i)\}$ and $\{\delta(i)\}$ are appropriately chosen time-varying weight sequences. Algorithm $\mathcal{DU}$ is distributed since at node n it involves only the data from the nodes in its communication neighbourhood Ω_n. In order to implement the $\mathcal{DU}$, each node stores and updates two states, $\bar{\mathbf{s}}_n(i)$, the estimate of $J(\widehat{\mathbf{x}})$, and $\widetilde{\mathbf{x}}_n(i)$, an auxiliary state used for the update of $\bar{s}_n(i)$. In fact, the auxiliary sequence evolves as the $\mathcal{M} - \mathcal{CSE}$ scheme and converges to the WLS estimate of $\widehat{x}$ of x.

We refer to the recursive estimation algorithm in (7.23) and (7.24) as $\mathcal{DBD}$. The following assumption on the connectivity of the inter-node communication network is assumed:

(E.2) *Connectivity*: The inter-node communication network determined by the communication neighbourhoods Ω_n is connected.[2]

Further, the time-varying weight sequences $\{\alpha(i)\}$ and $\{\delta(i)\}$ are assumed to satisfy:

(E.3) *Time-varying weights*: The sequences $\{\alpha(i)\}$ and $\{\delta(i)\}$ are of the form

$$\alpha(i) = \frac{c}{(i+1)^{\tau_1}}, \quad \delta(i) = \frac{d}{(i+1)^{\tau_2}}, \tag{7.25}$$

where $c, d > 0$ are constants and the exponents τ_1 and τ_2 satisfy

$$0 < \tau_1 \leq 1, \quad 0 < \tau_2 \leq 1. \tag{7.26}$$

Also, the constant step sizes a and b satisfy

$$0 < a, b < 2/\lambda_{\max}. \tag{7.27}$$

The following result characterizes the convergence of the individual node estimates $\{\bar{s}_n(i)\}$ to the desired statistic $(1/N)J(\widehat{x})$:

THEOREM 7.1 *Consider the $\mathcal{DBD}$ under* **(E.1)–(E.3)**. *Then, for each n, the estimate sequence $\{\bar{s}_n(i)\}$ converges a.s. to the statistic $(1/N)J(\widehat{x})$, i.e.*,

$$\mathbb{P}\left(\lim_{i \to \infty} \bar{s}_n(i) = (1/N)J(\widehat{x}), \ \forall n\right) = 1. \tag{7.28}$$

7.4 Case study

In this section, the proposed distributed bad-data processing algorithm $\mathcal{A} - \mathcal{BD}$ is illustrated in the standard IEEE 14-bus system. We assume a DC state-estimation model with 1.0 per unit (p.u.) voltage magnitudes at all buses and $j1.0$ p.u. branch impedance. The system has four non-overlapping administrative control areas as shown in Figure 7.1. Administrative local control areas A_1, A_2, A_3, and A_4 contain $n_1 = 3$, $n_2 = 4$, $n_3 = 4$, and $n_4 = 3$ buses, with a total of $m_1 = 5$, $m_2 = 8$, $m_3 = 7$, and $m_4 = 6$ measurements, respectively. The chi-square test thresholds are chosen for a 95% confidence level. The largest normalized residual measurement is identified as bad data if it is greater than 3 (with a 99.7% confidence level). Measurement noises are assumed to be Gaussian with zero mean and finite variance. In addition, the physical and communication network graphs among the control areas are assumed to be identical in this system.

---

[2] We note here that the cyber communication neighbourhood could be significantly different from and sparser than the physical neighbourhood determined by electrical connections.

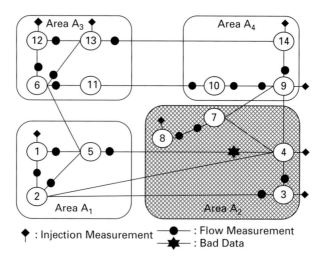

Figure 7.1 The IEEE 14-bus system.

In this section, two cases are simulated for the proposed $\mathcal{A} - \mathcal{BD}$ algorithm:

- Case 1: Single bad data exists at the boundary flow measurement in area A_2.
- Case 2: Multiple bad data exist at the internal flow measurement in area A_3 and boundary-injection measurement in area A_4.

For both cases, we assume that (i) A_2 (shaded) is locally unobservable (i.e., the local measurement Jacobian matrix of A_2 is rank deficient); and (ii) the system is globally observable (i.e., the global measurement Jacobian matrix is full-rank).

To investigate the impact of bad data on the convergence of the $\mathcal{DBD}$ algorithm, we have the following two performance indices corresponding to the chi-square test and estimation accuracy:

$$D_n(i) = J_n(i) - \frac{J}{4}, \qquad (7.29)$$

where J_n and J represent the value of the estimated objective function for the nth control area and the central coordinator, respectively, and

$$g_{j,k}(i) = \theta_{j,k}^{(d)}(i) - \theta_{j,k}^{(t)}, \qquad (7.30)$$

where $\theta_{j,k}^{(d)}(i) = |\theta_j^{(d)}(i) - \theta_k^{(d)}(i)|$ and $\theta_{j,k}^{(t)} = |\theta_j^{(t)} - \theta_k^{(t)}|$ represent the absolute values of bus j's and bus k's phase-angle differences in distributed estimation and true state value difference, respectively. Note that performance indices (7.29) and (7.30) correspond to distributed iterative equations (7.23) and (7.24), respectively.

7.4.1 Case 1

In this case, a single bad datum is introduced to the boundary flow measurement $P_{4,5}$ in area A_2. The true value $P_{4,5} = 0.0647$ is replaced by $P_{4,5} = 0.1772$ using a gross error of 20σ. We assume that all the power flow and injection measurements are corrupted by additive Gaussian noises with equal variances $\sigma^2 = 0.0001$. The IEEE 14-bus system has a total of 26 measurements, including 8 power-injection and 18 power-flow measurements.

The parameters a, d, and τ_2 in equation (7.23) for the convergence of the distributed chi-square test are chosen with $a = 0.05$, $d = 1.75 \times 10^{-7}$, and $\tau_2 = 1$, respectively. The parameters b, c, and τ_1 in equation (7.24) for the convergence of the distributed state estimate are chosen with $b = 0.1$, $c = 0.1$, and $\tau_1 = 0.2$, respectively. For both equations, the tolerance of simulation is set to $\epsilon = 0.00001$. We randomly illustrate four pairs $g_{1,2}$, $g_{5,3}$, $g_{7,8}$, and $g_{9,10}$ in the figure. Figure 7.2 shows two convergence plots in the distributed state estimation, corresponding to when $P_{4,5}$ has bad data and after $P_{4,5}$ is deleted. As shown in Figure 7.2(a), bad datum $P_{4,5}$ leads to an inaccurate estimation solution. After deleting $P_{4,5}$ (we assume that all the local control areas are able to perform the largest normalized residual test, and they verify the largest normalized residual $r_{4,5}^N = 41.78$ corresponding to $P_{4,5}$), we can see from Figure 7.2(b) that the distributed estimation results converge with the true state variable values. Figure 7.3

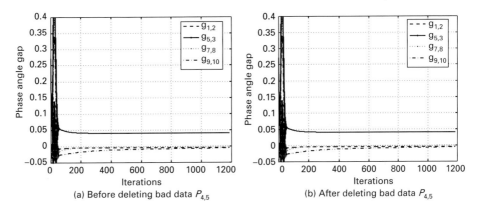

Figure 7.2 Convergence of the $\mathcal{M} - \mathcal{CSE}$ algorithm in Case 1.

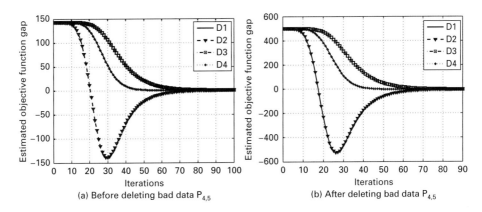

Figure 7.3 Convergence of the $\mathcal{DBD}$ algorithm in Case 1.

Table 7.1. Chi-square test when $P_{4,5}$ has bad data

PI	Central	Area A_1	Area A_2	Area A_3	Area A_4
$J(\widehat{\mathbf{x}})^{(1)}$	1427.5	1424.12	1424.12	1424.12	1424.12
$J(\widehat{\mathbf{x}})^{(2)}$	12.3	7.16	7.16	7.16	7.16

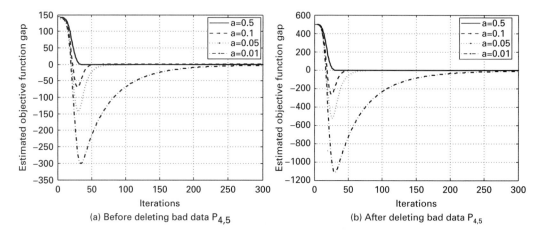

Figure 7.4 Convergence of estimated objective function J_2 with varying values a in Case 1.

shows the convergence of the $\mathcal{DBD}$ algorithm together with an initial value $J_n(0) = 500$. From these figures, it is observed that the value of the distributed estimated objective function J_n for the nth local control area converges exponentially to the centralized estimated objective function. Table 7.1 shows the results of the chi-square test when $P_{4,5}$ has bad data. We denote by $J(\widehat{\mathbf{x}})^{(i)}$ the value of the estimated objective function after the ith round of estimation. The second and third rows of Table 7.1 are compared to the chi-square test thresholds 21.02 and 19.7, respectively. Before $P_{4,5}$ is eliminated, the bolded values of estimated objective function of all the local control areas as well as the central coordinator are above their threshold. After $P_{4,5}$ is eliminated, the central coordinator and all the local control areas detect no bad data, as shown in the second row of Table 7.1.

Next, we investigate the sensitivity of the convergence rate of the $\mathcal{DBD}$ algorithm with respect to the constant step size a. Figure 7.4 shows the performance of the estimated objective function J_2 in area A_2 with varying $a = 0.01, 0.05, 0.1, 0.5$. As shown in these figures, it is observed that the convergence rate increases as a increases. This is due to the fact that as the feedback gain coefficient a increases, the proposed algorithm described in (7.23) will have a higher exponential convergence rate. However, the aforementioned sensitivity analysis does not hold true for the constant d and exponent τ_2 in the time-varying weight sequence.

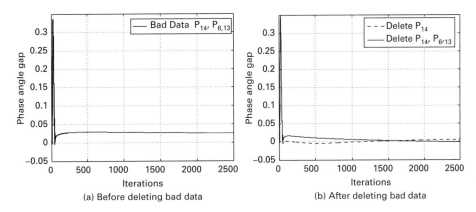

Figure 7.5 Convergence of $g_{9,14}$ in Case 2.

7.4.2 Case 2

In this case, multiple bad data are introduced to both the internal-flow measurement $P_{6,13}$ and the boundary-injection measurement P_{14}. A gross error of 20σ is applied to both measurements. Therefore, the true values $P_{6,13} = 0.0462$ and $P_{14} = 0.0401$ are replaced with $P_{6,13} = 0.116$ and $P_{14} = 0.0693$. This system is assumed to have the same measurement configuration as the IEEE 14-bus system illustrated in Case 1.

The parameters of the two distributed iterative algorithms in Case 2 are identical to those in Case 1, except $d = 3.8 \times 10^{-7}$. Figure 7.5 shows the convergence of $g_{9,14}$ before or after deleting bad data. Similar to Figure 7.2(a) in Case 1, Figure 7.5(a) shows that the proposed distributed algorithm affected by multiple bad data P_{14} and $P_{6,13}$ does not converge well to a true parameter. As multiple bad data are sequentially eliminated (the largest normalized residuals are $r_{14}^N = 28.13$ and $r_{6,13}^N = 16.95$ with the second and third round estimation, respectively), the performance of the proposed distributed state estimation becomes more satisfactory, as shown in Figure 7.5(b). Figure 7.6 shows that the $\mathcal{DBD}$ algorithm converges well no matter whether bad data exist or not. Similar to Figure 7.4, Figure 7.7 also shows that large a leads to higher convergence rate at the expense of potentially more oscillations, and vice versa. Table 7.2 summarizes the performance of the chi-square test with multiple bad data. The second, third, and fourth rows of Table 7.2 are compared to the chi-square test thresholds 21.02, 19.7, and 18.3, respectively. After the first round of state estimation, $J(\widehat{\mathbf{x}})^{(1)}$ is greater than its threshold since there exist bad data. After P_{14} is eliminated, the third row of this table shows that $J(\widehat{\mathbf{x}})^{(2)}$ for all the local areas as well as the central control centre are still above their threshold. Therefore, another round of bad-data processing is initiated. After the third round of state estimation, $J(\widehat{\mathbf{x}})^{(3)}$ becomes less than the corresponding threshold and hence no bad data are suspected.

While a discussion of bad-data identification in a distributed manner is not within the scope of this chapter, it is nevertheless worth mentioning several possible approaches to distributed bad-data identification following the proposed detection algorithm. One possible approach is to compute the normalized residual directly based on (7.13). Since

Table 7.2. Chi-square test when $P_{14}, P_{6,13}$ have bad data

PI	Central	Area A_1	Area A_2	Area A_3	Area A_4
$J(\hat{x})^{(1)}$	941.5	930.4	930.4	930.4	930.4
$J(\hat{x})^{(2)}$	333.9	332.24	332.24	332.24	332.24
$J(\hat{x})^{(3)}$	12.7	17.6	17.6	17.6	17.6

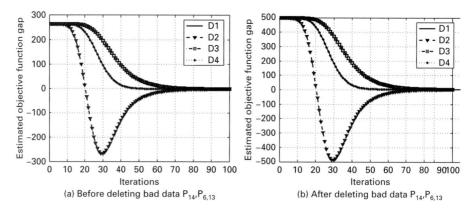

Figure 7.6 Convergence of the $\mathcal{DBD}$ algorithm in Case 2.

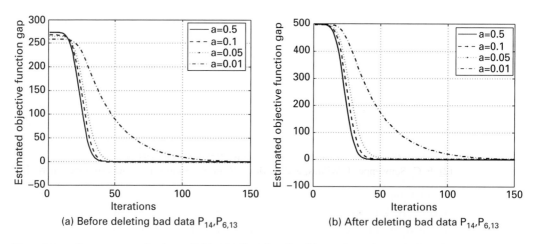

Figure 7.7 Convergence of estimated objective function J_1 with varying values a in Case 2.

the system sensitivity matrix S and measurement covariance matrix R do not change continuously over time as long as the system topology stays unchanged, offline communication could be established among local control areas to obtain the system-wide information of S and R. Another approach is to develop a distributed joint-detection and identification algorithm such as in [27].

7.5 Conclusion

We have discussed a distributed approach to bad-data detection for smart grid state estimation. For a multi-control-area power system, each control area can exchange its own measurement data with designated neighbouring control areas to achieve chi-square statistics that converge to that of a centralized bad-data detection approach. As long as the overall system is observable and the communication topology is connected, the proposed algorithm does not require local observability. For the case of linearized DC state estimation, the distributed bad-data detection could achieve *provably* the same performance as the centralized algorithm. The proposed scheme can be implemented in a fully distributed smart grid operation. Numerical results for an IEEE 14-bus system have shown satisfactory performance of the proposed distributed bad-data processing framework.

While the proposed distributed bad-data detection algorithm performs well in the simulated DC state-estimation case studies, an important open question is how to extend the proposed algorithm in distributed bad-data *detection and identification* in non-linear AC power-flow models. Given the cost of communication and need for near-real-time monitoring, another key challenge is to achieve satisfactory bad-data processing performance by using *local information and limited information exchange among control centres*. Last but not least, large-scale simulation using real-world system configurations would provide greater insight into the proposed distributed bad-data processing scheme.

Acknowledgement

The work of the first two authors was supported in part by the US Power Systems Engineering Research Center, and in part by the Texas Engineering Experiment Station.

References

[1] V. Terzija, G. Valverde, D. Cai, P. Regulski, V. Madani, J. Fitch, S. Skok, M. M. Begovic, and A. Phadke, 'Wide-area monitoring, protection, and control of future electric power networks', *Proceedings of the IEEE*, vol. 99, no. 1, pp. 80–93, January 2011.

[2] F. C. Schweppe, J. Wildes, and A. Bose, 'Power system static state estimation, parts I, II and III', *IEEE Transactions on Power Apparatus and Systems*, vol. 89, no. 1, pp. 120–135, January 1970.

[3] F. F. Wu, K. Moslehi, and A. Bose, 'Power system control centers: past, present, and future', *Proceedings of the IEEE*, vol. 93, no. 11, pp. 1890–1908, November 2005.

[4] A. Bose, A. Abur, K. Y. K. Poon, and R. Emami, 'Implementation issues for hierarchical state estimators', *Final PSERC Project Report*, Document 10-11, August 2010.

[5] L. Xie, D.-H. Choi, and S. Kar, 'Cooperative distributed state estimation: local observability relaxed', in *Proceedings of IEEE Power and Energy Society General Meeting*, July 2011.

[6] A. Abur and A. G. Expósito, *Power System State Estimation: Theory and Implementation*. Marcel Dekker, 2004.

[7] T. V. Cutsem, J. L. Horward, and M. R. Pavella, 'A two-level static state estimator for electric power systems', *IEEE Transactions on Power Apparatus and Systems*, vol. PAS-100, no. 8, pp. 3722–3732, August 1981.

[8] T. V. Cutsem and M. R. Pavella, 'Critical survey of hierarchical methods for state estimation of electric power systems', *IEEE Transactions on Power Apparatus and Systems*, vol. PAS-102, no. 10, pp. 247–256, October 1983.

[9] G. N. Korres, 'A distributed multiarea state estimation', *IEEE Transactions on Power Systems*, vol. 26, no. 1, pp. 73–84, February 2011.

[10] T. Yang, H. Sun, and A. Bose, 'Transition to a two-level linear state estimator–part I: architecture', *IEEE Transactions on Power Systems*, vol. 26, no. 1, pp. 46–53, February 2011.

[11] T. Yang, H. Sun, and A. Bose, 'Transition to a two-level linear state estimator–part II: algorithm', *IEEE Transactions on Power Systems*, vol. 26, no. 1, pp. 54–62, February 2011.

[12] W. Jiang, V. Vittal, and G. T. Heydt, 'A distributed state estimator utilizing synchronized phasor measurements', *IEEE Transactions on Power Systems*, vol. 22, no. 2, pp. 563–571, May 2007.

[13] L. Zhao and A. Abur, 'Multiarea state estimation using synchronized phasor measurements', *IEEE Transactions on Power Systems*, vol. 20, no. 2, pp. 611–617, May 2005.

[14] A. G. Expósito, A. Abur, A. V. Jaén, and C. G. Quiles, 'A multilevel state estimation paradigm for smart grids', *Proceedings of the IEEE*, vol. 99, no. 6, pp. 952–976, June 2011.

[15] D. M. Falcao, F. F. Wu, and L. Murphy, 'Parallel and distributed state estimation', *IEEE Transactions on Power Systems*, vol. 10, no. 2, pp. 724–730, May 1995.

[16] L. Xie, D.-H. Choi, S. Kar, and H. V. Poor, 'Fully distributed state estimation for wide-area monitoring systems', submitted to *IEEE Transactions on Smart Grid* (initial submission: May 2011).

[17] T. A. Clements, G. R. Krumpholz, and P. W. Davis, 'Power system state estimation residual analysis: an algorithm using network topology', *IEEE Transactions on Power Apparatus and Systems*, vol. PAS-100, no. 4, pp. 1779–1787, April 1981.

[18] H. N. Korres and G. C. Contaxis, 'A reduced model for bad data processing in state estimation', *IEEE Transactions on Power Systems*, vol. 6, no. 2, pp. 550–557, May 1991.

[19] S. Y. Lin and C. H. Lin, 'An implementable distributed state estimator and distributed bad data processing schemes for electric power systems', *IEEE Transactions on Power Systems*, vol. 9, no. 3, pp. 1277–1284, August 1994.

[20] G. M. Huang and J. Lei, 'Measurement design of data exchange for distributed multi-utility operation', in *Proceedings of IEEE Power and Energy Society Winter Meeting*, January 2002.

[21] D.-H. Choi and L. Xie, 'Fully distributed bad data processing for wide area state estimation', in *Proceedings of 2nd IEEE International Conference on Smart Grid Communications*, October 2011.

[22] P. Billingsley, *Convergence of Probability Measures*. John Wiley & Sons, Inc., 1999.

[23] Fan R. K. Chung, *Spectral Graph Theory*. American Mathematical Society, 1997.

[24] B. Mohar, 'The Laplacian spectrum of graphs', in *Graph Theory, Combinatorics, and Applications*, (eds Y. Alavi, G. Chartrand, O. R. Oellermann, and A. J. Schwenk). John Wiley & Sons, 1991, vol. 2, pp. 871–898.

[25] B. Bollobas, *Modern Graph Theory*. Springer-Verlag, 1998.

[26] E. Handschin, F. C. Schweppe, J. Kohlas, and A. Fiechter, 'Bad data analysis for power system state estimation', *IEEE Transactions on Power Apparatus and Systems*, vol. 94, no. 2, pp. 329–337, April 1975.

[27] S. Dayanik, C. Goulding, and H. V. Poor, 'Joint detection and identification of an unobservable change in the distribution of a random sequence', *Proceedings of 41st Annual Conference on Information Sciences and Systems*, March 2007.

8 Distributed state estimation: a learning-based framework

Ali Tajer, Soummya Kar, and H. Vincent Poor

8.1 Introduction

The present-day electricity grid is rapidly evolving towards a complex interconnection of distributed modules equipped with a broad range of heterogenous sensing and decision-making functionalities. The proliferation of highly intermittent small energy resources and changing customer needs highlight more adaptive and responsive grid operability in terms of decision making and control. Conventional grid-controlling techniques are unable to cope with such dynamics, as manifested by the increasing reliance on faster time-scale control (such as FACTs devices which enable power electronics-based switching [1]) and system sampling techniques (such as PMUs [2]) to mitigate rapid system fluctuations. It is hard to overemphasize the role of reliable system state estimation on the efficient operability of current and future grid-control techniques.

The complexity in estimator design stems mainly from the fact that, unlike conventional scenarios, the smart grid state estimator needs to possess the attributes of being *distributed*, *adaptive*, and *accurate* over relatively short time intervals. The design of distributed inference and decision-making tasks is indeed key to sustaining the evolving demands and functionalities of the grid [3–9]. Due to the sheer size of the network (at both the transmission and distribution levels), it will no longer be feasible to communicate the entire raw measurement data from all points at all times to a centralized SCADA for state estimation and control; rather, the various substations or regional transmission organizations (RTOs) should use the cyber or information processing/exchange layer efficiently to compute estimates and controls in a distributed manner. Moreover, any meaningful state-estimation technique should be adaptive in the wake of rapid system fluctuations as a result of physical system reconfigurations or varying degrees of reliability and sampling rates in the measurement data.

In this chapter we introduce an efficient distributed approach for power system dynamic-state estimation that takes into account the uncertainties pertinent to the underlying physical and sensing models and is adaptive to the rapidly evolving structural dynamics. We provide a theoretical framework for the development and analysis of estimation/control techniques that (i) adapts to the dynamic reconfigurations of the physical grid model based on closed-loop learning, and (ii) effectively combines (fuses) the sensed data from different sensing sources (with varying methodologies, accuracies, and sampling rates) by properly weighting the value of information from these different sources.

We briefly describe the outline of the rest of the chapter. Section 8.2 provides a relevant survey of existing methods in power system state-estimation. The multi-agent state-estimation model is formulated in Section 8.3, whereas Section 8.4 introduces our learning-based adaptive estimation approach. Convergence analysis and theoretical performance guarantees associated with the proposed learning-based estimation approach are further established in this section. Finally, Section 8.5 concludes the chapter and discusses directions for future research.

8.2 Background

The problem of power system state estimation [10, 11] has been a very active field of research with an extensive literature. This literature may be classified broadly into dynamic and static scenarios. The former seek recursive filtering approaches that offer online prediction of the time-varying state process over time, whereas the latter address the more common problem of static parameter estimation. The key assumption in such available literature is that the statistical model of the system (at least in the estimation time horizon of interest) is known and is often assumed to be static. Under such assumptions, the system state evolution is generally linearized about the operating point and state estimation relies on linear filtering/estimation approaches. Note that this is possible due to perfect knowledge of the system Jacobian matrix (assuming the grid model is perfectly known *a priori* and does not change over the estimation time horizon). Moreover, many existing approaches are centralized, in which it is assumed that information from different parts of the system are fed to a centralized SCADA at all times and centralized computation/information processing is performed. Finally, it is assumed that the value of information (possibly measured in terms of sample variances) from different regions of the network is more or less constant over time and data fusion rules are fixed throughout the course of the procedure. This is of course with the exception of the so-called bad-data detection and localization procedures in static state estimation [12–14], in which anomalies in the proposed information value are measured based on small sample statistics. (See also Chapter 7.)

In general, the information content of a sensor may not be well predicted from a single sample and better results are likely to be obtained by studying its long-term trend or large-sample characteristics, especially in dynamic environments. Especially, the incorporation of more non-conventional sensing resources, for example PMUs with high sampling rates, suggests the use of more general sequential learning approaches in order to measure the value of information in dynamic circumstances. As emphasized earlier, in the face of the major transformations that the grid is undergoing currently, the classical approach of centralized state estimation/control with perfect physical and sensing model information may no longer be feasible, and one needs to resort to distributed approaches that utilize the cyber infrastructure efficiently and that are adaptive with respect to the rapidly varying underlying physical and sensing models. The major undertaking in this chapter is to define a theoretically rigorous learning-based framework for adaptive and distributed decision making. In this context we note that the problem of distributed power-state estimation has been addressed in prior work assuming time-invariant linearized system dynamics (see, for example, [15–31]). However, as has been mentioned earlier, these approaches lack adaptive self-tuning ability with respect to model perturbations and changes in the contextual information value over time and space.

8.3 State estimation model

We consider a smart grid network that is comprised of many interconnected and geographically distributed subnetworks (clusters). Each of the subnetworks has access to some partial (local) information about the state of the network. Obtaining the global state of the network, which serves as the basis for taking the globally optimal control actions in the network, necessitates that the regional subnetworks share their perceptions about the network dynamics with a central decision-maker entity through a backbone communication network. The subnetworks share only their estimate of the network state and do not exchange the full extent of the information that they collect from individual substations or local consumers. Such a level of information sharing has three main advantages:

- The communication channels are prone to be overheard by unauthorized, and possibly malicious, entities, and exchanging only the local estimates of the network state will hide the data of the individual consumers and generators from out-of-network entities.
- The computational load for obtaining the optimal control actions will be distributed among many processors, thereby reducing the computational load of the central entity, which consequently reduces the computation time.
- Full information exchange can potentially overburden the backbone communication channels, a problem that is mitigated by this approach.

The smart-grid network consists of N possibly overlapping clusters that cover the entire network. The state vector of the entire network at time t is denoted by the M-dimensional complex vector

$$\mathbf{x}_t := [\mathbf{x}_t[1], \ldots, \mathbf{x}_t[M]]. \tag{8.1}$$

The measurement that cluster n takes at time t is denoted by $\mathbf{z}_t^n$ and is related to the state vector $\mathbf{x}$ through a mapping

$$\mathbf{z}_t^n = f_t^n(\mathbf{x}_t), \tag{8.2}$$

where $f_t^n(\cdot)$ is a (generally) non-linear function that maps the effects of the network state, measurement errors, and potential adversarial manipulation of the network state (when the network state is compromised by attackers) to the measurements taken by cluster n at time t. Note that often in the literature the observations are related to the network state through a stationary stochastic process, where a transformation of the network state is contaminated by additive white Gaussian noise. In this chapter we assume that the mapping rule $f_t^n(\cdot)$ models an unknown and unspecified mechanism that could be deterministic, stochastic, or even adversarial.

Given the measurement $\mathbf{z}_t^n$, cluster n estimates a subset of the state parameters (elements of the state vector $\mathbf{x}_t$) that correspond to the power flow states in cluster n. We define $\mathcal{S}_i$ to be the set of the indices of all clusters whose flow states are functions of $\mathbf{x}_t[i]$ and estimate it, i.e.,

$$\mathcal{S}_i := \{n \mid \mathbf{x}_t[i] \text{ is estimated by cluster } n\}. \tag{8.3}$$

For each state parameter $\mathbf{x}_t[i]$ and for all clusters $n \in \mathcal{S}_i$, we define $\hat{\mathbf{x}}_t^n[i]$ as the estimate of $\mathbf{x}_t[i]$ provided by cluster n. Such cluster-level state estimates are related to the local measurements through the mappings

$$\hat{\mathbf{x}}_t^n[i] = g_t^n(\mathbf{z}_t^n), \tag{8.4}$$

where the mapping rule $g_t^n(\cdot)$ characterizes the estimator. The estimators are not necessarily identical across the clusters. Also, the estimators are unknown to the central entity. Hiding the estimators deployed from any other agent in the network, including the central entity, further facilitates protecting the data of the consumers and the generators within each cluster.

Upon obtaining the local estimates, each cluster n will report the estimates of its related state parameters to the central estimator of the network for computing the global state of the network. More specifically, at time t all clusters $n \in \mathcal{S}_i$ report the estimates $\hat{\mathbf{x}}_t^n[i]$ to the central estimator. Given all the local estimates, the central estimator obtains the global estimate of $\mathbf{x}_t[i]$, denoted by $\bar{\mathbf{x}}_t[i]$, as a function of the estimates $\{\hat{\mathbf{x}}_t^n[i]\}_{n \in \mathcal{S}_i}$.[1]

In order to assess the estimation quality in each cluster, we denote the *non-negative* estimation cost between the true parameter x and its estimate by $C(x, \hat{x})$. Two popular cost functions corresponding to the minimum mean-square error (MMSE) and maximum *a-posteriori* probability (MAP) estimation criteria are

$$\text{MMSE}: \quad C(\hat{x}, x) = \|x - \hat{x}\|^2$$

and

$$\text{MAP}: \quad C(\hat{x}, x) = \begin{cases} 0, & \|x - \hat{x}\| \leq \delta \ll 1 \\ 1, & \text{otherwise.} \end{cases}$$

Given this estimation cost function, corresponding to each state parameter $\mathbf{x}_t[i]$ and each cluster $n \in \mathcal{S}_i$, we define a *cumulative* estimation cost accumulated up to time T as

$$R_T^n[i] := \sum_{t=1}^{T} C(\mathbf{x}_t[i], \hat{\mathbf{x}}_t^n[i]). \tag{8.5}$$

Similarly, the estimation cost associated with the global estimate of $\mathbf{x}_t[i]$, which we denote by $\bar{\mathbf{x}}_t[i]$, is $C(\mathbf{x}_t[i], \bar{\mathbf{x}}_t[i])$. Based on this estimation cost, for each state

---

[1] In general, the state parameters can be correlated and exploiting such correlation can improve the estimation quality.

parameter we define the estimation cost of the global estimator accumulated up to time T as

$$R_T[i] := \sum_{t=1}^{T} C(\mathbf{x}_t[i], \bar{\mathbf{x}}_t[i]). \qquad (8.6)$$

8.4 Learning-based state estimation

8.4.1 Geographical diversity

Due to the geographical diversity of the clusters spread across a large area, it is very unlikely that the estimates of $\mathbf{x}_t[i]$ in all clusters $n \in \mathcal{S}_i$ are inaccurate. In other words, with a high probability at least one of the clusters $n \in \mathcal{S}_i$ provides a sufficiently accurate estimate on $\mathbf{x}_t[i]$. Therefore, by exploiting the inherent geographical diversity, the task of the central estimator reduces to tracking the *best* local estimator for each state parameter. For tracking the quality of the global estimates with respect to the local ones, we define a *relative* cost function that captures the difference between the estimation cost incurred by the global estimate and the local estimates. For each state parameter $\mathbf{x}_t[i]$ and for all clusters $n \in \mathcal{S}_i$, we define the relative estimation cost as

$$\Delta L_t^n[i] := C(\mathbf{x}_t[i], \bar{\mathbf{x}}_t[i]) - C(\mathbf{x}_t[i], \hat{\mathbf{x}}_t^n[i]), \qquad (8.7)$$

and accordingly we define the cumulative relative estimation cost up to point T as

$$L_T^n[i] := \sum_{t=1}^{T} \Delta L_t^n[i] = R_T[i] - R_T^n[i]. \qquad (8.8)$$

8.4.2 Side information

Two major obstacles for computing the network state in real time are the huge load of information exchange across the network and the pertinent computational complexities for processing the information of the entire network. These two obstacles, however, are less prohibitive in an offline setting as such settings can tolerate more delay in receiving information as well as in processing it. Therefore, the central estimator can acquire fairly accurate information about the previous states of the network through some offline processing. Such side information about the past states can be exploited to compare the relative cumulative estimation costs of the global estimator with respect to the local estimators and identify the more reliable ones, which are those that incur smaller long-term aggregate estimation cost.

8.4.3 Weighted average estimation

The core step of the estimation procedure at the central entity is how to use the real-time and offline information in order to combine the local state estimates provided by the clusters and obtain the global estimate for the global network state. One natural estimation strategy is to find the global state estimate as the *weighted* average of the local estimates $\mathbf{x}_t[i]$ according to

$$\bar{\mathbf{x}}_t[i] = \frac{\sum_{n \in \mathcal{S}_i} \alpha_t^n[i] \, \hat{\mathbf{x}}_t^n[i]}{\sum_{n \in \mathcal{S}_i} \alpha_t^n[i]}, \qquad (8.9)$$

where $\alpha_t^n[i]$ are the weights assigned to the local estimates of $\mathbf{x}_t[i]$, for which we have $\alpha_t^n[i] = 0$ for $n \in \mathcal{S}_i$. The side information about the past estimation costs incurred by different estimators can be exploited to dynamically adjust these weighting factors. Intuitively, at time t, if $L_{t-1}^n[i]$, which is the *relative* cumulative cost for estimator n up to time $t-1$, is large, we assign a larger weighting factor $\alpha_t^n[i]$ to that estimator. This will result in weighting the estimators n whose cumulative costs $R_{t-1}^n[i]$ are small. Therefore, the weights are increasing functions of the cumulative relative cost. For this purpose, for $n \in \mathcal{S}_i$ we write the weight $\alpha_t^n[i]$ as the *derivative* of a non-negative, convex, and increasing function $\phi : \mathbb{R} \to \mathbb{R}$ of $L_{t-1}^n[i]$, i.e.,

$$\alpha_t^n[i] = \phi'\left(L_{t-1}^n[i]\right), \qquad \forall n \in \mathcal{S}_i. \qquad (8.10)$$

Corresponding to each state parameter $\mathbf{x}_t[i]$, we define the relative cost vector $\mathbf{r}_t^i \in \mathbb{R}^N$ as

$$\mathbf{r}_t^i[n] := \begin{cases} \Delta L_t^n[i], & \text{if } n \in \mathcal{S}_i \\ 0, & \text{if } n \notin \mathcal{S}_i \end{cases} \qquad (8.11)$$

and the corresponding cumulative cost vector $\mathbf{R}_T^i \in \mathbb{R}^N$ as

$$\mathbf{R}_T^i := \sum_{t=1}^T \mathbf{r}_t^i. \qquad (8.12)$$

Also, corresponding to each state parameter, we introduce the *potential* function $\Phi_i : \mathbb{R}^N \to \mathbb{R}$ as

$$\Phi_i(\mathbf{u}) := \psi\left(\sum_{n=1}^N \phi(u_n)\right), \qquad (8.13)$$

where $\psi : \mathbb{R} \to \mathbb{R}$ is a non-negative, strictly increasing, concave, and twice differentiable auxiliary function. Based on this definition and (8.10) we find that for each i and for all $n \in \mathcal{S}_i$

$$\alpha_t^n[i] = \nabla \Phi_i(\mathbf{R}_{t-1}^i)_n = \frac{\partial \Phi_i(\mathbf{R}_{t-1}^i)}{\partial L_{t-1}^n[i]}. \qquad (8.14)$$

Therefore, by invoking the convexity of ϕ we can provide the following lemma:

LEMMA 8.1 (Blackwell condition) *If the cost function C is convex in its first argument, then $\forall i$*

$$\sup_{\mathbf{x}_t[i]} \mathbf{r}_t^i \cdot \nabla \Phi_i(\mathbf{R}_{t-1}^i) \leq 0, \qquad (8.15)$$

where '$\cdot$' denotes the inner product.

Proof. By using Jensen's inequality for all $\mathbf{x}_t[i]$, we have

$$C(\bar{\mathbf{x}}_t[i], \mathbf{x}_t[i]) = C\left(\frac{\sum_{n \in S_i} \phi'\left(L_{t-1}^n[i]\right) \hat{\mathbf{x}}_t^n[i]}{\sum_{n \in S_i} \phi'\left(L_{t-1}^n[i]\right)}, \mathbf{x}_t[i]\right)$$

$$\leq \frac{\sum_{n \in S_i} \phi'\left(L_{t-1}^n[i]\right) C(\hat{\mathbf{x}}_t^n[i], \mathbf{x}_t[i])}{\sum_{n \in S_i} \phi'\left(L_{t-1}^n[i]\right)},$$

where by rearranging the relevant terms we obtain for all $\mathbf{x}_t[i]$,

$$\sum_{n \in S_i} \left[C(\mathbf{x}_t[i], \bar{\mathbf{x}}_t[i]) - C(\mathbf{x}_t[i], \hat{\mathbf{x}}_t^n[i])\right] \phi'\left(L_{t-1}^n[i]\right) \leq 0 \quad \Rightarrow \quad \mathbf{r}_t^i \cdot \nabla \Phi_i(\mathbf{R}_{t-1}^i) \leq 0.$$

□

Given the Blackwell condition, the following theorem applies to any global estimator that satisfies the Blackwell condition:

THEOREM 8.1 *For any global estimator with the given potential functions $\Phi_i(\mathbf{u})$ for all $T \in \mathbb{N}$, we have*

$$\Phi_i(\mathbf{R}_T^i) \leq \Phi_i(\mathbf{0}) + \frac{1}{2} \sum_{t=1}^{T} Q_i(\mathbf{r}_t^i),$$

where

$$Q_i(\mathbf{r}_t^i) := \sup_{\mathbf{u} \in \mathbb{R}^N} \psi'\left(\sum_{n=1}^{N} \phi(u_n)\right) \sum_{n=1}^{N} \phi''(u_n) \mathbf{r}_t^i[n].$$

Proof. By using the Taylor expansion of $\Phi_i(\mathbf{R}_T^i)$ we have

$$\Phi_i(\mathbf{R}_T^i) = \Phi_t(\mathbf{R}_{T-1}^i + \mathbf{r}_T^i)$$

$$= \Phi_i(\mathbf{R}_{T-1}^i) + \mathbf{r}_T^i \cdot \nabla \Phi_i(\mathbf{R}_{t-1}^i) + \frac{1}{2} \sum_{n=1}^{N} \sum_{n'=1}^{N} \left.\frac{\partial \Phi_i}{\partial u_n \partial u_{n'}}\right|_\xi \mathbf{r}_t^i[n] \mathbf{r}_t^i[n']$$

$$\leq \Phi_i(\mathbf{R}_{T-1}^i) + \frac{1}{2} \sum_{n=1}^{N} \sum_{n'=1}^{N} \left.\frac{\partial \Phi_i}{\partial u_n \partial u_{n'}}\right|_\xi \mathbf{r}_t^i[n] \mathbf{r}_t^i[n'],$$

where ξ is some vector in $\mathbb{R}^N$. Also, some simple manipulations show that

$$\sum_{n=1}^{N} \sum_{n'=1}^{N} \left.\frac{\partial \Phi_i}{\partial u_n \partial u_{n'}}\right|_\xi \mathbf{r}_t^i[n] \mathbf{r}_t^i[n'] \leq Q_i(\mathbf{r}_t^i).$$

□

8.4.4 Estimation performance

We evaluate the *asymptotic* performance of the estimator in the asymptote of large T for the following specific potential function:

$$\Phi_i(\mathbf{u}) = \left(\sum_{n=1}^{N}(u_n)_+^p\right)^{\frac{2}{p}} = \|\mathbf{u}_+\|_p^2, \tag{8.16}$$

where $p \geq 2$ and $(u)_+$ denotes the positive component of u. For this given potential function the weights assigned to the estimators are given by

$$\alpha_t^n[i] = \nabla\Phi_i(\mathbf{R}_{t-1}^i)_n = \frac{2(\mathbf{R}_{t-1}^i[n])_+^{p-1}}{\|(\mathbf{R}_{t-1})_+\|_p^2}.$$

For the above given set of weighting factors, the following theorem establishes the asymptotic performance of the global estimator:

THEOREM 8.2 *When the estimation cost function C is convex in its first argument, and the potential function is $\Phi_i(\mathbf{u}) = \|\mathbf{u}_+\|_p^2$, the relative cumulative estimation cost of the global estimator with respect to the most reliable estimator (with the smallest cumulative cost) satisfies*

$$R_T[i] - \min_{n \in \{1,\ldots,N\}} R_T^n[i] \leq \sqrt{T(p-1)N^{\frac{2}{p}}}.$$

The theorem above demonstrates that the relative estimation cost per estimation round compared with the best estimator diminishes as $T \to \infty$. In other words, for all state parameters we have

$$\lim_{T \to \infty} \frac{1}{T}\left(R_T[i] - \min_{n \in \{1,\ldots,N\}} R_T^n[i]\right) = 0. \tag{8.17}$$

8.5 Conclusion

We have introduced a learning-based state-estimation approach in a smart-grid setting that consists of multiple possibly overlapping clusters. In each cluster, the state parameters that are related to the power flow in that cluster are estimated locally by a local estimator. All such estimates from across the network are fed to a central entity whose role is to collect all the local estimates for each state parameter and identify the most reliable local estimate for each state parameter. It has been shown that if the estimation criteria deployed by each cluster satisfy certain convexity conditions, the central entity is guaranteed to identify the best state estimators across the network that exhibit the smallest estimation error in the long-term.

We end the chapter by discussing some extensions of this basic estimation framework. A particularly interesting topic for future research concerns designing fully distributed fusion approaches, in which the various agents (subnetworks) interact over a sparse communication topology in the absence of a central coordinator. This would limit the undesirable issues associated with a central fusion-based architecture, leading to more scalable and reliable distributed information dissemination schemes. Another topic that would be worth investigating concerns the partial information case, in which the different subnetworks (agents) may only have partial state estimates (i.e., estimates corresponding to a local subregion and not the entire geographically distributed field of interest). The estimate fusion will be more challenging in this case as the local estimates provided by the different subnetworks may be significantly non-overlapping, raising the need to address intricate system-theoretic issues like observability, for example.

References

[1] F. D. Galiana, K. Almeida, M. Toussaint, J. Griffin, D. Atanackovic, B. T. Ooi, and D. T. McGillis, 'Assessment and control of the impact of facts devices on power system performance', *IEEE Transactions on Power Systems*, vol. 11, no. 4, pp. 1931–1936, 1996.

[2] R. F. Nuqui and A. G. Phadke, 'Phasor measurement unit placement techniques for complete and incomplete observability', *IEEE Transactions on Power Systems*, vol. 20, no. 4, pp. 2381–2388, 2005.

[3] M. D. Ilić, L. Xie, U. A. Khan, and J. M. F. Moura, 'Modeling of future cyber-physical energy systems for distributed sensing and control', *IEEE Transactions on Systems, Man and Cybernetics*, vol. 40, no. 4, pp. 825–838, July 2010.

[4] M. D. Ilić, J. W. Black, and M. Prica, 'Distributed electric power systems of the future: institutional and technological drivers for near-optimal performance', *Electronic Power Systems Research*, vol. 77, no. 9, pp. 1160–1177, July 2009.

[5] K. Tomsovic, D. Bakken, V. Venkatasubramanian, and A. Bose, 'Designing the next generation of real-time control, communication, and computations for large power systems', *Proceedings of the IEEE*, vol. 93, no. 5, pp. 965–979, May 2005.

[6] S. K. Mazumder, K. Acharya, and M. Tahir, 'Joint optimization of control performance and network resource utilization in homogeneous power networks', *IEEE Transactions on Industrial Electronics*, vol. 56, no. 5, pp. 1736–1745, May 2009.

[7] S. M. Amin and B. F. Wollenberg, 'Toward a smart grid: power delivery for the 21st century', *IEEE Power and Energy Magazine*, vol. 3, no. 5, pp. 34–41, September 2005.

[8] A. Bose, 'Smart transmission grid applications and their supporting infrastructure', *IEEE Transactions on Smart Grid*, vol. 1, no. 1, pp. 11–19, June 2010.

[9] F. F. Wu, K. Moslehi, and A. Bose, 'Power system control centers: past, present, and future', *Proceedings of the IEEE*, vol. 93, no. 11, pp. 1890–1908, November 2005.

[10] F. C. Schweppe, J. Wildes, and A. Bose, 'Power system static state estimation, parts i, ii and iii', *IEEE Transactions on Power Apparatus and Systems*, vol. 89, no. 1, pp. 120–135, January 1970.

[11] A. Monticelli, 'Electric power system state estimation', *Proceedings of the IEEE*, vol. 88, no. 2, pp. 262–282, February 2000.

[12] E. Handschin, F. C. Schweppe, J. Kohlas, and A. Fiechter, 'Bad data analysis for power system state estimation', *IEEE Transactions on Power Apparatus and Systems*, vol. 94, no. 2, pp. 329–337, April 1975.

[13] A. Monticelli, F. F. Wu, and M. Yen, 'Mutiple bad data identification for state estimation by combinatorial optimization', *IEEE Transactions on Power Delivery*, vol. 1, no. 3, pp. 361–369, July 1986.

[14] H. J. Koglin, T. Neisius, G. Beibler, and K. D. Schmitt, 'Bad data detection and identification', *International Journal of Electrical Power and Energy Systems*, vol. 12, no. 2, pp. 94–103, April 1990.

[15] T. V. Cutsem, J. L. Horward, and M. R.-Pavella, 'A two-level static state estimator for electric power systems', *IEEE Transactions on Power Apparatus and Systems*, vol. PAS-100, no. 8, pp. 3722–3732, August 1981.

[16] T. Cutsem and M. R.-Pavella, 'Critical survey of hierarchical methods for state estimation of electric power systems', *IEEE Transactions on Power Apparatus and Systems*, vol. PAS-102, no. 10, pp. 247–256, October 1983.

[17] H. Sasaki, K. Aoki, and R. Yokoyama, 'A parallel computation algorithm for static state estimation by means of matrix inversion lemma', *IEEE Transactions on Power Systems*, vol. 2, no. 3, pp. 624–632, August 1987.

[18] S. Iwamoto, M. Kusano, and V. H. Quintana, 'Hierarchical state estimation using a fast rectangular-coordinate method', *IEEE Transactions on Power Systems*, vol. 4, no. 3, pp. 870–880, August 1989.

[19] A. A. El-Keib, J. Nieplocha, H. Singh, and D. J. Maratukulam, 'A decomposed state estimation technique suitable for parallel processor implementation', *IEEE Transactions on Power Systems*, vol. 7, no. 3, pp. 1088–1097, August 1992.

[20] S. Y. Lin and C. H. Lin, 'An implementable distributed state estimator and distributed bad data processing schemes for electric power systems', *IEEE Transactions on Power Systems*, vol. 9, no. 3, pp. 1277–1284, August 1994.

[21] D. M. Falcao, F. F. Wu, and L. Murphy, 'Parallel and distributed state estimation', *IEEE Transactions on Power Systems*, vol. 10, no. 2, pp. 724–730, May 1995.

[22] R. Ebrahimian and R. Baldick, 'State estimation distributed processing', *IEEE Transactions on Power Systems*, vol. 15, no. 4, pp. 1240–1246, November 2000.

[23] L. Zhao and A. Abur, 'Multiarea state estimation using synchronized phasor measurements', *IEEE Transactions on Power Systems*, vol. 20, no. 2, pp. 611–617, May 2005.

[24] A. J. Conejo, S. de la Torre, and M. Canas, 'An optimization approach to multiarea state estimation', *IEEE Transactions on Power Systems*, vol. 22, no. 1, pp. 213–221, February 2007.

[25] W. Jiang, V. Vittal, and G. T. Heydt, 'A distributed state estimator utilizing synchronized phasor measurements', *IEEE Transactions on Power Systems*, vol. 22, no. 2, pp. 563–571, May 2007.

[26] W. Jiang, V. Vittal and G. T. Heydt, 'Diakoptic state estimation using phasor measurement units', *IEEE Transactions on Power Systems*, vol. 23, no. 4, pp. 1580–1589, November 2008.

[27] G. N. Korres and G. C. Contaxis, 'Application of a reduced model to a distributed state estimator', in *Proceedings of IEEE PES Winter Meeting*, January 2000.

[28] G. N. Korres, 'A distributed multiarea state estimation', *IEEE Transactions on Power Systems*, vol. 26, no. 1, pp. 73–84, February 2011.

[29] U. A. Khan, M. D. Ilic, and J. M. F. Moura, 'Cooperation for aggregating complex electric power networks to ensure system observability', in *Proceedings of First International Conference on Infrastructure Systems*, Rotterdam, November 2008, pp. 1–6.

[30] L. Xie, D. H. Choi, and S. Kar, 'Cooperative distributed state estimation: local observability relaxed', in *Proceedings of IEEE Power and Energy Society General Meeting*, Detroit, MI, USA, July 2011.

[31] L. Xie, D. H. Choi, S. Kar, and H. V. Poor, 'Fully distributed state estimation for wide-area monitoring systems', submitted to *IEEE Transactions on Smart Grid* (initial submission: May 2011).

Part III

Smart grid and wide-area networks

9 Networking technologies for wide-area measurement applications

Yi Deng, Hua Lin, Arun G. Phadke, Sandeep Shukla, and James S. Thorp

9.1 Introduction

A wide-area measurement system (WAMS) consists of advanced measurement technology, the latest communication network infrastructure, and integrated operational framework. The supervisory control and data acquisition (SCADA) infrastructure for energy-management system (EMS) has been widely used in power systems for a long time. Some of the functionalities of an EMS are system state monitoring, tie-line bias control, and economic dispatch [1]. However, in recent years, various deficiencies of the existing SCADA-based EMS (such as quasi-steady-state calculation, non-synchronized data acquisition, and relatively low data transmission rate) have been pointed out. These defects make it impossible to sample the global state of a power system in real time. As more and more wide-area blackouts are reported, it is clear that acquiring real-time or wide-area state information would be needed in the future. The state information in terms of phasors of voltages and currents from a distributed wide area in real time is therefore critical for avoiding large-area disturbances by effecting wide-area control based on wide-area measurements.

The main enabler of WAMS is phasor measurement unit (PMU) technology. With the innovation of PMU, the problem of measuring the phasor quantities simultaneously from a wide area of distributed substations, also called 'synchrophasor', has been solved. At present, the PMU technology is one of the essential enablers for WAMS. It utilizes the availability of high-precision synchronized clock sources – extracted from global positioning system (GPS) receivers and samples the instantaneous analogue – quantities of voltage and current magnitudes and phase angles. Many utilities are installing PMUs at very high voltage substations, but the trend is that eventually PMUs will be used to monitor all high-voltage transmission lines. There are debates as to whether the wide-area or long-distance communications that communicate phasor data, or control action messages, should only be done via proprietary networks, or the existing Internet infrastructure with quality-of-service (QoS) differentiated delivery could be used. These questions require further simulation studies as well as financial and policy decisions which are not considered in this chapter. We concentrate mostly on evaluating the network architecture, the specific network medium and technology, the appropriate protocols, etc. that will be required for making the latency and bandwidth requirements necessary for specific WAMS applications. Our work is based on simulation of realistic, but small-scale, PMU networks and profiling the transported data traffic for various applications during simulation. Our simulations also corroborate some of the real experimental latency studies done by Ken Martin in the recent past [2].

The communication infrastructure for WAMS applications can be considered as a real-time performance system. The question: how to construct the entire system in order to realize a variety of applications and satisfy the timing constraints, is the primary problem for system designers. We propose an integral solution that consists of device intelligence, hardware configuration, software design, tradeoff analysis of various choices of communication medium, specific application analysis, component modelling, and system simulation.

This chapter is organized as follows. In Section 9.2, a brief definition of WAMS and its constituents (including PMU and phasor data concentrator (PDC), actual hardware architecture and versatile software infrastructure) is presented. Section 9.3 discusses the communication needs for WAMS applications, available transmission medium, and widely used communication protocols. A comprehensive depiction of WAMS applications from polyphase monitoring to emergency protection to control is presented in Section 9.4. Section 9.5 addresses the WAMS modelling and network simulation results by utilizing the state-of-the-art simulation tool OPNET. We present an experimental framework that is used to explore the networking architecture, protocols, etc. and the latency and bandwidth requirements for WAMS applications. Finally, Section 9.6 concludes our current work and discusses some future directions.

9.2 Components of a wide-area measurement system

A typical WAMS consists of distributed PMUs, regional PDCs, centralized super-phasor data concentrator (SPDC), and hierarchically organized communication networks.

9.2.1 PMU and PDC

The first PMU was invented and built at the Power Systems Research Laboratory at Virginia Tech [3]. The main components and structure of a typical PMU are shown in Figure 9.1.

The phasor data is time-tagged at the source by the PMU, and the data is sent over the communication interface to a PDC. Once the PDC receives the measurements, it checks for transmission errors using the cyclic redundancy check (CRC) and restores the data into its internal data memory. When the data is written into its memory, the PDC accesses the internal binary counter that is synchronized with the universal time coordinated (UTC) clock. The PDC compares the time-tag recorded by various PMUs and rearranges the measurement data to align the data that belongs to the same UTC time. As shown in Figure 9.2, PDC often feeds data to various subscribed local applications.

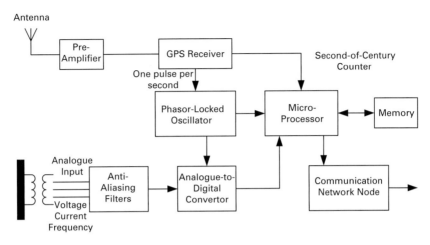

Figure 9.1   The inner structure of the phasor measurement unit [3].

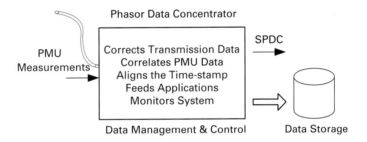

Figure 9.2   The function of the phasor data concentrator [4].

Some applications require the PDC to act as a regional decision maker. For example, it can calculate the existing local phasor measurements and send out protection messages to block or trip the protection relays. In most cases, for wide-area monitoring, the aligned time-stamp data will be sent over the network to one level up in the hierarchy – to the super PDC.

9.2.2   Hardware architecture

The hardware architecture of a practical WAMS is shown in Figure 9.3. This WAMS is composed of five main components: substations with PMUs, substations with PDCs, centralized SPDC, a relay office, and high-performance regional or wide-area networks.

Located at the lowest layer of the WAMS hierarchy, PMU-installed substations consist of basic measuring devices including PMUs, digital relays, and intelligent electronic devices (IEDs). Since the volume of transmission data among all the interconnected equipment is modest, all these devices within a substation are connected by the shared media access Ethernet. The communication protocols for working over such an Ethernet are defined in the IEC 61580 standard formulated by the International Electrotechnical

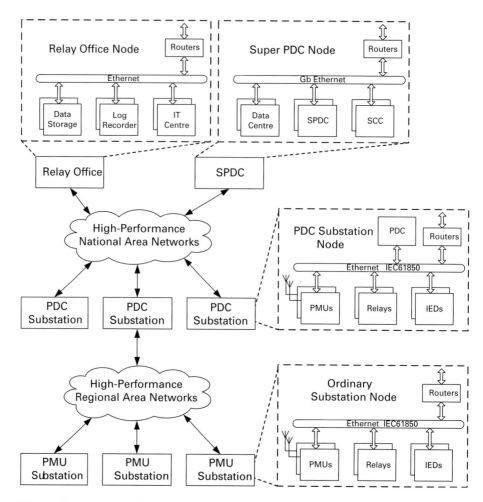

Figure 9.3 The specific hardware architecture of WAMS.

Commission (IEC) Technical Committee 57 (TC57) [5]. There are routers connecting the intra-substation Ethernet and the regional area networks.

In a PDC substation node, in addition to the existing three types of networked measuring equipment mentioned above, there is usually a PDC installed on the substation Ethernet. The first responsibility of a PDC is to gather all the PMU measurement data that is within the scope of its region. It uses the time-stamps to realign data. It may send the time-aligned data to the higher level of PDC such as a super PDC, over the network. It may also have the functionality to make certain regional control decisions.

The super PDC node, which has the capability to store, analyse, and illustrate measurement data stays at the top level of the overall architecture. It may be housed in a data centre, a phasor data processing centre, or a system control centre (SCC). Most of the monitoring functions, parts of the global protection schemes, and all of the controlling strategies are executed through SPDC.

As the observation node for relay engineers, the relay office node has the function of recording the relay's operation logs, dealing with relay alarms, and revising the relay parameters remotely.

The communication infrastructure of WAMS can be classified into three types: intra-substation local-area networks (LAN), high-performance regional networks, and wide-area fibre-optic networks. The communication standard IEC 61850 defines the mapping of data models to a series of protocols such as manufacturing message specification (MMS), generic object-oriented substation events (GOOSE), and sampled measured values (SMV). All the protocols can run over an Ethernet protocol directly. The large transmission bandwidth available on a gigabit Ethernet can guarantee the 4 ms response time required by some applications. The high-performance regional networks interconnect several distributed PMUs and one PDC. This kind of network may be regulated by an independent utility or it may be dedicated for the sole purpose of the power system utility. The measurement data sent to a PDC may be used for making regional protection decisions and then transferred to the SPDC through the wider-area fibre-optic networks. The highest level of communication network is the most congested network. All the phasor information gathered by PMUs should upload to the centralized system monitoring centre. The SPDC could be responsible for sending time-critical commands back to protection equipment, stability controllers, etc. It may so happen that the data volume will increase gradually as the number of PMUs increases – the communication backbone among PDCs could present a communication bottleneck for the WAMS.

9.2.3 Software infrastructure

The software infrastructure may be composed roughly of four layers, shown in Figure 9.4: the physical layer, the communication protocol layer, the middleware layer, and the application layer. This specific layered software architecture corresponds to the functions of different equipment installed in a WAMS. However, different equipment only needs to implement one or two layers of this infrastructure. For example, software designed for PMUs does not necessarily implement an application layer, but must contain a physical layer, a communication protocol layer, and a middleware layer.

The physical layer, which is situated at the bottom of the overall software architecture, implements the different driver functions. It isolates the software from the electrical interfaces that could be changed with the development of higher-rate transmissions. The access to the transmission medium, such as fibre-optic-based synchronous digital hierarchy (SDH), is wrapped in the software of in the physical layer. Except for the CRC error-detection algorithm normally implemented in hardware, the software in the physical layer can also define its own error correction schemes.

The second layer of software infrastructure is the communication protocol layer that is used mostly for data transmission and reception. The complexity of communication flows in WAMS applications requires the communication protocols to be diverse and efficient.

The middleware layer – in most cases implemented in all the devices – consists of a real-time operating system (RTOS) and other pre-verified libraries. The reason for

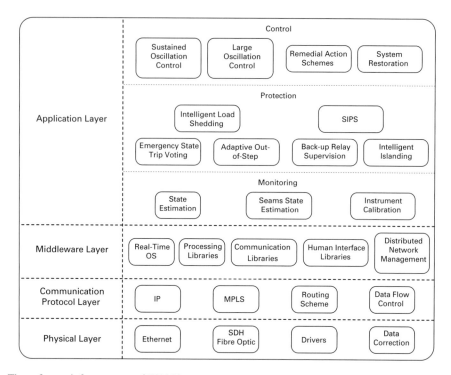

Figure 9.4 The software infrastructure of WAMS.

using RTOS in WAMS is that it allows for real-time reactions to various interrupts such as reception of data. The processing libraries are made up of various signal-processing algorithms such as digital filters and discrete Fourier transform implementations – which will be used widely in various applications such as state estimation. The communication libraries wrap all the communication protocols and provide application programming interfaces (APIs) to the application developers. The human interface libraries provide the basic functions for visualization – for example. Distributed network management-related libraries such as simple network management protocols (SNMP) [6] may be helpful for distributed management of network resources. Considered as a parallel and distributed system, WAMS may need optimization algorithms for optimal location of PMUs, and other resources.

The application layer defines a large number of commonly used applications in power systems. Some applications that will be implemented in PDC can act as a regional executor for supervising relays, special protection schemes, etc., whereas others with the functionality of wide-area monitoring and control might be installed in the SPDC.

9.3 Communication networks for WAMS

When implementing a realistic WAMS, the interconnection structure could be more sophisticated than we have depicted above. The WAMS architecture is considered as both a scalable integration of equipment such as PMUs, PDCs, SPDC, and SCC, and a judicious combination of systematic software and a large amount of embedded software [7, 8]. WAMS utilizes a multi-level hierarchical communication network to combine all these components working together. Therefore, the performance of the network is recognized as one of the most important factors for WAMS. The designing process of establishing a WAMS network could experience four stages. First, the communication requirements for the entire system should be proposed, and every communication need should be fully verified. These verifications are associated with specific applications. Second, system designers could select the appropriate transmission medium and communication protocols by considering the construction costs as well as latency, bandwidth requirements. Third, an optimal topology for WAMS communication would be proposed with the help of network simulation. Finally, the deployment of WAMS communication networks represents a compromise between enough reliability, robustness, and cost of novel technologies.

9.3.1 Communication needs

Although the communication needs for different applications differ [9–11], there are common characteristics for a unified WAMS network. These features are reliability, real-time responsiveness, scalability, and security.

- *Reliability*. Like any other monitoring and control applications in industrial control systems (ICS), WAMS acquires accurate measurement data from a large number of 'sensors' – especially PMUs. Consequently, the reliability of communication network for WAMS is crucial. If there are transmission errors in protection commands transmission, or no abnormal states are reported to the system monitoring centre, it will bring disaster to power grid operation and management.
- *Real-time responsiveness*. After the correctness of transmission, the property of real-time transmission becomes the second priority. Most of the applications used in WAMS are time-critical applications. In some cases, the initiator attempts to obtain a response in a limited period of time – typically on the order of milliseconds. If some information is not sent to the particular destinations on time, or not received timely, actions such as load shedding and islanding would be too late to execute. Therefore, the real-time performance is a vital parameter for WAMS networks.
- *Scalability*. In many cases, the communication needs of WAMS will increase dramatically, caused by unpredictable events. For instance, a region of power grids that previously belonged to different utilities will plan to merge into a unified WAMS. This requires that the communication network of WAMS has scalability. When adding a great deal of regional PMU measurements to a limited-capacity network, the transaction would be blocked, decreased, or even crashed. Scalability also defines that the transmission medium and network protocols used in WAMS could be upgraded to newer techniques conveniently and economically.
- *Security*. The safe operation of WAMS is related directly to the health of the entire grid, because many of the parameters used for controlling the power grid are acquired from

WAMS measurements. If the communication network is unsafe, or could be attacked easily, this will cripple the communication system and even destroy the power grid in the worst-case scenario.

9.3.2 Transmission medium

The transmission medium chosen by the power industry has experienced quite a long period of evolution. From the perspective of economy and performance, different kinds of transmission medium have their inherent advantages and disadvantages when implementing WAMS. Until now, there have been at least four commonly used transmission medium in power system communication design: power-line communication, satellite communication, microwave communication, and optical-fibre communication [12].

Power-line communication (PLC)

PLC is probably one of the most economic and widely used technologies for data communication in power systems. However, it is still in the process of development with several unresolved problems. PLC has been used since the 1950s. One of the limitations of power-line communication is that it can only be used in simplex mode (one direction) and it has much narrower data bandwidth than demanded by today's WAMS applications. These characteristics limit its use to only a few applications, such as remote relay control, non-critical public lighting, and regional home automation applications. In recent years, the communication speed over PLC has largely improved. Nevertheless, the various disadvantages of PLC still obstruct its utilization in WAMS. As the reliability and security are of primary concern for WAMS applications, noisy transmission environments, limited capacity, irreparable open circuit issues, signal distortion and weakness of cyber-security make widespread adoption of PLC in WAMS communication unrealistic.

Satellite communication

Two significant features of satellite communication – global coverage and convenient deployment – make the utilization of satellites as the communication solution in power systems attractive [13]. There exist some remote substations where no communication infrastructures exist. In some other cases, the remote substations have been destroyed by unpredicted artificial events or damaged by natural disasters. Practical experiences tell us that the satellite communication is quite appropriate for these cases. However, there are several problems to be tackled before it is widely used in WAMS. The transmission latency is a prime concern for protection applications, and satellite communications have been shown to have unsatisfactory performance measured under multiple round-trip paths. These relatively long transmission delays are caused by the long distances between satellite receivers and the geo-synchronous communication satellites. Some suggestions such as using lower-orbit satellites, could mitigate the delay to some extent, but other actors, such as the impact of terrains or electromagnetic environments, and the long-term rental fee should also be considered.

Microwave communication

Once considered as a prevailing communication mechanism, microwave communication provides the high-volume transmission channels, anti-interference ability, and portability. Nevertheless, some defects such as line-of-sight propagation distance, undesirable transmission loss, and low diffraction ability would reduce the communication performance. Moreover, if deployed in a wide area, the large number of microwave radio relays would make the entire system unreliable.

Optical-fibre communication

As one of the most promising communication options, optical-fibre communication has been growing rapidly in the power grid communication field. Several characteristics, such as immunity to electromagnetic interference, make it especially suitable for use in strong electromagnetic interference (EMI) and radio-frequency interference (RFI) environments. From a practical deployment point of view, the optical-fibre communication supports even longer-distance transmission due to the fact that light has very low transmission loss when propagating in fibre. Other key features for optical-fibre are unsurpassed channel capacity and high-speed data transfer rates. Although the communication needs for WAMS are not as heavy as some other particular transmission-intensive storage applications, the real-time performance (especially the protection and control processes) require that the WAMS networks are latency-sensitive as well. In addition, considering the low radiation in fibres, optical-fibre communication is also known as a more secure infrastructure against side-channel information leakage issues than any other alternatives. Small size and light weight lead to the manageable deployment. In order to implement the integration of system monitoring, protection, control, and even office automation, the high-bandwidth optical-fibre-based communication backbone infrastructure provides the margin of expansion.

9.3.3 Communication protocols

SONET/SDH technology

Taking the real-time performance and reliability into account, synchronous digital hierarchy (SDH) in optical-fibre networks is becoming the mainstream for WAMS. Developed from synchronous optical network (SONET), SDH was adopted by the International Telecommunication Union Telecommunications Standardization Sector (ITU-T) as an international standard for the second generation of digital transport technology in 1988. In the frame structure of SDH, the overhead section occupies about 2.96% of the frame size, so that the payload transmission efficiency is quite high. Furthermore, maintenance and management sections within the frame structure help to form a robust network for WAMS communication. Due to their higher level, IP-family protocols are popular in packet-switching networks, which can be supported by SDH datalink layer protocols.

MPLS technology

Compared to circuit-switching networks, packet-switching networks are more flexible and efficient. Also, in order to introduce differentiated QoS guarantees, the multiprotocol label-switching (MPLS) scheme – an Internet Engineering Task Force (IETF) standard based on Cisco's tag switching – has been established [14]. This is a virtual connection-oriented structure integrated into the otherwise connectionless IP network. One major function of MPLS is resource reservation protocol – traffic engineering (RSVP-TE), which manages the flowing pass for every IP package and avoids data aggregation at the congested nodes. Another function of MPLS is the priority scheme to supply the IP precedence or label-based QoS [14]. An MPLS virtual private network (VPN) helps divide the different WAMS application services into prioritized channels. Conclusively, the SONET/SDH with MPLS scheme seems to be the most appropriate candidate for WAMS.

9.4 WAMS applications

Conventionally, the applications of WAMS can be classified into three major categories according to their purposes: power-system monitoring, protection, and control [15–17]. Power-system monitoring applications provide first-hand observations of the system state data for system dispatchers. Distributed measuring equipment or IEDs located in each substation periodically sample the voltage, current, and frequency values and send them to the system's monitoring centre for visualization or logging.

Power-system protection applications work on the basis of collective measurements obtained from monitoring applications. They analyse the measurement data and judge the status of the power system so as to guarantee the safe operation of the system. In case a fault happens, the wide-area protection applications can help to trip or block protective devices at critical junctures from the systematic perspective. In current power system protection applications, most of the decision-making and protection actions are executed by local protectors, such as circuit breakers and various kinds of relays. The introduced WAMS enhances the dependability and security of system protections. Some of the applications are time-critical, and require the end-to-end delay or response time to be measured within one hundred milliseconds; the communication needs for different protection applications may be diverse.

Similarly, power-system control applications maintain system stability based on wide-area measurements also provided by monitoring applications. Combining real-time PMU measurements with on-line planning schemes, these control applications are especially useful for damping a broad range of oscillations.

9.4.1 Power-system monitoring

State estimation
Conventional static-state estimation relies on measurements of complex power flows to calculate bus voltages. The remote terminal units (RTUs), composed of microprocessors and communication modules, take charge of the measurement scanning. The RTU measurement are exchanged on the SCADA infrastructure. The SCADA data do not have unified time-tags. Therefore, system dynamics cannot be captured due to unsynchronized

measurements. The WAMS equipped with PMUs measures the voltage phasor directly so that the dynamic system state can be obtained without estimation. However, the initial use of WAMS only involves a limited number of deployed PMUs. Hence, there should be a large amount of SCADA data mixed with a small amount of phasor measurement data. In a conventional static-state estimation SCADA system, the refresh rate of the data is as low as 0.1 to 0.25 Hz. There should be no communication problems even if a small amount of phasor data is injected into the system. However, the perspective of more potential phasor measurements or even an all-phasor estimators application will require more communication bandwidth than the traditional ones. The communication infrastructure should be rebuilt to satisfy the new needs.

All-PMU estimators are usually used on the highest voltage level in the power system and the estimation of the real and imaginary parts of the voltages and currents is linear and non-iterative. As shown in Figure 9.5, the bus voltage and line current of a 500 kV network are measured. The approximate number of installed PMUs for estimating all the 500 kV voltages is one-third of the number of 500 kV substations [18]. In a high-voltage system, suppose there are N buses and nearly $n \sim (N/3)$ measurements, then the number of phasor measurements will be $N \sim 2N$. A conservative value of $1.5N$ can be chosen. In future implementation, all the high-voltage buses will be equipped with PMUs. Then the total number of phasor measurements will increase up to $4N$.

Seams between state estimates

One typical problem that is commonly encountered in system-wide monitoring is how to use PMUs to concern the interface between two or more large state estimators. In order

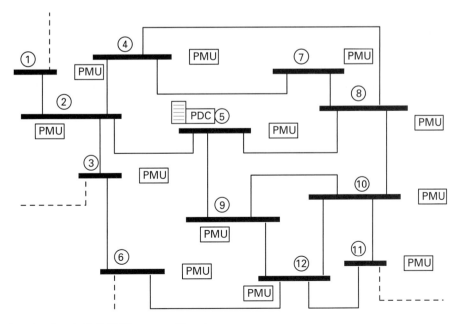

Figure 9.5 A regional part of all-PMU power grid system.

to monitor wider area state information and produce a larger estimator, the independent system operator (ISO) would like to share its own PMU measurement data with other adjoining ISOs. By doing so, the number of bus measurements increases by about n times. Take two adjoining ISOs as an example, if the individual ISO monitoring centre can find a common reference phasor angle between the separate references, the simplest way is to install two PMUs in this area. Then the measured phase-angle difference between the two PMUs can be used to adjust the measurement data transfer between the two ISOs.

Another more sophisticated but accurate method is to utilize the measurement data and concentrate on all the buses located at the boundary between the two IOSs. From Figure 9.6, we can see that the boundary buses both in ISO1 and ISO2 are covered by both estimators. The two estimators can use the differences $(\theta_{(i,1)} - \theta_{(i,2)})$ as the reference angles. In [3], another more refined way to get the differences in reference angles is utilizing the flows on the tie-lines.

Instrument transformer calibration

Instrument transformers installed in the power system, such as magnetic core multi-winding current transformers (CTs), capacitive voltage transformers (CVTs), and precision potential transformers (PTs), are shown in Figure 9.7.

The network equation: $E_t = Z \cdot I_t$, represents the relationship between true voltages and currents. The actual measurements are related to true values by the ratio correction factor k. Therefore, by obtaining a set of actual measurements over several testing points: $I_m = k_i \cdot I_t$ and $E_m = k_j \cdot E_t$, the instrument ratio correction factor can be calibrated.

The data collection period for this application should be 12 hours. All the PMU measurement data which are pre-stored in the distributed PDCs will be uploaded to the system monitoring centre.

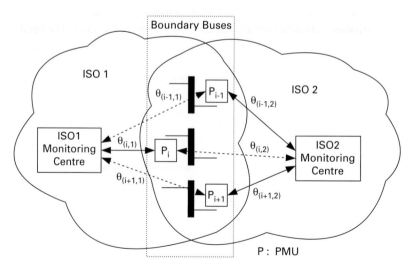

Figure 9.6   The boundary buses monitoring for seams state estimates.

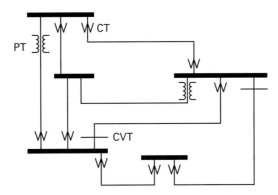

Figure 9.7 CT, CVT, and PT calibration network.

9.4.2 Power-system protection

Adaptive reliability

At critical locations in the power system, the protection schemes for independent relays are preset based on off-line solutions. In the normal system state shown in Figure 9.8(a), to ensure high dependability, the relays (R_1, R_2, and R_3) located at critical places should be set on a more sensitive mode for fault detecting. That means the trip output logic is in the 'or' manner: if any one of the relays detects the fault, the output will execute the tripping action. In addition, if the system is in the emergency state shown in Figure 9.8(b), the decision will be made by a 'voting' procedure in the system control centre. For example, two-thirds of total relays (R_4, R_5, and R_6) should agree to trip before a real trip takes place.

The system control centre makes decisions for the next state of the system (normal state or emergency state) based on the current state information and the predefined experimental plans. If the emergency state is reached, the system control centre will send a message to the critical relay indicating that the current state is emergency and a tripping action has been taken over by a voting procedure. The bandwidth need for this application is modest, and the respond requirement is not critical.

Adaptive out-of-step

Within the protection applications, the adaptive out-of-step function can be enhanced efficiently by PMU-based WAMS [19, 20]. This problem can be described in a relatively easy example where the interface between the two large areas is modelled as a two-generator machine system shown in Figure 9.9.

The work flows of the adaptive out-of-step mechanism can be described in the following two major stages: the detection stage and the action stage. In the detection stage, the PMUs which are installed in the generators monitor the rotor angles in real time. Rotor angles which can be calculated from generator terminal phasor measurements and machine-equivalent circuits will be transmitted to the control centre in the period of 30 times a second. After the data is gathered in the control centre, the angles are tracked automatically. In almost all cases of disturbance, the related two groups of rotors will

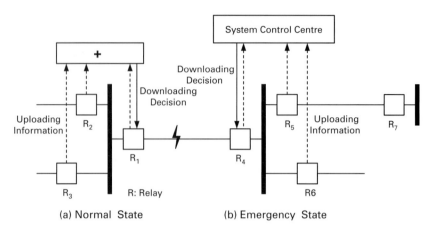

Figure 9.8 Critical relay actions in two-system state.

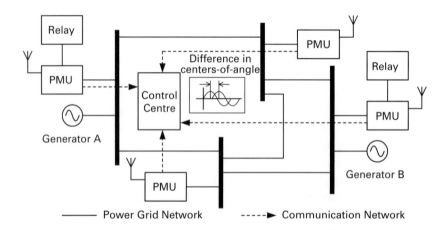

Figure 9.9 Two-generator machine system.

become coherent. The centre-of-angle of each group can be recorded and compared during the operation time. If the difference of the two centres-of-angle reaches an upper limit of about 120–150 degrees, a major transient stability alarm will be declared. There are two major decision algorithms that are used for judging the stability or instability of a system. One is called the time-series prediction method, and the other is called the extended equal area criterion. If the disturbance is unstable, the protection scheme of islanding coherent groups of machines will be performed. The control centre will play a key role in rotor angle tracking, determining coherent groups, and performing time-series prediction.

Supervision of backup zones

From the investigation of recent power system blackouts, mis-operation of backup protection zones is the main contributing factor of the disaster. In the traditional protection system, the distance relays and over-current relays are used for backup protection. When a major disturbance event occurs, unpredictable load flows and non-directional power swings could trick the backup relays to get a wrong apparent impedance value, and trip the power line sequentially. This kind of cascading failure can lead to a wide-area blackout [12].

One effective method to solve this problem is to use PMUs which are located in the backup zones to monitor the apparent impedance shown in Figure 9.10. If there is a fault occurring over the power line between the zone 1 protection relays R_1 and R_2 and the zone 1 protection relay R_2 fails to trip for some reason, the backup relay R_5 should see this fault and trip within 300 ms. Before tripping, R_5 should check the information of PMU P_5 or PMU P_2, which are located in zone 1. If none of the PMUs see the event as a zone 1 fault, there must be errors which may be caused by load incursion in R_5. The trip action should not be executed. Adding PMU measurement data to the decision making will prevent false trips at critical junctures in the progression of a cascading failure.

Intelligent load shedding

In order to balance the affordable generation with the system load, power utilities use the under-frequency load-shedding relays to operate in urgent cases when a system island is formed and frequency is decaying. The system engineers could design a pre-calculated load-shedding scheme which adjusts the available generation of the system constantly by monitoring the tie-line power flows. When the values of the tie-line flows change to

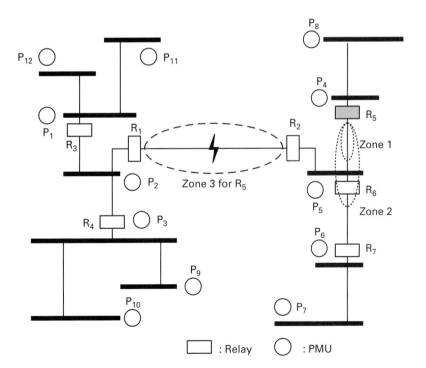

Figure 9.10 Using PMUs to control the mis-operation of backup relays.

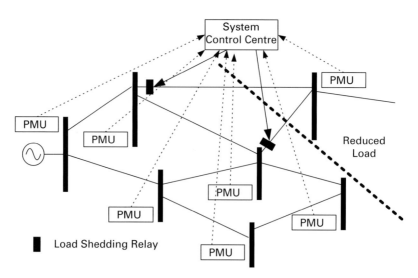

Figure 9.11 The conceptual of intelligent load shedding.

a high level, the phenomenon of 'generation insufficient' is indicated. If the measured tie-line values reach above the predefined limit, the system control centre should send reducing commands to load-shedding relays so as to prevent breaking up and formation of an islanded system as shown in Figure 9.11. In addition, the amount of load to be shed should be the same as the tie-line flow changes.

The WAMS provides the computing and communication architecture for system supervisory control. The tie-line flows in the power grid network are calculated by the PMUs using the current and voltage values at the tie buses. Then the distributed values will be transmitted to the system control centre where the load-shedding message is generated.

Intelligent islanding

To avoid system instability, an intelligent islanding plan is required. As shown in Figure 9.12, after gathering all the real-time PMU measurement data, the system control centre can accurately determine whether the power system is on the edge of an unstable state and system islanding is necessary to avoid a blackout. Then, the system control centre will calculate the optimal islanding boundaries for the current occasion. The islanded system should meet the balance requirements between generation and load. The entire system in this application is in a chaotic environment, several individual relays will trip the lines according to their own settings. So, one of the urgent controlling tasks is both to block undesirable trips and to trip the necessary lines or transformers. Without WAMS, these visions cannot be achieved.

System-wide implementation and integration of SIPS

The system integrity protection schemes (SIPS), also known as remedial action schemes, are formulated as a set of strategies to solve the integrity problems of a power system. These schemes, reported by the Power System Relaying Committee in 2002, can

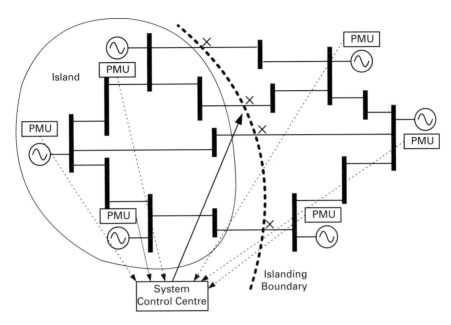

Figure 9.12   Islanding architecture in WAMS.

address the specific control and protection issues such as load shedding before the frequency begins to decay (under frequency load shedding), and before the voltage becomes unstable (under voltage load shedding). Other problems such as out-of-step tripping and blocking, congestion mitigation, static var compensator (SVC)/static compensator control, dynamic braking, generator runback, black start of gas turbines, and system separation are all mentioned in the scheme report [12]. As the number of SIPS increases, a new added scheme should be evaluated carefully before implementation, so that it will not disrupt the other existing schemes due to the complexity of power system events. The WAMS provides a platform and gives an opportunity for the system control centre to communicate with the overall power system and acquire the SIPS response in real time.

9.4.3   Power-system control

Sustained oscillations

Using PMU-based WAMS to reduce the sustained inter-area oscillations which are low-frequency small signal oscillations is a typical application in power systems. As shown in Figure 9.13, the two low-frequency oscillation centres which belong to different adjoining ISOs are hundreds of miles apart. When making damping controls, the PMU measurements will travel through a long distance in the region to the system control centre, and the feedback controlling messages will go through a greater transmission delay than in state applications.

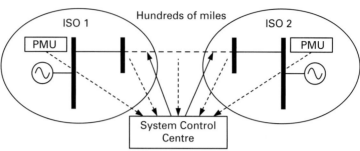

Figure 9.13 Sustained oscillations control architecture.

Large oscillations

In order to avoid unstable transient events, which can lead to islanding or even blackouts, some artificial intelligences are introduced in power system control. Some artificial intelligent controls for damping the inter-area oscillations are handy to prevent the system from islanding. Therefore, the system control centre needs to monitor a larger number of PMUs and utilize many novel decision algorithms such as expert systems, genetic algorithms, neural networks, or fuzzy set theories to make precise statements. The communication infrastructure for control of large oscillations will be more complicated than small signal ones, because of not only the growth of measurement data, but also the real-time transmission needs in a nationwide area.

Remedial action schemes

Considered as a control application, remedial action schemes are used to mitigate the large-area instability issues. The traditional protection schemes are mainly concentrated on specific transmission devices such as transmission lines, transformers, bus bars, and generators. Nowadays, the overall protection schemes which are commonly used in utilities have proved to be more effective for system monitoring and control.

System restoration

When the unexpected islanding or blackout occurs, the most important thing is how to re-energize the islanding area, reconnect with the other networks, and restore the power system [21]. The restoration strategy is a step-by-step procedure as shown in Figure 9.14. There are four steps that reconnect sequentially from no. 1 to no. 4 to merge the islanding area to the system. Through the whole process, the PMUs will measure the phasor information in every substation which is located in the junction of the island. The system control centre which monitors and controls the entire process will take the PMU measurements before, during, and after the disturbance into consideration for a restoration plan.

In the literature, there are a number of restoration methods being proposed, and three of them are discussed most: automate restoration, computer-aided restoration, and cooperative restoration [3]. In these restoration techniques, the communication systems are

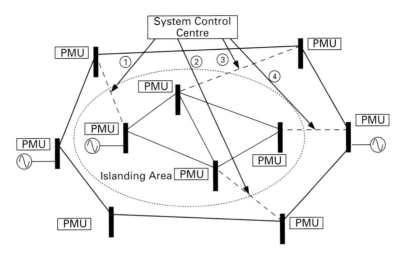

Figure 9.14 Illustration of the power system restoration plan.

SCADA-based controlling systems. Being upgraded to WAMS, the communication performance can be enhanced. The system architecture for using a power-system restoration plan is essentially the same as those of the all-PMU state estimator applications.

9.5 WAMS modelling and network simulations

Before implementing a realistic communication system of WAMS, the performance of networks should be assessed by using simulation software. Network simulation technologies supply an effective way to find out the bottlenecks and redundancies of a large complex communication system [22, 23]. The simulation results in this section reveal the feasibility that one full-featured communication architecture can satisfy the WAMS requirements.

9.5.1 Software introduction

The network modelling and simulation tool we use is the OPNET Modeler [24], a powerful and comprehensive modelling and simulation software which is dedicated to communication network research and development. The hierarchical network modelling architecture which corresponds to actual protocol layers, device layers, and network layer can enable the accurate simulation of WAMS-like, end-to-end, system-level network architecture.

9.5.2 System infrastructure modelling

We model the entire WAMS communication networks in a hierarchical manner. It consists of local last-mile access networks, intra-substation networks, and wide-area backbone networks. A skeletal network composition is shown in Figure 9.15.

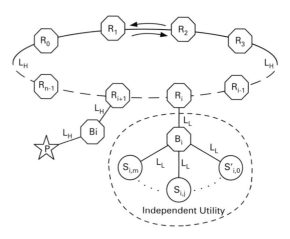

Figure 9.15 The structure of the backbone network.

The octagon nodes marked with R_i $(i = 0, 1, 2, ..., n-1)$ represent one of the n routing nodes (RNs) installed on the backbone network. These RNs interconnect with nearby RNs to form a ring topology. Each routing node is composed of a various number of high-performance networking routers. The octagon nodes marked with B_i represent the LAN routers. There are two types of circle nodes marked with $S_{(i,j)}$ and $S'_{(i,j)}$ that represent the PMU-equipped substation node without PDC and with PDC, respectively. Usually, there is only one $S'_{(i,j)}$ substation node in a specific region. The main function of the PDC is to aggregate and realign the PMU data based on time-stamps. Substations in the same regional area are connected to one of the routing nodes. Figure 9.15 only shows the subsidiary substation for the routing node, R_i, but in fact other routing nodes also have their own group of subsidiary substations. The star node P represents the SPDC centre in WAMS. In this particular design, there is only one SPDC in charge of system monitoring and control from the highest level.

Physically, this ring topology backbone network is an optical-fibre network using SDH. The communication links between any two routing nodes $(R_i \leftrightarrow R_{i+1})$ denoted by L_H are modelled as 155.520 Mbps SDH STM-1. This type of interconnection link is also used to connect between the SPDC node and the routing node $(P \leftrightarrow R_i)$. The communication links between substations and routing node $(S \leftrightarrow R_i$ or $S' \leftrightarrow R_i)$ denoted by L_L are modelled as 2.048 Mbps E1 links which have the potential to upgrade to L_H if necessary.

A more detailed WAMS communication infrastructure is shown in Figure 9.16. In total, 120 PMUs are placed in the simulation framework which can cover most areas of the eastern and central US grid. Within each substation, we configure two PMUs, various relays, and circuit breakers over the 100 Mbps Ethernet. In OPNET, the paradigms of this equipment are modelled by reconfigurable workstations and servers. The SPDC node in the system is composed of server-based SPDC and workstation-based SCC. The subnet of the routing node R_i is composed of six high-performance realistic routers

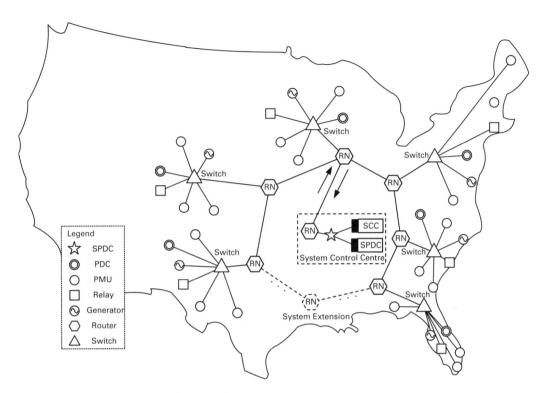

Figure 9.16 A nationwide WAMS construction.

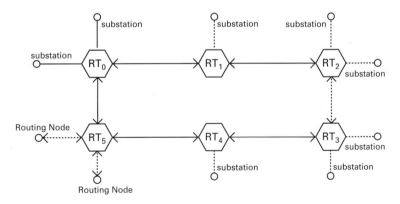

Figure 9.17 The topology of the subnet of routing nodes.

RT_i $(i = 0, 1, 2, ..., n-1)$. As we can see in Figure 9.17, the routers RT_i will be responsible for connecting with substations located in its region, and connecting with other regions' routing nodes. The dotted line between RT_2 and RT_3 indicates that the routing node subnet has the capability to expand to a larger scale by increasing the number of routers.

Table 9.1. Application category

No.	WAMS application types	Application profiles
1	Periodic transfer without acknowledgements	Video conference
2	Large amount of burst data transfer without acknowledgements	FTP data transfer
3	Small amount of burst data transfer without acknowledgements	Print operation
4	Burst transfer with acknowledgement required	Remote login with response

9.5.3 Application classification

WAMS supports various operations of a power system, such as monitoring, protection, and control. In the literature, there are 10 frequently used WAMS applications [9]. By analysing the communication needs, these applications can be classified into four different data transmission profiles: periodic transfer without acknowledgements, a large amount of burst transfer without acknowledgements, a small amount of burst transfer without acknowledgements, and burst transfer with acknowledgement. Mapping to OPNET's predefined transaction profiles, these four communication profiles can be modelled as: video conference, file transfer protocol (FTP) data transfer, print operation, and remote login with response. The corresponding relationship between WAMS application types and OPNET application profiles is listed in Table 9.1.

This classification can distinguish the time-critical applications from other applications. Take the application 'supervision of backup zone' as an example – designed to prevent mis-operation of backup protection zones – the PMUs on remote buses need to monitor the apparent impedance of the transmission lines. When a false fault is picked up by backup relays, the PMUs around the backup relays should send messages to the backup relays to prevent the false trips. The whole processing flow belongs to the fourth type: 'Burst transfer with acknowledgement required' of application. Only by acquiring the acknowledgement from backup relays, can the sender make sure that this critical action has been executed correctly.

9.5.4 Monitoring simulation

Although there are three independent monitoring applications discussed above, we can find that their communication features are similar. In fact, these three applications can be captured by using the same network simulation profile. For 'seams between state estimates' application, the SPDC can reuse the uploaded all-PMUs measurement data including boundary reference buses data in both areas, and distinguish the two state estimators and their reference angle differences. Regarding 'the instrument transformer calibration' application, all the needed information has been transmitted and stored to

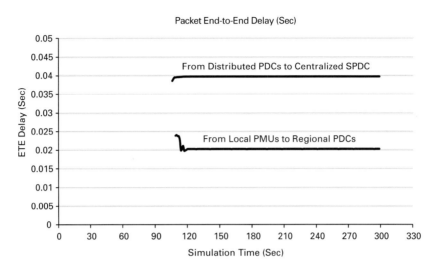

Figure 9.18 ETE delays for monitoring applications.

the SPDC periodically. These pre-stored data will be re-fetched every 12 hours when carrying out calibration calculations. Hence, the communication data flows in these two applications are the same as in the all-PMUs state estimation application, so that only the 'all-PMUs state estimation' simulation profile is needed.

The packet end-to-end (ETE) delays for the power system monitoring applications are plotted in Figure 9.18. The entire transmission delay can be divided into two stages. In the first stage, the aggregation delay will be generated from local measurement PMUs to regionally realign PDCs. In the second stage, the gathering delay will be calculated from distributed PDCs to the centralized SPDC. From the curves shown, we can see that the regional transmission delays – (from local PMUs to regionally PDCs) are stable at 20 ms with the simulation time increased from 0 to 300 s. The wide-area transmission delays (from distributed PDCs to centralized SPDC) are kept at 40 ms. Therefore, we can infer that the whole transmission delays (from distributed PMUs to centralized SPDC) are approximately 60 ms. Compared to the 100 ms communication requirement [9], the simulation results not just meet the need, but acquire a 40 ms margin.

In Figure 9.19 and Table 9.2, the network throughputs and link utilizations of critical pass are illustrated. From the data shown, the highest throughput channel is the communication link between SPDC and its connecting routing node. The other throughput values between two routing nodes increase proportionally with the increasing number of PMUs. The number of local PMUs can be controlled in a predictable manner. Therefore, the communication throughput between PMUs and PDCs is controllable. In this case, it is the lowest value of all the statistics. If more PMUs are installed in the WAMS infrastructure, the throughputs of these communication channels will increase accordingly. Therefore, the capacity issues of the existing backbone network are likely to become a bottleneck for the WAMS network.

Table 9.2. Throughputs and utilizations for different transmission stages

Segment	Direction	Throughput (Mb/s)	Utilization (%)
Regional	From RN to PDCs	0.428	21
Wide area	Between routing nodes	0.514–2.568	0.33–1.65
Wide area	From RN to SPDC	5.135	3.3

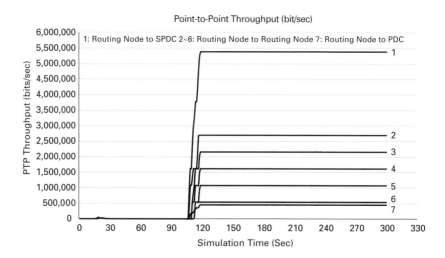

Figure 9.19 The throughput of three communication link types.

9.5.5 Protection simulation

The simulation results of the maximum end-to-end delays and the maximum response time for the major five protection applications are listed in Table 9.3. The processing flows of the five applications can be described as follows. In the adaptive reliability (AR) application, the system control centre decides the current state of the system, and then sends voting messages to nine distributed critical relays asking for acknowledgements. In the adaptive out-of-step (AOS) application, the PMUs, which are installed outside the generators, upload the measurements and track rotor angles. To determine coherent groups, the PMUs send the information to the PDC with no acknowledgements. In the supervision of backup zones (SBUZ) application, PMUs which fall inside the backup zones can monitor the apparent impedance and send decisions to backup relays with acknowledgements. Through joint decisions, the relays may eventually prevent mis-operations. In the intelligent load-shedding (ILS) application, the PMUs, which are used to monitor the tie-line power flows, will take charge of the supervisory control, and send measurements to PDC without acknowledgements. In the intelligent islanding (II) application, since instability is inevitable, the system control centre has to make decisions to trip or block the related circuit breakers with acknowledgements needed.

Table 9.3. Delays and response time of protection applications

Applications	Direction	Maximum ETE delay (ms)	Maximum response time (ms)
AR	SCC to 9 critical relays	40	79
AOS	2 Generator PMUs to PDC	44	N/A
SBUZ	10 PMUs to backup relay	25	55
ILS	4 Tie-line PMUs to PDC	44	N/A
II	SCC to 9 circuit breakers	46	90

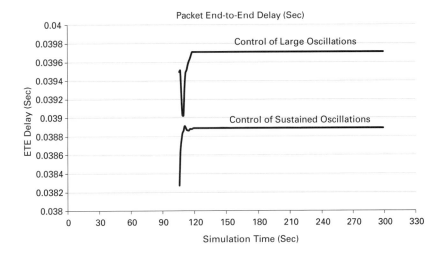

Figure 9.20 The ETE delay for power system control applications.

The maximum end-to-end delays and maximum response time of all the applications are shown in Table 9.3. In [9], the allowable timing constraints for ETE delay and response time are 50 ms and 100 ms, respectively. The simulation results show that our designed WAMS can satisfy the communication requirements of protection applications.

9.5.6 Control simulation

The power system control is a widespread concern application. We will take two realistic applications – the control of sustained oscillations and the control of large oscillations – as simulation examples. The simulation results for these two applications are shown in Figure 9.20.

In the sustained oscillations control application, using PMUs to damp the low-frequency inter-area oscillations is one of the most attractive methods in WAMS. In our simulation, we assume that there are 5 control devices and a total of 25 remote PMUs installed in the system. The simulation results show that the ETE delay for sustained oscillations control is below 50 ms. For the large oscillations control, the number of distributed remote PMUs will increase to 50. This application involves wide-area communication where measurement data might travel hundreds of miles. This application is used mainly for preventing transient instability. Although the ETE delay value is higher than with sustained oscillations control, the application delay below 50 ms can still make sure of the timely control of WAMS.

9.5.7 Hybrid simulation

In a realistic WAMS communication system, all the time-critical applications, which are mentioned in system monitoring, protection, and control together with other non-critical applications, may work over the same measurement system, as illustrated in Figure 9.21.

In practice, all the applications mentioned above are going to run simultaneously, no matter what the transaction types are. In order to guarantee the accuracy and effectiveness of PMU measurement data, these time-critical applications should have constraints on end-to-end transmission delay. In the worst-case scenario, we simulated these independent applications working at the same period of time. The hybrid simulation results of end-to-end delays are shown in Figure 9.22. The end-to-end delays between distributed

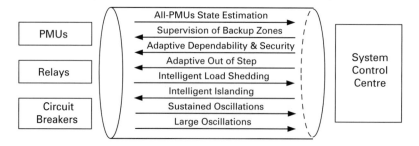

Figure 9.21 Applications running in the same communication architecture.

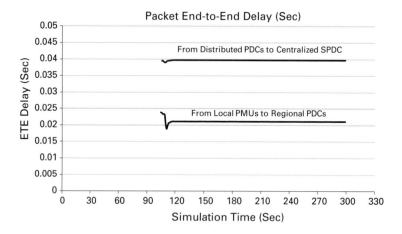

Figure 9.22 The hybrid simulation results of packet end-to-end delays.

PDCs and centralized SPDC are about 40 ms, which are the same as the delays of the all-PMUs state estimation application. The end-to-end delays from local PMUs to regional PDCs are a little higher than 20 ms, that represents the heavier traffic in regional area data transmission.

9.6 Conclusion

Historically, considered to be the most complicated electrical system, the power grid has evolved slowly over the recent half century. Starting from the invention of the PMU in 1988, many academic researchers and power engineers have paid much attention to the development and utilization of the PMU to promote the power grid in a future revolution. Until now, PMU constructed WAMS have been seen as the way to the future power grid, and the mentioned wide-area measurement technologies are promising to be the mainstream in smart grid operation at the transmission level.

In this chapter, we have defined and depicted the construction of realistic WAMS in detail and shown the special communication constraints imposed by the wide-area power system. A variety of WAMS applications in terms of state estimation, adaptive reliability, and damping of sustained oscillation, etc. combined with their communication issues are discussed. We have constructed a realistic WAMS in a simulation platform to study the nature of different communication infrastructures which are suited for the power grid. The simulation results have shown that with a reasonable fibre-optic backbone architecture and appropriate bandwidth protocols, all WAMS applications' network bandwidth and latency requirements can be met easily in the current scale. The constructed system model can leave room for further expansions through addition of more PMUs to monitor and control low-voltage lines as well.

Meanwhile, there remain some issues which can be improved. In order to gather more accurate data for deployment, the simulation model needs to be improved. More sophisticated models for PMU and PDC should be implemented. A step-by-step, pipelined time-scale processing representing every detail of the input and output data transmission is needed. In addition, almost all parameters of the transmission network can be adjusted. The topology of networks should also be more convenient to change.

From the emerging investments in WAMS, we predict that this is just the very beginning of PMU deployment. Every independent large-scale power grid operator will install hundreds of PMUs or even more in their high-voltage transmission lines. How the fully meshed communication network can accommodate increasing data transmission associated with WAMS applications will become an inevitable challenge. The diversity and practicality of WAMS applications are also worth researching. The merging of digital relays and PMUs into single devices will continue, and the processing for adaptive protection and WAMS-based control will also be distributed around the network rather than concentrated in the control centre. The designed applications will control the entire system intelligently, from measurement to computation to execution. In the meantime, the highlighted WAMS security issues will need to be addressed continuously.

References

[1] M. Shahidehpour and Y. Wang, 'Communication and control in electric power systems, applications of parallel and distributed processing', IEEE Press, John Wiley & Sons Inc., 2003.

[2] K. E. Martin, 'Summary of BPA phasor measurement system communication timing tests', personal communications, October 2005.

[3] A. G. Phadke and J. S. Thorp, 'Synchronized phasor measurements and their applications', Springer, 2008.

[4] K. E. Martin, 'Phasor measurement systems in the WECC', *IEEE Power Engineering Society General Meeting*, 2006.

[5] K. Kalam and D. P. Kothari, 'Power system protection and communications', New Age Science Limited, 2010.

[6] D. R. Mauro and K. J. Schmidt, *Essential SNMP*, 2 edn, O'Reilly Media, Inc., 2005.

[7] A. G. Phadke, 'The wide world of wide-area measurement', *IEEE Power & Energy Magazine*, pp. 52–65, September/October 2008.

[8] A.-R. A. Khatib, 'Internet-based wide area measurement applications in deregulated power systems', PhD thesis, Virginia Tech, VA, USA, 2002.

[9] A. G. Phadke and J. S. Thorp, 'Communication needs for wide area measurement applications', *Critical Infrastructure (CRIS) 5th International Conference*, September 2010.

[10] M. Shahraeini, M. H. Javidi, and M. S. Ghazizadeh, 'Comparison between communication infrastructures of centralized and decentralized wide area measurement systems', *IEEE Transactions on Smart Grid*, vol. 2, no. 1, pp. 206–211, March 2011.

[11] B. Naduvathuparambil, M. C. Valenti, and A. Feliachi, 'Communication delays in wide area measurement systems', in *Proceedings of 34th Southeastern Symposium on System Theory*, 2002.

[12] S. H. Horowitz and A. G. Phadke, *Power System Relaying*, 3rd edn, Research Studies Press Limited, John Wiley & Sons Inc., 2008.

[13] V. C. Gungor and F. C. Lambert, 'A survey on communication networks for electric system automation', *Computer Networks*, pp. 877–897, February 2006.

[14] H. G. Perros, *Connection-Oriented Networks: SONET/SDH, ATM, MPLS and OPTICAL Networks*, John Wiley & Sons Ltd, 2005.

[15] A. G. Phadke, 'Synchronized phasor measurements in power systems', *IEEE Computer Applications in Power*, vol. 6, no. 2, pp. 10–15, April 1993.

[16] J. Bertsch, C. Carnal, D. Karlsson, J. Madaniel, and K. Vu, 'Wide-area protection and power system utilization', *Proceedings of the IEEE*, vol. 93, no. 5, pp. 997–1003, May 2005.

[17] V. Terzija, G. Valverde, D. Cai, P. Regulski, V. Madani, J. Fitch, S. Skok, M. M. Begovic, and A. G. Phadke, 'Wide-area monitoring, protection, and control of future electric power networks', *Proceedings of the IEEE*, vol. 93, no. 1, pp. 80–93, January 2011.

[18] A. G. Phadke and J. S. Thorp, *Computer Relaying for Power System*, 2 edn, Research Studies Press Limited, John Wiley & Sons Ltd, 2009.

[19] J. D. L. Ree, V. Centeno, J. S. Thorp, and A. G. Phadke, 'Synchronized phasor measurement applications in power systems', *IEEE Transactions on Smart Grid*, vol. 1, no. 1, pp. 20–27, June 2010.

[20] M. G. Adamiak, A. P. Apostolov, M. M. Begovic, C. F. Henville, K. E. Martin, G. L. Michel, A. G. Phadke, and J. S. Thorp, 'Wide area protection – technology and infrastructures', *IEEE Transactions on Power Delivery*, vol. 21, no. 2, pp. 601–609, April 2006.

[21] A. S. Bretas and A. G. Phadke, 'Artificial neural networks in power system restoration', in *IEEE Transactions on Power Delivery*, vol. 18, no. 4, pp. 1181–1186, October 2003.

[22] M. Chenine, E. Karam, and L. Nordstrom, 'Modeling and simulation of wide area monitoring and control systems in IP-based networks', *IEEE Power & Energy Society General Meeting*, July 2009.

[23] K. Hopkinson, X. Wang, R. Giovanini, J. Thorp, K. Birman, and D. Coury, 'EPOCHS: a platform for agent-based electric power and communication simulation built from commercial off-the-shelf components', *IEEE Transactions on Power System*, vol. 21, no. 2, pp. 548–558, May 2006.

[24] OPNET Technologies Inc., http://www.opnet.com

10 Wireless networks for smart grid applications

David Griffith, Michael Souryal, and Nada Golmie

10.1 Introduction

By connecting the various entities in the grid and enabling a two-way flow of information related to the production and distribution of energy, communication networks, and more specifically wireless networks, are poised to play a significant role in the modernization of the electric grid. In fact, most functions related to making energy production and consumption more efficient require some form of collecting and sharing information to monitor current usage levels, predict future demands, balance loads, and tightly control the production, distribution, and transmission of power.

While there are many advantages for using wireless networks in the grid, there are many concerns as well. The benefits include providing untethered and universal access to information, and deploying and maintaining infrastructure easily and at low cost. Drawbacks are mostly related to signal propagation properties in what could be a noisy environment and are inherent to the use of a shared medium, which may be prone to interference.

This chapter discusses the use of wireless networks in the context of the smart grid. First, it provides an overview of the various applications envisaged in the smart grid and discusses their communication requirements. Sifting through thousands of communication requirements leads to an identification of different classes of applications based on their traffic and quality-of-service (QoS) requirements as well as an identification of the communication actors and their logical topologies. Subsequently, there is a discussion of the main factors to consider in designing and deploying wireless networks, including spectrum, coverage, resilience, and security. The chapter closes with a discussion of performance trends and tradeoffs related to coverage, capacity, latency, and reliability.

10.2 Smart grid application requirements

Before one can design a network to support smart grid communications, the traffic that the network will carry must be well understood. This section examines the types of applications that run over the smart grid communications network and classifies them according to their performance requirements and their traffic characteristics.

10.2.1 Application types

The smart grid as envisioned, will support many functions. It will allow utilities to collect fine-grained use statistics from large numbers of residential, commercial, or industrial customers. It will allow customers to determine when per-kWh prices are lowest and either schedule the operation of their appliances accordingly or allow smart appliances to do the scheduling themselves. Customers will also access use and price information to schedule when they charge their plug-in hybrid electric vehicles, and they may allow the utility to use the vehicle battery as an emergency power source. The utility may also allow customers to sell power back to the grid if they have on-site generation capacity such as roof-mounted solar panels or wind turbines.

As a result of its many functions the smart grid must support a varied set of applications. Examining the characteristics of these applications is an important step in determining the amount of traffic that the smart grid communications network must support. The Open Smart Grid (OpenSG) working group, which is a creation of the UCA International Users Group (UCAIug), has developed a matrix of application requirements for smart grid communications. These requirements are extensions of the use cases that have been tabulated by the National Institute of Standards and Technology (NIST) in the interoperability knowledge base (IKB) [1], by the Electric Power Research Institute (EPRI) in their Use Case Repository [2], and by Southern California Edison (SCE) in their use case collection that they developed with EPRI [3]. OpenSG identified distinct use cases for encoding into an applications requirements matrix. Each entry gives the communications endpoints associated with the requirement, the use case, how often messages are generated, the message size, and the performance requirements. The matrix includes not only the end-to-end application layer flow requirements but all of the individual link requirements along the flow's path.

10.2.2 Quality-of-service (QoS) requirements

The performance requirements for smart grid applications depend on the type of application and the location of the endpoint actors. The performance requirements in the OpenSG matrix are given in terms of latency and reliability. The latency is the total processing time and network transit time, measured from the time of message departure from the source actor to the time of message reception at the destination actor. Note that the latency is an end-to-end metric. The reliability is the probability that the message is successfully received within the latency limit. If we are examining the performance of a single link or group of links, we need to consider how the latencies associated with each link affect the total latency budget for the connection.

The OpenSG matrix presents these requirements in the form of a maximum latency and a minimum reliability. There are varying levels of complexity associated with these requirements. For example, an on-demand meter reading request from the meter data-management system (MDMS) to the two-way electrical meter must reach the meter successfully in at most 15 s at least 98% of the time. Note that the reliability requirement is defined so that a successful delivery of the message with a delay that is greater than the threshold is still considered a failure. Formally, we require that

$$P_f + P_s \int_L^\infty f_{D|s}(t)\, dt \leq 1 - R, \quad \text{or} \quad P_s \int_0^L f_{D|s}(t)\, dt \geq R, \tag{10.1}$$

where R is the reliability requirement, L is the latency requirement, P_f is the probability that the message is lost somewhere in transit, $P_s = 1 - P_f$ is the probability that the message reaches the destination actor, and $f_{D|s}(t)$ is the density of the delay of a message, conditioned on successful delivery.

The matrix also includes more complex reliability and latency criteria. For example, for the case of the two-way meter sending multi-interval data to the advanced metering infrastructure (AMI) headend, we have a latency requirement equal to the time between message transmissions, which is usually 4 hours or 6 hours. The corresponding reliability requirement is

$$R = \begin{cases} 90\%, & \text{success every 4–6 hours} \\ 98\%, & \text{success over a day} \\ > 99.5\%, & \text{over two days.} \end{cases}$$

Here, we require that the probability that a single message reaches the destination within the allotted latency limit 90% of the time, while the probability that at least one message reaches its destination over the course of a day is 98%, and 99.5% over two days.

10.2.3 Classifying applications by QoS requirements

The applications supported in the smart grid vary considerably with respect to their QoS requirements, which are encoded in the reliability and latency requirements that we discussed in the previous section. In addition, the various types of messages associated with a single use case can have very different QoS requirements, depending on the message type.

In Figure 10.1, we show a graphical representation of the relative QoS requirements for a set of applications identified in the OpenSG smart grid requirements matrix, with an accompanying legend that defines each of the acronyms in the figure. Following the classification scheme in the requirements matrix, the applications are partitioned into use cases such as meter reading, plug-in hybrid electric vehicles (PHEVs), or price. For each use case, the figure compiles the end-to-end QoS requirements for each message and encodes them as (reliability, latency) ordered pairs. Based on their values, each of the hundreds of ordered pairs was assigned to one of the 16 bins shown in the figure. For example, the reliability and latency requirements for all 20 messages associated with the voltage/volt-ampere-reactive (VAR)-centralized control (VVC) use case are '> 99.5%' and '< 5 s', respectively. Thus, the VVC use case is shown as a single entry in the lowest, rightmost bin in the figure. For any use case that appears in the figure as a single acronym with no surrounding shaded region, all its associated reliability and latency requirements fall into the bin where the use case name is located.

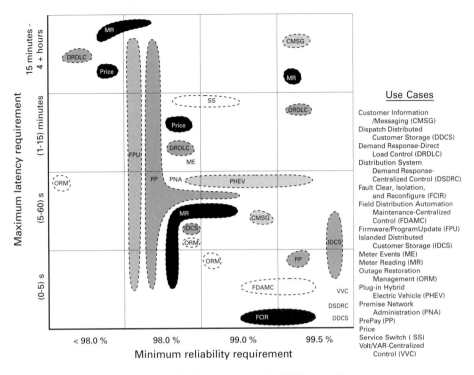

Figure 10.1 Major smart grid use cases, categorized by latency and reliability requirements.

The reliability requirements for all messages never fall below 90%, with the single exception of the requirement for outage restoration management (ORM) that two-way meters should be able to send outage notification to the outage management system (OMS) within 20 s, 30% of the time during large outages. Aside from that one case, we essentially have a four-tier, 'good, better, best, very best' system for rating reliability requirements. As smart grid deployment proceeds, these requirements may become more nuanced and diversified.

The figure also shows use cases whose QoS requirements occupy multiple bins, such as PHEV. By examining the requirements matrix, we found that, out of 11 messages, 8 had respective reliability and latency requirements of '> 98%' and '< 15 s', and the remaining three messages' requirements were '> 99%' and '< 10 s', '> 99%' and '< 15 s', and '> 99.5%' and '< 10 s'. This results in the requirements for PHEV being spread across three bins, as shown in the figure. Likewise, shaded regions that stretch across multiple bins indicate that QoS requirements associated with that use case exist in each bin. Also, multiple 'islands' of the same shade indicate the presence of QoS requirements for a use case in multiple non-adjacent bins.

From the figure, one can make the observation that the use cases' requirements, with some exceptions, fall largely along the diagonal bins. In the lower right-hand corner, there is a clearly delineated set of high-QoS use cases (i.e., low latency and high reliability) such as distribution system demand response-centralized control

(DSDRC), dispatch distributed customer storage (DDCS), field distribution automation maintenance-centralized control (FDAMC), fault clear, isolation, and reconfigure (FCIR), islanded distributed customer storage (IDCS), and VVC. There is also a large group of intermediate-QoS use cases in the centre bins, including meter events (ME), meter reading (MR) (multi-interval and on-demand commands, responses, and errors), ORM, PHEV, premise network administration (PNA), and service switch (SS). The upper left-hand corner of the figure contains a set of relatively low-QoS use cases that include demand response-direct load control (DRDLC), price, and bulk meter read messages.

Because of the existence of this hierarchy of use cases, and because there are broad variations in QoS requirements within use cases, the network will need to implement QoS-based treatment of messages based on their delay and loss tolerances. This is similar to the need for differentiation of service for packets in the commercial Internet based on traffic type. However, there are differences between smart grid traffic and the traffic that the Internet carries. For example, real-time voice and video traffic has stringent latency requirements but can tolerate occasional packet losses (strict latency, low reliability), while file transfers may have more relaxed latency requirements but no tolerance for packet loss. Figure 10.1 shows that some smart grid applications have strict latency and strict reliability requirements (e.g., FDAMC) while others have loose reliability and latency requirements. The network will have to give priority to messages associated with VVC, DSDRC, and DDCS, while, for example, meter event messages can be given lower priority. For other cases, such as pre-pay (PP) messages, the priority level given to a message will depend on its type; some PP messages must be delivered within seconds while others can be delayed for much longer without causing problems.

Example: firmware/program update (FPU)

Several use cases defy easy classification because their requirements are spread over many bins. An example is the FPU use case, which consists of a set of applications that push software or firmware updates out to equipment in the field (two-way meters and demand aggregation points (DAPs)). These updates can be sent from entities in the network (e.g., the AMI headend) or from field tools. Updates and update confirmation messages have relatively low reliability requirements, and the tightest latency requirements are associated with update confirmation messages. Longer latencies, on the order of minutes to hours, are acceptable for the updates themselves. The lowest delay tolerances are associated with the DAP, which is a relatively critical piece of equipment, while longer latencies are allowed for the two-way meters, up to 7 days per 100,000 meters. The allowed latencies for firmware updates are greater than those for program updates, but this is because the firmware updates are expected to be larger.

Example: field distribution automation maintenance-centralized control (FDAMC)

The FDAMC use case involves communications between the distribution management system (DMS) and various field devices such as capacitor bank controls (CBCs), reclosers, switches, voltage regulators, and field sensors. The messages consist of regularly transmitted status request and sensor data request messages from the DMS to the various devices, status response messages and sensor data messages from the devices back to the DMS, alarms and operation failure reports from the devices to the DMS, and alerts indicating deviations or status change, again from the devices to the DMS. The timely and accurate reception of all these messages is critical to ensuring that the electrical network functions well, and all of these message types have high reliability and low latency requirements.

Example: plug-in hybrid electric vehicle (PHEV)
As shown in Figure 10.1, the PHEV use case is an example of one whose requirements fall across a contiguous set of bins, where the reliability requirements vary but the latency requirements are all roughly the same. The three PHEV messages with high reliability requirements are error messages associated with power charging rate messages and vehicle identification number (VIN) information requests, which are transmitted from the AMI headend to the PHEV and operations entities such as the load management system (LMS). The remainder of the messages, including other types of error messages, have relatively low reliability requirements, so that the PHEV use case overall is a relatively low-QoS one.

Example: customer information/messaging (CMSG)
Customer messaging QoS requirements are split into two non-adjacent bins in Figure 10.1. One group of messages, with maximum latencies of 15 s and 30 s and associated reliabilities of 99%, consists of customer account requests that originate from a web browser and the associated data responses from the appropriate web portal, which can belong to the utility or a retail electric provider (REP), or which can be a common web portal. Also included in this group are on-demand meter-reading error messages, which the customer's web browser forwards to the utility's web portal from the in-premises device (IPD) or energy-management system (EMS), depending on how the customer's premises are configured. Because the latency and reliable delivery of these messages directly impacts customer satisfaction, it is important that they be delivered promptly.

The other group of CMSG messages can be delayed by up to 1 hour but must arrive within that time period 99.5% of the time. These messages contain bulk customer account information, meter information, and outage information that are sent to an operational data warehouse (ODW) from the customer information service (CIS)/billing-utility, AMI headend/MDMS, and OMS actors, respectively. The ODW forwards the information to the various web portals in a secondary wave of bulk data-transfer messages. These transfers take place at night, and are not time-critical in the same way that information exchanges are that involve the customer as a direct participant.

Example: meter reading (MR)
The meter reading use case incorporates several types of applications that run at different times of the day, and which have very different QoS requirements. Figure 10.1 shows three clusters of meter reading requirements; one is in a relatively low-latency bin while the other two are in high-latency bins. The messages in the lowest-latency bins, which span the 98% and 99% reliability ranges, are associated with on-demand meter reading

events, which can originate from the CIS/billing-utility actor or from the customer, and with multi-interval meter reading events, in which the MDMS sends read requests to individual meters. Both of these applications are active during daytime hours.

The high-reliability, high-latency application that resides in the upper right-hand bin in the figure is bulk meter reading, which occurs during night hours. Commands to send bulk data originate in the CIS/billing-utility actor and pass to the AMI headend via the MDMS. These message transactions have allowed latencies of up to 1 hour, but must complete within that time 99.5% of the time.

The AMI headend responds to bulk meter-read commands by sending the data that its subordinate meters sent to it during the preceding day. The responses, which can be on the order of many megabytes in size, have loose latency requirements, typically with delivery windows of 2–4 hours, and reliability requirements that are relatively low with respect to those for use cases such as VVC. These requirements allow for a 90% success rate for a single message, with higher success rates over the course of multiple days, as described in the previous section.

10.2.4 Traffic requirements

In addition to their QoS requirements, we can classify smart grid applications and use cases according to their traffic characteristics. The OpenSG matrix lists message frequency and message size for each use-case message, and we can collect these together for each use case to create a composite set of requirements that indicate the use case's contribution to the network load.

In this subsection, we consider four of the example use cases whose QoS characteristics are described in Section 10.2.3. We catalogued the arrival rate and corresponding message-size information for all the messages associated with each use case, and plotted the resulting (rate, size) ordered pairs using a log–log scale. Because rates in the matrix are typically given on a per-device basis, we use the same convention in the graphs. When either the rate or the size entry denotes a range of values (e.g., if the 'How Often' column entry is, '1–3 times per day'), we plot the resulting set of ordered pairs as a horizontal line if the message size is a constant or as a vertical line if the arrival rate is a constant. If both the rate and the size are intervals, we plot the resulting set of ordered pairs as a shaded region on the graph. The resulting plots for the FDAMC, PHEV, CMSG, and MR use cases appear in Figure 10.2(a), (b), (c), and (d), respectively.

Example: FDAMC

We start by considering the FDAMC use case which, as we recall from Section 10.2.3, involves status and sensor data requests and responses and alarm messages between the DMS and various field devices, which have strict latency and reliability requirements. Operation failure alarms occur about once per month for each voltage regulator, capacitor bank, recloser, and switch. All of these messages are 50 bytes in size except for a 100 byte message associated with the voltage regulator. Messages that occur about once a week per device include alarms from voltage regulators, capacitor banks, reclosers, and switches,

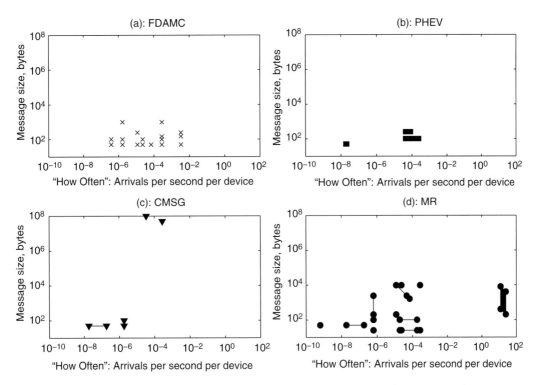

Figure 10.2 Plot of message size vs. arrival rate for use cases: (a) FDAMC, (b) PHEV, (c) CMSG, (d) MR.

as well as status requests and reports between the DMS and field sensors. These messages range in size from 50 bytes to 1 kB. Reclosers and switches also generate alarms and status-change alerts once per day per device on average; these messages are 50 bytes long, except for the recloser status change message, which is 250 bytes in length.

The most frequent message exchanges occur from twice per day per device to once every 5 min per device, and involve the exchange of status information with the DMS. These messages range in size from 50 bytes to 1000 bytes. The resulting traffic characteristic appears in Figure 10.2(a). Looking at the figure, we can see that the message sizes associated with this use case tend to be small and that the load generated by a single field device is rather light, on the order of a few bits per second. However, a DMS typically communicates with multiple feeder lines, each of which has multiple devices associated with it, so that there may be hundreds of switches, reclosers, and other devices generating FDAMC traffic. In such a case, the most frequent messages will result in an offered load of a few hundred bytes per second.

Example: PHEV

The PHEV use case consists of a relatively small number of message types. PHEV messages are sent only during daylight hours, in a 15-hour span. There are three message sizes: 50 bytes, 100 bytes, and 255 bytes. The 50-byte messages are all error messages, with the exception of a PHEV VIN information request message from the PHEV to the

MDMS. We assume that the error messages occur at a rate of 1 message per 1000 PHEV meters per day; the VIN data command message rate is defined as one transaction per PHEV meter connection per day; we note that not every PHEV-connected meter connects to the network every day. The 100-byte messages are rate-negotiation messages and charging-status message that are exchanged by the PHEV and the LMS-utility actors. The negotiation message exchange occurs 2–4 times per PHEV-connected meter per day; the status message goes out 2–4 times per charging event per day. Finally, the 255-byte message contains price-rate data, and goes from the LMS to the PHEV once per PHEV meter connect per day. We plot the resulting rate/size profile in Figure 10.2(b), which includes the ranges in the arrival rates.

Example: CMSG

In the customer-messaging use case, there are three message sizes: 50 bytes, 100 bytes, and an unspecified large message. The 50-byte messages are customer account information requests that the customer web browser sends to the appropriate web portal and that occur at a rate of 100 messages for every 1000 meters, and on-demand meter-read error messages. We assume that these error messages are generated once per 1000 meters per day. The 100-byte messages are all response messages that the web portal sends back to the customer browser and these messages are generated on a one-to-one basis for each request, so they also are generated at a rate of 100 messages per 1000 meters per day. The 50-byte and 100-byte messages are all sent during daytime hours. The large messages are sent at night during an 8-hour period. These are batch messages, in which the data for groups of meters is reported out at regular intervals so that all meter information is reported by the end of the 8-hour period. We assume that the message generation rates during the nighttime period vary from one per hour to one per day, and we assume the amount of data in the messages is 100 bytes per meter for the single-batch transaction, and 400 bytes per meter for the hourly batch transaction. We also assume a population of 1,000,000 meters, which gives the plot shown in Figure 10.2(c).

Example: MR

The meter-reading use case, like the customer-messaging use case, consists of two sets of messages that are sent at different times of the day. The meter-reading use case consists of a large number of message types, and the corresponding message-generation rates depend on network settings, as well as the number of meters. The resulting plot of message size versus message arrival rates per device is shown in Figure 10.2(d).

The smallest messages associated with this use case are 25 bytes in length, and are all meter read requests. As we discussed in Section 10.2.3, the requests associated with on-demand meter reading and multi-interval meter reading occur during the day, while the bulk meter read commands occur only at night. Meter read commands from utility entities occur at a rate of 25 commands per 1000 meters per day, while similar commands from the customer EMS or IPD occur more frequently, at a rate of 1–10 commands per meter per day. The frequency of bulk meter read commands depends on how the smart grid communications network is configured, but we assume that they occur over a range of rates, from a rate of one per day to 12 per day (once per hour during the night-time period when bulk-data transfers occur). The 50-byte messages associated with this use case are all error messages; the exact rate is unspecified, and is given in the matrix as one error message per x read requests, so we will assume that $x = 1000$ in this case.

All 100-byte meter-read messages contain on-demand meter-read response data, and these occur at the same rate as the 25-byte request messages that generated them. The two-way water meter sends a 200-byte data packet to the AMI headend once per day, while the gas and electric meters send multi-interval response data to the AMI headend 25 times per 1000 meters per day, in a data packet whose size ranges from 200 bytes to 2400 bytes.

The remaining message types are response messages that contain metering data. Their size depends on the interval between responses and the amount of data that is collected for each interval. Gas and electric meters send multi-interval metering data to the AMI headend at regular intervals over the whole day; this information is bundled by the AMI headend and sent to the MDMS and CIS/billing-utility actors at night, when the load on the smart grid communications network is lighter. The gas meter data is sent out 1–6 times per day, and contains from 4 hours of data (if sent 6 times per day) to 24 hours of data (if sent once per day). The OpenSG matrix lists the corresponding message sizes as 1600 bytes for 4 hours of data and 2400 bytes for 6 hours of data, so 1 hour of data requires 400 bytes. This means that the message that is sent once per day is 9600 bytes long, and that in general we need a message size of $9600/N$ bytes to encode data that is sent N times per day. We can represent this with a line segment in the rate/size plot.

Electric meters also send multi-interval data to the AMI headend; commercial/industrial meters send the data 12–24 times per day, while residential meters send reports 4–6 times per day. This data is taken at intervals of 15 minutes to 1 hour, and consists of 4–20 data points per interval for commercial/industrial meters and 4–8 data points per interval for residential meters. If a meter sends data every hour, with 4 points per interval, the resulting message size is 200 bytes, which implies that each data point is associated with 50 bytes of data. The largest messages use 20 data points at 15-minute intervals, which requires 4000 bytes to encode an hour of data, and 8000 bytes to encode 2 hours of data. This results in a parallelogram-shaped region on the plot, as shown in the figure.

Finally we have the collected bulk meter-read data that flows from the AMI headend to the CIS/billing-utility via the MDMS. These flows occur between 6:00 pm and 6:00 am, and their size is not specified. We assume that they are sent out from once per night to once per hour (i.e., 12 times per night). We assume that each meter sends 400 bytes of data per hour so that the overnight reports contain 9600 bytes of data per meter.

10.3   Network topologies

The smart grid network exists in the physical world as an arrangement of devices that communicate over wired or wireless links. This physical network topology has an associated logical topology, which describes the message flows between the various actors in the network. Examining the logical network topology associated with a single-use case in combination with the use-case QoS requirements and traffic characteristics gives insight into network behaviour and can provide information that in turn can be used to optimize the placement of communication devices in the physical network topology.

10.3.1 Communication actors

An actor in the smart grid is any device or system whose application layer acts as the source or destination for a communications network flow that is required to support some function of the smart grid. A large number of actors have already been defined as part of the work on smart grid standards by NIST and other groups [4]. The OpenSG matrix has considered a subset of these actors in defining use case requirements. Actors can be classified according to what part of the smart grid they occupy; one possible mechanism for doing this is to use the set of domains that were used in [4] as part of the conceptual model that was developed to aid in identifying interactions between different parts of the network.

If we use this approach, we can sort the OpenSG actors into a number of domains. The most familiar of these domains for most people is the customer domain. This is typically a residential, commercial, or industrial premises that contains electrical service end-users. At its edge are a variety of two-way meters (water, gas, and electric), which monitor electricity use and report it out so that the customer can be billed. Associated with the meter is an energy systems interface (ESI), which may be inside the meter or housed in a separate device. The customer domain also encompasses smart appliances, which communicate with the meter in order to take advantage of changes in billing rates or smooth out aggregate customer demand, and PHEVs, which also communicate with the electricity provider to negotiate rates and log charges. This domain also may include a programmable communicating thermostat (PCT), a customer web browser, and, depending on the customer's premises configuration, an IPD or a customer EMS.

Elements of the distribution domain are also familiar to many people; this domain includes the local feeder lines, transformers, and substations that are visible in many neighbourhoods. The distribution domain's actors communicate with actors in the customer domain via the AMI network; at the edge of this network are DAPs and Field Network Gateways. The actors in this domain include feeder capacitor banks, fault detector sensors, reclosers, sensors, switches, and voltage regulators. The distribution domain also includes circuit breakers and sectionalizers, and can possibly include distribution storage devices. In addition, there are field tools that are used by field crews; these are not associated with a particular domain, but they connect to the communications network by interfacing to the customer meter, a DAP, or the AMI network directly.

The distribution operations domain contains actors that are responsible for network monitoring and control, as well as recording and storing network data, and performing data analysis. This domain contains the control entities such as the MDMS, the network management system (NMS), the LMS, the distribution supervisory control and data acquisition (SCADA) front-end processor (FEP), the demand-side manager (DSM), the

DMS, and the AMI headend, which acts as the communications gateway for the actors in this domain when they interact with actors in other domains.

10.3.2 Connectivity

In the smart grid network, each actor does not communicate with every other actor. Using the OpenSG matrix and considering only the parent requirements that identify the application layer endpoints for individual message flows, we developed a connectivity matrix that identifies the use cases that are associated with flows from one actor to another. The resulting matrix is rather sparse; many actors are sources or destinations for only a small number of message types. Using this information, we can develop a logical topology for each of the use cases. In contrast to a network's physical topology, which is the real-world arrangement of the network devices and the connections between them, a logical topology is defined by the set of communications pathways between actors. If two actors are endpoints for a message transmission, then they are directly connected in the logical topology, even though there may be multiple links and devices between them in the physical topology.

We show an example in Figure 10.3. The physical topology, consisting of network nodes A through F, is at the bottom of the figure, with physical links between nodes shown as black lines. In this example network, there are application-layer message flows between nodes A and C, B and C, and E and F. The resulting logical topology appears at the top of the figure. Node D is not part of the logical topology since it does not exchange any application-layer messages with any of the other nodes in the network. Also, the logical topology is partitioned into two islands, even though the physical topology is fully connected; furthermore, nodes A and B are not connected in the logical topology even though they share a physical connection.

Consider the FDAMC use case. As we noted in Section 10.2.4, the various feeder devices such as switches and voltage regulators typically communicate only with the DMS. Reclosers and switches also communicate with the distributed application controller (DAC), but in connection with the FCIR use case. This results in the star-shaped

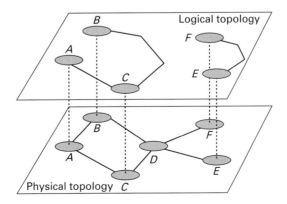

Figure 10.3 An example of a logical topology overlaid on a network physical topology.

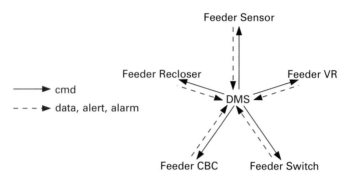

Figure 10.4 End-to-end flows between actors for the FDAMC use case.

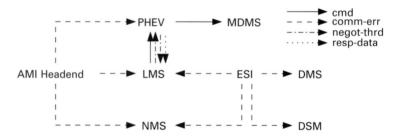

Figure 10.5 End-to-end flows between actors for the PHEV use case.

logical topology shown in Figure 10.4. Recalling the relative sizes of the messages, the figure also reveals that the bulk of the traffic flows upstream from the feeder devices to the DMS.

Next, we consider the logical topology for the PHEV use case, shown in Figure 10.5. The bulk of the traffic in this use case is between the PHEV and the LMS, with a single command (cmd) message to the MDMS to request VIN information. The communication error (comm-err) messages that originate from the AMI headend and ESI are generated only in the event of communications failures involving cmd messages, rate-negotiation messages (negot-thrd), or data (resp-data); these messages go to multiple distribution operations domain entities as part of the network's error tracking and response functions.

We show the virtual topology for the CMSG use case in Figure 10.6. From the figure, we can easily see that the primary interactions between actors are the requests and responses that flow between the customer browser, which can be in a traditional computer or a mobile device, and either a utility or REP web portal, or a common web portal. Behind the scenes, from the customer's point of view, the web portals are stocked with account information from the centralized ODW, which receives regular data updates from the CIS/billing, MDMS, and AMI headend actors, as well as information on outages and failures from the OMS actor.

Finally, we consider meter reading, which is one of the more complex use cases. The logical topology for this use case appears in Figure 10.7. Command-type messages originate from the IPD or customer EMS in the case of customer-initiated on-demand

Wireless networks for smart grid applications 247

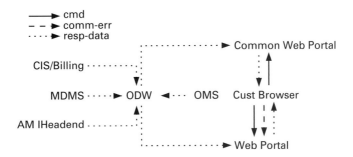

Figure 10.6 End-to-end flows between actors for the CMSG use case.

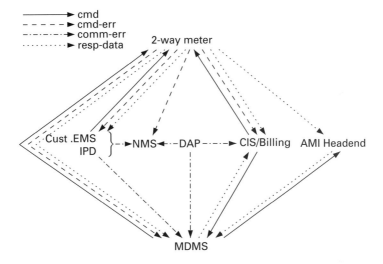

Figure 10.7 End-to-end flows between actors for the MR use case.

meter reading events, from CIS/billing or the MDMS to the two-way meter in the case of operations-based multi-interval meter reading, and from CIS/billing to the MDMS and from the MDMS to the AMI headend in the case of night-time bulk meter-data transfers. Meter data (the resp-data flows) originates at the meter and travels either to the IPD/customer EMS in the case of customer-initiated on-demand meter reading or to the requesting operations domain entity in the case of all other on-demand meter reading. Data also flows to the AMI headend in the case of multi-interval meter reading, and at night the AMI headend forwards this data in bulk form to the MDMS, which forwards it to CIS/billing. Error messages associated with this use case can be of two basic types. Command error (cmd-err) messages are notifications of a failure in responding to a meter-reading request, and originate from the meter and return to the requesting actor. Communication error (comm-err) messages are all sent to the operations domain entities that monitor faults (MDMS, NMS, and CIS/billing), and are sent in response to meter-reading faults. Note that the DAP is an origination point for these messages in addition to the IPD and customer EMS; meter-read requests and data responses traverse the

DAP, which is responsible for reporting communications failures on the interface to the customer domain.

Using these logical topology diagrams, we can obtain a clearer picture of the activity associated with each of the use cases that the smart grid must support, and we can clearly see which actors are involved. Combining this information with the QoS and traffic characteristics of the various use cases will let us do additional analysis to determine what resources are needed to support all the network's traffic requirements.

10.4 Deployment factors

This section highlights some of the key factors to consider in the deployment of wireless networks in general and smart grid wireless networks in particular. They include the choice of spectrum, the coverage and capacity of the network, and its resilience to failure. Given the rather stringent requirements imposed by smart grid applications on coverage and reliability, this section concludes with a brief discussion of opportunities for sharing wireless network resources to reduce costs.

10.4.1 Spectrum

One of the most important factors to consider when deploying smart grid wireless networks is the choice of spectrum. A fundamental choice is that between a licensed and an unlicensed spectrum. The major advantage of a licensed spectrum is that the utility would have exclusive use of the spectrum and, therefore, be better able to manage RF interference in the allocated band. This benefit, of course, comes at the cost of purchasing or leasing the licence to use the spectrum.

An unlicensed spectrum, on the other hand, while not having any associated licensing costs, presents other issues, chief among them being the need to coexist with other users sharing the band. The use of an unlicensed band implies the lack of direct control over the level of in-band RF interference and, therefore, a greater degree of uncertainty in performance. To control and mitigate interference, an unlicensed band may also have more stringent limitations on radiated power, impacting range, as well as limitations on the access scheme (e.g., mandated use of spread spectrum techniques), impacting spectral efficiency.

Another decision facing network planners is the location of the spectrum. Lower-frequency bands tend to offer longer propagation distances than higher-frequency bands, but bandwidth there may be more scarce and in greater demand. Thus, the choice of spectrum has implications for both the coverage of a wireless network as well as for its capacity. The following sections elaborate on propagation issues, as well as factors to consider in coverage and capacity planning of smart grid wireless networks.

10.4.2 Path-loss

An important factor in the design of a wireless network is the effect of the wireless channel itself. As the wireless signal propagates from its source to its destination, it experiences *path-loss*; in other words, various effects cause the amount of signal energy that impinges on the receiver's antenna to be much less than what was launched from the transmitter. Physical obstructions such as trees, buildings, or terrain features can absorb some of the signal's energy, or block it altogether, and environmental phenomena such as the amount of water vapour in the air can cause signal attenuation as well. Finally, the signal can reflect off various objects; these reflections are often delayed with respect to the component of the signal that travels along the line of sight from the transmitter to the receiver. As a result, the various signal components can add constructively or destructively, introducing another element to the path-loss. Because of the effect of obstructions, the height of both the transmitting and receiving antennas impacts path-loss as well. Determining the dependence of path-loss on transmitter–receiver distance is an important part of wireless channel modelling.

Ignoring multipath, the simplest expression of path-loss is the Friis equation, which expresses the loss as a function of d, the distance from the transmitter to the receiver, as follows:

$$PL(d) = PL(d_0) + 10n \log_{10}(d/d_0) + \chi, \qquad (10.2)$$

where d_0 is the reference distance, n is the path-loss exponent, and χ is the shadowing loss. Usually one models the shadowing as a log-normally distributed random variable (i.e., a random variable that is normally distributed in the dB domain), whose standard deviation in dB is σ_χ.

The value of the path-loss exponent is important for determining channel behaviour. In an obstruction-free environment, known as free space, $n = 2$ since signal strength decreases with the square of the distance from the transmitter. In most environments, we will have $n > 2$.

Since the 1960s, various people have developed more sophisticated path-loss models. The Hata model is one of the most well known, and characterizes path-loss in urban environments at carrier frequencies ranging from 150 MHz to 1500 MHz [5]. It features a closed-form expression that captures the results of previous work by Okumura, which presented the path-loss solely as plotted curves. The Hata model was extended by Action 231 'Evolution of Land Mobile Radio (Including Personal Communication)', which operated under the aegis of Cooperation in Science and Technology (COST). COST is an intergovernmental instrument that is directed by the European Union. The extension allows the Hata model to be used for frequencies in the range of 1500–2000 MHz and distances in the range of 1–10 km. The valid ranges for the transmitter and receiver antenna heights are 30–200 m and 1–10 m, respectively.

10.4.3 Coverage

In many smart grid applications, such as advanced metering or distribution automation, it is desirable to reach as many actors with as little communication infrastructure as possible, especially under coverage constraints.[1] While coverage is typically discussed in

---

[1] Under capacity constraints, additional infrastructure may be needed to increase capacity, as discussed in Section 10.4.4.

the context of point-to-multipoint topologies, the same underlying issues apply to point-to-point links; the transmitter–receiver range determines how far a point-to-point link can go without the additional infrastructure, and potential throughput loss, of a repeater.

The prior discussion on spectrum pointed to the relationship between the choice of spectrum and propagation characteristics, with lower-frequency bands offering longer propagation distances than higher-frequency bands. The effect of carrier frequency on coverage can be illustrated with the Hata path-loss model.

Figure 10.8 shows the path-loss versus transmitter–receiver distance at three different carrier frequencies according to the COST231–Hata model, assuming a transmitter antenna height of 30 m, a receiver antenna height of 1.5 m, and a suburban environment. At a transmitter–receiver distance of 2 km, for example, the model predicts a median path-loss of 148 dB at 2 GHz compared with only 125 dB at 700 MHz. Alternatively, a system which could only reach 1 km at 2 GHz would be able to reach 5 km at 700 MHz, according to this model.

While the choice of spectrum influences coverage, another factor which has a large impact on coverage – but one over which the network planner has little control – is the type of environment needing coverage. The Hata model is based on an urban area but includes corrections for suburban and open rural areas. Figure 10.9 illustrates path-loss in urban, suburban, and rural environments at 900 MHz, according to the Hata model. At this frequency, the model predicts the path-loss in an urban area to be approximately 10 dB higher than in a suburban area and nearly 29 dB greater than in a rural area. In terms of range, a signal that would reach 1 km in an urban area would reach nearly 2 km in a suburban area and over 6 km in a rural area.

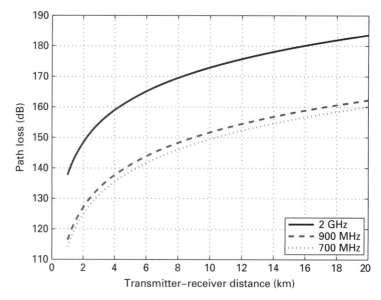

Figure 10.8 Path-loss (dB) vs. distance (km) at various carrier frequencies according to the COST231–Hata model.

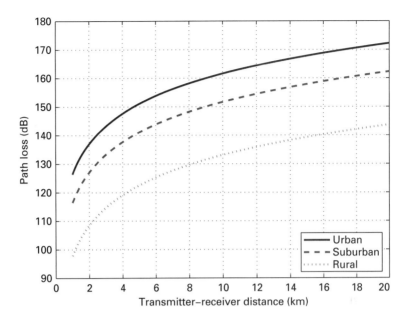

Figure 10.9 Path-loss (dB) vs. distance (km) in various environments at 900 MHz according to the Hata model.

Conventional wireless networks will require a minimum number of infrastructure elements (base stations or access points) to provide the necessary coverage in a utility's territory. The required infrastructure will roughly scale inversely with the footprint of an element. As each infrastructure element requires backhaul to the core (typically wired) communication network, it is desirable to minimize the number of installed elements.

One approach being pursued by some utilities to reduce the number of infrastructure elements in coverage-limited networks is to use a mesh architecture to extend the range of each infrastructure element. In a mesh network, end devices such as smart meters or sensor nodes communicate with one another on a peer-to-peer basis to reach the nearest infrastructure element. While a sensor node within the coverage footprint of a base station would have a direct link to the core network, nodes out of range of the base station would use other sensor nodes as relays to establish a connection with the nearest base station. An added benefit of a mesh architecture is that if a base station were to fail, the end devices could reroute their communications to a neighbouring base station. However, the mesh approach is only viable when the last-hop links to the base station have excess capacity to support the additional relayed traffic. Furthermore, when a mesh approach is applied to advanced metering, security concerns associated with relaying a customer's information through other customers' meters must be adequately addressed.

10.4.4 Capacity

The capacity of a smart grid wireless network is the maximum level of traffic it can reliably carry. Since most smart grid applications are data oriented, capacity is most generally measured in bits per second. For specific applications, such as advanced metering, it could also be measured in terms of the maximum number of smart meters an infrastructure element can support, assuming some level of traffic per meter.

At the level of an individual link or infrastructure element, capacity depends on the spectral efficiency of the wireless technology in use, measured in bits per second per hertz. Multiplying the technology's spectral efficiency by the amount of spectrum in hertz gives a rough measure of the raw capacity of an infrastructure element. To arrive at an application-specific measure of capacity, one would need to account for the overhead of intervening layers between the radio access network and the application layer, such as the overhead for addressing, encryption, and network and session management. Capacity is also influenced by the quality of the RF channel, including path-loss, multipath fading, noise, and interference. Techniques that a technology may use to enhance capacity on an individual link include spatial multiplexing with multiple antennas and link adaptation.

At a network-wide level, capacity can also be increased through careful network planning. For example, when deploying a cellular network, extensive planning of carrier frequency use and cell-site antenna configurations is performed to reduce inter-cell interference, thereby increasing link quality. Furthermore, when a cell's traffic load nears capacity, the cell can be split, introducing another infrastructure element to carry some of the load.

Planning of a smart grid wireless network will need to ensure that capacity is sufficient to handle expected traffic loads. During the earlier phases of a service roll-out, infrastructure requirements will be driven primarily by coverage constraints. A minimal level of infrastructure will be needed to provide the necessary coverage. However, as a deployment matures and the number of end devices increases, capacity constraints may begin to take hold. Additional communications infrastructure (e.g., access points or base stations) may be needed in order to increase the communications capacity of the smart grid. Capacity planning should take into account the expected load of a range of scenarios. According to a study sponsored by the Electric Power Research Institute, capacity requirements of a smart grid wireless network will be driven by the communication needs during disaster recovery rather than the needs of day-to-day operations [6].

10.4.5 Resilience

Given that smart grid networks will be expected to operate during disaster recovery, hardening of the communications infrastructure for resilience to natural and man-made incidents will be necessary. Besides hardening equipment for physical integrity, fault-tolerant designs can include overlapping coverage provided by different infrastructure elements, as well as backhaul facilities with alternate routes. In some cases, the wireless network as a whole may serve as secondary backup to a primary wireline system. A truly resilient design will also include backup power via battery or local generation. In addition to measures protecting the permanent network, a disaster recovery plan may also include the use of portable, truck-mounted wireless infrastructure that can be deployed to incident areas on a temporary basis with microwave or satellite links to the core network.

10.4.6 Security

The broadcast nature of wireless communication renders it somewhat more susceptible to security threats than wired communication. For example, rogue servers may spoof client devices and launch man-in-the-middle attacks, allowing adversaries to steal sensitive information or inject malicious software into the system. Furthermore, consumers have expressed concerns about the privacy of their usage information, particularly because appliances have unique signatures that make metering data a trove of information about the activities and routines of the members of a household [6].

Consequently, security in the wireless context must be thoroughly addressed at various layers of the protocol stack. The OpenSG requirements address this need to a certain extent by requiring additional protocol and traffic events in support of security signalling, such as authentication and authorization, as well as additional overhead for encryption. Furthermore, the requirements specify for each event security objectives in confidentiality, integrity, and availability (CIA). Mapping security protocols at the various layers to the CIA levels (low/moderate/high) will be needed to adequately address security.

10.4.7 Resource sharing

The deployment and operation of a highly reliable, resilient, and secure wireless network for smart grid communications constitute a substantial financial investment. There are costs associated with the use of a licensed spectrum, providing coverage throughout the utility's service territory, building in ample capacity for disaster operations, and deploying redundant facilities for resilience to disasters. Considering that utilization of the network is likely to be low during normal operations, there may be incentives to share network resources with other users.

Sharing arrangements can range in the required level of coordination, from simply sharing sites, to sharing spectrum, to sharing a common network. At the site level, the utility would own and operate its own communications infrastructure, and potentially its own spectrum, but share sites (i.e., towers) and power resources with other operators. In a spectrum-sharing arrangement, the utility would still operate and maintain its own network infrastructure but coordinate its use of the spectrum with another operator. In a full network-sharing scenario, the utility would share both spectrum and network facilities with other entities.

As an example, in its National Broadband Plan [8], the US Federal Communications Commission envisions the latter scenario for utilities and public safety agencies. Citing common requirements between the two user groups for near-universal coverage and a resilient network able to function in emergency situations, the plan proposes that utilities have access to 700 MHz public safety broadband networks.

10.5 Performance metrics and tradeoffs

When evaluating wireless network alternatives, a common set of performance metrics is needed by which to quantitatively compare one solution with another. This section presents a minimal set of metrics that can be used to quantify the coverage capability, capacity, reliability, and latency of a smart grid wireless network. It also highlights some of the inherent tradeoffs between these quantities.

10.5.1 Coverage area

The coverage of a wireless network deployment may be expressed in terms of the geographic area (e.g., square kilometres) where wireless communications are available, or in relative terms as a percentage of the utility's service territory. Since the RF channel is often characterized in stochastic terms, it is customary to quote coverage subject to a minimum level of reliability or coverage probability. At any given location in the coverage area, there is some probability that service will be unavailable, due for example to severe shadowing by an obstruction, a deep multipath fade, or a high level of interference. Furthermore, these phenomena, especially fading and interference, may change over time. Thus, a coverage area is often referenced with respect to a certain probability of coverage, or conversely a probability of outage. For example, a deployment may be characterized as serving 98% of the territory with 99% reliability. In the context of a point-to-point link, the range of the link connecting the two points is likewise quoted with reference to a reliability level or outage probability.

The coverage radius, or range, of a wireless technology can be expressed, therefore, as the maximum transmitter–receiver distance that satisfies a minimum reliability level (or a maximum outage level) as follows:

$$R_{\text{max}} = \max\{R : P_{\text{out}}(R) \leq P_{\text{out,max}}\},$$

where $P_{\text{out}}(R)$ models the outage probability at distance R, and $P_{\text{out,max}}$ is the maximum outage probability that can be tolerated. The outage probability, in turn, is often modelled as the probability that the received signal strength or signal-to-noise ratio is below some minimum value needed for reliable communication. This probability can be quantified with the use of a link budget and the statistics of the RF channel.

Simple path-loss models are monotonically decreasing functions of distance, translating to an outage probability that increases with the transmitter–receiver distance, R. Examples of outage probability curves based on such models for home-area and neighbourhood-area networks are illustrated in Figure 10.10, obtained from an analysis of an IEEE 802.11 network for smart metering [9]. This analysis computed the average outage probability in a circular area of radius R_{max} assuming the signal experiences lognormal shadowing and Rayleigh fading. Three cases illustrate the impact of the type of environment (e.g., indoor home-area network vs. outdoor neighbourhood metering) as well as the transmission power. Given an outage criterion, such curves can be used to obtain the approximate footprint of an access point. For instance, this model predicts that the maximum coverage radius satisfying a 1% average outage criterion in an outdoor urban environment with moderate transmission power is 220 m.

Analogous coverage results for a wide-area cellular network are reproduced in Figure 10.11, based on an analysis of 3GPP Long Term Evolution (LTE) [10]. In this

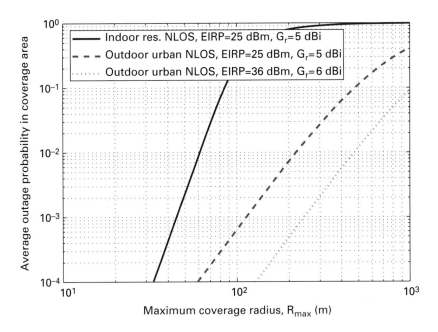

Figure 10.10 Outage probability vs. distance of an IEEE 802.11 network operating in the unlicensed 2.4 GHz band [9].

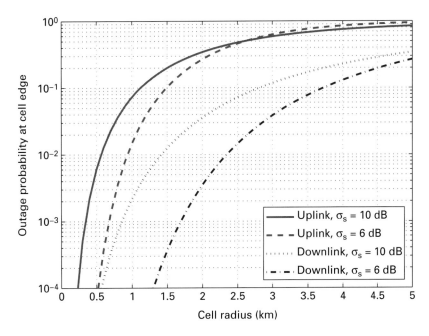

Figure 10.11 Outage probability vs. cell radius of a 3GPP LTE network operating at 700 MHz, by lognormal shadowing parameter σ_s [10].

analysis, the coverage criterion was defined in terms of the outage probability at the cell edge. Comparing the uplink and downlink curves, it is clear that coverage is limited by the uplink, which has a lower transmission power than the downlink. The analysis also illustrates the effect of different channel statistics, specifically the standard deviation of the lognormal shadowing, σ_s. For example, while a cell radius of 1.5 km can provide 90% coverage probability in 6 dB shadowing, a smaller cell radius (1.1 km) is required to provide the same coverage probability in a more cluttered environment with 10 dB shadowing. The preceding analysis assumes coverage is provided by a single-cell site. Assuming a terminal at the cell edge has independent channels to two cell sites, a 90% cell edge coverage probability increases to 99%.

A coverage analysis can be used to provide an estimate of initial infrastructure requirements. For example, if deploying a network DAPs to communicate with customer smart meters, the coverage capability of the chosen radio technology translates to the footprint each DAP can cover with some reliability. The footprint of each DAP, in turn, determines the approximate number of DAPs required to service a territory.

10.5.2 Capacity

Capacity can be measured in multiple ways. In an information-theoretic sense, capacity is the maximum rate at which information can be transmitted reliably – that is, with vanishing error probability as the block size increases. In practice, with finite message sizes, capacity may be defined as the maximum rate at which a device can send or receive information with a tolerable error rate. For a point-to-point link, this practical definition is sufficient.

In point-to-multipoint topologies, the wireless communications channel is shared by multiple actors. In this context, capacity may be measured in terms of the maximum number of actors that can be supported, with a specified data rate per actor. Alternatively, for a fixed number of actors, capacity may be defined in terms of the maximum data rate that each actor can support. Thus, in many-to-one scenarios, a network's capacity places limits on both the number of actors and the rate of each actor.

Given that both the number of actors sharing the channel and the data rate of each actor are limited by the capacity, one might be tempted to assume that capacity may simply be measured as the largest product of the two (number of actors × data rate per actor). However, depending on the wireless technology, there is often some overhead associated with mediating access by the actors to the shared channel, an overhead that grows with the number of devices contending for the channel. Furthermore, where multiple access interference is a factor, additional actors impact capacity through mutual interference. As a result, one cannot simply divide the 'capacity' of a technology by the number of actors to arrive at a maximum data rate per actor, in general.

Capacity can be assessed in the planning stages using mathematical and/or simulation models. Using a model, the offered load can be increased until signs of congestion are observed. Technologies based on a shared, random access channel will display congestion in the form of high access delays and a throughput which no longer increases with the offered load. For example, Figure 10.12 illustrates the throughput

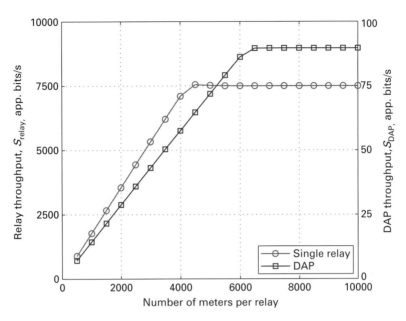

Figure 10.12 Throughput between a smart-meter relay and a DAP vs. the number of smart meters aggregated per relay over an IEEE 802.11 network operating in the unlicensed 2.4 GHz band [9].

on a neighbourhood-area network connecting a DAP to smart meter relays over an IEEE 802.11 network [9]. With increasing traffic load, measured in terms of the number of smart meters aggregated per relay, the relay–DAP link reaches capacity just beyond 4000 meters per relay on the uplink (relay to DAP) and 6000 meters per relay on the downlink (DAP to relay). The network is uplink-limited since more traffic is generated on the uplink (i.e., by the smart meter relays) than on the downlink in this case.

Other technologies, such as wide-area cellular networks, may rely on a centralized scheduler to mediate access to the wireless channel. In that case, once the network reaches capacity, the scheduler will reject requests for new connections. An example of the capacity of such a network is shown in Figure 10.13, which plots the capacity per smart meter (in messages per second) as a function of the cell radius in an urban LTE deployment supporting wide-area advanced metering [10]. This example also illustrates the impact of inter-cell interference. As the downlink utilization (v) and the uplink noise rise (Λ) increase, due to increasing traffic in other cells, the capacity of the cell of interest decreases.

The previous example illustrates a fundamental tradeoff in most wireless deployments between coverage and capacity. As the coverage footprint of a base station or access point increases, naturally so does the number of devices (e.g., smart meters) that are served by that cell. Hence, the finite capacity of that cell is shared by a larger number of devices. Another factor is the impact of longer transmitter–receiver distances. Most wireless digital communications systems adapt the rate of the link to the quality of the RF channel. Larger cell sizes imply longer distances and weaker signals at the edges of

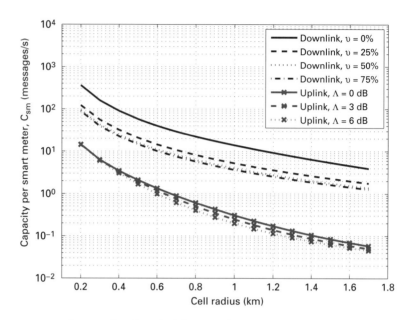

Figure 10.13 Smart-meter capacity vs. cell radius of a 3GPP LTE field-area network operating at 700 MHz [10].

the cell. These longer links consume, on average, more time and/or frequency resources to achieve a required data rate, leaving fewer resources for other devices and further depressing the cell's capacity.

Initial roll-outs are likely to be coverage-limited, because capacity requirements will be minimal at first. Networks which remain coverage-limited may be good candidates for resource sharing, in order to more efficiently utilize network resources. As network traffic increases, the network may transition from being coverage-limited to capacity-limited, at which point additional infrastructure may be needed to keep pace with capacity requirements, maintain reliability, and limit latency.

10.5.3 Reliability

Of particular importance to a control and measurement-oriented network such as the smart grid is reliability. For example, faults must be detected and power control commands received with a high degree of reliability.

Two aspects of reliability can be considered, each in the context of a different timescale: service availability (long-term) and transmission reliability (short-term). Service availability refers to the percentage of time that a communications link or network is up and running. Certain aspects of service availability are common to all communications networks (wireless and wireline) and include the reliability of associated hardware, software, power, and transmission lines.[2] Unique to wireless communications is an aspect

---

[2] Even a wireless deployment will include some transmission line facilities for backhaul.

of service availability related to RF impairments. A link may be down as a result of high signal attenuation due to transmitter–receiver distance, building penetration loss, precipitation, or clutter. Given that most smart grid communication links will be between fixed-location (not mobile) devices, these sources of signal attenuation tend to vary slowly over time but can vary widely from one location to another. Thus, the outage probability measure discussed in Section 10.5.1 can be interpreted as a measure of how likely the RF link will be unavailable at a randomly chosen location.

While RF attenuation is unlikely to vary greatly between two fixed points, RF interference can be very dynamic and cause short-term transmission losses. Sources of RF interference can be other devices within the same network as well as emitters from other systems. Regulatory mechanisms can help control other-system interference, however in-network interference may be unavoidable, especially under high load conditions. For example, cellular systems must contend with inter-cell (and sometimes intra-cell) interference, and random access systems are subject to collisions between simultaneous transmissions. If the data traffic is bursty in nature, such interference can be transient, and retransmission protocols can be used to recover from transient losses. But if the interference persists long enough or collisions occur frequently enough, the maximum number of transmission attempts may be reached without successful reception, resulting in message loss.

Thus, transmission reliability in smart grid communication networks will be managed primarily by maintaining loads at or below acceptable limits. An analysis like that of the aforementioned IEEE 802.11 metering network example can be used to identify those limits; in Figure 10.14 they can be seen as the levels of load on the horizontal axis at which

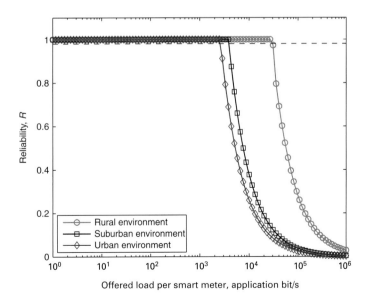

Figure 10.14 Transmission reliability vs. offered load in an IEEE 802.11 metering neighbourhood-area network [11]. The horizontal dashed line is the 98% reliability requirement.

reliability starts to drop rapidly. The load, in turn, can be managed through appropriate provisioning of resources and scheduling of traffic. For example, firmware upgrades can be staggered in time such that any given infrastructure element (base station or access point) is not overloaded.

10.5.4 Latency

Also related to the quality of service of a wireless link is latency, the length of time between when the sender's link layer receives a message from the higher layer and when the destination's link layer correctly receives the message from the physical layer. The latency is made up of multiple components, including buffering (i.e., time spent in a queue) at the transmitter, any handshaking or control exchange that may be part of the transmission protocol (such as a request-to-send/clear-to-send, or a resource request/resource grant), time for the channel to become available, the time to transmit the message including any overhead, propagation time (which is usually negligible in terrestrial communications), and processing time. If the message received from the higher layer needs to be fragmented by the link layer, then the total latency is the sum of the latency of each fragment. Likewise, if the route between the sender and receiver is made up of multiple wireless links (such as in a mesh network), then the total latency is the sum of the latency on each link.

If on any link the message, or a message fragment, is received in error, then the protocol may retransmit it up to some maximum number of attempts. Retransmissions are key to improving transmission reliability. However, each transmission attempt incurs additional latency as described above. Reliability can also be improved with increased redundancy in the forward error correction code, which also increases latency through longer transmission times. These two factors point to a tradeoff between reliability and latency: the greater the reliability required, the larger the latency.

Latency can also be related to how reliability is defined: if a message is not received within a specified time limit, it is deemed lost. In such a case, it is assumed that beyond that maximum tolerable latency, the message is no longer meaningful.

Like reliability, latency is a function of the load on the network. The more congested a network is, the longer a message needs to wait in queue before it can be transmitted, and the more likely retransmissions are needed to recover from collisions. To illustrate this relationship on a random access network, Figure 10.15 plots the latency in each direction on a neighbourhood-area network connecting DAPs to smart meter relays, as a function of the number of meters aggregated per relay. Just as the reliability drops when the traffic load reaches a critical level, latency jumps several orders of magnitude at the network's maximum sustainable load.

As the wireless network is likely to be but one component of the larger smart grid communications architecture, so the latency discussed herein is one component of a larger latency budget. Other, wireline networks that serve the application flow will contribute their own delays to the end-to-end latency. But given the more limited bandwidth and harsher channel that are typical of wireless links, the delay on the wireless portion of the network is likely to be the dominant component.

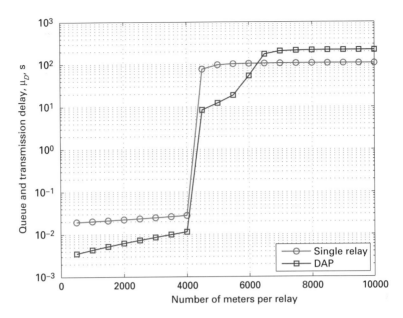

Figure 10.15 Average latency in each direction between a smart-meter relay and a DAP, vs. the number of smart meters aggregated per relay over an IEEE 802.11 network operating in the unlicensed 2.4 GHz band [9].

10.6 Conclusion

While the motivation for using wireless networks in the context of the smart grid is obvious, there are many considerations to factor in when designing and deploying such networks to support smart grid applications. This chapter first reviewed the plethora of applications, classifying them according to their QoS requirements and traffic characteristics. A view of the logical topology of the network vis-à-vis the various use cases provides a better understanding of where the major traffic flows are. A number of factors that pertain to the deployment of wireless networks must be considered, such as the choice of spectrum, the limits of coverage and capacity, and building in resilience and security. Key performance tradeoffs specific to wireless networks must then be balanced in designing solutions that meet smart grid communication requirements.

References

[1] M. Burns (May, 2010) Interoperability knowledge base [online]. Available at: http://collaborate.nist.gov/twiki-sggrid/bin/view/SmartGrid/InteroperabilityKnowledgeBase

[2] (2010) Smartgrid resource center: Use case repository [online]. Available at: http://smartgrid.epri.com/Repository/Repository.aspx

[3] (2011) Edison smartconnect™ – industry resource center: 2008–2009 smart grid use cases [online]. Available at: http://www.sce.com/CustomerService/smartconnect/industry-resource-center/smartgrid-usecase.htm

[4] 'NIST framework and roadmap for smart grid interoperability standards, release 1.0', NIST Special Publication 1108, Technical Report, January 2010.

[5] T. S. Rappaport, *Wireless Communications: Principles & Practice*. Prentice Hall PTR, 1996.

[6] 'Wireless field area network spectrum assessment', Technical Update 1022421, Electric Power Research Institute, Palo Alto, CA, December 2010.

[7] P. McDaniel and S. McLaughlin, 'Security and privacy challenges in the smart grid', *IEEE Security & Privacy*, vol. 7, no. 3, pp. 75–77, May–June 2009.

[8] *Connecting America: the national broadband plan*, Federal Communications Commission, March 2010 [online]. Available at: http://www.broadband.gov/

[9] M. Souryal, C. Gentile, D. Griffith, D. Cypher, and N. Golmie, 'A methodology to evaluate wireless technologies for the smart grid', in *Proceedings of IEEE SmartGridComm*, October 2010.

[10] M. Souryal and N. Golmie, 'Analysis of advanced metering over a wide area cellular network', in *Proceedings of IEEE SmartGridComm*, October 2011.

[11] 'Guidelines for assessing wireless standards for smart grid applications', NISTIR 7761, July 2011 [online]. Available at: http://collaborate.nist.gov/twiki-sggrid/pub/SmartGrid/PAP02 Wireless/NISTIR7761.pdf

Part IV

Sensor and actuator networks for smart grid

11 Wireless sensor networks for smart grid: research challenges and potential applications

Dilan Sahin, Vehbi Cagri Gungor, Gerhard P. Hancke Jr, and Gerhard P. Hancke

11.1 Introduction

The electrical grid is a critical infrastructure that could have a major impact on human lives, economics, and politics [1]. Hence, any instabilities related to the structural and operational characteristics of the existing power grid, equipment failures, blackouts, poor communication, and lack of effective monitoring of the infrastructure, create additional challenges to the power utilities due to the prospect of vast economic losses, customer dissatisfaction, inefficient electricity usage, and the huge amount of CO_2 emissions. The rising costs of new infrastructure and maintaining existing ones, increasing energy demand, and a declining number of skilled personnel, will drive utilities to operate systems more dynamically and efficiently. The need for real-time monitoring and management of transmission and distribution systems will become increasingly important. Such systems can be realized by the utilization of various types of sensors, and possibly actuators (actors).

Sensor systems will help provide the required information to utilities to achieve the goal of dynamic efficiency. The real-time information acquired from these sensors can be analysed to diagnose problems early, serve as a basis for taking remedial action, and thereby reduce service outages. This will reduce lost revenue and minimize person hours required to locate and rectify faults. For example, the Northeast blackout of 2003 in the United States was widespread and adversely affected over 50 million people with a huge economic loss [2]. There are many components of the electrical grid that must be monitored by advanced sensor technologies based on a combination of measurements, such as voltage, current, overhead conductor sag, temperature profile of conductors, power quality disturbances, system frequency, etc. [3–6]. On the other hand, there are millions of assets and hundreds of thousands of miles of power lines geographically distributed over millions of square miles needed to be monitored and utilized [7].

Recently, wireless sensor networks (WSNs) have increasingly been recognized as a promising technology to achieve seamless, energy-efficient, reliable, and low-cost remote monitoring and control in smart grid applications, from generation to consumer sites [3, 7, 8]. The existing and potential applications of WSNs in power grid span a wide range, including advanced metering, remote power system monitoring and control, electricity fraud detection, fault diagnostics, demand response, dynamic pricing, load

Table 11.1. Wireless sensor network applications in smart grid environments

Applications	Power grid sides
Wireless automatic meter reading (WAMR)	Consumer side
Residential energy management (REM)	Consumer side
Automated panels management	Consumer side
Building automation	Consumer side
Demand-side load management	Consumer side
Process control monitoring	Consumer side
Properties control monitoring	Consumer side
Equipment management and control monitoring	Consumer side
Equipment fault diagnostics	T&D side
Overhead transmission line monitoring	T&D side
Outage detection	T&D side
Underground cable-system monitoring	T&D side
Conductor temperature and dynamic thermal rating systems	T&D side
Overhead and underground fault circuit indicators	T&D side
Cable, conductor, and lattice theft	T&D side
Conductor temperature and low-hanging conductors	T&D side
Insulators	T&D side
Fault detection and location	T&D side
Animals and vegetation control	T&D side
Real-time generation monitoring	Generation side
Remote monitoring of wind farms	Generation side
Remote monitoring of solar farms	Generation side
Power quality monitoring	Generation side
Distributed generation	Generation side

control, energy management, and power automation, etc. However, the realization of these WSN-based smart grid applications depends directly on efficient communication capabilities among electric power system elements.

The remainder of the chapter is organized as follows. The potential applications of WSNs in smart grid are introduced in Section 11.2. Specifically, WSN-based applications are explored for power generation systems, transmission and distribution networks, and consumer facilities as depicted in Table 11.1. The section also gives an overview of the existing status of relevant technologies, i.e., sensors, communication systems, and tools that can be used by business in order to gather information about physical conditions and parameters that are critical for improving and optimizing the performance of transmission and distribution systems. Section 11.3 describes research challenges for WSN-based smart grid applications. Finally, the chapter concludes with Section 11.4.

11.2 WSN-based smart grid applications

WSN technology is a promising technology for achieving seamless, energy-efficient, reliable, and low-cost remote monitoring and control in smart grid applications. The required information can be provided to utilities by sensor systems to enable them to achieve dynamic efficiency. The real-time information gathered from the sensors can be analysed to diagnose problems early and serve as a basis for taking remedial action. In general, WSN-based smart grid applications can be divided into three groups: consumer side, transmission and distribution (T&D) side, and generation side, as in Table 11.1. In the following section, WSN-based smart grid applications are briefly described.

11.2.1 Consumer side

Consumer-side WSN-based smart grid applications can be divided into two groups: residential-side and industrial-side applications. Consumer-side applications have a direct relationship with the end-users; hence it is an important layer, which consists of wireless automatic metering, residential energy management, automated panel management, building automation, and demand-side load management.

Residential side

- *Wireless automatic metering*. Meter reading techniques, such as direct physical access to a meter or visual meter reading, may not be cost-effective considering the large scale of the metering infrastructure. Recent wireless sensor network platforms can offer several advantages, including decreased utility operational costs by eliminating the need for human readers and the prevention of meter tampering. With the integration of wireless metering systems into the grid, real-time and dynamic pricing, which provides different charging techniques during the peak hours of a day, can be realized.
- *Residential energy management*. WSN-based applications have formed an indispensable part of our daily lives, as they have an extensive diversity from the energy conservation domain, to the health, safety and comfort domains. Since a major goal of the power utilities is to have more control in reducing the peak demands and provide a balance between the supply and demand, many applications have been developed for industrial and residential customers to shift the demands to off-peak hours [9]. Energy-related applications provide real-time feedback about energy consumption behaviour to the customers, which has a significant effect on reducing overall energy consumption during the peak hours or off-peak hours. The realization of WSNs in monitoring and managing power consumption is one of the most popular solutions in the residential energy management sector [9].
- *Automated panel management*. The generation of solar energy from solar panels will be more efficient with the integration of sensor nodes into the system. According to [10], the sun's rays can be captured in a more efficient way if sensor nodes are used to track the sun's rays.
- *Building automation*. Building automation aims to control various appliances' energy consumption and enable a communication network to connect these appliances to act more efficiently and prevent redundant energy use. Lighting, heating, ventilation, and air conditioning (HVAC) are some of the smart appliances that are actively participating in reducing the use of energy. WSNs reduce redundant cabling costs and complexity of the installation process of building automation systems. Recent studies

have shown that it is possible to save up to 30% of energy consumption of buildings with efficient energy management [11].
- *Demand-side load management.* Sustainable systems focus on providing a variety of energy services from low-risk energy sources. Since demand-side management is the key parameter of sustainable energy systems, the optimization of demand-side management with efficient utilization of end-use energy should be performed to decrease the energy demand and energy cost. The challenge here is that the demand–supply is not sustainable, however, using advanced efficiency technologies, innovative management methodologies, integration of end-use energy efficiency, and renewable energies, the energy demand could be reduced [12]. WSNs can play a key role in realizing such systems. For instance, Kantarci *et al.* propose a load-shifting mechanism by using wireless sensor networks to reach the energy-management units that schedule the appliances [9].

Industrial side
- *Process control monitoring.* With continuous monitoring using wireless sensor nodes, real-time data transfer creates an efficient production process in industry by providing efficient energy usage, fault-minimized goods production, early-fault detection, reduced-deficient goods [13], product consistency, and reduced-process time [10].
- *Properties control monitoring.* The control of the physical properties in production processes, the integration of WSN technology enabling advanced monitoring with smart sensor nodes and the availability of different resources, measurement of different properties, energy savings during production and reduction of pollutants, are just some of the consequences of WSN technology [13].
- *Equipment management and control monitoring.* Temperature, pressure, humidity or vibration values of the industry machines that provide information about the health of the machine are measured and analysed by the sensor nodes and when potential problems are detected, the necessary signals are sent to make predictive maintenance possible [10].

11.2.2 Transmission and distribution side

Transmission and distribution-side WSN-based smart grid applications play a significant role in the system, since they are responsible for successful power transmission. Overhead transmission line monitoring, outage detection, conductor temperature and dynamic thermal rating systems, underground cable system monitoring, equipment fault diagnostics, overhead and underground fault circuit indicators, cable, conductor and lattice theft, fault detection and location, and animals and vegetation control are some of the WSN applications on the transmission and distribution side.

- *Equipment fault diagnostics.* For the generation and transmission side of the power grid, reliable performance of equipment such as power transformers is very important [14]. Failures result in the discontinuity of power flow, unavailability of equipment with a resulting revenue loss. Equipment fault diagnostic systems with the integration of digital information technology and intelligent techniques are needed to increase the performance of electric equipment and reduce electrical system failures of the power grid. The combination of equipment fault diagnostic systems with a cost-effective, scalable nature of WSNs will provide a reliable and efficient performance of the power grid.

- *Overhead transmission line monitoring.* The transmission line is the most critical component of the power grid. There exist many threats that will influence the safety, reliability, and security features of transmission lines, which have a direct effect on the economy of the country and safety and welfare of the citizens. Lightning strikes, icing, hurricanes, landslides, bird damage, and overheating of transmission lines are some possible threats that transmission lines may face. On the other hand, the distributed nature of the transmission lines also creates difficulties in maintaining it easily. Hence, an intelligent monitoring system for overhead transmission lines is required with the integration of advanced, low-cost, and durable technologies. Wireless sensor network technology best fits this kind of application as the scale of the grid is expanding continuously, which will cause other technologies to be costly and inefficient. Hence, with WSNs technology, automatic energy transmission monitoring with fast response will be possible. Hung *et al.* point out these important issues in designing the network model to support overhead transmission line monitoring applications. Delay, reliability, and energy efficiency are some of the factors that should be considered carefully when designing the network [15]. In overhead transmission line monitoring applications, sensors deployed near the towers/poles collect the information and send it to the relay node, which is deployed on the pole. Hung *et al.* also present a network model solution that is based on traffic characteristics and resource constraints in which sensor nodes transmit the data in a hop-by-hop manner to the relay node, which is a GSM/GPRS/UMTS-enabled device and will switch on when needed. Hence, each relay node sends the collected data to the data collector via GSM towers.

- *Outage detection.* In the USA, the estimated cost of outages in 2002 was in the order of 79 billion dollars, which is equivalent to a third of the total electricity retail revenue of 249 billion dollars [16]. Hence, outages have both social and economic consequences. The lack of automated analysis and poor visibility are the basic reasons that outages in the electric system cannot be detected. Advanced sensors and monitoring systems are needed to reduce outages and increase the reliability of the power system.

- *Underground cable system monitoring.* There are many failures in joints and terminations of underground cable systems, as well as overhead transmission line systems. However, monitoring and maintenance of underground cables is much harder due to the harsh characteristics of the underground environment. WSNs will be well suited for underground cable system monitoring to reduce the maintenance costs and provide more accurate status information of the underground cables [7].

- *Conductor temperature and dynamic thermal rating systems.* Power utilities pay attention to the temperature values of the cables because it is an essential measurement to get the optimum cable use. The load capacity of the cables has a direct relation with the cable conductor temperature ratings and since it is one of the key values of the power system, measuring the cable conductor temperature with smart sensor nodes will be a cost-effective, reliable, and flexible solution [7].

- *Overhead and underground fault circuit indicators (FCI)*. FCIs are used to identify where faulty equipment on distribution circuits are located. It is important to quickly and accurately pinpoint fault locations to reduce power restoration times. Some of the system variables that can cause false readings affecting FCI operations are [18]: inrush current, cable discharge, proximity effect, back-feed voltages and currents, etc. These variables need to be monitored and the communication capability of some FCI products [18–20] can be used in a WSN for this purpose.

- *Cable, conductor, and lattice theft*. There continue to be frequent reports of cable theft, both of power cables and of telecommunication cables, from many parts of the world [21]. Cables, especially copper cables, are being stolen for the scrap value of their metal content and this has become an increasingly serious issue due to the rise in metal prices. In addition, the stealing of lattice members (steel support structures) of the lower end of the tower on the transmission and distribution network is a huge risk. Outages due to damaged high-voltage transmission lines and associated structures are time-consuming to fix and cost billions of dollars in lost revenue. Combating theft is a very challenging problem as transmission lines cross many miles of remote country through dedicated corridors with no effective means of physical security detection or protection [22].

- *Conductor temperature and low-hanging conductors*. Sagging in high-voltage (HV) lines is mainly due to conductor temperature, which in turn is a function of current loading. The monitoring of weather conditions is also related to conductor temperature monitoring. Sagging in low and medium-voltage lines is not due to temperature but rather due to tilting of the wooden poles arising from geological reasons. This type of sagging is predominant in rural areas and poses fatal hazards to man and animals. Various temperature and sag measurement techniques for high-voltage lines have been described in [23–25].

- *Insulators: partial discharge and leakage current*. Partial discharge (PD) and leakage current monitoring is crucial for preventing or minimizing system failures due to the breakdown of cable insulation and insulators. The benefits of on-line PD field measurements are that it is a predictive, non-intrusive [26] and non-destructive test that is relatively inexpensive when compared to off-line testing. A PD measurement system could also provide the location [27] of the PD in real time [28]. In medium and high-voltage applications, a leakage current is the current that flows either through the body or over the surface of an insulator and may lead to flashover of the insulation. Leakage current measurement as a diagnostic tool for outdoor insulation has been described in [29–31]. When the leakage current is continuously monitored, it is possible to generate an alarm about the probability of a dangerous flashover-risk level and recommend the necessary maintenance actions.

- *Fault detection and location*. Traditionally, the short-circuit faults in power distribution lines were located by a trial and error method. This is time-consuming, especially if it involves driving to remote locations, and newer more intelligent methods are required to quickly and accurately determine the location of the fault [7]. Various automatic means of implementing management functions for faults using WSNs exist and have demonstrated value in reducing outage times. A selection of techniques proposed for fault detection and location is presented in [32–36].

- *Animal and vegetation control.* Animal and vegetation control is necessary to achieve expanded, safe, and reliable operations of the power grid. Reducing avian interactions, and taking precautions to prevent animals from damaging cables will reduce the blackouts, short-circuit problems and WSN technology will be the perfect choice to detect animal and avian interactions [7]. Pre-fault detection of relevant phenomena is vital for predictive maintenance and some work has been done, e.g., the effect of trees on the nearby electric field [37]; and the identification of trees or animals, which caused faults [38].

11.2.3 Generation side

Most of the generation-side applications are based on monitoring tasks. Some examples are as follows: real-time generation monitoring, remote monitoring of wind farms, remote monitoring of solar farms, power quality monitoring, and distributed generation.

- *Real-time generation monitoring.* In the existing power grid, some methods are used to store the energy, such as pumped hydro, compressed air and flywheel, however, these are inconvenient to store the renewable energy generated from solar and wind farms [39]. Making energy storage decisions is quite possible with real-time generation monitoring systems in which WSNs will be a preferred solution due to their low-cost characteristics [39].
- *Remote monitoring of wind farms.* Wind farms are one of the most important renewable energy resources, whose performance can easily be affected by some external conditions, such as outdoor pressure and temperature values, the orientation of wind, bird collisions, etc. [39]. These external factors may have a lesser effect on the performance of wind farms, if they can be monitored wisely. Capturing the audio and visual data in a cost-effective manner will enable the identification of external parameters, which influence the performance of the wind farms [39].
- *Remote monitoring of solar farms.* The temperature value, radiation, DC voltage, and weather conditions are some of the external parameters that have a direct effect on the performance of the renewable energy generation of the solar farms [39]. WSN-based remote monitoring systems will evaluate the external effects on solar farms and improve their performance.
- *Power-quality monitoring.* The quality of power is critical for the safe operation of control units and electrical appliances and the deregulation of the electrical power market, etc. Power quality monitoring, which enables a continuity and increase in power quality, collects the voltage and the current data and sends this data to remote centres for further decision-making processes. WSNs provide a low-cost, efficient, and reliable data communication system for power quality monitoring applications.
- *Distributed generation.* In [40], an application of a low-cost IPv6-based wireless sensor network in distributed generation is proposed. The IEEE 802.15.4-based link

layer technology is used and all the sensor nodes are capable of communicating with other IP-based devices. The main focus is to improve the power management process by correcting the distributed generators' reference signals.

11.3 Research challenges for WSN-based smart grid applications

In the following paragraphs some research challenges and design objectives for WSN-based smart grid applications (as listed in Table 11.2) are described.

- *Interoperability*. Energy generation units, distribution networks and energy consumers all are important parts of the smart grid that need advanced communication techniques among each component to exchange information. To provide such a complex communication infrastructure will be very challenging if standard-based and interoperable communication protocols are not used [41].
- *Memory consumption*. Sensor nodes have limited memory capacity. Available memory capacity often limits the functionality of the system. The system software running on the sensor nodes, the communication protocol and complex computations should be chosen wisely to decrease memory consumption.
- *Power management*. Power management is a challenging task since the sensor nodes are battery powered. Minimizing the energy consumption is very important since performing computations, sensing the environment, and communication with other sensor nodes are quite complex processes, which may increase energy consumption. To this end, power-efficient communication protocols and advanced sleep schedules can be used to prolong the network lifetime [42].
- *Dynamic pricing and configuration updates*. Energy-management systems pose another challenging task, which is the process of dynamically updating price information. In [9] it is stated that better billing can be accomplished by using dynamic price rates according to energy demand, which could result in load oscillation.
- *Security*. The ability to provide secure communication between a remote control centre and a device by protecting the communication from external denial of service (DoS) attacks, can be called security [43]. In the smart grid, sensitive and confidential data can be generated from smart meters and smart home appliances. This data should be transmitted safely to the power utility's data servers to prevent unauthorized access. Hence, secure end-to-end communication protocols should be used to protect the confidential data against cyber and physical attacks [41]. Security for wireless sensor networks is influenced by a number of factors, such as deployment strategy, system architecture, underlying communication infrastructure, the node platform and the application. It is likely that a new WSN deployment would not be able to use existing solutions without some degree of customization and further evaluation.
- *Quality-of-service (QoS) requirements*. WSN-based smart grid applications can have different QoS requirements and specifications in terms of reliability and communication delay. For example, in the case of alarm conditions and dynamic pricing notifications, it is important to receive the data in a timely manner. Data with long

Table 11.2. Challenges vs. design objectives in WSN-based smart grid applications

Challenges	Design objectives
Interoperability	Standard-based WSN protocols and products
Memory consumption	Low-overhead and simple protocols
Power management	Energy-efficient protocols and energy harvesting solutions
Dynamic pricing and configuration updates	Adaptive protocols
Security	Secure design and protocols
Quality-of-service (QoS)	QoS-aware protocols and cross-layer designs
Unreliability of wireless links	Link quality-aware routing and MAC protocols
Data management	Data aggregation and compression
System integration	Data aggregation and integration techniques
Large scale	Scalable protocols
Heterogeneous communication techniques	Cross-layer design and hybrid protocols
Transmission line conductor galloping	Anti-vibration or damping schemes
Mechanical strength of towers and poles	Durable towers and poles design
Energy harvesting for powering sensors	Energy harvesting

delay due to processing can be outdated and result in wrong decisions. Hence, assigning appropriate QoS requirements for WSN-based smart grid applications is essential for providing a reliable monitoring system [44].

- *Unreliability of wireless links.* The significant levels of unreliability and asymmetry of wireless links adversely affect the communication performance of WSNs. Signal attenuation by the distance, asymmetry in wireless links, non-uniform radio signal strength, fading and multipath effects are some of the causes of the unreliable nature of the wireless links [45]. Most of the proposed routing protocols work well for ideal conditions [39, 46]. However, the harsh environmental conditions in the smart grid environment cause these mechanisms to perform very poorly. Link quality estimation provides an important value to choose the best route for the data packets in WSN [47]. The measurement, characterization, and utilization of the wireless link quality with less energy consumption of the sensor nodes [48] is a great motivation for researchers to find the link quality of the sensor networks. Real-time decision making processes require on-time packet delivery, hence, any latency related to this issue can lead to some serious problems in the power grid [39]. Routing and MAC protocols should be implemented wisely for mission-critical WSN-based smart grid applications.
- *Data management.* A huge amount of data is generated from smart meters and smart home appliances. This confidential and sensitive data should be transmitted to the power utility centres securely. The communication network should be capable of performing the complex tasks related to transmission, collection, storage, and maintenance of this huge amount of data [49].

- *System integration.* Compression and aggregation of data and thus preventing data overload, data extraction to create information from disparate data sources, and integration with the existing SCADA system are critical requirements.
- *Large scale.* Covering wide geographical areas with large numbers of sensor nodes creates scalability challenges, which necessitate the use of intelligent and efficient aggregation and summarization techniques to manage the extensive data gathered from the sensor nodes [50]. The large-scale sensor networks may lead to some delay-related problems for some mission-critical applications. Hence, the choice of the routing protocols should be done wisely.
- *Heterogeneous communication techniques.* WSN-based smart grid applications require a reliable, resilient, secure, flexible, cost-effective communication system [8]. The challenge is that there is no single communication technique that provides all these requirements simultaneously. Hence, a combination of communication techniques should be applied. However, the heterogeneity will create some additional problems, such as the interoperability among these techniques.
- *Transmission line conductor galloping.* A 'galloping' condition is defined as a low-frequency vibration of the conductor in the range of about 0.1 Hz to 1 Hz for a predetermined length of time (e.g., between 0.1 and 300 s, or several cycles or more). Effective detection of conductor galloping in overhead lines is important, as galloping can cause mechanical failure of the conductor or structure, or breakdown of the insulation between conductors on different phases. Research efforts have focused on anti-vibration or damping schemes [51, 52], i.e., avoidance, but not detection of galloping.
- *Mechanical strength of towers and poles.* Failures of poles, towers, and structures may lead to power outages, high repair costs, and are potentially very dangerous. Therefore, inspecting and maintaining them in a timely manner, and preferably continuously, is essential to system integrity and maximizing service life of the equipment [7]. Several measurement techniques are proposed such as drilling or chipping, stress wave, sonic or ultrasonic, electrical resistivity, infrared, radar, and tomography. These techniques are normally destructive, and/or only test a local area of the structure rather than evaluating the state of the entire structure.
- *Energy harvesting for powering distributed sensors.* Sensor nodes require an energy source. The typical power supply for a stand-alone sensor, i.e., batteries, is not a viable option. A solution being researched is energy harvesting from any available sources near to a sensor node such as solar, thermal, vibrations, magnetic, or electric fields, as discussed in [17]. Other solutions are offered by methods utilized under HV conditions, using an optical source [53] and a current transformer source [54].

11.4 Conclusion

The industry is recognizing the fact that the implementation of advanced sensor networks and systems could potentially have a huge social and economic impact by contributing towards improving the efficiency of transmission and distribution systems and related components of the electrical energy-providing smart infrastructure. One of the most important criteria used to measure the quality-of-service in the electric utility industry is the uninterrupted service supplied to customers. Preventative maintenance is required to avoid failures, but other factors such as theft of critical components, asset management in general, and safety considerations also play a major role. A large volume of relevant research outputs indicates that the research community is also aware of these needs. However, many challenges still remain to be resolved.

In this chapter, the potential applications of WSNs in smart grid have been introduced along with the related technical challenges. Specifically, WSN-based applications have been explored for power generation systems, transmission and distribution networks, and consumer facilities. It has identified and discussed suitable sensor systems for the monitoring and management of infrastructure components, and has given an overview of the existing status of relevant technologies, i.e., sensors, communication systems, and tools that can be used by business in order to gather information about physical conditions and parameters that are critical for improving and optimizing the performance of the smart grid.

Acknowledgements

The work of D. Sahin and V. C. Gungor was supported by the European Union FP7 Marie Curie International Reintegration Grant (IRG) under Grant PIRG05-GA-2009-249206. The work of G. P. Hancke Jr and G. P. Hancke was supported by Eskom, South Africa.

References

[1] Z. Wang, A. Scaglione, and R. J. Thomas, 'Generating statistically correct random topologies for testing smart grid communication and control networks', *IEEE Transactions on Smart Grid*, vol. 1, no. 1, pp. 28–39, June 2010.

[2] Northeast Blackout 2003 [online]. Available at: https://reports.energy.gov/BlackoutFinal-Web.pdf

[3] V. C. Gungor, L. Bin, and G. P. Hancke, 'Opportunities and challenges of wireless sensor networks in smart grid,' *IEEE Transactions on Industrial Electronics*, vol. 57, no. 10, pp. 3557–3564, October 2010.

[4] E. F. Livgard, 'Electricity customers' attitudes towards smart metering,' *IEEE International Symposium on Industrial Electronics (ISIE 2010)*, pp. 2519–2523, 4–7 July 2010.

[5] R. Moghe, Y. Yi, F. Lambert, and D. Divan, 'A scoping study of electric and magnetic field energy harvesting for wireless sensor networks in power system applications', in *Proceedings of IEEE Energy Conversion Congress and Exposition (ECCE 2009)*, pp. 3550–3557, 20–24 September 2009.

[6] V. K. Sood, D. Fischer, J. M. Eklund, and T. Brown, 'Developing a communication infrastructure for the smart grid', *IEEE Electrical Power and Energy Conference (EPEC 2009)*, pp. 1–7, 22–23 October 2009

[7] Y. Yang, F. Lambert, and D. Divan, 'A survey on technologies for implementing sensor networks for power delivery systems', *IEEE Power Engineering Society General Meeting, 2007*, pp. 1–8, 24–28 June 2007.

[8] S. Ullo, A. Vaccaro, and G. Velotto, 'The role of pervasive and cooperative sensor networks in smart grids communication', *IEEE Mediterranean Electrotechnical Conference (MELECON 2010)*, pp. 443–447, 26–28 April 2010.

[9] M. E.-Kantarci and H. T. Mouftah, 'Wireless sensor networks for domestic energy management in smart grids', in *Proceedings of 25th Biennial Symposium on Communications (QBSC 2010)*, pp. 63–66, 12–14 May 2010.

[10] Smart sensor networks: technologies and applications for green growth, December 2009 [online]. Available at: http://www.oecd.org/dataoecd/39/62/44379113.pdf

[11] X. Guan, Z. Xu, and Q.-S. Jia, 'Energy-efficient buildings facilitated by microgrid', *IEEE Transactions on Smart Grid*, vol.1, no. 3, pp. 243–252, December 2010.

[12] Sustainable Energy Systems [online]. Available at: http://www.wupperinst.org/uploads/txwibeitrag/ sustainableenergysystems.pdf

[13] Department of Energy, United States, Energy Technology Solutions Public Private Partnerships Transforming Industry, Washington, DC [online]. Available at: http://www1.eere.energy.gov/industry/pdfs/itp_successes.pdf

[14] B. Lu and V. C. Gungor, 'Online and remote energy monitoring and fault diagnostics for industrial motor systems using wireless sensor networks', *IEEE Transactions on Industrial Electronics*, vol. 56, November 11, November 2009.

[15] K. S. Hung, W. K. Lee, V. O. K. Li, K. S. Lui, P. W. T. Pong, K. K. Y. Wong, G. H. Yang, and J. Zhong, 'On wireless sensors communication for overhead transmission line monitoring in power delivery systems', in *Proceedings of IEEE First International Conference on Smart Grid Communications (SmartGridComm 2010)*, pp. 309–314, 4–6 October 2010.

[16] K. Moslehi and R. Kumar, 'Smart grid: a reliability perspective', *Innovative Smart Grid Technologies (ISGT 2010)*, pp. 1–8, 19-21 January 2010

[17] Energy Harvesting Forum, 'Energy harvesting electronic solutions for wireless sensor networks and control systems' [online]. Available at: http://www.energyharvesting.net/

[18] Cooper Power Systems [online]. Available at: http://www.cooperpower.com

[19] Horstmann GmbH [online]. Available at: http://www.horstmanngmbh.com

[20] Remote Monitoring Systems [online]. Available at: http://cable-fault.com

[21] ICF News (January 2008) [online]. Available at: http://www.icf.at/en/6050/cable_theft.html

[22] C. Scott, G. Heath, and J. Svoboda (April 2006) Preprint: 'Vibration monitoring of power distribution poles' [online]. Available at: http://www.inl.gov/technicalpublications/Documents/3493254.pdf

[23] C. Bernhauer, H. Bohmer, S. Kornhuber, S. Markalous, M. Muhr, and T. Strehl (May 2007) 'Temperature measurement on overhead transmission lines (OHTL) utilizing surface acoustic wave (SAW) sensors' [online]. Available at: http://tinyurl.com/395seso

[24] Doble Lemke, RITHERM [online]. Available: http://tinyurl.com/335hsvo

[25] Sensa – a Schlumberger Company. Power-circuit condition monitoring [online]. Available at: http://sensa.org/user/files//sensa_power_brochure_.pdf

[26] P. J. Moore, I. Portugues, and I. A. Glover, 'A nonintrusive partial discharge measurement system based on RF technology', *IEEE Power Engineering Society General Meeting*, Toronto, Canada, 2003, pp. 629–633.

[27] B. Quak, E. Gulski, J. J. Smit, F. J. Wester, and P. N. Seitz, 'PD site location in distribution power cables', *IEEE 2003 International Symposium on Electrical Insulation*, Boston, MA, USA, 2002, pp. 83–86.

[28] P. Wagenaars, P. A. A. F. Wouters, P. C. J. M. van der Wielen, and E. F. Steennis, 'Technical advancements in the integration of online partial discharge (PD) monitoring in distribution cable networks', in *IEEE Conference on Electrical Insulation and Dielectric Phenomena*, CEIDP 2009, Eindhoven, Netherlands, 2009, pp. 323–326.

[29] T. P. Hong, P. Thinh, D. G. Trong, and D. V. Hoang, 'Leakage current analysis for predicting flashover in distribution network', *IEEE Electrical Insulation and Dielectric Phenomena* (CEIDP 2009), Virginia Beach, VA, USA, 2009, pp. 462–465.

[30] G. Montoya, I. Ramirez, and R. Hernandez, 'The leakage current as a diagnostic tool for outdoor insulation', in *Proceedings of IEEE PES Transmission and Distribution Conference and Exposition: Latin America*, 2008, pp. 1–4.

[31] S. C. Oliveira, E. Fontana, F. J. M. M. Cavalcanti, R. B. Lima, J. F. Martins-Filho, and E. Meneses-Pacheco, 'Fiber-optic sensor system for leakage current detection on insulator strings of overhead transmission lines', in *IEEE MTTS International Conference on Microwave and Optoelectronics*, 2005 SBMO, Brasilia, DF, Brazil, 2005, pp. 368–373.

[32] C. M. Fuk and M.S. Demokan, 'On-line fault location system for overhead power transmission lines using passive quasi-distributed fibre-optic sensing', in *Proceedings of IEEE 2nd International Conference on Advances in Power System Control, Operation and Management*, Hong Kong, 1993, pp. 243–251.

[33] M. M. Nordman and T. Korhonen, 'Design of a concept and a wireless ASIC sensor for locating earth faults in unearthed electrical distribution networks', *IEEE Transactions on Power Delivery*, vol. 21, no. 3, pp. 1074–1082, July 2006.

[34] E. C. Senger, G. Manassero, C. Goldemberg, and E.L. Pellini, 'Automated fault location system for primary distribution networks', *IEEE Transactions on Power Delivery*, vol. 20, no. 2, pp. 1332–1340, April 2005.

[35] O. Vähämäki, S. Sauna-aho, S. Hänninen, and M. Lehtonen, 'A new technique for short circuit fault location in distribution networks', *IRED 18th International Conference on Electricity Distribution*, Turin, Italy, June 2005.

[36] T. Welfonder, V. Leitloff, R. Feuillet, and S. Vitet, 'Location strategies and evaluation of detection algorithms for earth faults in compensated MV distribution systems', *IEEE Transactions on Power Delivery*, vol. 15, no. 4, October 2000.

[37] M. Suojanen, 'Effect of spruce forest on electric fields caused by 400 kV transmission lines', in *International Conference on Power System Technology*, Perth, Australia, 2000, pp. 1401–1405.

[38] L. Xu and M.-Y. Chow, 'A classification approach for power distribution systems fault cause identification', *IEEE Transactions on Power Systems*, vol. 21, no. 1, pp. 53–60, February 2006.

[39] M. E. Kantarci and H. T. Mouftah, 'Wireless multimedia sensor and actor networks for the next generation power grid', *Ad Hoc Networks*, vol. 9, no. 4, pp. 542–551, 2011.

[40] G. Bag, R. Majumder, and K. K.-Hyung, 'Low cost wireless sensor network in distributed generation', in *Proceedings of IEEE First International Conference on Smart Grid Communications (SmartGridComm 2010)*, pp. 279–284, 4–6 October 2010.

[41] Z. Fan and G. Kalogridis, 'The new frontier of communications research: smart grid and smart metering', in *Proceedings of 1st International Conference on Energy-Efficient Computing and Networking, e-Energy*, 2010, pp. 115–118.

[42] F. Osterlind, E. Pramsten, D. Roberthson, J. Eriksson, N. Finne, and T. Voigt, 'Integrating building automation systems and wireless sensor networks', in *Proceedings of IEEE Conference on Emerging Technologies and Factory Automation (ETFA 2007)*, pp. 1376–1379, 25–28 September 2007.

[43] V. C. Gungor and F. C. Lambert, 'A survey on communication networks for electric system automation', *Computer Networks*, vol. 50, pp. 877–897, May 2006.

[44] V. C. Gungor and G. P. Hancke, 'Industrial wireless sensor networks: challenges, design principles, and technical approaches', *IEEE Transactions on Industrial Electronics*, vol. 56, no. 10, pp. 4258–4265, October 2009.

[45] J. Shin, U. Ramachandran, and M. Ammar, 'On improving the reliability of packet delivery in dense wireless sensor networks', in *Proceedings of 16th International Conference on Computer Communications and Networks* (ICCCN 2007), pp. 718–723, 13-16 August 2007.

[46] G. Zhou, T. He, S. Krishnamurthy, and J. A. Stankovic, 'Impact of radio irregularity on wireless sensor networks', in *Proceedings of ACM MobiSys*, New York, 2004, pp. 125–138.

[47] M. Krogmann, T. Tian, G. Stromberg, M. Heidrich, and M. Huemer, 'Impact of link quality estimation errors on routing metrics for wireless sensor networks', in *Proceedings of 5th International Conference on Intelligent Sensors, Sensor Networks and Information Processing (ISSNIP 2009)*, pp. 397–402, 7–10 December 2009.

[48] D. Lai, A. Manjeshwar, F. Herrmann, E. U.-Biyikoglu, and A. Keshavarzian, 'Measurement and characterization of link quality metrics in energy constrained wireless sensor networks', in *Proceedings of IEEE Global Telecommunications Conference, GLOBECOM 2003*, vol. 1, pp. 446–452, 1–5 December 2003.

[49] S. S. S. R. Depuru, W. Lingfeng, V. Devabhaktuni, and N. Gudi, 'Smart meters for power grid challenges, issues, advantages and status', *IEEE/PES Power Systems Conference and Exposition (PSCE 2011)*, pp. 1–7, 20–23 March 2011.

[50] D. Pendarakis, N. Shrivastava, L. Zhen, and R. Ambrosio, 'Information aggregation and optimized actuation in sensor networks: enabling smart electrical grids', *IEEE 26th International Conference on Computer Communications (INFOCOM 2007)*, pp. 2386–2390, 6–12 May 2007.

[51] G. Diana, M. Bocciolone, F. Cheli, A. Cigada, and A. Manenti, 'Large wind-induced vibrations on conductor bundles: laboratory scale measurements to reproduce the dynamic behavior of the spans and the suspension sets', *IEEE Transactions on Power Delivery*, vol. 20, no. 2, pp. 1617–1624, April 2005.

[52] L. Wang, Y. Yin, X. Liang, and Z. Guan, 'Study on air insulator strength under conductor galloping condition by phase to phase spacer', in *Conference on Electrical Insulation and Dielectric Phenomena*, 2001 Annual Report, Kitchener, Ontario, Canada, 2001, pp. 617–619.

[53] C. Svelto, M. Ottoboni, and A. M. Ferrero, 'Optically-supplied voltage transducer for distorted signals in high-voltage systems', *IEEE Transactions on Instrumentation and Measurement*, vol. 49, no. 3, pp. 550–554, June 2000.

[54] Z. Gang, L. Shaohui, Z. Zhipeng, and C. Wei, 'A novel electro-optic hybrid current measurement instrument for high-voltage power lines', *IEEE Transactions on Instrumentation and Measurement*, vol. 50, no. 1, pp. 59–62, February 2001.

12 Sensor techniques and network protocols for smart grid

Rong Zheng and Cunqing Hua

12.1 Introduction

Spread over the grid, sensors and sensor networks monitor the functionality and the health of grid devices, monitor operation conditions, provide outage detection, and detect power quality disturbances [1]. Control centres can thus immediately receive accurate information about the actual conditions of the grid. Applications of sensor and actuator networks in power systems can be categorized based on the subsystems they are employed in, namely, generation systems, power transmission systems, distribution systems, and consumers.

In *generation systems*, sensor networks can be utilized to improve the efficiency of and monitor the conditions of the generation systems. For example, automated panels managed by sensors track the Sun's rays to ensure that solar power is gathered in a more efficient manner [2]. Sensors can report the structural health condition of wind turbines [3] and hydro dams continuously, as well as wind speed and direction for monitoring wind power generation [4].

On the *transmission side*, since active power flow in a power line is nearly proportional to the sine of the angle difference between voltages at the two terminals of the line, phasor measurements are important to the planning and operational considerations in a power system. Phasor measurement units (PMUs) have been deployed at substations that facilitate such measurements. The IEEE standard on synchrophasors (C37.118) governs the format of data files created and transmitted by PMUs, allowing interoperability among multiple device vendors. In addition to PMUs, other types of sensors have been used in practice. For instance, fibre Bragg grating (FBG) strain sensors and temperature sensors have been proposed to measure the temperature and icing conditions on transmission lines [5]. Non-contact sensors have been considered to measure three-phase power transmission line parameters [6].

On the *distribution side*, many of the same types of sensors in the transmission system can be utilized. Additionally, in substations, critical components such as transformers and circuit breakers must be continuously monitored to reduce the possibility of expensive and disruptive power outages [7,8].

On the *consumer side*, smart meters at customers' homes play a crucial role. They allow for real-time determination and information storage of energy consumption and provide the possibility to obtain energy consumption both locally and remotely [9]. Further, they also provide the means to detect fluctuations and power outages, permit remote limitations on consumption by customers, and permit the meters to be switched off. This results in important cost savings and enables utilities to prevent electricity theft. Electricity providers get a better picture of customers' energy consumption and obtain a precise understanding of energy consumption at different points in time. As a consequence, utilities are able to establish demand-side management (DSM) and to develop new pricing mechanisms. Energy can be priced according to real-time costs taking peak power loads into account, and price signals can be transmitted to home controllers or customers' devices, which in turn evaluate the information and make scheduling decisions accordingly.

Finally, *smart buildings* form their own eco-system in energy generation, distribution, storage, and consumption. Smart buildings rely on a set of technologies that enhance energy efficiency and user comfort as well as monitoring the safety of the buildings. The information and communication technologies (ICTs) are used in: (i) building management systems which monitor heating, lighting, and ventilation, (ii) software packages which automatically switch off devices such as computers and monitors when offices are empty, and (iii) security and access systems. The key characteristics of the newest generation of smart buildings are the ability to learn from the building and adapt their monitoring and controlling functions. Sensors and sensor networks are used in multiple smart building applications. These include: heating, ventilation, and air-conditioning systems (HVAC), lighting, shading, air quality, and window control; systems switching off devices, metering, security, and safety (access control). Sensors embedded in HVAC systems, for example, monitor the temperature and the status of parts of the buildings such as open or closed windows. In the field of air quality, new gas sensors, micro electrical–mechanical systems (MEMS), measure the content of CO_2 in rooms.

Traditionally, sensors are either involved in standalone local control loops or rely on data logger or wired lines for the collection of sensing information. Wireless sensor networks (WSN), since their inception in the early 2000s [10], have been shown to facilitate real-time, low-cost solutions in many application domains including structural health monitoring for civil structures, environment monitoring, and health care. While common features can be identified in the design of WSNs for different applications, requirements and specific characteristics of smart grids warrant in depth investigation.

In this chapter, we provide a detailed exposition of sensor technologies and sensor network protocols for smart grids. In Section 12.2, an overview of common sensors in power systems and the basic sensing principles are discussed. Specifically, we survey sensors used in metering and power quality monitoring, and power system status and health monitoring. In Section 12.3, communication protocols for smart grids are discussed in depth with focus on medium-access control (MAC), routing, and transport protocols.

12.2 Sensors and sensing principles

Sensors in power grids differ in the physical phenomena that they measure. Basic sensors are transducers that convert physical parameters into electronic signals, which can be either analogue or digitized. In the latter case, an analogue-to-digital converter is used. Standardized digital interfaces (e.g., RS232, SPI) can be used to communicate sensing data to computers. On-board storage may be available for data logging. Many of today's smart sensors are equipped with microprocessors for on-board processing and wireless transmission modules to send sensing data to remote stations in a real-time manner.

In this section, we review major types of sensors and sensing principles. We classify the basic sensors into two broad categories: (i) metering and power-quality sensors, and (ii) power system status and health-monitoring sensors.

12.2.1 Metering and power-quality sensors

As the names suggest, metering and power-quality sensors measure the usage and quality of electricity. The basic parameters measured by electrical sensors for metering and assessing power quality include voltage, current, harmonic phase, frequency, and many derived parameters such as apparent power, power factor, and volt–amps–reactive (VAR) etc.

Current measurements

Current sensing can be performed by three means, namely, resistive shunt, the current transformer, and the Hall effects [11] (Figure 12.1). The *resistive shunt* is simply a resistor placed in series with the load. According to Ohm's law, the voltage draw is proportional to the current flowing through the load. If the resistive shunt is manufactured to minimize its inductance, it is termed an AC shunt. The main advantages of DC resistive shunts are their low cost and high reliability. Both DC and AC resistive shunts require insertion into the power line and may lead to voltage drop and heat (energy dissipation).

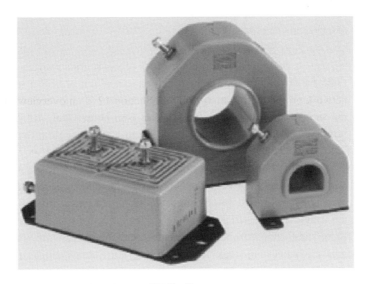

Figure 12.1　Current sensing using resistive shunt and Hall effects.

The *current transformer* is based on the fact that for a given flow, a proportional magnetic field is produced in accordance with Ampere's law. The current transformer couples this magnetic field into the secondary, providing a proportional current output. The operation of the device is identical to that of any voltage step-up transformer. The sensed, or aperture, current forms the primary turn, while the large number of turns wound on the magnetic core forms the secondary. The turns ratio determines the current output. The main advantages of the current transformer are its low cost in measuring AC currents, voltage isolation, and high reliability. However, it is only suitable for AC currents and the output is frequency dependent.

Hall-effect current sensors incorporate Hall generators, four-terminal solid-state devices that output a voltage proportional to the normal magnetic field, and the magnitude of the input control current. These detectors are either open loop or closed loop. Apart from their use of a Hall generator, core and amp, the two technologies are markedly different. In the open-loop setup, the Hall generator is mounted in the air gap of a magnetic core. A current-carrying conductor placed through the aperture of the core produces a magnetic field proportional to the current. The core concentrates the magnetic field, which is measured by the Hall generator. The signal from the Hall generator is low and is therefore amplified to a useful level, which becomes the output of the sensor. Closed-loop Hall-effect current sensors have five basic building blocks: the Hall generator, the magnetic core, the amplifier, a driver circuit, and a coil wound in series opposition around the magnetic core. The term 'closed-loop' is used because the magnetic field generated by the current-carrying conductor is nulled within the magnetic circuit of the core, thus closing the magnetic loop. This technique allows great improvements in sensor performance. The Hall generator senses the magnetic field generated by the conductor under measure and concentrated by the magnetic core. The output of the closed-loop sensor is therefore proportional to the aperture current and the number of turns of the coil. The main advantages of Hall-effect sensors are the ability to measure both large DC and AC currents, providing electrical isolation, and high reliability. However, they require external power supply.

Phasor measurements

A phasor is a complex number that represents both the magnitude and phase angle of the sine waves found in electricity. It is well known that active (real) power flow in a power line is nearly proportional to the sine of the angle difference between voltages at the two terminals of the line. Phasor measurements that occur at the same time are called 'synchrophasors', as are the PMU devices that allow such measurements. Synchrophasors measure voltages and currents at diverse locations on a power grid and can output accurately time-stamped voltage and current phasors. Because these phasors are truly synchronized, synchronized comparison of two quantities is possible, in real time. These comparisons can be used to assess system conditions. Report rates of phasor measurements range from 10–50/s for 50 Hz systems and 10–60/s for 60 Hz systems, allowing real-time decision and control. In addition to positive sequence voltages and currents, these systems also measure local frequency and rate of change of frequency, and may be customized to measure harmonics, negative and zero sequence quantities, as well as individual phase voltages and currents.

Smart meters

Smart meters will allow the central distribution system to monitor in real time the energy usage of each individual home on the grid and allow information about outages to be transported back to the utility (Figure 12.2). Smart meters are part of the advanced metering infrastructure (AMI) that measures, collects and analyses energy usage, and communicates with metering devices such as electricity meters, gas meters, heat meters, and water meters, either on request or on a schedule.

Smart meters have been deployed in many places in the world. In the USA, several major cities have smart meter projects underway. California's energy regulators approved a programme to roll out conventional meters retrofit with communications co-processor electronics to 9 million gas and electric household customers in the Northern California territory of Pacific Gas and Electric (PG&E). The Los Angeles Department of Water and Power (LADWP) has chosen to expand its advanced metering infrastructure (AMI) serving its commercial and industrial (C&I) customers. Austin Energy, the nation's ninth largest community-owned electric utility, with nearly 400,000 electricity customers in and around Austin, Texas, began deploying a two-way RF mesh network

Figure 12.2 Smart meter.

and approximately 260,000 residential smart meters in 2008. More than 165,000 two-way meters have been installed by spring 2009, and integration with AE's meter data management system is underway. Centerpoint Energy in Houston, Texas is currently in the deployment stage of installing smart meters to over 2 million electricity customers in the Houston-Metro and Galveston service locations. Current estimated completion of Centerpoint Energy's smart meter deployment is 2012.

12.2.2 Power system status and health monitoring sensors

Circuit-breaker sensors

Circuit breakers (CB) are electromechanical devices used in the power system to connect or disconnect the power flow at the generator, substation, or load location. The circuit breakers are capable of making, carrying, and breaking currents under normal circuit conditions and also making, carrying for a specific time, and breaking currents under specified abnormal circuit conditions such as those of a short circuit. As part of the supervisory control and data acquisition (SCADA), system circuit-breaker contacts are used to determine power system topology. By monitoring the open/closed status of all circuit breakers, it is possible to create a bus/branch topology configuration of the power system. This information is essential for several power system applications used to improve the reliability of the power system, such as power flow, state estimation, and alarm processing. Two types of information are important, namely, CB status and CB condition. CB status is measured by voltage on CB auxiliary contacts. CB condition monitoring needs to determine the condition of a specific circuit breaker and the condition of the circuit-breaker support and control functions. When a current is interrupted, an arc is generated. This arc must be contained, cooled, and extinguished in a controlled way, so that the gap between the contacts can again withstand the voltage in the circuit. Sulphur hexafluoride (SF_6) is often used in high-voltage circuit breakers to quench the stretched arc generated due to the interrupted current. Thus, sensors can be adopted to detect the leakage of SF_6 gas.

Transformers

Transformers represent one of the more expensive assets of local utility companies. There are many types of transformers used in practice, predominantly falling into two categories: dry-type and liquid-filled transformer. Dry-type transformers are environmentally safe, high-temperature insulation systems cooled by normal air ventilation, and thus they are suitable to be deployed close to loads in high rise buildings, underground tunnels, etc. For large transformers used in power distribution or electrical substations, the core and coils of the transformer are immersed in oil which cools and insulates. Oil circulates through ducts in the coil and around the coil and core assembly, moved by convection. To ensure the safe operation of transformers, hot-spot temperature measurement is needed. Fibre-optic thermometers use various sensing materials such as phosphors, semiconductors or liquid crystals, connected to the free end of fibre-optic links. They have the advantage of a wide range of operating conditions, from $-100°C$ to $400°C$. For liquid-filled transformers, hydrogen gas sensors can be used to measure H_2 and C_2H_2 in the headspace and oil. Acoustic sensors can detect leakage in the oil tank.

Transmission-line monitoring

For better grid situation awareness, transmission-line monitoring is an important topic in smart grid. Transmission-line monitoring aims to monitor and display the actual situation of overhead and underground lines, including thermal stability, line resistance, line current, active and reactive losses, erosion, tilt and vibration on the power lines. Some commercial systems such as the overhead line monitoring by Siemens [12] can also provide forecasting capability for contingency response, dispatch, generation, and outage scheduling.

Most recently, Yang *et al.* [13] described the design of a power-line sensornet, which consists of four core functions: sensing and measurement of critical line parameters in the immediate vicinity of the sensor; continuous online estimation of line status and identification of incipient faults in a changing environment; operation as a node of the sensornet communication system; power and fault management while allowing sensing and communication functions. In power-line sensors, line segmentation is needed to limit the sensing to the length of the line in the immediate vicinity of the sensor. To estimate the dynamic thermal capacity for line segments (defined as the maximum allowable short-term overload of a line), real-time information regarding the conductor and ambient temperature can be measured and plugged into a neural network-based model.

Table 12.1 gives a summary of sensors that can be utilized in power systems based on their functionalities [14].

12.3 Communication protocols for smart grid

In the past years, wireless sensor networks (WSNs) have been studied extensively for potential applications in a wide range of areas, including military, environment, health, home, industry, and many other commercial areas. Recently, WSN has been recognized as a promising technology for enhancing the functionality of different components in smart grid, from generation and transmission to utilization. Although many communication architectures have been proposed for WSNs in the literature, and different algorithms and communication protocols have been developed for each layer according to some general or specific application requirements, there are still many challenging issues in smart grid due to the harsh and complex electric-power system environments, which introduce many unique research problems for applying sensor network technologies in this area.

In this section, we provide an overview of the challenging issues for the design of MAC, routing and transport protocols for WSNs in smart grid, which are critical in connecting electric-power devices with an IP-based infrastructure [15]. Since smart grid shares many common features with other WSN applications, we start with a brief survey of existing algorithms and protocols developed for general WSN applications, followed

Table 12.1. Transmission and substation sensors [14]

Area	Component	Sensor	Applications
Substations	Substation-wide	Antenna array	Location and identification of discharging components
		On-line infrared	Automated processing of video thermal images of components
	Transformer	MIS gas sensor	Low-cost sensor to measure H_2 and C_2H_2 in headspace and oil
		3D Acoustics	Location and analysis of discharge activity in transformer
		Acoustic fibre-optic	Identifying low-level internal discharges in high-risk regions
		Gas fibre-optic	Identifiying gassing in high-risk regions
		On-line FRA	Continuously monitor frequency response using natural transients
	Load tap charge	LTC gassing	Identifying overheating or coking or worn contacts
	Post and bushing external insulation	RF leakage current	Identification of high-risk contacts wirelessly
	Disconnect	RF disconnect	Identifying high-risk contacts wirelessly
	Current and potential transformers	RF acoustic emissions	Using wireless mesh to identify internal discharges wirelessly
	Breaker	RF SF_6 density	Using wireless mesh to send SF_6 density wirelessly
Underground lines	Oil	MIS sensor	Low-cost sensor to measure H_2 and C_2H_2 gases in oil
	Underground cable system	Various	Identifying potential breakage
Overhead lines	Compression connector	RF temperature and current	Measures connector temperature and current to determine risk and identify high-risk components
	Conductor	RF temperature and current	Measures connector temperature and current for rating
	Insulator	RF leakage current	Identification of high-risk insulation requiring maintenance
	TLSA	RF leakage current	Assesses condition and number of operations
	Shield wire	RF fault magnitude and location	Determine location and magnitude of fault current
		RF lightning	Distribution of lightning current magnitudes
	Structure	Sensor system	Integrates RF and image recognition sensors to investigate transmissionline issues

by a discussion of the applicability of these schemes in smart grid, and lastly a more detailed introduction of the protocols specifically designed for smart grid.

12.3.1 MAC protocols

The design of energy-efficient MAC protocols has been the primary objective in the process of standardization for WSNs, such as ZigBee (IEEE 802.15.4), Bluetooth (IEEE 802.15.1), ultra wide band (UWB), and other custom-defined technologies. These short-range communication systems are mainly tailored to provide high energy efficiency for applications with relaxed throughput and latency requirements in wireless personal-area networks (WPANs). However, some emerging applications (such as smart grid) have imposed more stringent requirements and specifications for WSNs in terms of reliability, latency, throughput, fairness, etc., which has motivated the studies and investigations of new MAC protocols for WSNs.

As shown in Figure 12.3, existing MAC protocols for WSNs can be classified into three categories according to the channel access and collision resolution mechanisms: *contention-based protocol*, *contention-free protocol*, and *hybrid protocol*. The characteristics of each category and existing protocols are as follows:

- The *contention-based* MAC protocols are normally based on the carrier sense multiple access (CSMA) mechanism and its variations to avoid collision and hidden terminal problems, such as virtual and physical carrier sensing, RTS/CTS mechanisms, etc. In addition to the energy-efficiency issue, these protocols have been designed for different objectives. For example, the sensor-MAC (S-MAC) protocol [16] attempts to maintain good scalability and collision avoidance by locally managing the synchronization and periodic sleep–listen schedules; the DS-MAC protocol [17] is a variation of the S-MAC protocol with a dynamic duty cycle, which achieves a good tradeoff between energy consumption and latency by adjusting its sleep–wakeup cycle time based on the energy consumption level and the average latency it has experienced; MS-MAC [18]

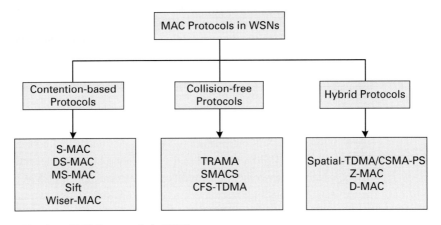

Figure 12.3 Classification of MAC protocols in WSNs.

is an adaptive mobility-aware MAC protocol that addresses the mobility issue in mobile sensor applications. It is similar to S-MAC in stationary scenarios in order to conserve energy. However, in highly mobile scenarios, it switches to an operating mode similar to the IEEE 802.11 MAC and dynamically adjusts the frequency of mobility handling actions based on the presence of mobile nodes and their moving speeds. Sift protocol [19] is motivated by the observations that sensor networks are usually event-driven and have spatially correlated contention. It uses a non-uniform probability distribution function for picking the contention slot, which is based on the shared belief of the current number of competing nodes; wise-MAC protocol [20] combines non-persistent CSMA with synchronized preamble sampling to mitigate idle listening, which leads to lower power consumption than the power-save scheme in the IEEE 802.11 and IEEE 802.15.4 MAC under low-traffic conditions.

- The *collision-free* MAC protocols are in general based on some time-division multiple access (TDMA)-like mechanisms, which can provide energy-efficient collision-free channel access while maintaining good throughput, acceptable latency, and fairness. For example, the traffic-adaptive medium access (TRAMA) protocol [21] assumes a single time-slotted channel. It adopts a transmitter-election algorithm that is inherently fair and promotes channel reuse as a function of the competing traffic around a given source or receiver; the self-organizing medium access control (SMACS) protocol [22] enables a collection of nodes to discover their neighbours and establish TDMA-like schedules for communicating with them without any local or global coordination. The contention-free scheduling TDMA (CF-TDMA) protocol [23] uses a periodic message model to construct a contention-free schedule for transmitting and receiving the messages. It is possible to combine the message scheduler with the task scheduler in that node. Therefore, this MAC protocol is highly scalable to large sensor networks. The MMAC protocol [24] is a schedule-based protocol. It can dynamically adapt the frame-time, transmission slots, and random-access slots according to the mobility patterns, which makes it suitable for mobile sensor networks under both *weak mobility* and *strong mobility* scenarios.

- The *hybrid* MAC protocols combine the features of both *contention-based* and *collision-free* protocols. For example, the spatial TDMA and CSMA with preamble sampling protocol [25] adopts two communication channels: a data channel and a control channel. In the data channel, a spatial TDMA protocol is used to transmit periodic and frequent data, while a low-power CSMA protocol is used in the control channel to transmit sporadic signalling traffic, which allows a node to sleep most of the time when the channel is idle, and thus improves energy efficiency and prolongs network lifetime. The Zebra-MAC (Z-MAC) [26] uses the CSMA protocol as the basic channel-access mechanism, while at the same time it uses a TDMA schedule to improve contention resolution. Specifically, under low-contention conditions, it behaves like a CSMA protocol and can achieve high channel utilization and low latency. Under high-contention conditions, it behaves like a TDMA protocol and can achieve high channel utilization and reduce collisions among two-hop neighbours at low cost. The D-MAC protocol [27] addresses the data-forwarding interruption problem in multihop data delivery by staggering the schedule of the nodes on the multihop

Table 12.2. Comparison of MAC protocols in WSNs

Category	Protocol	Access protocol	Time synchronization	Adaptivity	Mobility support
Contention-based	S-MAC	CSMA		Good	
	DS-MAC	CSMA		Good	
	MS-MAC	CSMA		Good	✓
	Sift	CSMA/CA		Good	
	Wiser-MAC	CSMA		Good	
Collision-free	TRAMA	TDMA/CSMA	✓	Good	
	SMACS	TDMA/CSMA	✓	Good	
	CF-TDMA	TDMA	✓	Weak	
	MMAC	TDMA	✓	Good	✓
Hybrid	Spatial-TDMA/CSMA-PS	TDMA/CSMA	✓	Good	
	Z-MAC	TDMA/CSMA	✓	Good	
	D-MAC	TDMA/Slotted Aloha	✓	Weak	

path so that the nodes will wake up sequentially like a chain reaction, which can enable continuous data forwarding on a multihop path.

A comparison of these MAC protocols in terms of channel access mechanism, time synchronization requirement, adaptivity, and mobility support is summarized in Table 12.2. We can see that although most of the above MAC protocols have not been designed specifically for smart grid application, many of them have the salient features, which are easy to adapt according to the application requirements in some components of smart grid. For example, for transmission-line monitoring, some *collision-free* MAC protocols (e.g., TRAMA [21], SMACS [22]) can be applied since the topology of WSN in this scenario is normally organized in a chain, which is easy to design a TDMA-like scheduling policy. Some *contention-based* protocols, for example, the D-MAC protocol [27] can also be applied because it has been designed to enable multihop data delivery. On the other hand, the network topology and environments are more complex in other application scenarios in smart grid, for example, substation automation and AMI networks. In this case, the *contention-based* or *hybrid* MAC protocols are more suitable since they have better adaptability under dynamic network conditions and can scale to a large network size.

In addition to these existing algorithms and protocols designed for WSNs, some researchers have studied the unique requirements of smart grid for MAC protocols.

For example, Sun *et al.* [28] propose a QoS-enhanced MAC protocol targeted for smart grid based on the IEEE 802.15.4 protocol, which can provide differentiated service for data traffic with different priorities. Their basic idea is to store the incoming traffic in different queues based on their priority, whereby high-priority data will not only have higher probability of channel access, but also can interrupt the service to the low-priority traffic by forcing it to backoff. The authors also provide theoretical models to analyse the network delay, goodput, and collision ratio. This scheme is well suited for the power distribution network-monitoring system, which carries two classes of traffic: operational and emergency data, where the emergency data normally requires higher priority of service than the operational data.

12.3.2 Routing protocols

Wireless sensor networks are normally deployed to monitor the interesting events specified by the target applications, and the collected data from each sensor node has to be processed and transmitted back to the sink nodes, which are often beyond the communication range of sensor nodes. Therefore, some special multihop wireless routing protocols between the sensor nodes and the sink node are needed.

Similar to the MAC protocol, the energy efficiency is one of the most important considerations in the design of routing protocols for WSNs. In addition, a unique feature of WSNs is that data collected by neighbouring nodes are spatially correlated, which leads to a 'data-centric' paradigm whereby the routing function should be combined with data aggregation to improve the energy efficiency. Many routing protocols have been proposed for WSN accordingly to meet these unique requirements, which are generally classified as *flat routing, hierarchical routing*, and *location-based routing* depending on the network structure (Figure 12.4). A brief survey of routing protocols in each category is as follows:

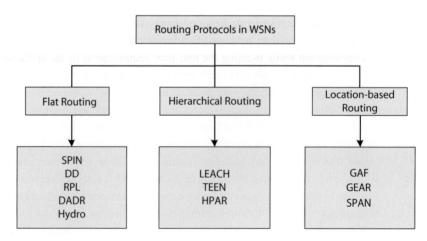

Figure 12.4 Classification of routing protocols in WSNs.

- In *flat-routing* protocols, every node performs similar functionality, and neighbouring nodes can collaborate in the event-sensing task. To reduce the amount of redundant information in the neighbouring nodes, a data-centric paradigm is adopted whereby each node can perform data fusion (known as in-network processing) in addition to the conventional routing function. For example, in the sensor protocols for information via negotiation (SPIN) protocol [29], each node performs a meta-data negotiation with its neighbouring nodes, which can remove redundant information from the data before it travels through the network to the sink node. The protocol can also adapt the routing path according to the residual energy levels of the nodes, which can prolong the network lifetime by balancing the energy consumption across nodes. The directed diffusion (DD) protocol [30] is another data-centric flat-routing protocol, which sets up routing paths from multiple source nodes to a single sink node using the information gradient, which allows the data generated by different sources belonging to the same interest to be aggregated at the intermediate nodes to reduce redundancy and minimize the number of transmissions.
- In *hierarchical routing*, the sensor network is partitioned into multiple clusters, each consisting of a single (not necessarily fixed) cluster head and a set of member nodes. The functionality of the network is also divided into two levels accordingly: one layer is to select cluster heads which are responsible for collecting and aggregating data from member nodes, the other is for the member nodes to report sensing data to the cluster head according to the schedules. By performing data aggregation and fusion within the cluster, it reduces the number of transmitted messages to the sink and energy consumption. For example, the low-energy adaptive clustering hierarchy (LEACH) [31] is a cluster-based protocol. It randomly selects a subset of nodes and designates them as the cluster heads. This role is periodically rotated among the nodes so as to balance the energy consumption. LEACH uses a TDMA/CDMA MAC to reduce inter-cluster and intra-cluster collisions. However, data collection is centralized and is performed periodically. Therefore, this protocol is most appropriate when there is a need for constant monitoring by the sensor network.
  In the threshold-sensitive energy-efficient sensor network (TEEN) protocol [32], the sensor nodes can generate data constantly, but they transmit using a lower frequency according to the threshold specified by the cluster head, which can reduce the energy consumption while meeting the real-time requirements of the application. Therefore, this protocol is suitable for time-critical applications. The HPAR (hierarchical power-aware routing) protocol [33] divides the network into groups of sensors. Each group of sensors in geographic proximity is clustered together as a zone and each zone is treated as an entity. Routing is performed at the zone level, whereby each zone decides how it will route a message hierarchically across the other zones such that the lifetime of the nodes in the network is maximized.
- In the *location-based routing*, the location information of sensor nodes is utilized in routing and data aggregation for energy-efficiency and reliability purposes. For example, in the geographic adaptive fidelity (GAF) protocol [34], the network area is divided into fixed zones and forms a virtual grid; each node is associated with a point in the virtual grid according to its location. Nodes within the same zone cooperate with

Table 12.3. Comparison of routing protocols for WSNs

Category	Protocol	Power usage	Overhead	Scalability	Multipath	Data aggregation
Flat routing	SPIN	Limited	Low	Limited	Yes	Yes
	DD	Limited	Low	Limited	Yes	Yes
	RPL	Limited	Low	Good	Yes	Yes
	DADR	Limited	Low	Good	No	Yes
	Hydro	High	High	Limited	No	Yes
Hierarchical routing	LEACH	High	High	Good	No	Yes
	TEEN	High	High	Good	No	Yes
	HPAR	Limited	Low	Good	No	No
Location-based routing	GAF	Limited	Low	Good	No	No
	GEAR	Limited	Low	Limited	No	No
	SPAN	Limited	Low	Limited	No	No

each other to share the responsibility for monitoring and routing data to the sink node. The geographic and energy aware routing (GEAR) protocol [35] takes into account the location and energy-consumption level of each node when selecting the next-hop nodes. The key idea is to restrict the number of interests in directed diffusion by only considering a certain region rather than sending the interests to the whole network. By doing this, GEAR can conserve more energy than directed diffusion. The SPAN [36] protocol selects a set of nodes as coordinators according to their positions. A node will become a coordinator if two neighbours of a non-coordinator node cannot reach each other directly or via one or two coordinators. The set of coordinators will act as the backbone nodes to forward for other nodes.

In addition to the above classification (Table 12.3), routing protocols can be considered as adaptive if they can adapt to the current network conditions and available energy levels. The above-mentioned protocols can also be categorized as *multi-path routing, query, negotiation, or quality-of-service*-based routing protocols depending on the actual operations. Alternatively, they can be broken down into *proactive, reactive*, or *hybrid* protocols according to the way the source finds the destination.

In smart grid, safety and reliability are the most critical requirements. Therefore, in addition to the problem due to the resource constraints (energy, memory, processing, etc.), the harsh environmental conditions and stringent quality-of-service requirements (reliability, throughput, latency, packet errors) pose great challenges to the design of reliable routing protocols for WSNs in smart grid application. To this end, a working group has been set up in IETF to draft the RPL (routing protocol for low power and lossy networks) for low-power and lossy networks (LLN) [37]. RPL maintains the network topology using a directed acyclic graph (DAG) rooted at the sink node (or gateway) as shown in Figure 12.5. Each node is associated with a rank value, which represents its 'position' relative to other nodes. In this way, the position of each node in the DAG can be determined using this ranking property. The RPL protocol is designed for a wide range

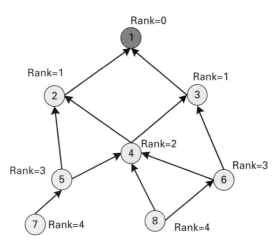

Figure 12.5 Directed acyclic graph (DAG) structure and node ranking in RPL protocol.

of WSN applications, such as environmental monitoring, building automation, health care, asset tracking, etc., and smart grid is also an important application of this protocol.

Note that RPL is still under the draft process; many detailed issues have been left unsolved due to the diverse requirements of application scenarios. For example, it is not specified how to modify and implement the RPL protocol in smart grid applications. Wang et al. [38] present a practical implementation of RPL that can support bi-directional traffic in smart grid. The authors adopt the expected transmission time (ETX) as the link metric to quantify the quality of wireless links, and propose an ETX-based rank computation method for DAG construction and maintenance, which tends to use links with smaller ETX, and therefore provides good end-to-end reliability. To support outward traffic (from the gateway to sensor nodes), a lightweight reverse-path recording mechanism is proposed for setting up the reverse routing path, whereby each node records the source and the last-hop nodes from the inward packets passing through the node. In this way, additional protocol overheads can be avoided since it exploits the information from the regular traffic, which is valid for AMI networks in smart grid since meter readings have to be sent back to the gateway periodically.

Tripathi et al. [39] present a simulation study of the RPL protocol based on the measurement data from an outdoor smart grid substation network. Since the RPL protocol is based on the DAG structure rooted at the sink, the major concern is the link repair problem when a link is broken due to node or link failure. There are two repair mechanisms for this problem, one is the 'global repair' scheme whereby each disconnected child node has to wait for a new DAG SequenceNumber to trigger the link repair process, which may keep the node out of service for a time on the order of minutes or hours. The other scheme is the 'local repair', whereby a node finding path failure can start the link-repair process and notify its children for alternate parents, which significantly reduces the repair waiting time for the global repair scheme. Simulation results suggest that the RPL protocol has the desired advantages in terms of control overhead, delay, and quick repair of local link outage.

Xie *et al.* [40] study the routing loop problem arising in the DAG-based mechanism adopted by the RPL routing protocol. In the DAG structure, the acyclic property (i.e., loop-free) is guaranteed if each node always has a higher rank than any of its parents. However, a node may change its rank when the rank or cost of existing parents changes, or it needs to add new parents. For example, a node A tries to increase its rank so that it can add a new node B (whose rank is higher than node A's current rank), then this may result in routing loops if node B is a descendant of node A in the DAG. In the worst case, this can result in the *count to infinity* problem, when the increase of rank for one node forces the increase of rank for other nodes, leading to excessive exchange of control messages. Through simulation study, it is found that most routing-loop problems can be resolved quickly. However, some routing-loop problems can lead to multiple routing loops and affect a large number of nodes. Some mechanisms for dealing with this problem – such as loop prevention, loop avoidance, and loop detection – are discussed. The authors also compare the performance for the case with and without loop avoidance strategy with simulation study. However, simulation results suggest that the use of loop avoidance is not necessarily beneficial for reducing the maintenance overheads in the RPL protocol.

Iwao *et al.* [41] focus on the effect of unreliable links on the routing protocol in WSNs. The conventional wisdom is that wireless links can be abstracted as either 'connected' or 'disconnected' according to the link quality (such as SNR). However, wireless links are dynamic and unpredictable in nature, and make no clear distinction between these two states. As a result, routing paths are subject to frequent failures, which requires constant route repairs and introduces much control overhead. To address this problem, the authors propose a DADR (distributed autonomous depth-first routing) protocol, which decouples the traditional one-phase routing protocol into two planes, the 'control plane' and 'data forwarding plane'. The *control plane* uses a lightweight proactive routing protocol to keep a 'soft' routing table for each node, which consists of a set of redundant next hops for each destination. When a routing path failure occurs, the data-forwarding algorithm finds an alternate next hop to search for the routing path to the destination. Therefore, there is no need to perform routing maintenance or routing repair in the control plane, which can avoid the control overhead in traditional proactive routing protocols. The *data-forwarding plane* employs a depth-first search approach to find an alternate routing path in case of routing failure by using the routing table provided by the control plane. It consists of a number of steps, including the *basic forwarding, loop detection, backtrack*, and *route avoidance* that attempt to avoid the routing loop problem in searching the alternate paths. The authors have implemented the proposed algorithm in several environments, including the AMI networks in Japan and the USA. Through indoor and outdoor experiments, it is shown that the proposed scheme can restore the reliability of routing paths quickly in the data-forwarding phase.

Haggerty *et al.* [42] propose Hydro, a hybrid routing protocol for low-power and lossy wireless networks, which is motivated by the observation that in addition to the conventional many-to-one (data collection) and one-to-many (data dissemination) traffic, point-to-point traffic is emerging in many wireless sensor applications, for initiating data transfer, sending end-to-end acknowledgments, or sending commands to nodes for time synchronization, etc. Hydro intends to support both robust data collection and point-to-point communication by combining the centralized and distributed mechanisms. The *distributed DAG formation* builds a DAG for routing data from sensor nodes to border routers (gateway nodes) using a distributed algorithm. It allows each node to maintain a list of neighbouring nodes in the *default route table* that are in the direction of a border router. Therefore, data collection can go through multiple paths and end-to-end reliability can be improved significantly. In addition to data collection, the DAG provides a basic triangle point-to-point routing by allowing nodes to forward the data to its destination via the border routers. *Global topology construction* builds a global view of network topology by aggregating the *topology reports* from each node in the network, which provides the information for a subset of bi-directional links in the network. The border routers can exploit this global information to optimize the routing paths for the active point-to-point flows by updating the flow table in the corresponding nodes, which can significantly reduce the extra hops for point-to-point flows. The authors have implemented Hydro on top of a low-power IPv6 stack using 6LoWPAN and evaluated its performance on two network testbeds. It is shown through experimental studies that Hydro can provide high reliability for the data collection, while reducing the routing stretch for point-to-point communications without overloading constrained nodes with excessive state or traffic requirements.

12.3.3 Transport protocols

In addition to tolerable end-to-end delay and low network capacity, many WSN applications require a reliable data delivery guarantee between the sensor nodes to the sink node (and vice versa). Traditionally, reliable data transmission is performed at the transport layer. For example, TCP (transmission control protocol) is a connection-oriented end-to-end transport protocol for the Internet. It sets up a connection between a sender and a receiver. The sender maintains a congestion window for each connection, and congestion control is realized by adjusting the congestion window size (decreasing when congestion is found, increasing otherwise).

However, TCP cannot be adopted directly for WSNs due to the unique characteristics of WSNs. First, the data volume for each node in WSN is normally very low (on the order of a few bytes). As a result, the three-way handshake process in the TCP protocol constitutes a non-negligible burden for the data traffic. Second, the time to set up a TCP connection depends on the round-trip time between two end nodes, which makes it difficult for the sensor nodes far from the sink node to get fair shares of the bandwidth. Last but not the least, the end-to-end approach adopted by the TCP protocol leads to a longer response time in case of packet losses, which in turn leads to much lower throughput since it relies on a retransmission strategy to achieve the reliability.

To address the above problems, several transport protocols have been developed for WSNs. Due to the unique traffic pattern of WSNs, these transport layer protocols have been designed to guarantee the reliability for either upstream traffic (from sensor nodes to the sink node) or downstream traffic (from the sink node to the sensor nodes). Therefore, they can be classified into two categories: upstream congestion control and downstream reliability guarantee.

- The *upstream congestion control* scheme tries to provide reliable transmission for data originated from the sensor nodes and avoid congestion closer to the sink node. Due to the unique feature of WSNs, 'reliability' of data transmission can be characterized in different ways. For example, ESRT (event-to-sink reliable transport) [43] attempts to achieve reliable *upstream event* detection and congestion control while maintaining the minimum energy expenditure. It guarantees only the end-to-end reliable delivery of individual events, not individual packets from each sensor node. The notion of reliability is defined with respect to the number of data packets originated by any event that are reliably received at the base station. ESRT includes a congestion control component that serves the dual purpose of achieving reliability and conserving energy, and the reliability of event detection is controlled by the sink which has more power than sensors. CODA (congestion detection and avoidance) [44] adopts a congestion detection mechanism by checking the buffer occupancy for each sensor node. Once the value is found to be beyond a certain level, an open-loop *hop-by-hop* back-pressure mechanism is used to notify the neighbouring nodes to reduce the transmission rate. The CODA protocol also adopts a closed-loop end-to-end scheme for the sink node to regulate the transmission rate from multiple sensor nodes. The RMST (reliable multi-segment transport) protocol [45] provides upstream delivery reliability using a hop-by-hop approach. In this protocol, packets generated by the sensor nodes are fragmented into multiple segments. In case of losses, any nodes along the path to the source nodes can initiate retransmission for the missing segments if they have a cached copy.
- In the *downstream reliability guarantee* scheme, the focus is on the recovery of loss packets disseminated from the sink node to the sensor nodes. For example the, PSFQ (pump slowly fetch quickly) protocol [46] employs a hop-by-hop error-recovery mechanism by caching the fragments at the intermediate nodes. When a node experiences packet loss, it can recover missing packets from the immediate neighbours quickly. GARUDA [47] provides downstream reliable delivery for control codes and query-metadata. It uses a three-phase approach to select a subset of nodes as the core nodes that can cache the packets, and other non-core nodes can recover any lost packets from the core nodes using out-of-sequence NACK. This approach ensures the highest availability of the lost fragments to the non-core sensors and reduces channel contention and congestion.

In addition to the aforementioned protocols that intend to provide either upstream or downstream reliable transmissions, a few transport protocols have been proposed for both traffic patterns (Table 12.4). For example, the ART [48] protocol provides upstream end-to-end event reliability, downstream end-to-end query reliability, and upstream congestion control. It selects a subset of 'essential nodes (E-nodes)' to cover the entire sensing area, and only those E-nodes participate in the reliable upstream and downstream data transfer, and recovering the lost fragments. RAP [49] is another protocol that provides a real-time upstream and downstream communication protocol for large-scale WSNs. It supports both periodic and event-based data flow.

Table 12.4. Comparison of transport protocols for WSNs

Protocol	Upstream	Downstream	Congestion control	Loss recovery	End-to-end	Hop-by-hop
ESRT	✓		✓	✓	✓	
CODA	✓		✓		✓	✓
RMST	✓			✓		✓
SA-TCP	✓		✓	✓	✓	
PSFQ		✓		✓		✓
GARUDA		✓		✓		✓
ART	✓	✓	✓	✓	✓	
RAP	✓	✓	✓		✓	✓

Note that smart grid shares many common features and requirements with other WSN applications in the design of transport protocols in terms of congestion control and reliability guarantee. Therefore, many aforementioned transport protocols can be applied for the smart grid applications with certain modifications. For example, Khalifa et al. [50] show that the traditional TCP protocol is ineffective in smart grid because individual meters in AMI networks generate and transmit data at very low data-rate (a packet every few minutes). If the conventional TCP protocol is applied, the congestion window will be reset to the initial value after each unsuccessful transmission, which essentially invalidates the functionality of congestion-control mechanism. On the other hand, due to the large number of meters in each AMI network, it will create a large amount of traffic that does not follow the TCP-friendly rule, and thus consumes more bandwidth than its fair share. In addition, the degree of congestion gets higher as it gets closer to the sink node due to the traffic aggregation; packets can be dropped without proper congestion control, leading to the large delivery latency, retransmission traffic and energy loss. To address this problem, the authors proposes a split and aggregated TCP (SA-TCP) protocol, which is based on the idea of adding a TCP aggregator node between the data sources (meter nodes) and the data collection server, which is responsible for collecting data from source nodes and forwarding data to the collection server. The aggregator will set up a TCP connection with the data collection node on behalf of all the meters. The reliability of data transmission and congestion control is thus performed between the aggregator and the data collection node. It is shown that this aggregation scheme is effective in improving the throughput, reducing the packet drop rate, and improving the fairness.

12.4 Challenges for WSN protocol design in smart grid

As has been discussed in previous sections, although many different algorithms and protocols have been developed for general application scenarios, the design of network protocols for smart grid is more challenging than the general WSN applications for the following reasons:

- *Complex and heterogeneous electric-power environments.* The most challenging issue for applying WSNs in smart grid is the complex environmental conditions in electric-power systems. The major components of smart grid are located in different environments, the network structure of WSNs should be designed according to the features of these components. For example, the power generating plants and substations are normally located in the concentrated area, with many densely distributed power devices and equipments. Therefore, dense WSNs with complex network topologies should be designed accordingly to monitor the target area. In contrast, transmission lines are geographically dispersed over hundreds of miles, and thus the linear network topology is more appropriate for assessing the components along the transmission line. In addition, the network topology and wireless channels may change dynamically due to RF interference, temperature, humidity level, and other environmental conditions, which requires the communication protocols to be adaptive and robust.
- *Reliability and availability requirements.* Both are common requirements for almost all communication systems, that is, nodes should be reliable and reachable under all conditions. However, maintaining reliability and availability is challenging in smart grid since the wireless channels are subject to dynamic change because of the larger-scale and small-scale fading, and sensor nodes may undergo failure in harsh environments due to high temperature, high humidity levels, dirt and dust in the open fields. Therefore, network topology should be designed carefully such that the connectivity of the network can still be maintained even with some of the sensor nodes or wireless links under failure conditions. Meanwhile, robust mechanisms should be provided in the communication protocols (MAC, routing, and transport protocols) to guarantee reliable transmission in the lossy and complex smart grid environments.
- *Diverse quality-of-service (QoS) requirements.* Unlike many WSN applications, smart grid has to support a number of data classes for metering, control commands, monitoring, and emergency responses. These data have diverse QoS requirements expressed in one or multiple metrics, including end-to-end throughput, delay and loss rates. For example, some time-critical information is normally carried in the control command and emergency response data, therefore they should be delivered in a timely manner, while the video surveillance data for monitoring transmission lines, transformers, and substations should be provided with the minimum bandwidth guarantee. Therefore, it is important to provide certain QoS mechanisms in the communication protocols according to the requirements of different data types.
- *Scalability issues.* In smart grid, a primary substation can provide service to tens of thousands of nodes, in particular in the densely populated area. Even though the meter reading data is small for each node, the overall volume of data traffic to be aggregated at the primary substation can be huge. Therefore, the scalability of WSNs is required not only from the network dimension perspective, but also from the network capacity perspective.
- *Security issues.* The integrity and privacy of the data in smart grid are crucial since the sensors are deployed for collecting information from power plant, wind farms, transmission lines, distribution units, and substations. Therefore, smart grid requires the highest level of security guarantees, which includes not only PKI technologies, but

also trusted computing elements in the networks. Assessment of the impact of security attacks on the stability of smart grid and electricity markets is another important research issue.

In summary, the unique features of electric-power systems pose many new challenges to the application of WSNs in smart grid. However, it also gives arise to research opportunities for communication engineers to design new network and communication protocols that are customized for smart grid environments. We therefore envisage that reliable and secure network protocol design in smart grid will become an exciting research area.

12.5 Conclusion

In this chapter, sensor technologies and sensor network protocols for smart grid have been discussed. With the advancement of smart grid, traditional sensors deployed in different subsystems of the power grids are connected to form a highly distributed and heterogeneous network of networks. Challenges arise in non-intrusive and low-cost sensing, scalable and reliable sensing data collection, and command control as well as effective processing of the large amount of sensed information to ensure safe, stable, and efficient grid operations.

References

[1] OECD, Smart sensor networks: technologies and applications for green growth, OECD Digital Economy Papers, No. 167, OECD Publishing.

[2] R. Atkinson and D. Castro, 'Digital quality of life – understanding the personal & social benefits of the information technology revolution', The Information Technology and Innovation Foundation, Washington, DC, October 2008.

[3] R. A. Swartz, J. P. Lynch, B. Sweetman, R. Rolfes, and S. Zerbst, 'Structural monitoring of wind turbines using wireless sensor networks', *Journal of Smart Structures and Systems*, in Press.

[4] L. Lin and S. Ming-Xia, 'Design of a wind power generation monitoring system based on wireless sensor network', in *Proceedings of International Conference on Intelligent System Design and Engineering Application (ISDEA)*, 2010.

[5] L. Li, X. Sun, X. Meng, Z. Zhang, X. Chen, and M. Zhang, 'Real-time ice monitoring on overhead power transmission lines with FBG sensor', in *Proceedings of Second IITA International Conference on Geoscience and Remote Sensing (IITA-GRS)*, 2010

[6] M. Greitans, E. Hermanis, and A. Selivanovs, 'Sensor based diagnosis of three-phase power transmission lines', Electronics and Electrical Engineering. *Kaunas: Technologija*, vol. 6, no. 94, pp. 23–26, 2009.

[7] Department of Energy, United States, Grid 2030 – A Vision for Electricity's Second 100 Years, Washington, DC, 2003.

[8] A. Nasipuri, R. Cox, J. Conrad, L. Van der Zel, B. Rodriguez, and R. McKosky, 'Design considerations for a large-scale wireless sensor network for substation monitoring', in *Proceedings of 5th IEEE International Workshop on Practical Issues in Building Sensor Network Applications (SenseApp 2010)*, Denver, CO, October 2010.

[9] H. P. Siderius and A. Dijkstra, 'Smart metering for households: cost and benefits for the Netherlands', http://www.saena.de/media/files/Upload/smart_metering/PDF/Smart_Metering_NL.pdf

[10] B. Warneke, M. Last, B. Liebowitz, and K. S. J. Pister, 'Smart dust: communicating with a cubic-millimeter computer', *Computer*, vol. 34, no. 1, pp. 44–51, January 2001.

[11] DST control, Current Sensing Technology, http://www.dst-solar.com/DC-Current-Sensing-Technology.html.

[12] Siemens, ISCM – overhead line monitoring for maximum power-throughput and highly effective planning.

[13] Y. Yang, D. Divan, R. G. Harley, and T. G. Habetler, 'Power line sensornet – a new concept for power grid monitoring', *IEEE Power Engineering Society General Meeting*, 2006.

[14] Electric Power Research Institute, Sensor technologies for a smart transmission system, White Paper, 2009

[15] F. Baker and D. Meyer 'Internet protocols for the smart grid', draft-baker-ietf-core-13.txt, March 2011.

[16] W. Ye, J. Heidemann, and D. Estrin, 'Medium access control with coordinated adaptive sleeping for wireless sensor networks', *IEEE/ACM Transactions on Networking*, vol. 12, no. 3, pp. 493–506, June 2004.

[17] P. Lin, C. Qiao, and X. Wang, 'Medium access control with a dynamic duty cycle for sensor networks', in *Proceedings of 2004 IEEE Wireless Communications and Networking Conference (WCNC'04)*, Atlanta, GA, March 2004, pp. 1534–1539.

[18] H. Pham and S. Jha, 'An adaptive mobility-aware MAC protocol for sensor networks (MS-MAC)', in *Proceedings of 2004 International Conference on Mobile Ad Hoc and Sensor Systems (MASS'04)*, Fort Lauderdale, FL, October 2004, pp. 558–560.

[19] K. Jamieson, H. Balakrishnan, and Y. C. Tay, 'Sift: a MAC protocol for event-driven wireless sensor networks', LCS Technical Reports, May 1, 2003.

[20] A. El-Hoiydi *et al.*, 'WiseMAC: an ultra low power MAC protocol for the WiseNET wireless sensor network', in *Proceedings of 1st ACM Conference on Embedded Networked Sensor Systems (SenSys'03)*, Los Angeles, CA, November 2003, pp. 302–303.

[21] V. Rajendran, K. Obraczka, and J. J. G. L. Aceves, 'Energy-efficient, collision-free medium access control for wireless sensor networks', in *Proceedings of 1st ACM Conference on Embedded Networked Sensor Systems (SenSys'03)*, Los Angeles, CA, November 2003, pp. 181–192.

[22] K. Sohrabi *et al.*, 'Protocols for self-organization of a wireless sensor network', *IEEE Personal Communications*, October 2000, pp. 16–27.

[23] T. W. Carley *et al.*, 'Contention-free periodic message scheduler medium access control in wireless sensor/actuator networks', in *Proceedings of 24th IEEE International Real-Time Systems Symposium (RTSS'03)*, Cancun, Mexico, December 2003, pp. 298–307.

[24] M. Ali, T. Suleman, and Z. A. Uzmi, 'MMAC: a mobility-adaptive, collision-free MAC protocol for wireless sensor networks', in *Proceedings of 24th IEEE IPCC'05*, 2005.

[25] A. El-Hoiydi, 'Spatial TDMA and CSMA with preamble sampling for low power ad-hoc wireless sensor networks', in *Proceedings of 2002 IEEE Symposium on Computers and Communications (ISCC '02)*, Taormina, Italy, July 2002, pp. 685–692.

[26] I. Rhee, A. Warrier, M. Aia, and J. Min, 'Z-MAC: a hybrid MAC for wireless sensor networks', in *Proceedings of 3rd ACM Conference on Embedded Networked Sensor Systems (SenSys'05)*, San Diego, CA, November 2005.

[27] G. Lu, B. Krishnamachari, and C. S. Raghavendra, 'An adaptive energy-efficient and low-latency MAC for data gathering in wireless sensor networks', in *Proceedings of 18th International Parallel and Distributed Processing Symposium (IPDPS' 04)*, Santa Fe, NM, April 2004, pp. 224–231.

[28] W. Sun, X. J. Yuan, J. P. Wang, D. Han, and C. W. Zhang, 'Quality of service networking for smart grid distribution monitoring', in *Proceedings of First IEEE International Conference on Smart Grid Communications (SmartGridComm)*, Gaithersburg, MD (NIST), October 2010.

[29] W. R. Heinzelman, J. Kulik, and H. Balakrishnan, 'Adaptive protocols for information dissemination in wireless sensor networks', in *Proceedings of 5th Annual ACM/IEEE International Conference on Mobile Computing and Networking (MobiCom'99)*, pp. 174–185.

[30] C. Intanagonwiwat, R. Govindan and D. Estrin 'Directed diffusion: a scalable and robust communication paradigm for sensor networks', in *Proceedings of 6th Annual International Conference on Mobile Computing and Networking (MobiCom'00)*, pp. 56–67.

[31] W. R. Heinzelman, A. Chandrakasan, and H. Balakrishnan, 'Energy-efficient communication protocol for wireless microsensor networks', in *Proceedings of 33rd Annual Hawaii International Conference on System Sciences*.

[32] A. Manjeshwar and D. P. Agrawal, 'TEEN: a routing protocol for enhanced efficiency in wireless sensor networks', in *Proceedings of 15th International Parallel and Distributed Processing Symposium*, pp. 2009–2015.

[33] Q. Li, J. Aslam, and D. Rus, 'Hierarchical power-aware routing in sensor networks', in *Proceedings of DIMACS Workshop on Pervasive Networking*, May, 2001.

[34] Y. Xu, J. Heidemann, D. Estrin, 'Geography-informed energy conservation for ad-hoc routing', in *Proceedings of Seventh Annual ACM/IEEE International Conference on Mobile Computing and Networking 2001*, pp. 70–84.

[35] Y. Yu, R. Govindan, and D. Estrin, 'Geographical and energy aware routing: a recursive data dissemination protocol for wireless sensor networks', UCLA, Technical Report, 2001, CSD-TR-01-0023.

[36] B. Chen *et al.*, 'SPAN: an energy-efficient coordination algorithm for topology maintenance in ad hoc wireless networks', *Wireless Network*, vol. 8, no. 5, pp. 481–494, 2002.

[37] T. Winter and P. Thubert, 'RPL: IPv6 routing protocol for low power and lossy networks', draft-ietf-roll-rpl-04.txt, October 2009.

[38] D. Wang, Z. Tao, J. Zhang, and A. Abouzeid, 'RPL based routing for advanced metering infrastructure in smart grid', in *Proceedings of IEEE International Workshop on Smart Grid Communications*, May 2010.

[39] J. Tripathim, J. C. de Oliveira, and J. P. Vasseur, 'Applicability study of RPL with local repair in smart grid substation networks', in *Proceedings of First IEEE International Conference on Smart Grid Communications (SmartGridComm)*, Gaithersburg, MD (NIST), October 2010.

[40] W. Xie, M. Goyal, H. Hosseini, J. Martocci, Y. Bashir, E. Baccelli, and A. Durresi, 'Routing loops in DAG-based low power and lossy networks', in *Proceedings of 2010 24th IEEE International Conference on Advanced Information Networking and Applications (AINA'10)*, 2010.

[41] T. Iwao, K. Yamada, M. Yura, Y. Nakaya, A. A. Cardenas, S. Lee, and R. Masuoka, 'Dynamic data forwarding in wireless mesh networks', in *Proceedings of First IEEE International Conference on Smart Grid Communications (SmartGridComm)*, Gaithersburg, MD (NIST), October 2010.

[42] S. D. Haggerty, A. Tavakoli, and D. Culler, 'Hydro: a hybrid routing protocol for low-power and lossy networks', in *Proceedings of First IEEE International Conference on Smart Grid Communications (SmartGridComm '10)*, October 2010.

[43] Y. Sankarasubramaniam, B. Akan, and I. F. Akyildiz, 'ESRT: event to sink reliable transport in wireless sensor networks', in *Proceedings of ACM MobiHoc'03*, Annapolis, MD, June 2003.

[44] C. Y. Wan, S. B. Eisenman, and A. T. Campbell, 'CODA: congestion detection and avoidance in sensor networks', in *Proceedings of First International Conference on Embedded Networked Sensor Systems (SenSys'03)*, Los Angeles, CA, 2003, pp. 266–279.

[45] F. Stann and J. Heidemann, 'RMST: reliable data transport in sensor net-works', in *Proceedings of First IEEE International Workshop on Sensor Network Protocols and Applications*, Anchorage, AK, 2003, pp. 102–112.

[46] C. y. Wan, A. T. Campbell, and L. Krishnamurthy, 'PSFQ: a reliable transport protocol for wireless sensor networks,' in *Proceedings of ACM International Workshop on Wireless Sensor Networks and Applications*, Atlanta, GA, 2002, pp. 1–11.

[47] S. J. Park, R. Vedantham, R. Sivakumar, and I. F. Akyildiz, 'A scalable approach for reliable downstream data delivery in wireless sensor networks', in *Proceedings of International Symposium on Mobile Ad Hoc Networking and Computing (MobiHoc)*, Tokyo, Japan, 2004, pp. 78–89.

[48] N. Tezcan and W. Wang, 'ART: an asymmetric and reliable transport mechanism for wireless sensor networks', *International Journal of Sensor Networks*, vol. 2, no. 3/4, pp. 188–200, 2007.

[49] C. Lu, B. Blum, T. Abdelzaher, J. Stankovic, and T. He, 'RAP: a real-time communication architecture for large-scale wireless sensor networks', in *Proceedings of IEEE RTAS*, 2002.

[50] T. Khalifa, K. Naik, M. Alsabaan, A. Nayak, and N. Goel, 'Transport protocol for smart grid infrastructure', in *Proceedings of Second International Conference on Ubiquitous and Future Networks (ICUFN)*, 16–18 June 2010, pp. 320–325.

13 Potential methods for sensor and actuator networks for smart grid

Victor O. K. Li and Guang-Hua Yang

13.1 Introduction

In the past few years, smart grid (SG) has attracted much interest from governments, power companies, and research institutes [1–3]. Compared to the traditional power grid, by employing advanced information technologies (IT), SG can achieve better reliability and stability, higher energy efficiency, higher penetration of renewable energy (RE), and lower greenhouse gas emission [3, 4]. Sensor and actuator networks (SANETs) play a key role in realizing these advantages. Compared to sensor networks, SANETs can not only sense the environment, but also react to it. This characteristic makes SANETs an essential enabling technology for various monitoring and control applications. However, to properly design an effective SANETs for SG, we must overcome many challenges.

A SANET is a network of nodes which sense and react to their environment. Compared to traditional sensor networks, which focus on sensing, SANETs can be used for both monitoring and control purposes. With a SANET, closed-loop control can be achieved to support more powerful applications.

Major actors in a SANET include sensors, actuators, controllers, and communication networks. Sensors are components or devices to measure and convert physical properties into electrical signals and/or data. Controllers perform calculations on the sensed data and make control decisions. Actuators execute the control decisions, convert electrical signals into physical phenomena (e.g., displays) or actions (e.g., switches). Different actors in a SANET may be physically separated or located in a single device. Actors in a SANET communicate with each other through communication networks, operating diverse kinds of protocols and media, to enable collaboration among nodes and interaction between nodes and the surrounding environment. SANET actors and the closed-loop control are shown in Figure 13.1.

SANET design is application-centric, which means the major design requirements are determined by the specific application. Given the application requirements, the following questions must be considered:

- To realize the application, which functions are required? The requirements on physical properties, physical phenomena or actions, control logic and communication networks, need to be determined.

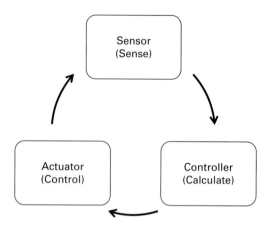

Figure 13.1 SANET actors and closed control loop.

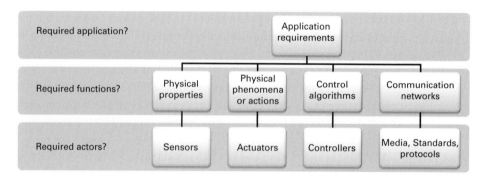

Figure 13.2 SANET design flow.

- To realize the function requirements, which actors are required? Specifically, sensors, actuators, controllers, communication media, standards, and protocols, need to be determined.

Figure 13.2 shows the simplified design flow, which leads to the determination of the actors for the SANET to achieve the specific application requirements.

In this chapter, we will introduce the composition and characteristics of SANETs, identify the major applications of SANETs in SG, highlight the major design and implementation challenges, and propose some innovative mechanisms to address these challenges. We also use a case study of a home energy-management system (HEMS) in the customer domain to demonstrate the effectiveness of the proposed mechanisms.

The rest of this chapter is organized as follows. In Section 13.2, the basics of information and energy flow in SG are discussed. In Section 13.3, the major applications of SANETs in SG are identified. The design issues are discussed, and the major design challenges are highlighted. In Section 13.4, some innovative mechanisms are proposed to address the design challenges. In Section 13.5, a HEMS is introduced as a case study. Section 13.6 concludes the chapter.

Table 13.1. A comparison of traditional power grid and smart grid

	Traditional grid	Smart grid
System reliability	Low reactive control, slow response and recovery	High proactive control, fast and effective response and fast auto-restoration
Power source	Centralized fossil fuel	Centralized + distributed fuel + RE
Energy flow	One way from grid to customer	Two way grid $<->$ Customer
Information flow	None or quite limited	Pervasive, two-way, broadband
Greenhouse gas emissions	High	Low (via increasing penetration of RE)
Energy efficiency	Low	High (via better balance of supply and demand)
Cyber security	No	Resilient against cyber attack
Customer participation	No participation	Better customer awareness and active participation

13.2 Energy and information flow in smart grid

As introduced in the previous chapters, compared to traditional power grid, SG enjoys various advantages. A comparison of traditional power grid and SG is shown in Table 13.1. There are three major driving forces of smart grid:

- improving security and reliability;
- enhancing energy efficiency;
- enhancing penetration of RE and reducing greenhouse gas emission to achieve sustainability.

These objectives are achieved by utilizing RE sources and employing modern power electronics and advanced information technologies.

As shown in Figure 13.3, the conceptual reference model of smart grid proposed by NIST [5] divides a smart grid into seven domains, specifically, customer, markets, service provider, operations, bulk generation, transmissions, and distribution. Connections among the different domains are two kinds of flows: energy flow and information flow.

Energy flow

The major energy flow is sourced at the bulk generation domain, delivered through the transmissions and distribution domains, and consumed at the customer domain. This flow is similar to what has existed in traditional power grid for decades. Besides the major energy flow, in an SG, there is energy flow in the reverse direction, from the distributed generators in the customer domain to the distribution networks.

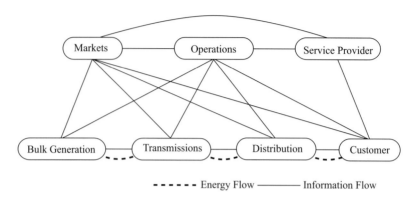

Figure 13.3 Conceptual reference model of smart grid (adapted from [5]).

The energy flow can be measured by energy measurement devices, such as power meters and power gauges. Meanwhile, the energy flow can be manipulated by actuators, such as breakers and switches.

Although the power grid is a homogeneous network in terms of the way energy is distributed, monitored, and controlled, the energy flow is quite dynamic in terms of quantity and quality. The dynamicity is due to variations in supply and demand, dynamic user behaviour, and continuously changing environments. In an SG, increasing usage of RE sources, such as wind turbines and solar panels, makes the problem even more challenging.

Information flow

In the traditional power grid, there is limited information flow. The most significant improvement of smart grid compared to traditional power grid is the deployment of a full-fledged SANET infrastructure, which carries all the information generated and consumed in a smart grid, such as real-time measurements, historical data, external events, control decisions, etc.

The information is exchanged among distributed actors within or among domains through diverse kinds of communication channels. The communication channels form a communication network, which is generally a heterogeneous and distributed network. The distributed nature is due to actors being physically distributed throughout the space, while heterogeneity is inevitable because different actors may follow different communication protocols, use different media (wired or wireless), and have different communication capabilities.

13.3 SANET in smart grid

From the energy flow and the information flow point of view, smart grid applications can be viewed as energy-flow management and optimization by utilizing the information

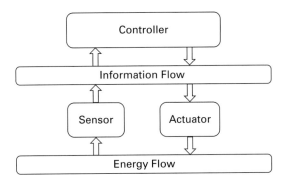

Figure 13.4 Relation of SANET and smart grid.

flow. This processing requires the capability of physical parameter sensing, decision making, and physical device control.

A high-level description of SANET in smart grid is shown in Figure 13.4. By employing SANET in smart grid, the energy flow and its supporting infrastructures are monitored by distributed sensors. Through information flow, the sensed data is transmitted to controllers for decision making. Controllers make control decisions and issue control commands to the actuators, also through the information flow. On receiving the control commands, actuators execute the control tasks.

13.3.1 Applications of SANET in SG

As introduced in the previous section, the three major driving forces of smart grid include improving security and reliability, enhancing energy efficiency, and reducing greenhouse gas emissions via increased penetration of renewable energy. Below, we will elaborate on the major smart grid applications to show why it is necessary to deploy SANET in smart grid and how SANET helps to achieve the three major objectives.

Grid monitoring and control (GMC)

Reliability is critical in the electricity network. However, the big blackout in the USA in 2003 indicated that the traditional electricity grid is still unreliable. A recent report showed that the US power grid is becoming less reliable over the years [6].

GMC is essential for reliable, secure, and high-quality electricity services. A survey conducted by Xie *et al.* [7] indicates that real-time monitoring and operating control systems may help in a very high percentage of the occurrences of power grid disturbances. Hauser *et al.* [8] point out that utility operators have inadequate situational awareness and are blind to disturbances in neighbouring areas, rendering them unable to limit the spread of disturbances. This leads to the call for new GMC mechanisms. With the capability of continuous system monitoring and control, SANET plays a key role in GMC.

The core duties of SANET in GMC include preventive and corrective functions. Specifically, SANET is required to monitor equipment health, predict and detect disturbances, prevent potential failures, respond quickly to energy-generation and consumption fluctuations and catastrophic events, and enable fast auto-restoration or self-healing.

Diverse kinds of SANET have been invented and employed for GMC. A supervisory control and data acquisition (SCADA) system has been used for GMC for decades to monitor electricity facilities, deliver data to operators, transmit commands to apparatus, and perform control. SCADA suffers from low sampling and transmission rates (data sensing and gathering from the remote terminal unit occurs every 25 seconds), and does not carry time synchronization information. To solve the problems of SCADA, phasor measurement units (PMU) and the PMU-based wide-area measurement system (WAMS) have been developed and deployed to enable fast and synchronous power-system monitoring in finer time scales and to provide more methods for system analysis on a regional and even national scale [9]. Most recently, the enhanced wide-area monitoring system (EWAMS) [10] was introduced as an enhancement of WAMS in terms of processing capability, database capacity, and management efficiency.

Generation dispatch (GD)

A good balance between the power supply and the power demand is critical to power grid stability, effectiveness in improving energy efficiency, avoiding unnecessary generation of power, and reducing greenhouse gas emissions.

One of the major mechanisms for power balancing is GD, which is a monitor and control mechanism to actively manage electricity generation such that the amount of power generated meets the demands at any time. As a realization of GD, generation scheduling and regulation have been deployed and play an important role in traditional power grid. However, this function in a smart grid must overcome additional challenges, as it has to actively manage significant amounts of distributed energy resources, especially RE resources.

Renewable energy (RE) sources include non-variable ones and variable ones. Non-variable RE sources, such as hydro, have already been widely utilized in existing power grids for decades. However, due to their intermittent nature, penetrations of variable RE sources, such as wind and solar, in considerable amounts (more than 10%) may cause severe problems in power grid stability.

With the help of SANET, renewable forecasting (RF) and real-time grid frequency regulation (GFR) are two effective mechanisms to address the RE penetration problem in GD.

RF [11] requires real-time distributed energy resource (DER) information to be sensed and gathered at the control centres; and after fast analysis, proper commands issued to generation scheduling and regulation functions. Specifically, with the help of SANET, accurate and up-to-date environmental information, such as wind speed, solar intensity, can be obtained to predict the characteristics of the RE generators. Based on the measurements and predictions, compensation mechanisms can be employed to adaptively control the backup generators, advanced storages, or even customer power loads to address the fluctuations of the RE supplies.

Real-time GFR [12] continuously performs fast detection on variations of frequency and voltage level. In response to the detected variations, compensative actions, similar to those in RF, are employed for generation-side optimization. GFR requires very sensitive and responsive hardware and extremely high-speed data transmission.

Besides the above mechanisms addressing RE penetration for GD, more advanced generation scheduling schemes [13–16] must be developed to further optimize the generation scheduling and regulation through acquiring and analysing more up-to-date grid state information, such as loading, generation availability, and occurrences of emergencies, etc. This process also relies on very high-speed data sensing and communication.

Demand-side management (DSM)

As the counterpart of GD located in the generation domain, DSM, which is also known as demand response (DR), works primarily in the customer domain and interacts with the service provider, market, and operation domains. DSM manages demand-side power load in response to power supply constraints [17, 18]. By employing automatic load management, DSM can realize peak shaving through reducing power load at peak times, or shifting power load from peak times to non-peak times, as shown in Figure 13.5.

DSM is an important application of SANET and imposes some special functional requirements on the underlying SANET, such as capabilities of real-time load monitoring, two-way data exchanging between the demand side and utilities, data processing, and demand-side load control, etc.

Advanced metering infrastructure (AMI) is a SANET realization for DSM, in which smart meters, which incorporate sensing, calculating, transmission, and control functions, play a key role [19].

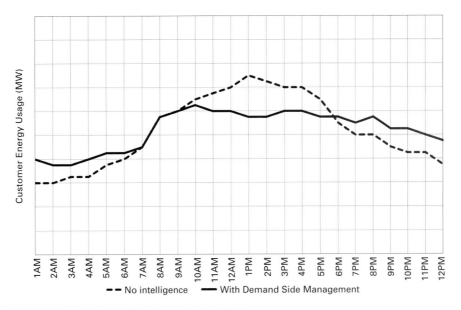

Figure 13.5 Peak shaving and load shifting by employing DSM.

13.3.2 Actors of SANET in smart grid

From the introduced SANET applications in smart grid we can find that the underlying idea of SG applications is to realize better management and optimization via better understanding of the up-to-date situation and performing the proper control actions. The realization of such purpose requires different actors. As introduced in the previous sections, SANET is composed of sensors, actuators, controllers, and communication networks. Below we highlight the major sensors and actuators commonly used in SG, and highlight the major requirements on controller and communication networks by the different smart grid applications.

Sensors

The basic requirement of sensors in a smart grid is to obtain the required state information of the energy flow, relevant infrastructures, and surrounding environment, in real time and with high accuracy. In Figure 13.6, based on the type of physical parameter measurements, we divide the major sensors commonly used in SG into three categories, namely, energy flow, environment, and working condition. Figure 13.7 shows a few sensor examples.

To support diverse kinds of SANET applications in smart grid, novel sensors are emerging. For example, novel magnetic sensors are proposed to measure transmission-line vibration [20] and sags in GMC. On the other hand, new SANET applications also impose special measurement requirements on sensors. For example, GMC requires highly durable temperature sensors to measure substation transformer temperature [21], and arc protection requires sensors to measure voltage and current of a power line under an extremely high sampling rate.

Actuators

The actuators in a smart grid react to the energy flow, relevant infrastructures, and the surrounding environment. In Figure 13.8, based on the type of actions, we classify the

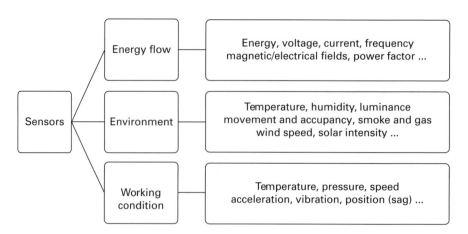

Figure 13.6 Sensor examples.

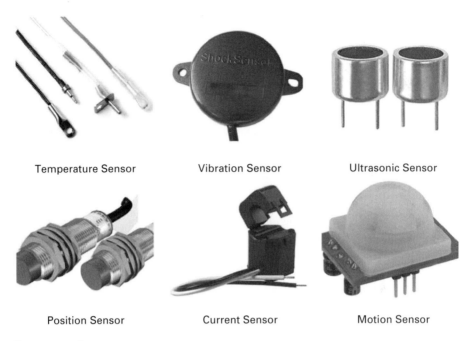

Figure 13.7   Sensor examples.

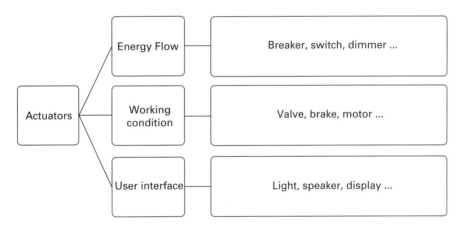

Figure 13.8   Actuator examples.

major actuators commonly used in SG into three categories. Figure 13.9 shows a few actuator examples.

In addition to the actuators for energy-flow control and working-condition control, SANET in smart grid may require other actuators, especially actuators to enable customer–grid interactions as well as grid–environment interactions. For example, energy in-home display, such as the one shown in Figure 13.9, can provide information on real-time energy consumption and tariff estimation for customers. Such customer

Figure 13.9   Actuator examples.

awareness is proved to be effective to motivate users to save power, reduce greenhouse gases, and realize sustainable lifestyles [24].

Controllers and control logic in smart grid

Depending on the application requirements, controllers in SANET may be complicated, powerful, centralized control centres, or simple, less-powerful, distributed microcontrollers. Normally, these two kinds of controllers work collaboratively to provide the monitor and control function in a single SANET application.

To handle tremendous data volume with high efficiency, data centres in smart grid require an extremely fast and large-capacity database. Meanwhile, to enable an effective utilization of the collected data, low-cost and secure access to the data is necessary for SANET applications.

Distributed controllers for SANET in smart grid need to achieve a good balance between capability and cost. Due to the great fluctuations in energy generation and consumption, SANET applications in smart grid may require computationally intense distributed control logics, such as fuzzy control and artificial intelligence (AI) control, to handle the dynamics, thus requiring powerful controllers. In addition, SANET applications in smart grid may require a large number of distributed controllers to work collaboratively. DSM is such a case, where thousands of smart meters and manageable customer loads are involved. This requires each controller to be low cost to enable a large-scale deployment.

Table 13.2. Requirements on SANET actors for different smart grid applications

SANET actors	SANET applications		
	GMC	GD	DSM
Sensors	Energy flow Environment Working condition	Energy flow Environment	Energy flow Environment
	Accuracy: high	Accuracy: medium to high	Accuracy: medium
	Sampling rate: extremely high	Sampling rate: low to high	Cost: low
Actuators	Energy flow Working condition	Energy flow Working condition User interface	Energy flow Working condition User interface
Controllers	Distributed & centralized	Distributed & centralized	Distributed & centralized
	Dynamic level: high Cost: medium to high	Dynamic level: high Cost: low to medium	Dynamic level: high Cost: low to medium
Communication networks	Reliability: extremely high	Reliability: high	Reliability: medium to high
	Bandwidth: extremely high	Bandwidth: low	Bandwidth: lower
	Delay: extremely low Coverage: local to wide	Delay: medium Coverage: local to wide	Delay: medium Coverage: wide

Communication network

To support the advanced features of SG, the data volume exchanged among different actors in SANET inevitably increases tremendously compared to traditional power grids. Meanwhile, different SANET applications in SG normally have different communication requirements, in terms of bandwidth, transmission delay, etc. When we design the communication network, we need to first consider the application requirements in terms of reliability, bandwidth, transmission delay, and coverage. In addition, technology availability, and costs in deployment and operation, must be considered to achieve a good balance. In Table 13.2, we summarize the characteristics and requirements on different SANET actors for the major SG applications.

13.3.3 Challenges for SANET in smart grid

Next we study the major design challenges of SANET in smart grid.

Heterogeneity and distributed operation. As pointed out before, heterogeneity and distributed operation are two major characteristics of the information flow in a smart grid. Since SANET relies on the information flow, the heterogeneity and distributed operation, which render the formation of a connected and efficient information flow, become the two major challenges of SANET in a smart grid.

Dynamics. The dynamicity is due to the variation of supplies and demands, dynamic user behaviour, continuously changing environments, and other random events. In a smart grid, increasing usage of RE sources, such as wind and solar, makes the problem even more challenging.

Scalability. A typical SANET application in a smart grid may cover hundreds of kilometers, and involves monitoring and control of thousands of pieces of equipment and devices. Scalability is a major challenge. It is necessary to employ protocols with low overhead and algorithms with linear complexity.

Flexibility. Smart grid is still evolving [22]. New technologies, policies, and user demands keep emerging. SANET is required to provide the flexibility to accommodate all the diversities and evolving factors.

Energy efficiency and cost efficiency. One of the driving forces of smart grid is to improve the efficiency of the power grid. Thus, SANET itself must be energy efficient. In addition, to lower the deployment barrier, it must be cost effective.

Cyber security. Another major challenge of SANET in smart grid is maintaining resiliency against cyber attacks. This issue will be elaborated in more detail in other chapters of this book.

13.4 Proposed mechanisms

In this section, we propose some effective mechanisms to address some of the major challenges of SANET in smart grid as outlined before. The proposed mechanisms and corresponding challenges addressed are shown in Table 13.3.

13.4.1 Pervasive service-oriented network (PERSON)

PERSON is a general framework to seamlessly integrate diverse kinds of actors and networks into a unified pervasive service-oriented network to address the challenges of heterogeneity and distributed operation, and bring lots of flexibility to develop diverse kinds of SANET applications.

PERSON has a three-layer structure as shown in Figure 13.10. The principle is to decompose the complexities into different layers and to have loose coupling among different layers. In the following we will introduce the three layers one by one.

Heterogeneous network platform (HNP)

The objective of HNP is to build a homogeneous communication infrastructure for the information flow. HNP provides simple APIs to the upper layer for information exchange. The upper layer does not need to care about how the information is delivered.

Table 13.3. Mechanisms to address the major challenges

Challenges	Mechanisms			
	Pervasive service-oriented network	Context-aware intelligent control	Compressive sensing	Device technology
Heterogeneity and distributed network	√			
Dynamics		√		
Scalability			√	√
Flexibility	√	√		√
Energy efficiency and cost efficiency			√	√

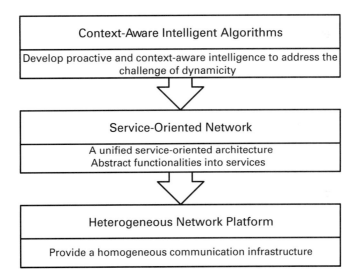

Figure 13.10 Three-layer structure of PERSON.

The underlying communication protocols, media, and communication capabilities are transparent to the upper layer. An implementation of such HNP can be found in [23].

Service-oriented network (SON)

The basic idea of SON is to achieve interoperability, modularity, and reusability by abstracting the functions provided by the actors into services. For a certain SANET application, a suite of services need to be defined. To support SON, mechanisms for service creation, registration, discovery, binding, and invoking are also required.

SANET applications

On this layer, the services provided by SON are exploited for specific SANET applications. The interoperability and service reusability provided by SON bring much flexibility to develop diverse kinds of applications, such as the applications introduced in the previous sections.

13.4.2 Context-aware intelligent control

Context-aware intelligent control is proposed to address the challenges of dynamics. The basic idea is to develop proactive and context-aware control logics to optimize the system performance under dynamic environment. Here, context has a very broad meaning, including but not limited to:

- energy flow states, such as voltage, current, power supply, and demand level;
- working conditions, such as temperature of transformers;
- environment parameters, such as wind speed and solar intensity;
- human behaviour, such as movement, preference on environment;
- economic incentives, such as tiered electricity rates;
- regulation schemes, such as DSM and RE penetration.

A simple example of context-aware intelligence is occupancy-based light control, where the context is whether the room is occupied or not, and light is turned on or off accordingly.

The context-aware intelligent algorithms exploit the contexts, obtained by exploiting the services of PERSON, to optimize the overall performance of a SANET application. More details of PERSON and context-aware intelligent control can be found in [24].

13.4.3 Compressive sensing (CS)

CS is proposed to address the challenges of scalability, energy efficiency, and cost efficiency. The basic idea of CS is to exploit data correlation in the time and space domains to avoid unnecessary sensing and data exchanging, and thus improve scalability, save energy, and reduce costs of hardware and communication.

Consider a distributed wind-power generator management system as shown in Figure 13.11, in which a number of micro wind-power generators are located in geographical proximity. The amount of power generated by each wind turbine depends greatly on the wind speed experienced. In order to get accurate information of DER for better GD, the wind speed needs to be measured in real time and the measurement results should be transmitted to the control centre. One option is to install wind speed meters at all the wind turbines, which is costly not only in hardware but also in data transmission. Since wind conditions in adjacent areas are similar, we can exploit such correlation in the spatial domain and deploy fewer sensors with less cost on hardware and data communication.

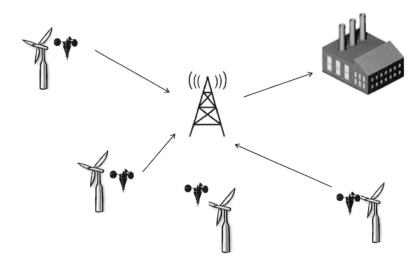

Figure 13.11 Compressive sensing in wind-power generator management.

The methodologies employed in CS include:

- complete continuous sampling to ensure reliability and complete understanding of the correlations;
- selective transmissions to ensure short delay and cost effectiveness;
- complete reconstruction based on sparse data.

An example of the application of CS can be found in [25].

13.4.4 Device technologies

Advanced device technologies can help improve energy efficiency and cost efficiency, and make a SANET more scalable and flexible to be employed for smart grid applications.

Low power-consumption design and power-harvesting technologies

The SANET itself inevitably consumes power. To reduce the total power consumption, low-power consumption design is required. In addition, mechanisms for power-harvesting are preferred.

In a SANET, all the major actors, such as sensor, actuator, controller, and communications, consume power. In Table 13.4, we list the potential mechanisms to reduce power consumption.

The employment of a mechanism for one actor may have an impact on others. For example, data compression and data aggregation may reduce the power consumption for data transmission, but increase the power consumption for regenerating the data. Therefore, power optimization needs to be considered from a system point of view.

Power harvesting is a process by which energy is derived from external sources, captured and stored. The major power-harvesting mechanisms applicable to SANET in a smart grid are listed in Table 13.5.

Table 13.4. Power-conserving mechanisms

Actor	Power-conserving mechanisms
Sensor	Compressive sensing to exploit correlations in time and space domains Sensing on demand to avoid continuous and unnecessary sensing
Actuator	Event-based control
Controller	Low-complexity algorithm
Communications	Compressive sensing Distributed data processing and control instead of centralized control Data compression and data aggregation Low-power data-transmission technologies

Table 13.5. Power-harvesting mechanisms

Energy type	Power-harvesting device (energy source)
Ambient radiation	Solar panel (solar energy) Antenna and transducer (RF energy)
Kinetic	Piezoelectric devices (mechanical strain, motion, vibration, noise) Micro wind turbine (wind power)
Thermal	Thermoelectric generator (thermal gradient)

Modular and compact design

Modular design can enhance the reusability of the modules to reduce the hardware design cost and development cost. Compact design can lower the development cost of the SANET actors, and improve the flexibility of employing the actors in diverse kinds of SANET applications. Figure 13.12 shows a compact wireless environment sensor applicable to many SANET applications.

13.5 Home energy-management system – case study of SANET in SG

13.5.1 Energy-management system

An EMS monitors, controls, and optimizes the performance of energy generation, transmission, distribution, and consumption. EMS is an important building block of an SG, and plays a key role in achieving the advantages of an SG. Among the seven domains, EMS for energy generation, transmission, and distribution has been studied for decades. Many models, standards, protocols, and systems have been proposed, implemented, and deployed in practical systems [26, 27]. However, EMS in the consumer domain is largely

Figure 13.12 Compact wireless environment sensor.

neglected in existing studies. In this section, we introduce the design and development of an EMS in the customer domain (for simplicity, we use EMS to denote EMS in the customer domain in the rest of this chapter) as a case study to demonstrate the effectiveness of the proposed mechanisms.

To support the advantages of smart grids, an EMS in the customer domain is required to support some basic features as follows:

- supports various existing actors as well as emerging actors;
- continuously monitors the energy consumption at different granularities, such as home level and appliance level;
- continuously monitors environment parameters, such as temperature and humidity, which can be exploited for context-based intelligent control;
- supports automatic and manual control of the actors;
- supports the integration of renewable power sources, such as solar and wind;
- interacts with actors in other domains to realize advanced features, such as DR.

Besides the basic features, some desired advanced features include but are not limited to the following:

- intelligence and efficiency – the ability to achieve optimized performance under dynamic situations;
- user friendliness – plug-and-play with self-configuration capability;
- high reliability and durability – robustness and self-healing capability after system failures;
- low cost and low power consumption.

Table 13.6. Major actors in EMS

Actor	
Sensor	Energy flow: voltage, current, energy
	Environment: temperature, humidity, luminance, motion
Actuator	Energy flow: dimmer, breaker, switch
	User interface: in-home display
Controller	Distributed: home gateway and control centre
	Centralized: data and service centre
Communications	Local area network: ZigBee, WiFi
	Wide area network: Internet

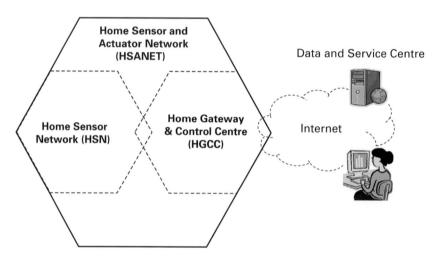

Figure 13.13 EMS architecture.

13.5.2 EMS design and implementation

The proposed EMS is based on a SANET composed of a home sensor and actuator network (HSANET) located in the customer domain, and a data and service centre (DSC) located in the service provider domain. The structure of the developed EMS is shown in Figure 13.13. The major actors employed in the EMS are listed in Table 13.6.

HSANET realizes the PERSON structure and provides the basic information infrastructure for EMS application. HSANET is composed of a ZigBee-based home sensor network (HSN) and a ZigBee–Internet home gateway and control centre (HGCC). HSN realizes the sensing and control of the energy flow. It also provides an in-home display, as a user interface to monitor the energy flow. HGCC hosts the local energy-management intelligence, provides interface between customers and service providers. Figure 13.14 shows the EMS prototypes. The device technologies introduced above are employed in the design and implementation of the devices.

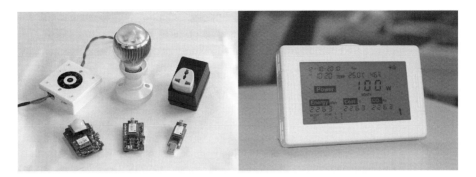

Figure 13.14 EMS prototypes.

To support EMS applications, we define a set of EM services and develop relevant mechanisms for service creation, registration, discovery, binding, and invoking. Based on the services and the supporting mechanisms, local context-aware intelligence is developed to optimize the energy consumption and handle the dynamics at the home level.

To further support the applications which require collaboration between the customer domain and other domains in an SG, we also develop and deploy a DSC at the service provider domain. DSC enables a broader and deeper energy awareness for a customer and a tighter customer–grid collaboration.

The mechanism of CS is deployed in the EMS to exploit data correlation in the time and space domains to avoid unnecessary sensing and data exchanging, and thus improve scalability, save energy, and reduce costs for hardware and communication.

As introduced above, DSM can help to realize peak shaving through managing demand-side power load in response to power supply constraints. With such capabilities as real-time load monitoring, two-way data exchanging between the demand side and utilities, data processing, and demand-side load control, etc., the EMS is an excellent platform for DSM. The effectiveness of the EMS has been demonstrated in a DSM application. More details can be found in [24].

13.6 Conclusion

In this chapter, we have studied the role of sensor and actuator networks in SG. We have discussed why SANET is important to SG and how SANET contributes to SG. We have highlighted the major design challenges and proposed some innovative mechanisms to address these challenges. Specifically, a pervasive service-oriented network framework has been proposed to seamlessly integrate diverse actors and networks into a unified pervasive service-oriented network to address the challenges of heterogeneity and distributed operation. Based on PERSON, context-aware intelligent control has been proposed to optimize the system performance under dynamic environment in a proactive manner. Furthermore, compressive sensing has been proposed to exploit data correlation in the time and space domains to improve scalability, save energy, and reduce the costs of hardware and communications. The effectiveness of the proposed schemes has been verified and demonstrated with a case study of an energy-management system.

References

[1] The American Recovery and Reinvestment Act of 2009, P.L.111-5, USA, February 2009.
[2] Federal Energy Regulatory Commission, Smart Grid Policy [Docket No. PL09-4-000], USA, July 2009.
[3] State Grid Corporation of China (SGCC), SGCC Green Development White Paper (in Chinese), April 2010 [online]. Available at: http://www.sgcc.com.cn/bps/index.shtml
[4] Electric Power Research Institute, The green grid: energy savings and carbon emissions reductions enabled by a smart grid, June 2008 [online]. Available at: http://my.epri.com
[5] National Institute of Standards and Technology (NIST), NIST Framework and Roadmap for Smart Grid Interoperability Standards, Release 1.0, USA, January 2010.
[6] S. M. Amin, 'U.S. electrical grid get less reliable', *IEEE Spectrum*, January 2011 [online]. Available at: http://spectrum.ieee.org/energy/policy/us-electrical-grid-gets-less-reliable
[7] Z. Xie, G. Manimaran, V. Vittal, A. G. Phadke, and V. Centeno, 'An information architecture for future power systems and its reliability', *IEEE Transactions on Power Systems*, vol. 17, no. 3, pp. 857–862, August 2002.
[8] C. H. Hauser, D. E. Bakken, and A. Bose, 'A failure to communicate: next-generation communication requirements, technologies, and architecture for the electric power grid', *IEEE Power and Energy Magazine*, vol. 3, no. 2, pp. 47–55, March/April 2005.
[9] M. Anjia, J. Yu, and Z. Guo, 'PMU placement and data processing in WAMS that complements SCADA', *2005 IEEE Power Engineering Society General Meeting*, 2005.
[10] B. Luitel, G. K. Venayagamoorthy, and C. E. Johnson, 'Enhanced wide area monitoring system', *2010 Innovative Smart Grid Technologies (ISGT)*, January 2010.
[11] T. Ackermann, G. Ancell, L. D. Borup, P. B. Eriksen, B. Ernst, F. Groome, M. Lange, C. Mohrlen, A. G. Orths, J. O'Sullivan, and M. de la Torre, 'Where the wind blows', *IEEE Power and Energy Magazine*, vol. 7, no. 6, pp. 65–75, November/December 2009.
[12] K. M. Rogers, R. Klump, H. Khurana, A. A. Aquino-Lugo, and T.J. Overbye, 'An authenticated control framework for distributed voltage support on the smart grid', *IEEE Transactions on Smart Grid*', vol. 1, no. 1, pp. 40–47, June 2010.
[13] P. P. Varaiya, F. F. Wu, and J. W. Bialek, 'Smart operation of smart grid: risk-limiting dispatch', *Proceedings of the IEEE*, vol. 99, no. 1, pp. 40–57, 2010.
[14] V. O. K. Li, F. F. Wu, and J. Zhong, 'Communication requirements for risk-limiting dispatch in smart grid', *Communications Workshop, 2010 IEEE International Conference on Communications*, May 2010.
[15] J. W. Bialek, P. Varaiya, F. F. Wu, and J. Zhong, 'Risk-limiting dispatch of smart grid', *2010 IEEE Power and Energy Society General Meeting*, July 2010.
[16] K. Cheung, X. Wang, B. C. Chiu, Y. Xiao, and R. Rios-Zalapa, 'Generation dispatch in a smart grid environment', *1st IEEE Conference on Innovative Smart Grid Technologies*, Washington, DC, US, January 2010.
[17] M. H. Albadi and E. F. El-Saadany, 'Demand response in electricity markets: an overview', *2007 IEEE Power Engineering Society General Meeting*, June 2007.

[18] F. Rahimi and A. Ipakchi, 'Demand response as a market resource under the smart grid paradigm', *IEEE Transactions on Smart Grid*, vol. 1, no. 1, pp. 82–88, June 2010.

[19] A. H. Rosenfeld, D. A. Bulleit, and R. A. Peddie, 'Smart meters and spot pricing: experiments and potential', *IEEE Technology and Society Magazine*, vol. 5, no. 1, pp. 23–28, March 1986.

[20] B. Garcia, J. C. Burgos, and A. M. Alonso, 'Transformer tank vibration modeling as a method of detecting winding deformations part I: theoretical foundation', *IEEE Transactions on Power Delivery*, vol. 21, no. 1, pp. 157–163, January 2006.

[21] M. K. Paradhan, 'Assessment of the status of insulation during thermal stress accelerated experiments on transformer prototypes', *IEEE Transactions on Dielectrics and Electrical Insulation*, vol. 13, no. 1, pp. 227–237, February 2006.

[22] C. H. Hauser, D. E. Bakken, and A. Bose, 'A failure to communicate: next-generation communication requirements, technologies, and architecture for the electric power grid', *IEEE Power and Energy Magazine*, vol. 3, no. 2, pp. 47–55, March/April 2005.

[23] V. O. K. Li, C. Li, Q. Liu, G.-H. Yang, Z. Zhao, and K.-C. Leung, 'A heterogeneous peer-to-peer network testbed', *1st International Conference on Ubiquitous and Future Networks*, Hong Kong, June 2009.

[24] G.-H. Yang and V. O. K. Li, 'Energy management system and pervasive service-oriented networks', in *Proceedings of 1st IEEE International Conference on Smart Grid Communications*, Gaithersburg, MD, October 2010.

[25] X. Yu, H. Zhao, L. Zhang, S. Wu, B. Krishnamachari, and V. O.K. Li, 'Cooperative sensing and compression in vehicular sensor networks for urban monitoring', in *Proceedings of 2010 IEEE International Conference on Communications*, South Africa, May 2010.

[26] A. J. Wood and B. F. Wollenberg, *Power Generation Operation and Control*, 2nd edn. John Wiley, 1996.

[27] K. Kato and H. R. Fudeh, 'Performance simulation of distributed energy management systems', *IEEE Transactions on Power Systems*, vol. 3, pp. 820–827, May 1992.

14 Implementation and performance evaluation of wireless sensor networks for smart grid

Nicola Bui, Angelo P. Castellani, Paolo Casari, Michele Rossi,
Lorenzo Vangelista, and Michele Zorzi

14.1 Introduction

This chapter focuses on the usage of wireless sensor and actuator networks to provide data connectivity in smart grids. In particular, we discuss the configuration adopted for the implementation of the sensor network test-bed deployed at the Information Engineering Department of the University of Padova, Italy. The test-bed has been designed to reproduce typical deployment scenarios in an urban network by mimicking diverse contexts such as dense building networks, sparse environmental scenarios, and linear deployments along streets.

The test-bed software has been realized taking full advantage of the most advanced solutions provided by the academic community and the standardization bodies by implementing a completely IP interoperable communication framework. Moreover, the latest solutions for the Internet of things [1] have been used to develop a lightweight modular architecture offering services and data sources through simple and efficient web services. All of this facilitates the integration of the test-bed functionalities into flexible web applications, capable of performing the needed monitoring and managing routines in the entire network as well as on single nodes.

The Internet-like approach, coupled with a variety of network configurations, has been used to verify the advantages brought by the usage of constrained wireless communication for smart grids. In particular, we have been able to quantify useful performance metrics, such as maximum throughput, delivery delay, and transmission reliability, in typical smart grid network scenarios. Specifically, these performance metrics were determined for linearly shaped multihop configurations, to address networks deployed along streets, such as those controlling the street lights, as well as for dense single- and multihop configurations to address small-to-medium-sized building deployments.

Wireless communication represents the final step for smart grids to reach the widest coverage. In fact, even though most of the nodes of the grid will be connected using high-end wired communication, this is neither cost nor energy efficient when applied to billions of devices belonging to millions of users. To give a few examples, constrained wireless networks will make it possible to detail the energy expenditure of every single appliance in the houses of the future and they will enable distributed monitoring and control of any recent renewable energy source, such as photovoltaic panels.

While the rest of the chapter will provide detailed technical considerations on the communication aspects linked to the usage of wireless sensor networks, it is worth noting here that quite a few practical advantages are achievable. First of all, with the constant growth of the market of wireless sensor nodes, devices are expected to become cheaper and cheaper; this, coupled with the low installation costs, will decrease the building costs of new smart grids and will make it possible to adapt grids already installed with minimal effort.

A second benefit derives from wireless sensor networks being highly energy efficient; in fact, it is of paramount importance for any energy monitoring network not to become a consumption burden itself. Finally, but the list of benefits could continue, wireless sensor networks allow the adoption of distributed self-optimizing routing. Such decentralization of the network management grants a double benefit: on the one hand failure recovery and other emergency routines can be performed in a timely manner by directly detecting the problem close to where it occurs, and on the other hand the communication overhead needed for letting the network adapt to different traffic conditions can be distributed on those parts which are affected by the problem without loading the core network.

However, all these advantages come at a price: constrained devices offer neither a high bandwidth nor a high computational power. This translates into a need to carefully design the communication, the data representation, and the control loop techniques adopted in the constrained network. This chapter will provide some insight into what is currently feasible with common wireless sensor devices, also trying to understand the boundaries of what will be achievable with state-of-the-art wireless technology. Finally, results obtained from the test-bed will be used to define guidelines for the realization of smart grids using constrained wireless networks.

The rest of the chapter is structured as follows: Section 14.2 provides an overview of the communication protocols adopted, explaining the rationale behind the choices. Section 14.3 describes the implementation of the communication solutions adopted, analysing the possible bottlenecks. Section 14.4 lists the outcome of the thorough experimentation campaign by also giving guidelines to overcome possible limitations. Section 14.5 summarizes the considerations related to the adoption of constrained wireless sensors.

14.2 Constrained protocol stack for smart grid

This section provides an overview of the protocol stack adopted in most smart grid realizations which exploit wireless sensor networks. As will become more evident by proceeding through this section, all the layers of the stack adopt the most efficient solution available in the state of the art. These solutions have been studied and developed during the last 10 years by researchers and engineers of different communities: starting from academic researchers who paved the way before standardization, to IEEE members for

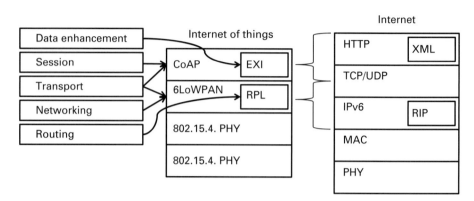

Figure 14.1 The protocol stack of the Internet of things compared with that of the standard Internet.

what concerns the PHY and MAC layer schemes of 802.15.4, to IETF members for network and application layers, and the W3C community for the presentation layer.

Figure 14.1 shows the whole protocol stack at a glance by comparing it with the current Internet protocol stack. Also, the figure highlights, on the left-hand side of each layer, the main function addressed by the specific protocol.

14.2.1 IEEE 802.15.4

Starting around the year 2000, wireless ad hoc networks were evolving to a different paradigm characterized by many more communicating entities, pure peer-to-peer communication, and very low power consumption. This growth went under the name of wireless sensor networks, since they were formerly intended to be used as distributed monitoring networks. In 2003, the IEEE released the first version of the 802.15.4 standard [2], intended to be the main driver for the standardization of PHY and MAC protocols for low-power and low-cost wireless communication.

Without analysing the entire standard in detail, we summarize here a few features of 802.15.4 that made it the main, if not the unique, standard for wireless sensor networks. First of all, to keep the power consumption low, transceivers have to be as simple as possible, and the data rate must be limited. For instance, typical modulation schemes of 802.15.4 are BPSK and O-QPSK for the $868/915$ MHz and the 2.4 GHz bands, respectively. Similarly, the maximum data rates in the two bands are $20/40$ kbps and 250 kbps. Finally, the packet size is limited to 128 bytes.

In order to achieve even higher energy efficiency, MAC protocols have to leverage on duty-cycling techniques, which let wireless nodes remain active for a small fraction of the time and sleep the rest of the time. In particular, 802.15.4 defines two possible MAC strategies: a coordinated version using beacon frames to synchronize nodes and imposing a hierarchical structure on the network, and an un coordinated version which adopts a pure CSMA/CA MAC protocol. Two considerations must be accounted for to understand the bandwidth limitation: first, of the available 128 bytes, 17 are used for PHY and MAC overhead, second, even without using any duty cycle, the error-prone nature of the wireless channel coupled with a CSMA MAC implies that the maximum single link throughput achievable is even lower than the nominal rate.

However, the benefits provided by the standard more than compensate the constraints introduced: compared with the main wireless competitors, Wi-Fi and Bluetooth, 802.15.4 offers a few orders of magnitude improvement in terms of energy expenditure, a degree of freedom in network topology which is not available in the other technologies, and the best coexistence performance of the three standards (up to 16 different 802.15.4 networks can be installed in the same area and still be able to operate).

Smart grids can exploit 802.15.4 for every application which is not time critical, but require distributed communication and coordination. In fact, even though this technology may not sound impressive for what concerns the single-link channel, the overall channel capacity, the flexibility, and the energy efficiency offered make it the first choice for low-cost wireless networks.

14.2.2 IPv6 over low-power WPANs

The Internet is composed of a heterogeneous set of subnetworks, possibly based on different lower-layer technologies (i.e., Ethernet), but interoperating with each other thanks to the Internet protocol. The total number of Internet hosts uniquely addressable using IPv4 has been known to be limited for many years, and this was the main design motivation of the IPv6 protocol. Such a protocol has never gained wide adoption in the Internet, mainly because there was no reason to deploy it that properly justifies the associated costs. In the smart grid use case, in the context of Internet of things applications, where many nodes are expected to be deployed as Internet hosts, the need for the larger address space provided by IPv6 is a strong requirement. IPv6 technology typically requires an 'IP-over-foo' RFC specification addressing the required link-layer adaptation required to run the IP service over the specified technology, which in general can be very simple but in the case of more challenging technologies may require more work. Transmitting IPv6 packets over the IEEE 802.15.4 link layer could be very complex, mainly because the frame payload available to the upper layer is small (ranging from 60 to 110 bytes depending on which MAC fields are present), especially when compared with the minimum MTU size required by IPv6. This led the RFC 4944 [3] specification to define 6LoW-PAN (IPv6 over low-power wireless personal-area networks), which mainly addresses the required adaptation to reduce the impact of IPv6 header overhead with respect to such small packets. In particular, the required adaptations are the following:

- *Encapsulation.* In IEEE 802.15.4 there is no field equivalent to the Ethernet ethertype; in order to enable the concurrent transmission of IPv6 and non-IPv6 packets, IPv6 packets are encapsulated with a dispatch byte header.
- *Header compression.* The IPv6 base header is 40 byte long, in the IEEE 802.15.4 small frames using uncompressed IPv6 header leads to a very high overhead impact. A header compression technique has been devised by exploiting the redundancy of fields in the MAC and IPv6 header.

- *Fragmentation.* IPv6 poses a requirement on the minimum MTU that has to be offered by the link layer in 1280 bytes. In order to provide such a service over the 15.4 link layer, a fragmentation technique very similar to IP fragmentation has been introduced at the adaptation layer.

The major novelty in the IPv6 adaptation layer is the header compression technique, especially for the fact that IETF standards in general avoid any kind of layer violations, whereas the 6LoWPAN adaptation is a cross-layer optimization. Compressing the IPv6 header is possible by exploiting the fact that using some fields available at the MAC layer (i.e., link-layer source and destination, payload length, checksum, etc.), some upper-layer fields can be derived from this information. A typical example is when a node is involved in a link-local communication with a neighbour, in this case the IPv6 source and destination addresses can be fully derived from the link-local addressing fields, thus the IPv6 adapted header is compressed, signalling this fact and omitting those fields.

Apart from the adaptation layer specification itself, other issues still have to be addressed to obtain an efficient realization of IPv6. To this end a set of optimizations to the IPv6 neighbour discovery procedures have been collected in a 6LoWPAN-specific ND draft [4], which focuses mainly on a set of assumptions and modifications that make the regular neighbour discovery work well in constrained network environments, where some features typical of regular cabled networks could be very costly (e.g., link-layer multicast).

14.2.3 Routing protocol for low-power and lossy networks

The routing protocol for low power and lossy networks (RPL) [5] is an IETF effort aimed at a comprehensive routing solution for constrained scenarios, where networked devices should provide a desired set of functionalities while possibly operating in a harsh environment, or subject to limited resources. Constrained networks of this kind can be found in many situations (e.g., at home, or inside commercial areas) as well as in factories and other industrial scenarios. Actually, many applications are conceptually similar across different scenarios. Such applications range from periodic monitoring of some parameters via specific sensors, to the prompt reporting of events of interest, the broadcast of unsolicited information, or the conveyance of alarm notifications to the appropriate entities to deal with exceptional events. In home scenarios, the specific case of a network of electric appliances and sensors, networked together to optimize power usage within the home as well as in larger areas such as a neighbourhood, is also known as a smart grid, and has recently been receiving substantial interest [6, 7].

In a general scenario, multiple applications may co-exist and require different guarantees in terms of network performance. A typical example is a house monitoring network, where a data centre aggregates the measurements of many sensors spread in the house: these measurements are generated continuously at a low rate, and routed over wireless or wired links to reach the data centre in a best-effort fashion. In other words, the delay incurred by the messages, or the probability of message loss, are not a major concern, insofar as they remain within very broad limits. As a house data centre does not usually need prompt decision making over general monitoring data, delay and loss constraints may be quite relaxed. On the contrary, alarm-detection equipment has completely different constraints, in that its event-based traffic must be delivered promptly to the appropriate entities. In view of this, delay and dependability are major requirements, and the network should operate in order to meet them. The above are two quite extreme examples: intermediate applications requiring a mix of reasonable delay and delivery guarantees include, e.g., reporting the completion of a factory process for re-initialization, real-time measuring and reporting of some physical quantity, remote detection and intervention by a human operator, and so forth.

RPL is designed to be configured so that different applications with different requirements can potentially co-exist in the same network. The basic RPL routing structure is the destination-oriented directed acyclic graph (DODAG). Every well-formed DODAG is rooted at a node which collects the traffic generated by the leaves: for instance, this node can be the home data centre cited before, or a home alarm centre providing alarm notifications to, e.g., firefighters or other safety operators. Every DODAG is characterized by an objective function (OF) which specifies the characteristics of the tree, and has a direct effect on how nodes connect to each other to form parent–children relationships. In its most basic implementation, the OF only sets up connectivity among nodes [8]: more refined objective functions can prescribe that a DODAG be created so that the hop-by-hop forwarding delay is less than a predetermined amount, or so that only links with a sufficiently low error probability be activated. Following the previous example, a general-purpose monitoring system would root a DODAG at a data centre and create a routing tree with little or no delay constraints, though possibly requiring a certain degree of reliability over each link. On the contrary, an alarm system would form a DODAG where hop-by-hop or end-to-end delays are of primary concern, and possibly require that the alarm be delivered to the root of the tree with near 100% reliability. Two such trees can co-exist in the same network, according to the RPL specifications: in fact, a node can choose to join different DODAGs where different OFs are in use.

DODAGs are formed by propagating messages that make nodes aware of the existence of roots and of the requirements of the DODAG, i.e., the OF. Such messages are called DODAG information objects (DIOs) and are initially propagated by the root. The other nodes downstream further replicate DIOs according to the rules dictated by the Trickle algorithm [9]. Such rules are designed to avoid useless message replication of DIOs that would propagate the same information. In particular, the Trickle protocol defines a listening interval I whose minimum length $I_{\min}$ is specified within the DIO sent by the root node. Every node willing to broadcast the DIO operates as follows: it picks a timer $T \in [I/2, I)$, and clears a counter c; then it listens to the channel until T expires, and increases c every time it receives information consistent with previous DIO messages. (The concept of 'consistency' is implementation-dependent and can be customized by the user.) When T expires, the node re-sends the DIO only if the counter c is less than a redundancy factor k: in this case the node resets the listening interval I to $I_{\min}$. Otherwise, the node suppresses retransmissions, and doubles the listening interval I (never exceeding a maximum value $I_{\max}$). The parameters of the Trickle algorithm (I, $I_{\max}$, and k) are typically DODAG-dependent, and are thus specified within the DIO itself: for example, a higher value of k triggers more retransmissions (thus facilitating the consistency of the information being broadcast) whereas a lower value saves more energy by suppressing more transmissions.

The concept of rank is very important for the formation, maintenance, and repair of a DODAG. The root has rank 1, whereas the nodes downstream in the tree are assigned higher ranks, so that to walk the direction of decreasing rank always means to get closer to the DODAG root. The way rank is computed is application-dependent, and may reflect topological distance (e.g., the node's hop count) or a mixture of this and other link-related metrics, such as the packet error rate or the expected transmission count. In any event, every node receiving a DIO and willing to join the DODAG computes its own rank based on the rank of the sender and its own measure of the link metrics concurring to rank calculation. This operation always assigns a DIO receiver a rank greater than the rank of the DIO sender.

While the propagation of DIOs is in process, nodes learn about the existence of the DODAG root and may decide to join the DODAG. This operation is performed via a destination advertisement object (DAO) message. After waiting for an implementation-dependent time, a node is likely to have received more than one DIO from prospective parents (note that the k factor of the Trickle algorithm has an impact on this process as well). The node can hence pick one or more parents from the parent set (again, as dictated by the OF) and unicast a DAO to each parent, indicating its own address and rank. This creates an association between the node and the parents. The association can also be employed to identify downward routes from the root to each node downstream. In more detail, RPL prescribes that either all nodes are 'storing' ones (i.e., they cache information about downward routes, typically in the form of {destination, relay, link metric} triples) or all nodes are 'non-storing', and thus the whole routing information is stored at the root node only. The latter case is an example of full-source routing, the former is completely stateful. No intermediate configurations are allowed. The amount of information in the DAO message varies depending on whether the nodes are storing or not: in fact, a network of non-storing nodes requires upward routes to be cached and sent along in DAOs, so that the root can reverse them and perform source routing correctly.

The configuration and topology of a DODAG must react to network changes (e.g., as nodes leave, join, move, or die). While the periodic retransmission of DIOs as prescribed by the Trickle timer rules is mostly sufficient, there are cases when the update of the parent–children associations must be re-triggered on demand. This happens, e.g., when a parent dies or moves away, leaving a portion of the network disconnected from the root node (i.e., a network partition occurs). RPL prescribes the usage of a DODAG information solicitation (DIS) message to correct this scenario and avoid prolonged partitions. The reception of a DIS by a neighbour triggers the transmission of a DIO, which in turn will allow the DIS sender to collect neighbourhood information and subsequently send a DAO to associate with a parent. The DIS can also be configured to have the DIO sender reset the Trickle timer: in this case, the topology change information is spread downstream faster.

The above describes how upward routes (towards the root) and downward routes are discovered and maintained in a DODAG. Peer-to-peer operations are also supported by travelling the DODAG from a sender, up to the first common ancestor (in the storing case) or to the root (in the non-storing case) and then downward to the destination of the peer-to-peer flow. The current specification of RPL does not support direct node-to-node communication; however, a specific DODAG could be created for this purpose, if the communication pattern is intense enough to justify its deployment and the related overhead.

14.2.4 Constrained application protocol

The inherent physical constraints limiting many typical embedded devices forming the smart grid justify the quest for a suitable communication paradigm enabling any general data exchange, while still being interoperable with regular computer networks. A recent IETF working group believes that such a protocol can be obtained using an approach based on web services mainly because of their flexibility and low requirements; moreover, web services are the most common communication paradigm adopted over the Internet.

The constrained application protocol (CoAP) [10] enables every smart network-connected device to offer web services relying on very low requirements and complexity on the embedded device, thus readily enabling a simple M2M communication language easily interoperable between constrained devices and the big Internet. CoAP specifies this web service protocol by defining two different internal protocol layers: (i) the 'messages' layer, describing a simple transport mechanism to obtain non-complex reliable communication over UDP, and (ii) the 'request/response' layer, providing a realization of the representational state transfer (REST) paradigm [11] using a subset of the features available in HTTP. REST is a pattern of resource operations that has emerged as a *de facto* standard for service design in Web 2.0 applications, which deals only with data structures and the transfer of their state.

The 'message' layer specified in CoAP defines four message types: (i) the CON message to transport messages requiring an acknowledgement; (ii) the NON message which instead is used to carry messages requiring no acknowledgement, (iii) the ACK message that is used to acknowledge a previously received CON message, and (iv) the RST message to abruptly interrupt the exchange, usually because an internal state to properly handle the message is missing.

Thanks to this transport layer, the rest of the specification can focus on a binary-encoded REST realization intended to scale up to very simple devices; this requirement is fulfilled by defining a very simple binary header easily parsable by constrained hosts, which implements a subset of the basic features very similar to HTTP. Moreover, the specification provides a generic binary option format, used to pass optional parameters required to complete the message itself.

A REST message exchange typically consists of one request from the client and one response from the server, supposing that the request is carried in a CON message, and the response can be piggybacked on the ACK message, or could arrive in a separate CON message after the ACK itself; the server decides whether to use a piggybacked or a separate response depending upon the timing required to retrieve the requested resource representation.

A very important property of the CoAP specification is that the protocol can easily be mapped to HTTP by proxy entities, so that the HTTP agents deployed in the regular networks can interoperate directly with CoAP devices without the explicit need to support this protocol. To this end a usual deployment scenario consists of CoAP endpoints in the constrained networks, and interoperation with the Internet via an HTTP–CoAP proxy on the edge of the constrained network. In this case both networks have no specific requirement, whereas the interoperability is completely delegated to the HTTP–CoAP proxy.

14.2.5 W3C efficient XML interchange

The World Wide Web Consortium (W3C) addressed the problem of reducing the size and complexity of the eXtensible Markup Language (XML); this activity led to the development of efficient XML interchange (EXI) [12], which defines a few solutions offering different levels of compression and complexity with respect to standard XML. Thanks to a design principle focused on simplicity and efficiency, EXI satisfies the implementation requirements of even the most constrained communication technology.

In order to support as much as possible the flexibility of the XML language, the EXI standard defines two operating modes: one, called schema-informed, relies on the *a priori* knowledge about a common schema; the other, called non-informed, is able to process any XML document. While the former offers better compression, the latter is preferable for versatility. Both operating modes aim to build a finite-state machine reproducing the possible structures of the XML documents; each state of the automaton uses a grammar for describing the organization of the contents within the document. The grammar information can then be used to define a set of short event codes. The EXI streams obtained using the schema-informed mode and these short event codes can be as small as $1/10$ of the original XML document.

For a smart grid deployment, EXI offers unprecedented reusability and versatility features: in fact, through EXI it is possible to leverage on the interoperability functionalities of HTTP, without its parsing complexity and memory demands. Also, straightforward advantages of adopting EXI are: the possibility of both publishing and requesting device capabilities by simply GETting or PUTting the XML schema; the ease of installing new services and functionalities by simply uploading new schemas to the node; finally, machine-to-machine interactions are natively supported, since data is semantically enhanced by XML tags.

14.3 Implementation

This section will provide information regarding the implementation of the protocols described in Section 14.2, taking into account the programming subtleties needed to make the whole stack efficient and reliable. The case study proposed in this chapter expands and improves one that appeared in previous papers [7, 13]. The main focus of these preliminary works was the deployment of a full-featured smart grid system leveraging on wireless sensor networks for realizing the constrained communication part of the grid.

From an architectural point of view, the constrained part of the smart grid will consist of a few different types of devices, which, in order to be interoperable, need to be developed using IP across the whole system. In particular, the main device classes are the following: (i) gateway nodes which are in charge of the internetworking with other networks (such as the Internet), (ii) router nodes tasked with establishing a wireless tree-shaped backbone, (iii) specialized nodes providing operational access to the actual sensors and actuators, and (iv) mobile nodes directly accessing the network and operating on its devices.

In our implementation the protocol stack has been implemented using TinyOS [14] (which is an operating system designed for the control of constrained devices) on telosb [15] sensor nodes, which are equipped with a TI MSP430 microcontroller and a CC2420 radio transceiver. Although focusing on a single OS for a single sensor platform, the rest of this section tries to differentiate between contingent challenges due to the particular choices and general challenges related to smart grid communication.

The central part of Figure 14.1 shows the TinyOS modules realized to develop our smart grid system. In order to enhance the memory usage efficiency of the protocol stack, a common memory manager is needed. In fact, thanks to this component it is possible to share a single memory pool, thereby reducing the memory needed for buffering at each layer. What is even more practical is the capability of the module to provide virtual memory allocation functionalities, through which other modules linked to it can allocate new memory areas, resize existing areas, and free older ones. Since this module associates actual memory pointers with static IDs, all other components just have to pass these IDs to one another in order to pass or share a given memory area, without the need to perform slow memory copy operations. Let us now describe each layer of the protocol stack.

14.3.1   802.15.4

The PHY and MAC layers of 802.15.4 used in our test-bed are those provided with TinyOS. In particular, for the experimental campaign of Section 14.4 we adopted a standard PHY layer in the 2.4-GHz band on top of which we run a simple CSMA MAC protocol, without implementing any duty cycle.

What is important to highlight in the design of this level is that the hardware choices are more likely to have an impact on other protocol layers. In fact, telosb nodes suffer from two main limitations: they lack a DMA module and the SPI is clocked too low. The former forces the microcontroller to actively manage the data transfer of packets to be transmitted and received; the latter implies that this transfer takes almost as much time as the transmission itself.

14.3.2   6LoWPAN

The 6LoWPAN component is the base component of the protocol stack and every node will include it. The 6LoWPAN module goes beyond RFC4944 [3] and other documents

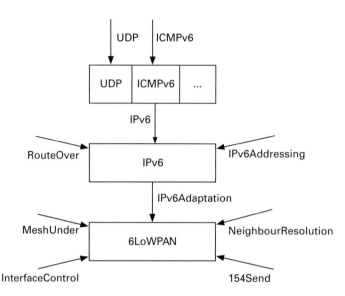

Figure 14.2 6LoWPAN protocol architecture.

related to the 6LoWPAN adaptation layer itself, because it needs to support several Internet standards, such as IPv6, UDP, ICMP, and possibly many others in order to enable Internet connectivity.

As shown in Figure 14.2, various software components contribute to build the high-level 6LoWPAN module, whereas the relevant adaptations of the upper-layer protocols for the constrained networks are isolated in the low-level 6LoWPAN component. The adopted 6LoWPAN software architecture is convenient because there is a clear correspondence between the software component and the implemented standard itself.

In particular, basic IPv6 support is obtained by implementing the base IPv6 header format as specified in Sec. 3 of RFC2460 [16]. The IPv6 component offers a modular extendible interface parameterized on the IPv6 next header field, thus enabling external modules to support any extension header or upper layer protocol.

Thanks to this interface, a UDP component has been bound to the IPv6 component fully implementing the UDP protocol as specified in RFC768 [17]; upper-layer protocols get access to the UDP protocol functions using a UDP interface parameterized on the local UDP port, thus enabling upper-layer components to process packets depending on the local application port specified in the packet.

ICMPv6 support as well has been delegated to a specific component handling ICMPv6 messages in general. Further processing of specific ICMPv6 messages can be performed by external modules using the ICMPv6 interface parameterized on the message type.

In order to transmit an IPv6 packet on an interface, IP standards require a technology-dependent adaptation layer for each IPv6 interface. That is where the various features offered by the 6LoWPAN standard for IEEE 802.15.4-based interfaces are implemented, and in particular: (i) next-header dispatch signalling the upper-layer protocol present in the data unit, required to overcome the fact that the 802.15.4 frame has no field equivalent to the Ethernet ethertype field, (ii) layer-2 fragmentation, similar to layer-3 fragmentation and enabling the transmission of a single layer-3 packet in multiple data units, and (iii) compression of upper-layer headers required to reduce the very high relative overhead used by IPv6 when sending the small data units typical of WS&AN traffic.

From a practical point of view, the adaptation process can be performed in the following order. At first the header compression is applied in a separate buffer and the compressed header size is computed, and the 'adapted' PDU is formed by combining the compressed header with the remainder of the packet. If the PDU does not fit in a single frame payload, layer-2 fragmentation is applied building fragments of size a multiple of 8 bytes fitting in the 802.15.4 frame.

A frequent issue found when implementing 6LoWPAN header compression is that the fragmentation header *datagram_size* and *datagram_offset* fields are computed with reference to the uncompressed IPv6 packet, whereas the fragmented data unit may have compressed headers. 6LoWPAN implementers tend to limit implementation size and may want to develop only and directly the 'adapted' PDU (compressed) encoding, however due to this specification constraint, this solution leads to a complex fragmentation process; our implementation has avoided merging IPv6 and 6LoWPAN components, enabling the straightforward fragmentation process described.

Additional interfaces required by the control plane must be offered by the 6LoWPAN/IPv6 stack, in particular: (i) IPv6 address management and (ii) forwarding next-hop signalling.

The IPv6 component provides network interface address-management features through an IPv6 addressing software interface, in particular IPv6 addresses of any class (link-local, global, multicast, etc.) can be added, deleted, and listed.

The forwarding process performs operations depending upon routing information provided by external modules. Two different software interfaces, RouteOver and MeshUnder, have been provided to perform the forwarding process, as it can happen at different layers of the protocol stack itself.

In the IP over low-power WPANs context, typically two different routing paradigms are available: (i) MeshUnder is often used to refer to link-layer mesh routing mechanisms, which received much attention in the past decade from the WSN research community, and (ii) RouteOver is used to refer to routing techniques more typical of the IP ecosystem, such as RIP, OSPF, etc. as well as the more recent RPL protocol.

Both approaches, even concurrently, are supported by the network layer through these interfaces.

14.3.3 RPL

Our implementation of RPL is directly dependent on the 6LoWPAN functionalities, as RPL messages are in fact ICMPv6 packets with a given type and code as deployed by IANA and reported in [18]. To make the integration more seamless, the node IDs contained in all messages are actually IPv6 addresses. As required for compression or internal translation purposes, the last two bytes are singled out to yield a 16-bit ID.

Our version of the DIO message (see also Section 14.2.3) includes the hop count of the sender, the RPL instance ID, the version (set to 0 by the root node), the rank of the sender and the control fields, which at this time only report that the DODAG is grounded (linked to a gateway device). In our implementation, we set the rank to be equal to the hop count of the node, and mitigate the possible connectivity problems that may arise [8] using a link quality criterion (see below). The parameters of the Trickle timer [9] regulating the replication of DIOs have been set to $I_{\min} = 1$ s, $I_{\max} = 32$ s, and $k = 5$ consistent replicas to be collected before a DIO transmission can be suppressed.

At reception of the first DIO carrying new information, a node waits for 4 s for further DIOs, and then associates with its chosen parent using a DAO. The DAO specifies the RPL instance ID read from the DIO and the address of the chosen parent, to which the DAO is unicast. The choice of the parent is driven by a modified hop-count criterion. In particular, every node chooses a single parent having lowest rank and whose DIO message has been received with the highest received signal strength indicator (RSSI). The optimization is carried out first over the rank, and then over the RSSI, throughout the nodes of equal rank. A minimum RSSI threshold of -80 dBm has also been set: those DIOs whose RSSI is below this threshold are automatically discarded. This solution avoids the usage of the hop count, as the only metric concurring with rank definition translates into poor connectivity and long, low-quality links [8]. In place of the RSSI, the link quality indicator (LQI) and the expected transmission count (ETX) have also been implemented, but will not be considered in the following discussion. The nodes periodically check the connectivity status over the link with their parent: if no DIO is received within a period twice as long as the maximum DIO transmission interval dictated by the Trickle timer, the node sends a DIS in order to trigger further DIOs.

We operate the network in a fully stateful, storing mode. This means that the root of the DODAG always has a complete routing table, whereas intermediate nodes store a partial routing table specific for the subtree rooted at them. Mobility and similar topology changes are supported using the prescribed DIS messages. In particular, a node detects movement via substantial changes in the average RSSI of known links, as well as by observing that some neighbours disappear and some previously out-of-reach nodes are now within communication range. As soon as movement is detected, the node triggers the rearrangement of the tree by sending a DIO with rank equal to infinity, so that all its children de-associate from it and rediscover new routes towards the sink. Also, the local routing tables of the parent nodes are reset for what concerns the subtree rooted on the moving node. The node itself sends a DIS in order to trigger DIOs and re-associate to a new parent; the DIS is configured not to reset the Trickle timer, as the connectivity changes are localized and do not require full-scale re-broadcasting.

14.3.4 CoAP

The CoAP module implements the lightweight web service functionalities in the protocol stack. In particular, the CoAP layer connects the application providing RESTful interfaces for any service offered as well as for hardware peripherals, such as sensors and actuators. The CoAP layer consists of two modules: one devoted to session management implementing the 'transport' part of the protocol, the other for REST parsing and resource registration implementing the 'message' part.

Designed to work on top of the UDP transport protocol, the session-management module of CoAP needs to implement a few functionalities to strengthen the reliability of the E2E communication, such as retransmission and separate response support. In particular, Figure 14.3 shows the finite-state machine of the 'transport' part. In the figure, the protocol states are represented with ellipses, the actions with rectangles, the choices with rhomboidal shapes, and the timers with squares. The transition triggers are represented with directed arrows whose label details the transition cause.

As can be seen in the picture, the automaton reacts to application commands, timeout interrupts and received packets, implementing a simple request/response paradigm. In addition to this, the protocol makes it possible to delay the transmission of a response to handle computationally heavy processes. In such a case, the client/server semantics is reversed as the server first replies to a request with an empty ACK, and then, after the application has provided the needed data, sends a request with the answer to the previous request; finally, the client acknowledges the reception of the request.

Since this separate response mechanism may block the transport layer for an arbitrarily long time, this module is implemented as a pool of sessions, each with a separate state variable. In this way, the CoAP server can still react to an incoming request, while others are waiting for the application to produce data for the response. In addition, each session can be cached in order to manage duplicate requests. The module is able to associate incoming requests and responses to sessions by means of the triple (host, port, token), where host and port connect the underlying UDP layer to the application while token is used to allow multiple sessions on the same port.

The CoAP 'transport' module is designed to reduce the interaction with the application; in fact, it is in charge of managing retransmissions, separate responses and duplicates automatically, only providing the application with error codes when anything abnormal happens.

On top of the 'transport' module, a 'REST' module is implemented, which takes care of the message parsing and the REST management. In addition, this module provides the application with interfaces to register the offered resources, such as sensors, actuators and other services, and interfaces for sending requests to other CoAP servers. Through these interfaces the application is able to install a CoAP server by specifying the URI paths describing the local addresses to access the resources as well as the implemented methods for each of them. Also, the application can behave as a client by issuing requests for other CoAP or HTTP servers.

This module behaves as a message parser, which supports the GET, POST, PUT, and DELETE methods to interact with resources and provides a set of status and error codes for M2M interactions. In particular, these error codes allow the server CoAP to behave very similarly to an equivalent web server, in terms of offered services. For a complete list of error codes, please refer to [10]. A few examples are: 2.05 specifies that the response carries the content, 4.04 is the well-known 'not found' error, while 5.03 reports a service unavailability.

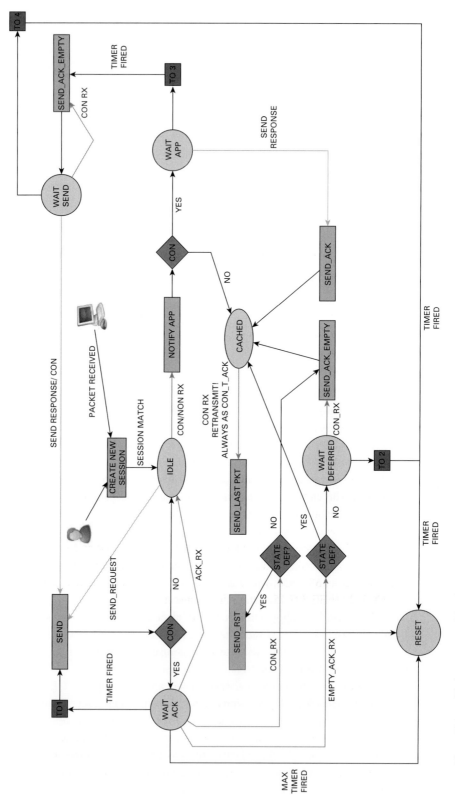

Figure 14.3 Finite-state machine for the 'transport' part of the CoAP protocol.

14.3.5 EXI

The EXI module is not a proper communication layer, but a fundamental block for the realization of a lightweight web service as it allows constrained nodes to leverage on the capabilities of expressive languages such as XML without implementing the complexity of the needed parser and the transmission overhead inherent to XML. To realize the EXI module, we opted for implementing coding and decoding functions between EXI streams and an intermediate memory format, which allowed us not to use (and store) XML documents on the nodes. This greatly improved the memory footprint of our implementation, as shown in [19].

Coding and decoding functionalities between EXI streams and the intermediate memory format were implemented using the *a priori* knowledge provided by a predefined XSD file. XSD files constrain the types and number of elements that can be used by XML requests and responses. Of course this trades off the flexibility and expressivity of the language for advantages in terms of overhead reduction and computational lightness.

From XSD files it is possible to generate grammar, which lists all the possible transitions among elements of an XML document. In the same way, grammars are used to interpret both the EXI streams and the intermediate memory format. In particular, this format contains one-to-many primitive types or pointers to similar memory structures, maintaining a tree-like organization, which is typical of XML documents.

14.4 Performance evaluation

The evaluation campaign of the implemented stack has been carried out in our test-bed deployed in the context of the SENSEI and WISE-WAI projects. The realization of the test-bed aimed to mimic a realistic wide-area deployment, thus it has been possible to evaluate the performance of our communication solutions in many different configurations. The test-bed coverage area spans several buildings in the Department of Information Engineering of the University of Padova, and contains 350 wireless sensor nodes, 40 gateways, and a main server. All nodes feature the telosb architecture, which consists of an MSP430, which is one of the most energy-efficient microcontrollers, and a CC2420 IEEE 802.15.4-compliant transceiver.

The network topology features several different areas: high-density areas containing up to 80 nodes within the same coverage range provided the environment for congestion tests; low-density linear areas have been used to test routing protocols and multihop forwarding; medium-density wide areas have been exploited to evaluate the average behaviour of a typical network.

The test-bed is deployed using a hierarchical configuration, where all nodes are connected through USB hubs to embedded computers, each of them connected via an Ethernet backbone to a central server coordinating the management and debugging activities in the test-bed. The dual connectivity mode made it possible to test the wireless communication performance while collecting statistics using the wired communication, thus not impacting the performance evaluation.

Other important features offered to the user are per-node power control, USB firmware reprogramming, USB serial connection and reservation backend; the wide set of features offered make extensive evaluation a rapid and simple process by which we can obtain realistic results.

The following subsections provide experimental results for the different layers of the protocol stack. Most of the tests have been targeted to determine the operating bounds of our implementation of the latest standards in constrained wireless communications applied to smart grid scenarios.

14.4.1 Link performance using IEEE 802.15.4

The first set of performance metrics, shown in Figure 14.4, addresses the throughput at the link layer. For this experiment, we only exploited the PHY and MAC layers as defined in the IEEE 802.15.4 standard, choosing the non-beacon-enabled mode MAC without implementing any duty cycle. This configuration has been installed in a varying number of sensor nodes, all of them within the same coverage range so that their transmissions have to share the same channel.

The objective of such a restrictive setup was to determine the bandwidth efficiency at the MAC layer. The figure plots the maximum throughput achievable by varying the physical frame size and the number of senders. We let the physical frame size vary from 60 to 110 bytes, which is near the maximum usable payload allowed by IEEE 802.15.4; also, we test this configuration using up to five sending nodes and a single receiver.

The first consideration to be made is that a single transmitter only achieves a throughput of up to 50 kbps, using the maximum physical frame size, which is equal to a fifth of the total bandwidth. This is due mainly to the telosb architecture that, not providing any DMA functions, forces the node firmware to pass packets from the micro to the radio using the SPI bus. Moreover, the SPI bus can only transfer up to 250 kbps at the working

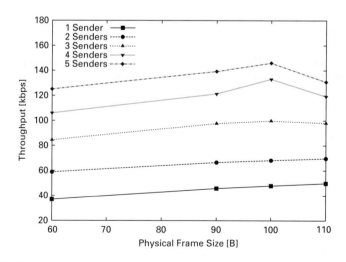

Figure 14.4 IEEE 802.15.4 throughput.

clock rate of the MSP430. However, this should only halve the achievable bandwidth, while our test results are far from this limit. This further performance reduction is due to the cooperative philosophy of TinyOS: whenever an interrupt is received by the micro, the operating system posts a task to manage the asynchronous event, however the task can be processed only when all the other tasks in the queue have been served.

The performance starts increasing from two transmitters onward: in fact, with multiple transmissions the CSMA MAC is able to use the bandwidth more efficiently. Figure 14.4 reports the maximum throughput achievable with 2, 3, 4 and 5 nodes transmitting, respectively. Three trends can be noticed here: the total throughput achievable at the link layer increases with the number of transmitters; it is bounded at about 150 kbps and increases with the physical frame size.

For what concerns smart grids, it is important to notice that, even though the maximum link data rate is limited, a simple CSMA MAC delivers acceptable performance. However, to achieve higher data rates a TDMA MAC is needed with the added complexity of maintaining the synchronization between devices. Also, better performance should be achievable by simply introducing a hardware DMA function in the node architecture.

14.4.2 Network throughput with 6LoWPAN

Adding the network layer to the protocol stack, the network is now able to exchange bigger packets exploiting fragmentation and reassembly. Since multiple fragments need to be sent per transmission, for this test we evaluate not only the network throughput, but also the average time needed to send a 6LoWPAN packet.

As in the previous test, one to five sensor nodes within the same coverage range have been used, however this time we implement 6LoWPAN on top of 802.15.4. This allowed us to let the packet size vary from 93 to 1272 bytes, however, for space constraints we only show the two boundary values. As before, we tested the performance varying the physical frame size from 60 to 110 bytes.

Figures 14.5 and 14.6 report the results obtained with 93-byte and 1272-byte packets, respectively. In both groups of figures we plotted on the left the throughput performance and on the right the average link delay.

Starting with Figure 14.5(a), it is possible to notice that the performance decreases as the frame size goes from 60 to 90 bytes and then reaches a maximum for 100 bytes. This behaviour is due to the fragmentation of the 6LoWPAN packet into smaller 802.15.4 packets that fit into the physical frame size.

Using a frame size bigger than the 6LoWPAN packet size achieves better performance, since there is no need to fragment the packet: hence, there are fewer contentions in the network and less processing time is needed in the node. However, the first throughput decrease, which happens when fragmentation is needed, is not obvious: this performance drop is due to fragments of very different size – the first packet uses the whole physical frame while the second packet just uses 3 bytes.

Figure 14.5(b) shows a similar behaviour, but for the average link delay. Again, the increase in the link delay for a 90-byte frame size is due to the unbalanced transmission of long and short fragments. By comparing the results obtained with small packets, it is

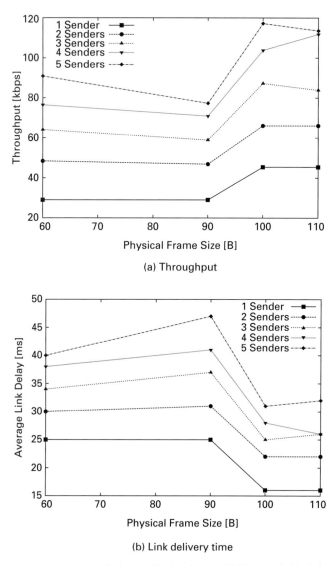

Figure 14.5 6LoWPAN performance with 93-byte packets.

worth noting that, even though using more senders in the same coverage range increases the average link-layer delay, the aggregate throughput increases as well.

Figure 14.6(a,b) shows the same metrics with 1272 bytes of physical frame size. With this frame size, performance, both in terms of throughput and in terms of delay, gets better with increasing frame size; this is due to a more balanced network behaviour, given that most of the packets are sent with a full frame size.

As a final note, comparing the performance achieved with small packets to that obtained with big packets, higher throughput is achieved with the maximum packet size at the price of the largest delay.

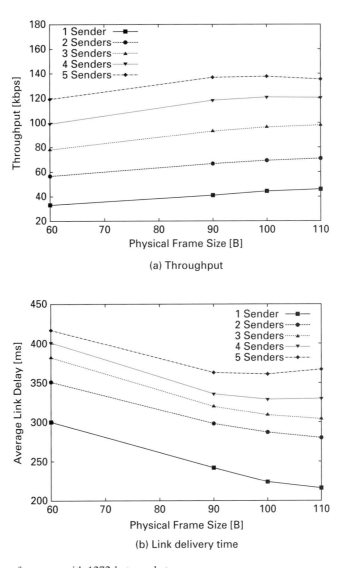

Figure 14.6 6LoWPAN performance with 1272-byte packets.

14.4.3 Network throughput with RPL in multihop scenarios

This section addresses the delivery rate of a complete network when RPL is used to route the packets towards the sink. Two tests have been performed with two network configurations using a portion of our test-bed: the first scenario, which we refer to in Figure 14.7(a) as *low density*, consists of 65 nodes deployed with a maximum hop-count distance from the sink of 6 hops; the second, called *high density* in Figure 14.7(b), uses 90 nodes over 8 hops. In this section, we are not addressing RPL performance in terms of setup time and control overhead; for these metrics the interested reader should refer to [7, 20, 21].

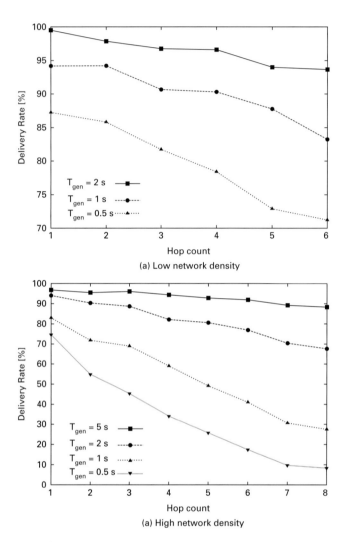

Figure 14.7 RPL delivery rate vs. hop count.

Curves are plotted, varying the average packet inter-arrival time, and show the average delivery rate for nodes at the same hop-count distance from the sink. A satisfactory successful delivery rate is achieved in both configurations: in the dense network a 5-s inter-arrival time is needed, while in the sparse environment 2 s is enough. Delivery rates have been obtained for inter-arrival times as short as 500 ms, however in these situations the network becomes congested and it is not possible to deliver all packets at the sink for of the nodes deployed in the first hop.

Smart grid networks must be deployed carefully according to these results; in fact, the maximum data rates achievable in the network are strongly related to both density and depth of the network. In order to achieve a given target throughput, it is of paramount importance to use the correct number of sink nodes to allow the network to have the maximum density and hop-count depth needed.

14.4.4 CoAP performance

This set of tests addresses the performance when the CoAP protocol is added to the protocol stack. The objective now is to obtain the communication bounds when a request/response paradigm is used in the network. In particular, we derived in Figure 14.8 the throughput achieved with a single CoAP server answering the request of a single (Figure 14.8(a)) and three (Figure 14.8(b)) CoAP clients, varying the node generation rate from 10 to 50 packets per second. Instead, Figure 14.9 reports the success rate.

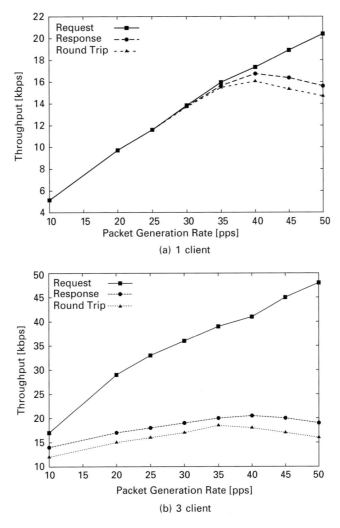

Figure 14.8 CoAP throughput performance in single-hop scenarios.

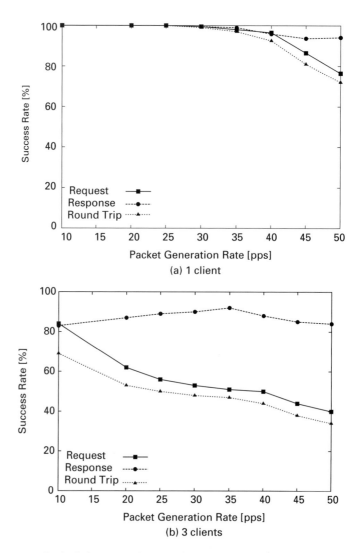

Figure 14.9 CoAP success rate in single-hop scenarios.

For these tests the throughput has been computed using a payload size of 70 bytes, which we assumed as a typical packet size for smart grid scenarios, where information such as the energy consumption at a particular point of the network is usually requested.

In the figures, curves have been plotted showing the throughput and the success rates achieved for request transmissions by the client(s), for request receptions by the server, and for response receptions by the client(s). A decrease in terms of performance was expected in these tests, since the bi-directional flow of packets congests the network faster than data dissemination and data gathering taken separately.

These results show that the web service capabilities offered by CoAP are obtained at a price: the bi-directional flow, coupled with the added processing complexity, strongly limits the network bandwidth. Hence, it is important to separate those services that can be provided using network aggregators from those that need to interact with a given node in the network: for instance, the average temperature reading in a building should not be obtained by asking every single node in the network, but rather by using a distributed procedure that first aggregates the average value at the sink and where users can then request the sink for that particular value.

14.4.5 CoAP multihop performance

The last performance metrics we tested imply the simultaneous use of the CoAP web service functionalities with RPL routing support. Figure 14.10 provides graphs for the

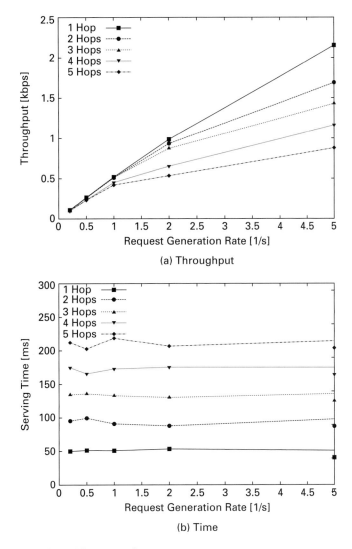

Figure 14.10 CoAP performance in multihop scenarios.

throughput achievable and the average round-trip time of communication on 1 to 5 hops from the client to the server, varying the request generation rate.

The first consideration is that the serving time increases linearly with the hop count distance, hence combining web services with routing is not impacting the performance more than expected. Also, the throughput degradation is due to the single-server scenario: here, both requests and responses share the same path, causing a higher collision probability and link congestion.

Finally, it is important to note the throughput achievable for processing a single request–response communication is very low. Hence, to exploit the full channel bandwidth, some data aggregation and multicast optimization is in order.

14.5 Conclusion

We have presented a case study of smart grid deployment using wireless sensor networks. In particular, we have described the standards used for the software implementation, highlighting the design subtleties needed to achieve the required efficiency in constrained environments. Subsequently, a section was devoted to the description of the actual implementation of the platform, and we discussed the different optimization techniques adopted in the development of both the network and the software used in it. Finally, a thorough experimental campaign was shown, where performance results were described for each of the protocol layers considered in the network.

From this case study it is possible to draw some general considerations about the feasibility of smart grid communication networks using constrained wireless devices. First of all, it is possible to integrate the Internet and the web service paradigms into smart grids. However, for any added functionalities or communication layer, the overall performance of the network decreases. Hence, smart grid engineers should select with care which functions can be offered, exploiting the full Internet of things protocol stack, and which are instead better implemented with optimized solutions working at the lower communication layers.

Moreover, since multihop communication round-trip time and packet-error probability increase with the number of hops, those services that require bounded and tight delay should be addressed locally as much as possible by distributing the control logic within the network and applying self-management functionalities. Thus, it is more efficient to remotely control these services by tuning their working parameters, but letting the actual M2M interactions happen within the constrained network.

As a final note, this case study testifies that the Internet of things communication solutions are mature enough to drive the integration of smart grids in the Internet, thus enabling the whole set of logistic and business applications to interact remotely with grid devices and to implement the complete supply chain from the energy producers to the consumers and through the suppliers.

References

[1] M. Zorzi, A. Gluhak, S. Lange, and A. Bassi, 'From today's INTRAnet of things to a future INTERnet of things: a wireless- and mobility-related view', *IEEE Wireless Communication Magazine*, vol. 17, no. 6, pp. 44–51, December 2010.

[2] 'IEEE 802.15.4 standard specification', downloadable at standards.ieee.org/getieee802/, 2006 [online]. Available at: http://www.ieee802.org/15/pub/TG4.html

[3] G. Montenegro, N. Kushalnagar, J. Hui, and D. Culler, 'Transmission of IPv6 packets over IEEE 802.15.4 networks', IETF Request For Comments, September 2007.

[4] Z. Shelby, S. Chakrabarti, and E. Nordmark, 'Neighbor discovery optimization for low-power and lossy networks', IETF draft-ietf-6lowpan-nd-15, December 2010 [online]. Available at: http://tools.ietf.org/html/draft-ietf-6lowpan-nd-15

[5] T. Winter, P. Thubert, A. Brandt, T. Clausen, J. Hui, R. Kelsey, P. Levis, K. Pister, R. Struik, and J. Vasseur, 'RPL: IPv6 routing protocol for low power and lossy networks', IETF internet draft draft-ietf-roll-rpl-12, 2010 [online]. Available at: https://datatracker.ietf.org/doc/draft-ietf-roll-rpl/

[6] F. B. Beidou, W. G. Morsi, C. P. Diduch, and L. Chang, 'Smart grid: challenges, research directions and possible solutions', in *Proceedings of IEEE International Symposium on Power Electronics for Distributed Generation Systems (PEDG)*, Hefei, China, June 2010.

[7] N. Bressan, L. Bazzaco, N. Bui, P. Casari, L. Vangelista, and M. Zorzi, 'The deployment of a smart monitoring system using wireless sensors and actuators networks', in *Proceedings of IEEE SmartGridComm*, Gaithersburg, MD, 2010.

[8] P. Thubert (ed.), 'RPL objective function 0', IETF draft, May 2011 [online]. Available at: http://tools.ietf.org/wg/roll/draft-ietf-roll-rpl/

[9] P. Levis, N. Patel, D. E. Culler, and S. Shenker, 'Trickle: a self-regulating algorithm for code propagation and maintenance in wireless sensor networks', in *NSDI*. USENIX, san Francisco, CA, 2004, pp. 15–28.

[10] Z. Shelby, B. Frank, and D. Sturek, 'Constrained application protocol (CoAP)', IETF internet draft draft-ietf-core-coap-06, 2010 [online]. Available at: http://datatracker.ietf.org/doc/draft-ietf-core-coap/

[11] R. T. Fielding, 'Architectural styles and the design of network-based software architectures', PhD dissertation, University of California, Irvine, CA, 2000 [online]. Available at: http://www.ics.uci.edu/ fielding/pubs/dissertation/top.htm

[12] J. Schneider and T. Kamiya, 'Efficient XML interchange (EXI) format 1.0', W3C working draft, 2008 [online]. Available at: http://www.w3.org/TR/2008/WD-exi-20080919

[13] A. P. Castellani, N. Bui, P. Casari, M. Rossi, Z. Shelby, and M. Zorzi, 'Architecture and protocols for the Internet of things: a case study', in *Proceedings of First International Workshop on the Web of Things (WoT-2010)*, Mannheim, Germany, March–April 2010.

[14] 'TinyOS community network' [online]. Available at: www.tinyos.net

[15] CrossBow, 'TelosB mote platform' [online]. Available at: http://www.xbow.com/Products/Product_pdf_files/Wireless_pdf/TelosB_Datasheet.pdf

[16] S. Deering and R. Hinden, 'RFC 2460: Internet protocol, version 6 (IPv6) specification', IETF RFC, December 1998 [online]. Available at: http://www.ietf.org/rfc/rfc2460.txt

[17] J. Postel, 'RFC 768: user datagram protocol', IETF RFC, August 1980 [online]. Available at: http://www.ietf.org/rfc/rfc768.txt

[18] T. Winter (ed.), 'RPL: IPv6 routing protocol for low power and lossy networks', IETF draft, March 2011 [online]. Available at: http://tools.ietf.org/wg/roll/draft-ietf-roll-of0/
[19] A. P. Castellani, M. Gheda, N. Bui, M. Rossi, and M. Zorzi, 'Web services for the Internet of things through CoAP and EXI', in *Proceedings of IEEE ICC RWFI Workshop*, Kyoto, Japan, June 2011.
[20] J. Tripathi, J. de Oliveira, and J. Vasseur, 'A performance evaluation study of RPL: routing protocol for low power and lossy networks', in *Proceedings of Conference on Information Sciences and Systems (CISS)*, Princeton University, NJ, March 2010.
[21] J. Tripathi, J. de Oliveira, and J. Vassew, 'Applicability study of RPL with local repair in smart grid substation networks', in *Proceedings of IEEE SmartGridComm*, Gaithersburg, MD, October 2010.

Part V

Security in smart grid communications and networking

15 Cyber-attack impact analysis of smart grid

Deepa Kundur, Salman Mashayehk, Takis Zourntos, and Karen Butler-Purry

15.1 Introduction

It is well known that the existing power grid represents a critical asset essential for the functioning and welfare of modern society. A movement to a *smarter* power grid promises to enable greater energy delivery, reliability, and efficiency. It also represents a critical foundation for reducing greenhouse gas emissions and transitioning to a low-carbon economy. The evolution from today's power grid to a smarter grid is only possible through greater dependency on information technology.

There are currently many working definitions for a smart grid. The North American Reliability Corporation (NERC) has defined the smart grid as 'the integration of real-time monitoring, advanced sensing, and communications, utilizing analytics and control, enabling the dynamic flow of both energy and information to accommodate existing and new forms of supply, delivery, and use in a secure and reliable electric power system, from generation source to end-user'. The movement towards cyber-enabled power systems increases the risk of attacks on information devices and communications systems for several reasons.

From an engineering perspective, there is increased opportunity for cyber attack in a smart grid because of the greater reliance on distributed advanced metering infrastructure (AMI), intelligent electronic devices (IEDs), and wireless and/or off-the-shelf communications components and systems. Such cyber infrastructure increases system connectivity and autonomous decision-making by employing standardized information protocols that often have (or will have in the future) publicly documented vulnerabilities. Motivations for attack also abound. Market deregulation and privatization of the energy industry has increased competition among energy providers to enhance consumer-centricity. Threats also exist in the form of dissatisfied utility insiders, consumers, and cyber terrorists. There have been more than 60 documented real-world cases of electronic attacks on control systems alone [1]. The Stuxnet worm targeting Siemens' supervisory control and data acquisition (SCADA) control software has recently received significant media attention.

There has been a recent focus, stemming in part from large-scale smart grid initiatives, by industrial parties, government organizations, and academia on developing and integrating cyber security solutions for emerging smart grid systems [2–9]. However, the problem of securing the power grid remains daunting for several reasons. First, the problem of properly training power system operators and engineers regarding security issues is challenging given that there is often a lack of a cyber security culture in the electric power sector. Second, for resource-constrained power utilities operating in a recently deregulated market, the task of overall system hardening is overwhelming given the need to enhance or add costly infrastructure. Last, the complex integration of cyber technologies with the electrical grid components and legacy system devices can create unforeseen emergent security vulnerabilities that should ideally be accounted for prior to grid redesign and system decision-making.

We assert that whether identifying or designing new smart grid devices or communication systems that promote cyber security, it is imperative to understand the various impacts of these new components, under a variety of attack situations, on the power generation, transmission, and distribution systems. This will enable a more comprehensive understanding of the security issues that exist, allow an evaluation of different technologies for a more detailed cost–benefit analysis, and facilitate the prioritization of system protection thrusts to address the most critical vulnerabilities first.

This chapter presents a work in progress towards the development of a framework for smart grid cyber-attack impact analysis. Our objective is to balance the needs of electric power utilities and smart grid system designers who aim to use such knowledge for resource prioritization as well as researchers who desire to glean fundamental knowledge about cyber-physical interactions and system security. Three main contributions are as follows:

1. Given the emerging nature of the area, a necessary background is provided to motivate and introduce fundamental research and development questions. The focus is on cyber-attack impact analysis.
2. We present a framework for cyber-attack impact analysis that employs a graph-theoretic dynamical systems approach for modelling the interactions between the cyber and the electricity networks. The focus is on the model synthesis stage.
3. A test case study using a system modified from the IEEE 13-node distribution test system is presented to demonstrate the potential for modelling future smart grid systems.

The next section provides a brief background of risk management and prior art in the field. Our cyber-attack impact analysis framework is provided in Section 15.3, where we focus on the model synthesis stage. Section 15.4, provides the results of model synthesis for a 13-node distribution test system demonstrating the potential of the approach for three attack scenarios. Final remarks are presented in Section 15.5.

15.2   Background

15.2.1   Risk management

In the field of information security, risk is a potential that a given *threat* will exploit one or more *vulnerabilities* of one or more *assets* and thereby cause damage or harm.

Classically, risk $\mathcal{R}$ of a system component is a product of three variables:

$$\mathcal{R} = P_{\text{threat}} \times P_{\text{vulnerability}} \times \mathcal{I}, \tag{15.1}$$

where P_{threat} represents the probability there is a threat, $P_{\text{vulnerability}}$ is the likelihood that the threat exploits system vulnerabilities, and $\mathcal{I}$ quantifies the potential *impact* of damage or harm caused by exploiting the vulnerabilities.

The objective of risk management in the context of cyber security is to identify vulnerabilities and threats to the cyber assets of a system and then make decisions on what countermeasures, if any, to apply to reduce risk to an acceptable level. The latter typically depends on the value of the cyber asset being considered. Risk management can be grouped into the following tasks:

1. Risk identification, where damages and their potential consequences are determined.
2. Risk assessment, in which the value of risk, for a concrete scenario and a recognized threat, is evaluated quantitatively and/or qualitatively.
3. Risk prioritization, whereby risk reduction measures are prioritized based on a strategy.

The integration of the traditional power grid with information devices and communications systems creates a host of new cyber-related threats for power systems. The complexity of integration, diversity of system vendors, and push for timely solutions will nonetheless lead to the existence of cyber vulnerabilities that can be discovered and exploited. While information technology can effectively be used to improve the reliability of power systems, there will be increased risk of cyber attack in contrast to traditional power systems. Every sensor, meter, and communication channel introduced will introduce an additional risk to cyber attack. Fundamental questions arise: *How can we mitigate such risks? How should we prioritize system hardening to address the most critical risks? Is the information available through advanced cyber infrastructure worth the increased security risk?*

Risk and vulnerability analysis approaches [10–12] aim to understand the answers to such questions. However, common strategies are as-of-yet ad hoc. It is well known that there is currently a lack of historical data to sufficiently estimate any of the quantities of (15.1) necessitating the development of appropriate analysis tools focused for emerging smart grid power systems. It may be possible to estimate P_{threat} and $P_{\text{vulnerability}}$ using conventional analysis methodologies for information systems security, but the impact $\mathcal{I}$ is difficult to assess given the complex system interdependencies that characterize a smart grid and the evolving nature of modern power systems. The impacts of a cyber attack on a smart grid are strongly related to the effects on power generation, delivery, and demand.

Before such evaluation can have practical significance, it is necessary to quantitatively study the potential severity of physical impacts of cyber attacks. This requires identifying cascading failures within and between the cyber and physical domains. Mathematical models of these interactive subnetworks are typically vague or often do not exist [13]. One of the stumbling blocks is the inability to formally measure the impact of a cyber attack on power-delivery metrics of importance to the power industry. To address this challenge, we study the development of a cyber security analysis methodology that accounts for the complex cyber-to-physical interactions.

15.2.2 Prior art

We consider the problem of *cyber-attack impact analysis* which involves quantifying the effects of given classes of cyber attack on the physical electrical grid, hence, providing information on the degree of disruption to power delivery that a class of cyber attacks can enable. This information is vital for risk management [14]. Furthermore, based on this information, sophisticated dependencies between the cyber and physical systems can be identified – also shedding light on behaviours of complex interdependent networks.

Existing research that has addressed the interaction between the cyber and physical aspects of emerging power systems can be categorized into a number of classes. *Static approaches* [15] focus on identifying system component dependencies to trace the sequence of steps an attacker would have to take in order to compromise a smart grid element. *Empirical approaches* [14, 16–18] exploit existing advances in communications and power systems simulators, combining them such that an attack is applied in the communication simulator that transfers data to the power systems simulator, which makes decisions based on this possibly corrupt or time-delayed information. Typical traditional power system reliability metrics are used to assess the impact of the cyber attacks. In *cyber–physical leakage approaches* [19–21], confidentiality of the cyber network is studied by identifying how physical observable measurements of the power generation and delivery system can be analysed for clues about cyber protocol activity. The latter information can be used to deduce, in part, what *state* (i.e., normal, emergency, or restorative) a grid is in. *Mechanistic techniques* [22–25] involve addressing the cyber security problem through developing new information protocols, algorithms, and architectures that account for the specific objectives of the power-system community. *Testbed systems research* addresses the exploration of practical vulnerabilities through SCADA testbed development and construction [26–28]. Finally, research on *attacks on control systems* [29–32] focuses on how data corruption of denial of information access can affect the control of the power grid.

The work presented in this chapter builds upon this body of research by mathematically representing grid component interactions to better identify non-cookie-cutter vulnerabilities, the relative physical impact of cyber attacks, and cost–benefit tradeoffs for potential countermeasures. Thus we aim to obtain a better compromise among computational complexity, generality, and modelling accuracy. Preliminary studies by the authors can be found in [33] and [34].

15.3 Cyber-attack impact analysis framework

Based on the requirements outlined above, we propose a paradigm for cyber-attack impact analysis that employs a graph-theoretic structure and a dynamical systems framework to model the complex interactions among the various system components. This approach is significant because it has the potential to provide timely tools and design insights essential for developing more reliable and secure smart grid infrastructure. Solutions for securing the smart grid are just beginning to emerge, and the proposed research provides a necessary framework in order to better assess and prioritize them.

15.3.1 Graphs and dynamical systems

A graph is a mathematical structure that represents pairwise relationships between a set of objects. A graph is defined by a collection of *vertices* (also called *nodes*) and a collection of *edges* that connect node pairs. Depending on the use of a graph, its edges may or may not have direction, leading to directed or undirected classes of graphs, respectively. Graphs provide a convenient and compact way to show relationships and relate dependencies within cyber–physical power systems, as witnessed by recent papers that employ this tool [15, 16, 18, 25, 35–40]. However, as cited in [39], purely graph-based approaches do not sufficiently model the state changes within the physical system. Moreover, they do not effectively account for the unique characteristics of the system at various time scales, nor provide a convenient framework for modelling system physics. We assert that modelling the electrical grid is a vital component to an effective impact-analysis framework.

One approach to physically modelling complex engineering interactions employs dynamical systems. A dynamical system is a mathematical formalization used to describe the time evolution of a *state* **x**, which can represent a vector of physical quantities. The reader should note that we include algebraic equations as trivial cases of this class. In continuous time the deterministic evolution rule describes future states from current states as follows:

$$\dot{x} = f(x, u), \qquad (15.2)$$

where $\dot{x}$ is the time derivative of x and u is an input vector. Dynamical systems theory is motivated, in part, by ordinary differential equations and is well suited to representing the complex physical interactions of the power grid [41].

We assert that a graph-based dynamical systems formulation is effective for a smart grid cyber-attack impact analysis framework for a variety of reasons. First, smart grid impact analysis necessitates relating the cyber attack to physical consequences in the electricity network. A dynamical systems paradigm provides a flexible framework to model (with varying granularity and severity) the cause–effect relationships between the cyber data and the electrical grid state signals and ultimately relate them to power-delivery metrics. Furthermore, secondary effects whereby the consequence of an attack itself influences the continued degree of attack can be represented.

Second, graphs enable a tighter coupling between the cyber and physical domains. For a smart grid, the cyber-to-physical connection is often represented through control signals that actuate change in the power system and the physical-to-cyber connection is typically due to the acquisition of power state sensor readings. These connections can conveniently be expressed as specifically located edges of the graphs. Furthermore, as we will discuss, the graphs induce a dynamical systems description of the overall smart grid, which conveniently expresses complex time-varying interrelationships. In this way, cascading failures and emergent properties from the highly coupled system can be represented. Mitigation approaches often involve islanding of the grid or partitioning of the core smart grid components from optimization functions [13], and a graph-based dynamical systems formulation can naturally portray such separation as well.

Last, a primary effect of including cyber attacks in traditional reliability analysis is that it increases the size of the system under study by several orders of magnitude [14]. Our proposed mathematical formulation has the potential to keep studies tractable because our granularity of detail can be tuned and the use of dynamics can enable sophisticated behaviours without a corresponding increase in complexity.

15.3.2 Graph-based dynamical systems model synthesis

An overview of our impact-analysis approach, which is currently a work-in-progress, is shown in Figure 15.1. The three stages of model synthesis, system analysis, and

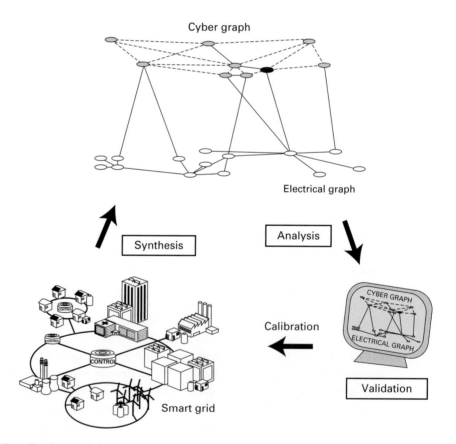

Figure 15.1 Overall cyber-attack impact analysis framework. The development of the framework employs the stages of model synthesis, model analysis, model and design validation. The results of the impact analysis can be used to calibrate a smart grid topology and enhance the synthesis model.

system validation are presented. In addition, the output of the validation stage is used to *recalibrate* the smart grid model to promote system resilience and security, or the accuracy of the synthesis approach.

The model synthesis stage takes an existing smart grid system and converts it to a mathematical construct that is intended to preserve the salient behavioural characteristics of the system. In this work we consider a combination of graphs and dynamical systems for this representation. As discussed previously, the use of graphs conveniently facilitates incorporating complex dependencies within and between the cyber and electric components. It also affords the flexibility to tune the granularity of detail.

After model synthesis, a convenient mathematical representation will enable analysis of system impacts to deduce global system properties such as stability and power disruption as a function of more local metrics in the face of various attack models. If appropriately modelled, the rigour afforded by the synthesis approach will enable some modularity, time-scale separation, and analysis of sensitivity to parameter variations in order to gain insight into topologies and mitigation robust to cyber attack. This leads to a security assessment and grid topology analysis phase. A validation state is required next. This stage is critical as it determines the relative accuracy of a smart grid impact analysis and dictates the possible analysis tools available to glean insights about vulnerabilities and strategies for system hardening.

In this chapter, we focus on results for a model synthesis approach. We have developed a general and systematic approach to begin modelling a smart grid system using a graph-based dynamical system approach. To elucidate our approach, we focus on a case study.

15.4 Case study

15.4.1 13-node distribution test system

We present the results of a case study on a 13-node distribution system that builds on the work of [33] and [34]. Our investigation is based on the IEEE 13-node test feeder system (http://ewh.ieee.org/soc/pes/dsacom/testfeeders/index.html), but has been modified in two significant ways to better characterize a smart grid. First, three distributed energy resources (DERs) have been added at nodes 634, 646, 680. Second, a switch has been added after node 650 in order to enable the overall distribution system to operate in islanding mode. In the study of this system, the switch is assumed to always be open. Therefore, the three DERs in the system are responsible for supplying as much of the system load as possible. The overall system is shown in Figure 15.2.

DER1 represents a 150-kW wind turbine that is connected to a synchronous generator. This generator is connected to node 634 through a power electronic interface. This DER is controlled in PQ mode such that the P and Q set points are provided for the generator and its controller maintains the associated active and reactive powers at these desired values. The wind speed is assumed to be 15 m/s and the P and Q set points are 150 kW and 120 kVAr, respectively.

DER2 is a 2000-kW small synchronous generator which is directly connected to the grid through node 680. DER2 is the system slack generator and is operated in isochronous

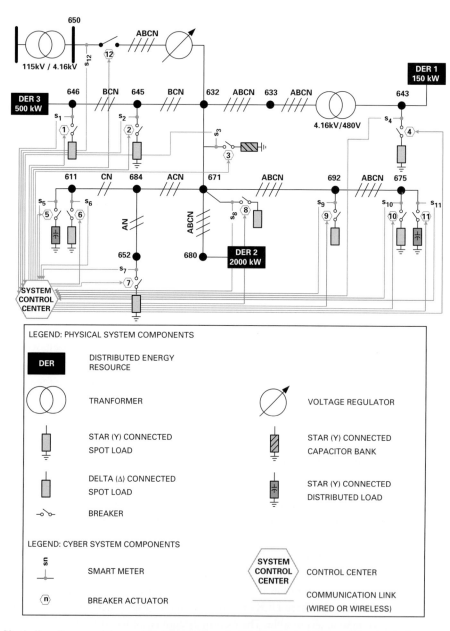

Figure 15.2 Single line diagram of the modified IEEE 13-node distribution test system. The physical electrical grid components are shown in black. The cyber components (communications links, smart meters, and actuator controls) are shown in grey.

mode such that its controller adjusts the output power to ensure that the system frequency remains around the nominal value of 60 Hz.

The third distributed generator, DER3, is a small 500-kW synchronous generator which is also directly connected to the grid and is controlled in droop mode. Thus, the

Table 15.1. System load specifications – the connection status column is 1 for connected and 0 for disconnected

Node	Priority	Connection status	Max demand (kW)	Demand at working pt (kW)	Loading (%)
671	1	1	1155	948	82.08
675	2	1	843	678	80.43
632–671	3	1	200	141	70.50
692	4	1	170	146	85.88
634	5	1	400	330	82.50
645	6	1	170	137	80.59
646	7	1	230	185	80.43
652	8	0	128	0	0
611	9	0	170	0	0
Total			3466	2565	

generator adjusts the power set point value (set to 470 kW in this example) to in part compensate for the generation–load mismatch when the frequency deviates from its nominal value of 60 Hz. The reader should note that the total generation–load mismatch will (in steady state) ultimately be compensated for by the slack generator DER2 and DER3 only participates in load sharing during the transients.

The test system also includes eight spot loads and one distributed load that are detailed in Table 15.1. The table provides information on the node that the load is connected to in Figure 15.2, its priority in the face of insufficient overall system power, its connection status (where 1 = connected, 0 = disconnected) under normal conditions (i.e., before an attack), its maximum power demand, its power demand at the current working point, and its percentege loading compared to the maximum power demand. Note that the maximum power column represents the total installed load at a node. Therefore, when the system is in operation the amount of power actually consumed at the node can be less than or equal to this installed capacity.

Microgrids typically have the ability to operate in one of two modes: (1) interconnected mode, and (2) islanded mode. The attack scenarios we study in this section assume operation in islanded mode; that is, switch number 12 in Figure 15.2 is open.

The total generation capacity of the three system DERs is 2650 kW (including loss). Since the maximum power demand of all loads is 3466 kW which exceeds the total generation capacity of DER1, DER2, and DER3, load-serving logic at the control centre must prioritize which loads can be served. Here, the control centre starts with the highest-priority unsupplied load and connects it to the system if enough generation margin is available for the load. Then it moves to the next load and applies the same rule to decide whether or not to connect the load. Note that according to this serving logic, if a higher-priority load cannot be supplied due to insufficient generation capacity but a lower-priority load can be supplied (because it is smaller), the control centre will serve the lower-priority load.

Table 15.2. Available generation margin

Total load (excluding loss)	2565 kW
System loss	26 kW
DER1 generation	150 kW
DER2 generation	1973 kW
DER3 generation	468 kW
Total load (including loss)	2591 kW
Available generation margin	59 kW

Each load is equipped with a smart meter which measures several quantities at the load point (including voltage, frequency, active and reactive power values) and sends them to the control centre with a predetermined transmission rate (for example every 15 minutes) through a communication network. Power demands for the loads at the current working point of the system (prior to any cyber attacks) are shown in Table 15.1. It can be seen that at this working point, the control centre supplies the loads with priorities 1 to 7 and the loads at nodes 652 and 611 are unsupplied. The power demands of the supplied loads add up to 2565 kW, with approximately 26 kW of system loss. Therefore, the total system generation is 2591 kW, which is distributed among the three generators as shown in Table 15.2. Therefore, at this working point the total available generation margin is 2650 kW − 2591 kW = 59 kW, which is insufficient to supply any of the unsupplied loads.

15.4.2 Model synthesis

We model each DER, switch, load, capacity bank, and cable as nodes in the electrical graph. Each meter, breaker actuator, and control centre is modelled as a distinct node of the cyber grid. The overall electrical and cyber graphs are shown in Figure 15.3. A 'composite' cable node is used to represent the five physical cables connected to nodes 632–671 of the test system. This has the effect of simplifying the graph while leading to the need for higher-dimensional dynamical system equations at this node.

The 13-node distribution test system is modelled in PSCAD, which is a transient power system simulation software. Each synchronous generator is modelled with its governor and exciter as discussed in [33]. DER2 and DER3 are tripped out if their frequencies vary $\pm 2\%$ from the nominal frequency of 60 Hz. A non-ideal Y–Y transformer model with leakage and magnetizing reactance is used for transformers. For power cables, a PI line model with non-symmetric self and mutual impedances is used. An appropriate model is used for each load according to its type (i.e., constant power, constant current, and constant impedance). For the distributed load between nodes 632 and 671, the 'exact distributed load model' is used in which two-thirds of the total distributed load is placed a quarter distance from the sending end and one-third of the load is placed at the end of the line [42]. Breakers are modelled with ideal models. Each breaker can be turned on or off by a signal from the control centre. In this example, smart meters measure the active and reactive powers of the node that they are connected to, and send them to the

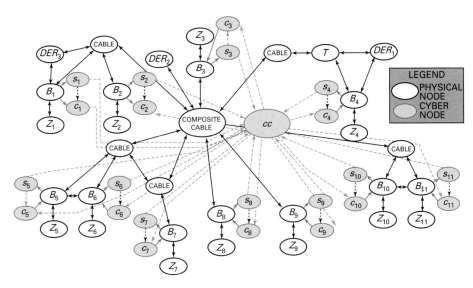

Figure 15.3 Electrical and cyber graphs for test system in islanding mode. Nodes are comprised of DERs, circuit breakers B_i, cables, a transformer T, and loads/capacitor banks Z_i of the electrical network and a control centre cc, smart meters s_i, and actuator controls c_i of the cyber network. The 'composite cable' graph node represents the five physical cables connected to nodes 632 and 671 of the IEEE 13-node distribution test system. Edges represent state dependencies for dynamical modelling. The cyber graph is distinguished with shaded nodes and dashed edges. Attack $\mathcal{A}$ targets the meter s_{10}.

control centre. Finally, the focus of this work is on the load-management functionality of the control centre and other control centre functionalities are neglected. Moreover, no bad data detection and identification method is assumed for the control centre.

We consider three attack scenarios outlined in the following sections.

15.4.3 Attack scenario 1

In this first attack scenario, we demonstrate how a coordinated cyber attack on the smart meters at specific nodes in the grid can cause incorrect decision-making at the control centre. In this scenario we demonstrate how the attack can create a series of cascading mishaps that result in a system overload situation and subsequent instability and blackout.

Here, the attacker alters the load measurements of the smart meters at nodes 671, 675, 634, and 645 with a bias of -100 kW, -70 kW, -50 kW, and -30 kW, respectively, at $t = 1$ s; this can be accomplished by, for example, corrupting the meter directly or attacking associated communication links to modify the information. Table 15.3 provides the system details in the presence of the attack before the control centre reacts to the biased information. Thus, the connection status column is the same as in Table 15.1. The right-most column provides the biased demand. Since the biases are all negative, the overall total power demand incorrectly *appears* to be lower than it is. The *perceived* available generation margin increases from 59 kW to 309 kW, as presented in Table 15.4.

Table 15.3. System load specifications in the presence of attack scenario 1 *before* control centre reactive decision-making

Node	Priority	Connection status	Power demand (kW)	Bias (kW)	Bias (%)	Biased demand (kW)
671	1	1	948	−100	−10.55	848
675	2	1	678	−70	−10.32	608
632–671	3	1	141	0	0	141
692	4	1	146	0	0	146
634	5	1	330	−50	−15.15	280
645	6	1	137	−30	−21.90	107
646	7	1	185	0	0	185
652	8	0	0	0	0	0
611	9	0	0	0	0	0
Total			2565	−250		2315

Table 15.4. Available generation margin for attack scenario 1 prior to control centre reactionary decision-making

Total load (excluding loss)	2315 kW
System loss	26 kW
DER1 generation	150 kW
DER2 generation	1973 kW
DER3 generation	468 kW
Total load (including loss)	2341 kW
Available generation margin	309 kW

Therefore, after a 0.5 s processing delay, at $t = 1.5$ s, the control centre decides to supply the two disconnected loads at nodes 652 and 611. In reality, this overloads the generators by 240 kW. As a consequence, the frequencies of DER2 and DER3 start to drop. Each generator is equipped with a protection system that disconnects the generator from the system in case the generator's frequency goes below 58.8 Hz or above 61.2 Hz. At approximately $t = 2.2$ s, the frequencies of DER2 and DER3 reach below 58.8 Hz and the protection systems disconnect them from the grid, resulting in a system blackout.

Figure 15.4 provides simulations of our synthesized model, demonstrating the predictable behaviour. Figure 15.4(a,b) shows the load measurements for the tree unbiased loads and biased loads, respectively. From Figure 15.4(b) it is clear that a bias is introduced at $t = 1$ s. From Figure 15.4(c) it is clear that the total system load is perceived to be below the total generation capacity due to the attack bias, however, the true load is above this value after $t = 1.5$ s. The total generation power is shown in Figure 15.4(d). The DER frequencies are shown in Figure 15.4(e), where it is clear that the overload situation drives the frequency down until DER2 and DER3 are tripped out. The biased

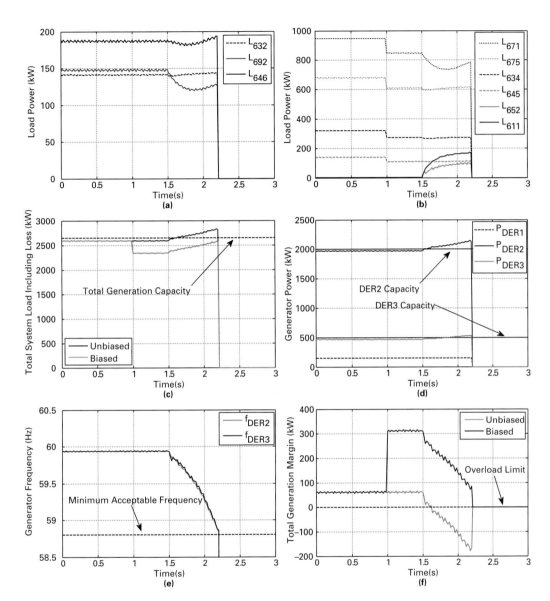

Figure 15.4 Simulation results for attack scenario 1. (a) Load measurements for unbiased loads, (b) load measurements for biased loads and newly prioritized loads, (c) total system load including loss, (d) generator power, (e) DER frequencies, (f) total system generation margin.

and unbiased generation margin is also shown in Figure 15.4(f). It is clear that the attack overestimates the generation margin, creating instability and subsequent blackout.

15.4.4 Attack scenario 2

In the second scenario, the load power measurements are once again biased but for nodes 671, 675, and 692 with values of +100 kW, +30 kW, and +20 kW, respectively.

Table 15.5. System load specifications in the presence of attack scenario 2 *before* control centre reactive decision-making

Node	Priority	Connection status	Power demand (kW)	Bias (kW)	Bias (%)	Biased demand (kW)
671	1	1	948	+100	+10.55	1048
675	2	1	678	+30	+4.24	708
632–671	3	1	141	0	0	141
692	4	1	146	+20	+13.70	166
634	5	1	330	0	0	330
645	6	1	137	0	0	137
646	7	1	185	0	0	185
652	8	0	0	0	0	0
611	9	0	0	0	0	0
Total			2565	+150		2715

Table 15.6. Available generation margin for attack scenario 2 prior to control centre reactionary decision-making

Total load (excluding loss)	2715 kW
System loss	26 kW
DER1 generation	150 kW
DER2 generation	1973 kW
DER3 generation	468 kW
Total load (including loss)	2741 kW
Available generation margin	−91 kW

Tables 15.5 and 15.6 provide the relevant details of the system load specification and generation margin for this scenario.

As a result of the misperceived −91 kW generation margin, after a half-second processing delay, at $t = 1.5$ s, the control centre decides to avoid an overload situation and according to load priorities, disconnects the lowest-priority load at node 646. This means that the attacker has achieved a denial-of-power-service attack on this load, even though it is possible to supply it in reality. Note that when the load at node 646 is disconnected, the (biased) generation margin is about 94 kW, which is not enough to supply any of the unsupplied loads.

The consequences of the load shedding are confirmed from the simulations of our model in Figure 15.5. The shedding of the load at node 646 creates unused generation capacity for DER2 and DER3. Therefore, their steady-state frequencies increase overall due to the attack, as shown in Figure 15.5(e). The overall generation margin from our model confirms the energy waste as presented in Figure 15.5(f).

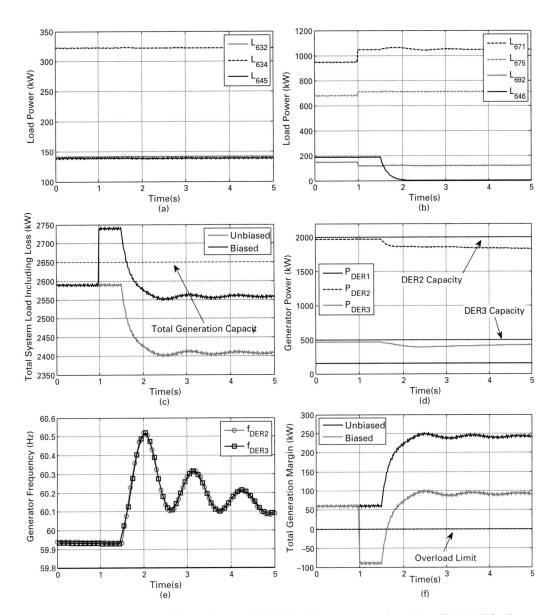

Figure 15.5 Simulation results for attack scenario 2. (a) Load measurements for unbiased loads, (b) load measurements for biased loads and newly prioritized loads, (c) total system load including loss, (d) generator power, (e) DER frequencies, (f) total system generation margin.

15.4.5 Attack scenario 3

In this third scenario that we present, the attacker alters the load measurements at nodes 675 and 692 with a bias of +80 kW and +20 kW, respectively. The details are shown in Table 15.7 and the generation margin changes are highlighted in Table 15.8.

Table 15.7. System load specifications in the presence of attack scenario 3 *before* control centre reactive decision-making.

Node	Priority	Connection status	Power demand (kW)	Bias (kW)	Bias (%)	Biased demand (kW)
671	1	1	948	0	0	948
675	2	1	678	+80	+11.80	758
632–671	3	1	141	0	0	141
692	4	1	146	+20	+13.70	166
634	5	1	330	0	0	330
645	6	1	137	0	0	137
646	7	1	185	0	0	185
652	8	0	0	0	0	0
611	9	0	0	0	0	0
Total			2565	+100		2665

Table 15.8. Available generation margin for attack scenario 3 prior to control centre reactionary decision-making

Total load (excluding loss)	2665 kW
System loss	26 kW
DER1 generation	150 kW
DER2 generation	1973 kW
DER3 generation	468 kW
Total load (including loss)	2691 kW
Available generation margin	−41 kW

Since the system generation margin decreases from 59 kW to −41 kW, after a processing delay of 0.5 s, the control centre will decide to disconnect the lowest priority load being served which is at node 646. By shedding this load, the generation margin will increase from −41 kW to 144 kW, which is enough to supply the 128 kW load connected to node 652. Therefore, at $t = 1.5$ s, the control centre disconnects the load at node 646 and connects the load at node 652, simultaneously.

Our model results are shown in Figure 15.6, which demonstrates how the attack can create an effective reprioritization of loads.

15.5 Conclusion

The overall objective of this chapter is to demonstrate how a graph-based dynamical systems concept works to describe smart grid system component interactions in order to predict cyber-attack impacts on the overall grid. We study the model synthesis component of our paradigm on a 13-node distribution test system. To remain close to the physics of the system components, the dynamical system models used for the nodes are highly variable for this study. Although this provides greater accuracy, which was our initial

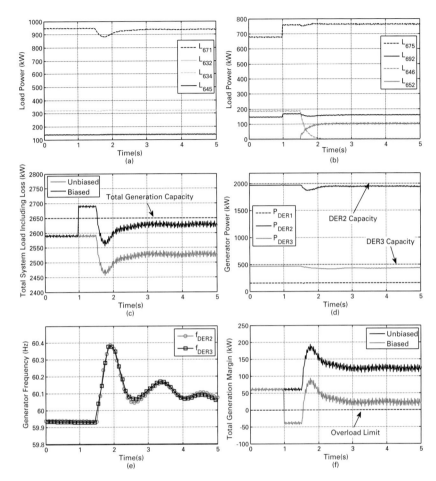

Figure 15.6 Simulation results for attack scenario 3. (a) Load measurements for unbiased loads, (b) load measurements for biased loads and newly prioritized loads, (c) total system load including loss, (d) generator power, (e) DER frequencies, (f) total system generation margin.

goal, it makes the subsequent task of analysis challenging. Analysis of such models will most likely lead to limited results in relating local system components to more global characteristics.

Future work will take two directions. First, we will focus on modelling the system components more homogeneously, perhaps by judiciously selecting what each node represents or by employing approximate techniques. If successful, this will possibly enable the application of consensus theory to our impact analysis framework. For example, the over/under-frequency problem of generators when a cyber attack occurs can be framed as a consensus problem, the goal of which is for all the generators to achieve consensus in frequency. Second, we will consider the use of switched-systems theory in order to model smart grid systems where attacks are applied to information systems that actuate breakers.

References

[1] E. Sandström and J. Weiss, 'Cyber security', in *Proceedings of 2005 CIGRE/IEEE Power Engineering Society International Symposium*, October 2005, pp. 282–289.

[2] D. Watts, 'Security and vulnerability in electric power systems', in *Proceedings of 35th North American Power Symposium*, Rolla, MO, October 2003, pp. 559–566.

[3] Substations Committee, 'IEEE standard for substation intelligent electronic devices (IEDs)', Standard IEEE Std 1686-2007, IEEE Power Engineering Society, 5 December 2007.

[4] L. Pietre-Cambacedes, C. Chalhoub, and F. Cleveland, 'IEC TC57 WG15 – Cyber security standards for the power systems,' Technical Report D2-02-C02, CIGRÉ Study Committee D2: Information Systems and Telecommunications, 2007.

[5] H. Endoh, 'Analyzing aspects of cyber security standards for M&CS', in *Proceedings of SICE Annual Conference*, August 2008, pp. 1478–1481.

[6] H. Falk, 'Securing IEC 61850', in *Proceedings of IEEE Power and Energy Society General Meeting – Conversion and Delivery of Electrical Energy in the 21st Century*, July 2008, pp. 1–3.

[7] R. McDonald, 'New considerations for security compliance, reliability and business continuity', in *Proceedings of IEEE Rural Electric Power Conference*, April 2008, pp. B1–B1–7.

[8] L. Piètre-Cambacédès, T. Kropp, J. Weiss, and R. Pellizzonni, 'Cybersecurity standards for the electric power industry – a "survival kit"', in *Proceedings of CIGRÉ Paris Session*, 2008, Paper D2–213.

[9] G. N. Ericsoon, 'Information security for electric power utilities (EPUs) – CIGRÉ developments on frameworks, risk assessment, and technology', *IEEE Transactions on Power Delivery*, vol. 24, no. 3, pp. 1174–1181, July 2009.

[10] J. Dagle, 'Vulnerability assessment activities', in *Proceedings of Power Engineering Society Winter Meeting*, 2001, vol. 1, pp. 108–113.

[11] J. Depoy, J. Phelan, P. Sholander, B. Smith, G. B. Varnado, and G. Wyss, 'Risk assessment for physical and cyber attacks on critical infrastructures', in *Proceedings of IEEE Military Communications Conference*, October 2005, vol. 3, pp. 1961–1969.

[12] Y. Jiaxi, M. Anjia, and G. Zhizhong, 'Vulnerability assessment of cyber security in power industry', in *Proceedings of IEEE Power Systems Conference and Exposition*, October–November 2006, pp. 2200–2205.

[13] S. M. Amin, 'Energy infrastructure defense systems', *Proceedings of the IEEE*, vol. 93, no. 5, pp. 861–875, May 2005.

[14] J. Stamp, A. McIntyre, and B. Ricardson, 'Reliability impacts from cyber attack on electric power systems', in *Proceedings of IEEE Power Systems Conference and Exposition*, March 2009, pp. 1–8.

[15] D. Conte de Leon, J. Alves-Foss, A. Krings, and P. Oman, 'Modeling complex control systems to identify remotely accessible devices vulnerable to cyber attack', in *Proceedings of First Workshop on Scientific Aspects of Cyber Terrorism*, Washington, DC, November 2002.

[16] D. D. Dudenhoeffer, M. R. Permann, S. Woolsey, R. Timpany, C. Miller, A. McDermott, and M. Manic, 'Interdependency modeling and emergency response', in *Proceedings of 2007 Summer Computer Simulation Conference*, July 2007, pp. 1230–1237.

[17] B. Rozel, M. Viziteu, R. Caire, N. Hadjsaid, and J.-P. Rognon, 'Towards a common model for studying critical infrastructure interdependencies', in *Proceedings of IEEE Power and Energy Society General Meeting – Conversion and Delivery of Electrical Energy in the 21st Century*, Pittsburgh, PA, July 2008, pp. 1–6.

[18] N. HadjSaid, C. Tranchita, B. Rozel, M. Viziteu, and R. Caire, 'Modeling cyber and physical interdependencies – application in ICT and power grids', in *Proceedings of IEEE Power Systems Conference and Exposition*, March 2009, pp. 1–6.

[19] H. Tan, 'Security analysis of a cyber-physical system', M.S. thesis, University of Missouri-Rolla, 2007.

[20] H. Tang and B. McMillin, 'Security property violation in CPS through timing', in *Proceedings of 28th International Conference on Distributed Computing Systems Workshops*, 2008, pp. 519–524.

[21] B. McMillin, 'Complexities of information security in cyber-physical power systems', in *Proceedings of IEEE Power Systems Conference and Exposition*, March 2009, pp. 1–2.

[22] S. Sheng, W. L. Chan, K. K. Li, D. Xianzhong, and Z. Xiangjun, 'Context information-based cyber security defense of protection system', *IEEE Transactions on Power Delivery*, vol. 22, no. 3, pp. 1477–1481, July 2007.

[23] D. Edwards, S. K. Srivastava, D. A. Cartes, S. Simmons, and N. Wilde, 'Implementation and validation of a multi-level security model architecture', in *Proceedings of International Conference on Intelligent Systems Applications to Power Systems*, November 2007, pp. 1–4.

[24] T. Mander, F. Nabhani, L. Wang, and R. Cheung, 'Integrated network security protocol layer for open-access power distribution systems', in *Proceedings of IEEE Power Engineering Society General Meeting*, June 2007, pp. 1–8.

[25] K. Xiao, N. Chen, S. Ren, L. Shen, X. Sun, K. Kwiat, and M. Macalik, 'A workflow-based non-intrusive approach for enhancing the survivability of critical infrastructures in cyber environment', in *Proceedings of Third International Workshop on Software Engineering for Secure Systems*, May 2007.

[26] C. M. Davis, J. E. Tate, H. Okhravi, C. Grier, T. J. Overbye, and D. Nicol, 'SCADA cyber security testbed development', in *Proceedings of 38th North American Power Symposium*, September 2006, pp. 483–488.

[27] A. Giani, G. Karsai, T. Roosta, A. Shah, B. Sinopoli, and J. Wiley, 'A testbed for secure and robust SCADA systems', *SIGBED Review*, vol. 5, no. 2, article no. 4, July 2008.

[28] G. Dondossola, F. Garrone, and J. Szanto, 'Supporting cyber risk assessment of power control systems with experimental data', in *Proceedings of IEEE Power Systems Conference and Exposition*, March 2009, pp. 1–3.

[29] A. A. Cárdenas, S. Amin, and S. Sastry, 'Research challenges for the security of control systems', in *Proceedings of 3rd USENIX Conference on Hot Topics in Security*, July 2008, Article 6.

[30] A. A. Cárdenas, S. Amin, and S. Sastry, 'Secure control: towards survivable cyber-physical systems', in *Proceedings of 28th International Conference on Distributed Computing Systems Workshops*, June 2008, pp. 495–500.

[31] A. A. Cárdenas, S. Amin, and S. Sastry, 'Secure control: towards survivable cyber-physical systems', in *Proceedings of First International Workshop on Cyber-Physical Systems*, June 2008.

[32] A. A. Cárdenas, T. Roosta, G. Taban, and S. Sastry, 'Cyber security basic defenses and attack trends', in *Homeland Security Technology Challenges*, G. Franceschetti and M. Grossi (eds), chapter 4, pp. 73–101. Artech House, 2008.

[33] D. Kundur, X. Feng, S. Liu, T. Zourntos, and K. L. Butler-Purry, 'Towards a framework for cyber attack impact analysis of the electric smart grid', in *Proceedings of IEEE International Conference on Smart Grid Communications (SmartGridComm)*, Gaithersburg, MD, October 2010, pp. 244–249.

[34] D. Kundur, X. Feng, S. Mashayekh, S. Liu, T. Zourntos, and K. L. Butler-Purry, 'Towards modeling the impact of cyber attacks on a smart grid', *International Journal of Security and Networks*, vol. 6, no. 1, pp. 2–13, 2011.

[35] D. D. Dudenhoeffer, M. R. Permann, and M. Manic, 'CIMS: a framework for infrastructure interdependency modeling and analysis', in *Proceedings of 38th Winter Simulation Conference*, December 2006, pp. 478–485.

[36] R. Dawson, C. Boyd, E. Dawson, and J. Manuel Gonzàlez Nieto, 'SKMA – A key management architecture for SCADA systems', in *Proceedings of Fourth Australasian Workshops on Grid Computing and E-Research*, January 2006, vol. 54, pp. 183–192.

[37] M. A. McQueen, W. F. Boyer, M. A. Flynn, and G. A. Beitel, 'Quantitative cyber risk reduction estimation methodology for small SCADA control system', in *Proceedings of 39th Annual Hawaii International Conference on Systems Sciences*, January 2006, vol. 9, pp. 226–236.

[38] W. Eberle and L. Holder, 'Insider threat detection using graph-based approaches', in *Proceedings of Cybersecurity Applications and Technology Conference for Homeland Security*, 2009, pp. 237–241.

[39] M. Ekstedt and T. Sommestad, 'Enterprise architecture models for cyber security analysis', in *Proceedings of IEEE Power Systems Conference and Exposition*, March 2009, pp. 1–6.

[40] H. Hadeli, R. Schierholz, M. Braendle, and C. Tuduce, 'Generating configuration for missing traffic detector and security measures in industrial control systems based on the system description files', in *Proceedings of IEEE Conference on Technologies for Homeland Security*, May 2009, pp. 503–510.

[41] X. Feng, T. Zourntos, and K. L. Butler-Purry, 'Dynamic load management for NG IPS ships', in *Proceedings of IEEE Power Engineering Society General Meeting*, Minneapolis, Minnesota, July 2010.

[42] W. H. Kersting, *Distribution System Modeling and Analysis*. CRC Press, 2002.

16 Jamming for manipulating the power market in smart grid

Husheng Li

16.1 Introduction

The beginning of the 2010s witnessed the rapid development of smart grid. The smart grid initiative (or similar concepts such as Intelligrid, utility of the future, and the Future-Grid) [1–5] is an attempt to modernize the current power grid with digital technology for communication, computing, and control to improve overall performance and to accommodate a high penetration of alternative energy sources and load responses. Among the many objectives related to smart grid, a key goal is to improve the electricity service to all end consumers such as residential houses, commercial buildings, and industrial loads. This calls for two-way communication in smart grid, i.e., power consumption/demand reports from the users to the centre (uplink) and price information from the centre to the users (downlink). Hence, the communication infrastructure for the power market is receiving intensive study, which also brings about a paradigm shift in the communities of communications and networking. Among many proposals for the communication infrastructure, wireless communication is a promising one due to its low cost, large coverage, and fast deployment [6–9].

Although the communication infrastructure can considerably improve the efficiency of the power market, it also brings significant vulnerabilities since malicious users can attack the communication system and thus cause various damages to the smart grid, or even result in large-area blackout. Hence, the security issue is of first priority in the study of smart grid and has attracted substantial attention in industry and academia. Various attack strategies have been proposed and studied [10–14]. For example, a false-data-injection attack can change the power sensor reports, thus perturbing the behaviour of the power market [15, 16]. The data privacy in smart grid is also studied, since the power consumption report contains much private information [10, 11]. Another possible attack is wireless jamming [17–19], i.e., the jammer uses high-power interference to suppress the information signal, if wireless technologies are used in smart grid. Such a jamming attack, the main purpose of which is to block the information flow in the enemy's communication network, has been studied widely in military communication networks. It can also be used to jam the sensor reports in smart grid, which has been studied in [20].

In this chapter, we propose a novel jamming-based attack scheme for manipulating the power market in smart grid. In contrast to other attacks like a false-data-injection attack, the jamming attack is much simpler since it only requires brute force interference and does not need to hack into the computer and communication systems in smart grid. The traditional security methodologies, like public/private key and hash function, cannot combat the jamming in the physical layer. However, one may have the following impression at first sight of the jamming attack: *The jamming attack can only* **block** *but cannot change the information; the worst case is that the information flow in the power market is completely blocked and the power market degenerates to the traditional one without communication infrastructures.* However, such a thought omits the other side of the jamming attack, which is also omitted in the traditional jamming attacks, i.e., the information **release** by stopping the jamming. As we will see, intelligently blocking and releasing the information in the power market via jamming the wireless communications can manipulate the power price, thus making profit for the jammer and causing damage to the power grid! It is similar to a dam, which can control the downstream water level by blocking and releasing the water in an intelligent manner, as illustrated in Figure 16.1.

The proposed attack is a severe threat to the power market in smart grid since it is easy to carry out a jamming attack and easy to operate the market manipulation if the designs of communication infrastructure and market mechanism are unaware of such an attack. Note that it is difficult to manipulate the traditional power market since one needs an extremely large capability of power consumption/generation to affect the market behaviour. However, when wireless communication is applied in smart grid, a jammer can affect the communication in a wide area, thus affecting the behaviour of many users in highly populated regions. Although the influence of each power user is negligible, the aggregated behaviour of many users will cause a significant impact on the power market. Hence, the market manipulation via jamming becomes feasible due to the introduction of wireless communication infrastructure in smart grid, thus yielding a pressing need to study such a novel and dangerous attack.

As will be shown, the proposed attack scheme can also cause an impulsive impact on the power grid, i.e., a sudden rise or drop of the power load. Such an impulsive impact may cause significant damage, e.g., the sudden drop of power load will accelerate the power generators, thus resulting in an increase in the frequency. It is well known that

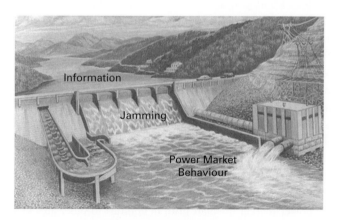

Figure 16.1 An illustration of blocking and releasing the information flow in smart grid by jamming.

the electricity frequency should remain very close to the reference value (e.g., 60 Hz in the USA), since a large frequency deviation may put many precise instruments out of operation. Hence, the proposed attack scheme has substantial impact on both the power market price and power grid stability.

To combat the jamming-based attack, various defence countermeasures can be taken in both the communication system design and the marketing policy. For the communication aspect, it is effective to carry out random frequency hopping over a wide frequency band such that the attacker cannot catch the wireless transmission. For the marketing side, we propose a random backoff scheme for the load adjustment in order to avoid the impulsive impact on the market price and power load. In this chapter, we will focus on only the countermeasure in the market mechanism.

The remainder of the chapter is organized as follows. In Section 16.2, we introduce a model of the power market, particularly the mechanism for pricing the electricity power. Then, we discuss the details of the attack scheme in Section 16.3. The countermeasure to combat the attack scheme will be given in Section 16.4. Finally, the conclusions are given in Section 16.5.

16.2 Model of power market

In power systems, a power market is used to balance the demand and supply of power via the mechanism of price [21–29]. Hence, the price is determined by the power demand, power generation cost, and constraints of transmission lines. An illustrative example of the power market is shown in Figure 16.2, in which the power consumers report their power demands to the power market via a communication infrastructure (the communication channels could be in time, frequency, or code); the power market decides the

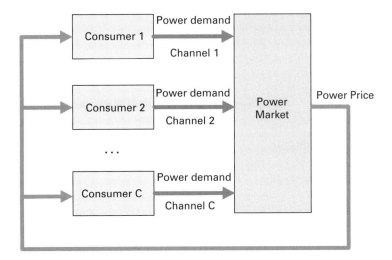

Figure 16.2 An illustration of the power market.

price and then feeds it back to the power consumers; the power consumers then adjust their demands according to the price.

Due to the power loss and the upper transmission limit of power lines, the power price could be different in different buses,[1] thus resulting in the locational marginal price (LMP) [30–38]. The LMPs can be obtained from the following optimization problem [31]:

$$\begin{aligned}
\min &\sum_{n=1}^{N} C_n \times G_n \\
\text{s.t.} &\sum_{n=1}^{N} G_n - \sum_{n=1}^{N} D_n = 0, \\
&\sum_{n=1}^{N} GSF_{n \to k} \times (G_n - D_n) \leq F_k^{\max}, \forall k = 1, ..., K, \\
&G_{\min} \leq G_n \leq G_{\max}, \quad \forall n = 1, ..., N,
\end{aligned} \quad (16.1)$$

where N is the number of buses, G_n is the power generation in bus n, C_n is the corresponding cost of power generation, D_n is the power demand at bus n, K is the number of transmission lines, $GSF_{n \to k}$ is the generation shift factor of bus n on transmission line k, $F_k^{\max}$ is the maximum transmission power of line k, and $G_{\min}$ and $G_{\max}$ are the minimum and maximum power generations, respectively. The LMP at bus n is given by $LMP_n = LMP^e + LMP_n^c$, where LMP^e can be considered as the common price of the energy, which is essentially the Lagrange factor of the first constraint in (16.1) and LMP_n^c is the price due to the congestion over the transmission lines, which is the Lagrange factor of the second constraint in (16.1). Note that the optimization problem in (16.1) is only a simplified model for determining the LMPs for the purpose of illustration. In practice, more constraints like power loss over transmission lines will be considered [23].

One of the main purposes of power price is to mitigate the power consumption peaks in certain times. Usually, power consumption peaks appear around the dinner period (lunch time or the noon of a hot summer day can also be peaks). Such power consumption peaks cause significant pressures on the power generators. Therefore, a high power price can be resorted to 'truncate' these peaks, thus making the power consumption more regular.

16.3 Attack scheme

16.3.1 Attack mechanism

Due to the fast deployment, large coverage, and large population of power users, wireless communications technology is an effective option for the communication infrastructure

---

[1] A bus means a common structure connecting multiple local electrical devices. Buses in different regions can be connected by transmission lines.

in smart grid. Although the wireless communication infrastructure can significantly enhance the agility of the power market and improve the efficiency of the power grid, it also brings vulnerabilities for the power market [29]. In particular, it provides an opportunity to manipulate the power market by simply jamming the wireless system. Essentially, wireless jamming can block or release the information flow in the power market, thus affecting the behaviour of many power consumers and causing a change in power market dynamics (predictable to the jammer). Such an attack is much simpler and more efficient (thus more dangerous) than that in traditional power markets, which requires a huge amount of money or a large power generation/consumption capability. Although the jamming attack cannot change the information in the power market, as will be seen in the example, the simple block/release of information can effectively manipulate the power price and cause many other negative side-effects.

The procedure of the proposed attack scheme is illustrated in Figure 16.3. At the beginning, the jammer jams the LMP signal in the downlink sent from the power market to a subset of power consumers, e.g., users within the range of a bus. Since the power consumers cannot receive the price information, it is reasonable for them to use the old LMP received before the jamming. The jammer keeps jamming until a significant change has occurred in the LMP, e.g., increasing from \$24/unit to \$25/unit. Then, the jammer suddenly stops jamming such that the jammed power consumers can receive the current power price. Subsequently, the power consumers will respond to the new power price. For example, they will decrease (increase) their power consumption if the new LMP is significantly higher (lower) than the old one they have used; then, the LMPs in the entire market will experience a significant price change since the total power consumption is

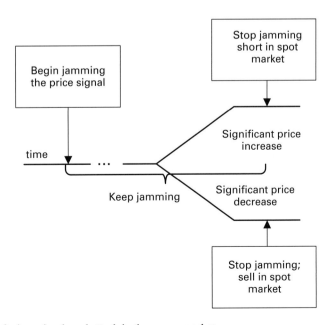

Figure 16.3 Illustration of a jamming-based attack in the power market.

substantially changed. Since the time of this significant price change is controlled by the jammer and the trend in price change can be predicted by the jammer, the jammer can make a profit from this predictable (only to the jammer) market dynamics. For example, if the price increases with a high probability after the jamming is stopped, the jammer can prepare power reservations with lower price in advance and sell them at the spot market when the change occurs.

Then, the procedure of the proposed attack is summarized below.

1. The attacker jams the price signalling in a highly populated area.
2. The jammed power users keep using the old power price since the new price is unknown. In practice, the power users can also use a predetermined default price.
3. The attacker keeps monitoring the power market, particularly the current market price, and jamming the price signalling.
4. When the true value of the price has changed significantly, compared with the price that the jammed power users are using, the attacker stops jamming.
5. The jammed power users adapt their power consumption according to the new power price. If the new power price is higher than the old one, the jammed power users will suddenly decrease their power consumption, thus making the power price drop with a large probability. If lower, the power consumption will increase, thus increasing the power price with a large probability.
6. If the power price is dropped, the attacker can make a profit by shorting in the power market. If increased, the attacker can prepare low-price power in advance and then sell it at the spot market, thus making profit.

The manipulation of the market can cause serious damage to the power market since the jammer can make a profit simply by jamming the price signal. Many other negative side-effects will also be incurred, which include:

- impulsive impact – the sudden and large-amplitude change of power consumption may cause an impulsive impact on the power grid dynamics and result in instability or even wide-area blackout;
- observation blocking – blocking the observations (i.e., the power demand reports) or the feedback control (i.e., the price) may make the market, as a dynamic system, unstable;
- market mechanism impairment – the jamming blocks the price signalling, thus seriously impairing the effect of 'peak truncation via dynamic pricing';
- communication congestion – when the power consumers have a sudden change in consumption, the communication volume may also increase (suppose that each user sends a report only when there is a significant change in the power consumption), thus causing congestion in the communication infrastructure. Note that, besides jamming the price signal in the downlink, the power consumption reports in the uplink can also be jammed, thus opening a new dimension of attack in smart grid.

16.3.2 Analysis of the damage

In this subsection, we analyse the damage caused by the proposed attack scheme, both analytically and numerically. We first introduce the PJM 5-bus model and the continuous pricing mechanism. Then, we propose a model for the power load response to the power price. The damage of the proposed attack scheme is subsequently discussed. Finally, we provide numerical simulation results.

PJM 5-bus model and pricing

To analyse the damage of the proposed attack scheme, we consider the PJM 5-bus model [39] for its simplicity. The topology of the 5-bus model is shown in Figure 16.4. The LMPs at different buses, as functions of the total load, are plotted in Figure 16.5. We observe that the LMPs are discontinuous functions of the power load. Such jumps in the power price, which result in sudden power-load changes, are undesirable in power systems; hence [31] proposed a continuous LMP scheme. In this chapter, we consider a simpler scheme which results in continuous LMP. For an arbitrary bus, we denote by $D_1, ..., D_n$ the power loads at which the LMP at this bus changes and denote by $p_1, ..., p_n$ the corresponding LMPs after the change. Given an arbitrary load D satisfying $D_i \leq D \leq D_{i+1}$, the continuous LMP is given by

$$p = \frac{D - D_i}{D_{i+1} - D_i} p_{i+1} + \frac{D_{i+1} - D}{D_{i+1} - D_i} p_i. \qquad (16.2)$$

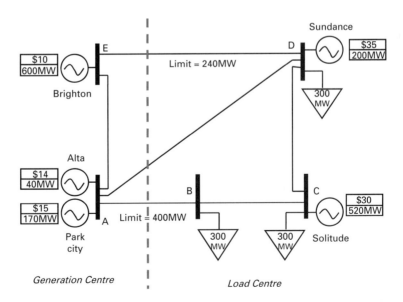

Figure 16.4 Configuration of the PJM 5-bus model.

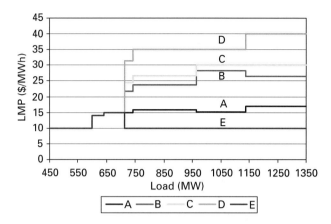

Figure 16.5 Price change with respect to the load for the PJM 5-bus model.

Power-load response

Since the automatic response to power price at electricity devices is still an open problem, there has not been a well-recognized mathematical model for the load response to power price in existing studies. We assume that the power consumption is determined by the utility function of the power consumer, which is denoted by $U(w,d)$, where w is a random parameter representing the internal power requirement and d is the power consumption. We assume that $U(w,d)$ is an increasing function of d and a decreasing function of w. Intuitively, this means that the utility function increases when more power is used and decreases when the internal power requirement is increased (e.g., the air conditioning needs to be used more intensively when the weather becomes hot).

We assume that the total reward of a power user is given by

$$R(w,d,p) = U(w,d) - dp, \qquad (16.3)$$

which is the utility function minus the payment for the power consumption. Hence, the optimal power consumption level should maximize the total reward, i.e.,

$$d^* = \arg\max_d \left[U(w,d) - dp \right]. \qquad (16.4)$$

For simplicity, we assume that all power users have the same utility function. We also assume that the optimal power consumption is a continuous, decreasing, and convex function of the power price, given a fixed power requirement index w.

The randomness of power consumption stems from the randomness of the internal power requirement characterized by the parameter w. We assume that w is a Markov chain, which has L values $w_1 \leq w_2 \leq w_3 \leq ... \leq w_L$. The state transition probability is denoted by p_{nm} for states n and m. Furthermore, we assume that the state transition occurs only between adjacent states w_i and w_{i+1}.

Damage analysis

We suppose that the power consumers in bus n are jammed by the attacker. We denote by p_o and p_n the old and new prices before and after ceasing the jamming (suppose that the time slots are t and $t+1$, respectively). We denote by π_i the proportion of users in the jammed bus that are in state i. When the number of users within a bus, i.e., M, is sufficiently large, the proportion π_i can be approximated by the stationary probability of the Markov chain.

Then, the change of the total power consumption after obtaining the power price is given by

$$\Delta D = \sum_{n=1}^{M} \left[d_{l_n(t+1)}(p_n) - d_{l_n(t)}(p_o) \right], \qquad (16.5)$$

where $l_n(t)$ is the power requirement index of user n at time t, d_i is the function of the optimal power consumption when the current power requirement index is i. Obviously, the expectation of the power consumption change is given by

$$E[\Delta D] = M \sum_{i=1}^{L} \pi_i \left[\sum_j P_{ij} d_j(p_n) - d_i(p_o) \right]. \qquad (16.6)$$

When M is sufficiently large, the power consumption change can be approximated by a Gaussian distribution, whose expectation is given in (16.6) and the variance can be computed similarly. We assume that $\Delta D > 0$. The analysis also applies in the case $\Delta D < 0$. Suppose that the attacker makes a satisfying profit from the market when $\Delta D > \Delta D_{\min}$. Then, according to the Gaussian approximation, the probability that the attacker succeeds in the attack is given by

$$P_s \approx 1 - Q\left(\frac{\Delta D_{\min} - E[\Delta D]}{\sqrt{Var[\Delta D]}} \right). \qquad (16.7)$$

Numerical results

In this section, we use the PJM 5-bus system [39] for the numerical simulation, as illustrated in Figure 16.4. We assume that there are two states for the power consumption, denoted by H and L. We set the transition probabilities $Q_{HH} = Q_{LL} = 0.8$ and assume $M = 200$, i.e., each bus supports 200 power users. The original LMP values (before being made continuous) with respect to the total power consumption are given in Figure 16.5. The corresponding parameters are given by $w_1 = 20$ and $w_2 = 5$. The utility function is assumed to be $U(w, d) = w \log d$. Therefore, the optimal power consumption is given by $d^* = \frac{w}{p}$.

We first assume that the attacker begins the jamming when the price at bus 2 is less than 24. The jamming is stopped when the price is more than 25. The cumulative distribution

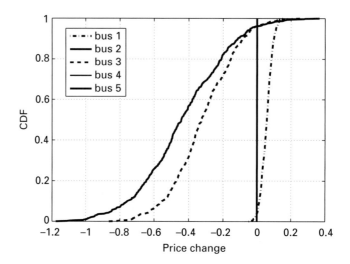

Figure 16.6 The CDF of price change after the attacker stops jamming.

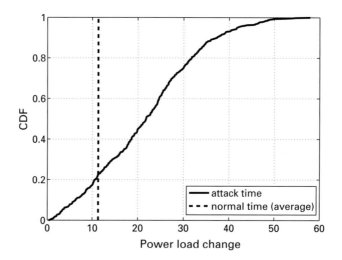

Figure 16.7 The CDF of load change after the attacker stops jamming.

function (CDF) of the price change after ceasing the jamming is shown in Figure 16.6. We observe that the prices for buses 2 and 3 will decrease with a very large probability. The price at bus 1 will increase with a large probability. Therefore, the attacker can use the jamming to manipulate the market. The load change after ceasing the jamming is also shown in Figure 16.7, as well as the average load change in normal time slots. We observe that the attack can also incur a significant change in the power load, thus causing much burden on the power generation.

16.4   Defence countermeasures

In this section, we discuss the defence countermeasure for the proposed attack strategy. We will focus on the random backoff in the power load response to price change. The random frequency hopping to combat the jamming has been discussed in many existing studies; hence we omit the corresponding discussion.

Due to the potential severe damage to the power market and power grid of the jamming attack, the price–response mechanism must be able to prevent such an attack. The essence of the countermeasure is to avoid changing the power consumption simultaneously. Meanwhile, the power price response should also be reasonably agile to the price change.

In this chapter, we propose a heuristic approach. The idea is borrowed from random access in communication systems, such as Aloha and carrier-sense multiple access (CSMA) protocols, in which different transmitters take random backoffs to avoid the collision incurred by simultaneous transmission. Motivated by the random access schemes, we proposed a random backoff scheme, in which each power consumer chooses a random time to change its power response. The random time T is a random variable depending on the difference between the new price and the old price, denoted by δp, as well as the time that the price signal has not been received, τ. We assume that the random time is exponentially distributed, i.e.,

$$p(T=t) = \left(e^\lambda - 1\right) e^{-\lambda t}, \qquad t = 1, 2, 3, ..., \tag{16.8}$$

where λ is a function of δp and τ. Obviously, λ should be an increasing function of δp and τ.

Then, if the attacker stops jamming at time 0, then the power load change at the jammed bus at time slot t, compared with the power load at time 0, is given by

$$\delta D(t) = \sum_{i=n}^{N} I(T_n < t) \left[\mathrm{d}_{l_n(t)}(p_n) - \mathrm{d}_{l_n(0)}(p_o) \right], \tag{16.9}$$

with expectation

$$E[\delta D(t)] = P(T \le t) M \sum_{i=1}^{L} \pi_i \left[\sum_j P_{ij}^t \mathrm{d}_j(p_n) - \mathrm{d}_i(p_o) \right], \tag{16.10}$$

where P_{ij}^t is the transition probability between power consumption levels i and j after t time slots.

Then, the criterion of choosing the function for λ is to spread the power consumption change into a much longer time interval and let the inherent randomness of the market counteract the change of the jammed users.

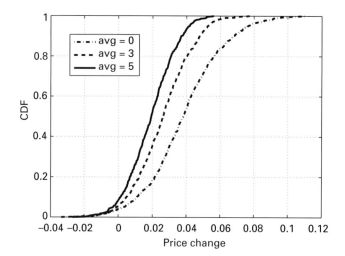

Figure 16.8 The CDF of price change when the countermeasure is taken.

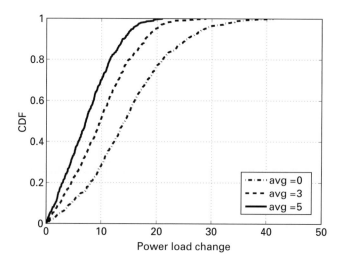

Figure 16.9 The CDF of load change when the countermeasure is taken.

The proposed method based on random backoff is tested in Figures 16.8 and 16.9, respectively, using the same configuration as that in Figures 16.6 and 16.7. We test the cases of the expected backoff time (denoted by 'avg') being 0, 3, and 5 time slots. We observe that, the longer the backoff is, the more damage is mitigated since the CDF curves are shifted to the left.

16.5 Conclusion

In this chapter, we have discussed the possibility of manipulating the power market by jamming the pricing signal. The basic idea of the attack scheme is to block and release the price information in the power market in order to result in sudden changes of power consumption and power price which are predictable by the attacker. Different from other attack schemes like the false-data-injection attack, the proposed attack strategy does not need advanced techniques to intrude on the computer system of a power grid and requires only jamming in the physical layer. The attacker can make a profit from the power market since it can manipulate the market with a large probability. Meanwhile, the attack can also incur an impulsive impact on the power grid stability. We have analysed the potential impact on the power load change and proposed a countermeasure to combat such an attack, motivated by the random access scheme in communication systems. Numerical results have shown that the attack strategy is valid to manipulate the power market while the countermeasure can effectively mitigate such an attack. Our future study will be focused on the feasibility of the attack; e.g., the risk and cost of the attacker, as well as the detailed jamming/anti-jamming approaches.

References

[1] US DOE Report, '20% Wind energy by 2030: increasing wind energy's contribution to US electricity supply', July 2008 [online]. Available at: http://www1.eere.energy.gov/windand-hydro/pdfs/41869.pdf, accessed in January 2010.

[2] R. Thresher, M. Robinson, and P. Veers, 'To capture the wind', *IEEE Power and Energy Magazine*, vol. 5, no. 6, pp. 34–46, November–December 2007.

[3] J. Blatchford, Participating Intermittent Resource Program (PIRP) 101, California ISO, January 2009 [online]. Available at: http://www.caiso.com/2343/2343d5d01ee50.pdf, accessed in January 2010.

[4] B. Parsons, M. Milligan, B. Zavadil, D. Brooks, B. Kirby, K. Dragoon, and J. Caldwell, 'Grid impacts of wind power: a summary of recent studies in the United States', NREL Report, June 2003 [online]. Available at: http://www.nrel.gov/docs/fy03osti/34318.pdf, accessed in January 2010.

[5] DOE smart grid website, http://www.oe.energy.gov/smartgrid.htm, accessed in January 2010.

[6] M. Anderson, 'WiMAX for smart grids', *IEEE Spectrum*, July 2010.

[7] R. H. Khan, T. F. Aditi, V. Sreeram, and H. H. C. Iu, 'A prepaid smart metering scheme based on WiMAX prepaid accounting model', *Smart Grid and Renewable Energy*, vol. 1, pp. 63–69, 2010.

[8] B. Heile, 'Smart grid for green communications', *IEEE Wireless Communication Magazine*, vol. 17, pp. 4–6, June 2010.

[9] K.-C. Chen, P.-C. Yeh, H.-Y. Hsieh, and S.-C. Chang, 'Communication infrastructure of smart grid', in *Proceedings of 4th International Symposium on Communications, Control, and Signal Processing 2010*, Limassol, Cyprus, 3–5 March 2010.

[10] H. K.-H. So, S. H. M. Kwok, E. Y Lam, and K.-S. Lui, 'Zero-configuration identity-based signcryption scheme for smart grid', in *Proceedings of 1st IEEE SmartGridComm*, October 2010.

[11] F. Li, B. Luo, and P. Liu, 'Secure information aggregation for smart grids using homomorphic encryption', in *Proceedings of 1st IEEE SmartGridComm*, October 2010.

[12] A. Bartoli, J. Hernández-Serrano, M. Soriano, M. Dohler, A. Kountouris, and D. Barthel, 'Secure lossless aggregation for smart grid M2M networks', in *Proceedings of 1st IEEE SmartGridComm*, October 2010.

[13] D. P. Varodayan and G. X. Gao, 'Redundant metering for integrity with information-theoretic confidentiality', in *Proceedings of 1st IEEE SmartGridComm*, October 2010.

[14] R. Berthier, W. H. Sanders, and H. Khurana, 'Intrusion detection for advanced metering infrastructures: requirements and architectural directions', in *Proceedings of 1st IEEE SmartGridComm*, October 2010.

[15] L. Xie, Y. Mo, and B. Sinopoli, 'False data injection attacks in electricity markets', in *Proceedings of 1st IEEE Conference on Smart Grid Communications (SmartGridComm)*, 2010.

[16] O. Kosut, L. Jia, R. Thomas, and L. Tong, 'Malicious data attacks on smart grid state estimation: attack strategies and countermeasures', in *Proceedings of 1st IEEE Conference on Smart Grid Communications (SmartGridComm)*, 2010.

[17] H. Hu and N. Wei, 'A study of GPS jamming and anti-jamming', in *Proceedings of 2nd International Conference on Power Electronics and Intelligent Transportation System (PEITS)*, vol. 1, pp. 388–391, 2009.

[18] C. Popper, M. Strasser, and S. Capkun, 'Anti-jamming broadcast communication using uncoordinated spread spectrum techniques', *IEEE Journal on Selected Areas in Communications*, vol. 28, no. 5, pp. 703–715, 2010.

[19] A. Mpitziopoulos, D. Gavalas, C. Konstantopoulos, and G. Pantziou, 'A survey on jamming attacks and countermeasures in WSNs', *IEEE Communication Surveys & Tutorials*, vol. 11, no. 4, pp. 42–56, 2009.

[20] H. Li, L. Lai, and R. C. Qiu, 'A denial-of-service jamming game for remote state monitoring in smart grid', in *Proceedings of Conference on Information Sciences and Systems (CISS)*, 2011.

[21] F. L. Alvarado, 'The dynamics of power system markets', Department of Electronics Computer Engineering, University of Wisconsin, Madison, WI, Technical Report PSERC-91-01, March 1997.

[22] J. Nutaro and V. Protopopescu, 'The impact of market clearing time and price signal delay on the stability of electric power markets', *IEEE Transactions on Power Systems*, vol. 24, pp. 1337–1345, 2009.

[23] S. Stoft, *Power System Economics – Designing Markets for Electricity*. IEEE/Wiley, 2002.

[24] R. Weron, *Modeling and Forecasting Electricity Loads and Prices*. John Wiley & Sons Ltd, 2006.

[25] K. M. Maribu, R. M. Firestone, C. Marnay, and A. S. Siddiqui, 'Distributed energy resources market diffusion model', *Energy Policy*, vol. 35, pp. 4471–4484, April 2007.

[26] J. Matevosyan and L. Soder, 'Minimization of imbalance cost trading wind power on the short-term power market', *IEEE Transactions on Power Systems*, vol. 21, pp. 1396–1404, August 2006.

[27] W. Reinisch and T. Tezuka, 'Market power and trading strategies on the electricity market: a market design view', *IEEE Transactions on Power Systems*, vol. 21, pp. 1180–1190, August 2006.

[28] A. Rabiee, H. A. Shayanfar, and N. Amjady, 'Reactive power pricing', *IEEE Power & Energy Magazine*, pp. 18–32, January/February 2009.

[29] K. Xie, Y. H. Song, J. Stonham, E. Yu, and G. Liu, 'Decomposition model and interior point methods for optimal spot pricing of electricity in deregulation environments', *IEEE Transactions on Power Systems*, vol. 15, pp. 39–50, February 2000.

[30] R. Bo and F. Li, 'Probabilistic LMP forecasting considering load uncertainty', *IEEE Transactions on Power Systems*, vol. 24, pp. 1279–1289, August 2009.

[31] F. Li, 'Continuous locational marginal pricing (CLMP)', *IEEE Transactions on Power Systems*, vol. 22, pp. 1638–1646, November 2007.

[32] F. Li and R. Bo, 'Congestion and price prediction under load variation', *IEEE Control System Magazine*, vol. 19, pp. 59–70, October 2009.

[33] E. Litvinov, T. Zheng, G. Rosenwald, and P. Shamsollahi, 'Marginal loss modeling in LMP calculations', *IEEE Transactions on Power Systems*, vol. 19, pp. 880–888, May 2004.

[34] J. E. Price, 'Market-based price differentials in zonal and LMP market designs', *IEEE Transactions on Power Systems*, vol. 23, pp. 1486–1494, November 2007.

[35] F. Li, 'Smoothing out step changes of LMP', *The Electricity Journal*, vol. 21, no. 7, pp. 43–49, August–September 2008.

[36] F. Li and R. Bo, 'DCOPF-based LMP simulation: algorithm, comparison with ACOPF, and sensitivity', *IEEE Transactions on Power Systems*, vol. 22, no. 4, pp. 1475–1485, November 2007.

[37] F. Li, Y. Wei, and S. Adhikari, 'Improving an unjustified common practice in ex post LMP calculation: an expanded version', in *IEEE PES General Meeting 2010*, Minneapolis, MN, 25–29, July 2010.

[38] F. Li and R. Bo, 'Small test systems for power system economic studies', in *IEEE PES General Meeting 2010*, Minneapolis, MN, 25–29, July 2010.

[39] PJM, *PJM Training Materials (LMP101)* [online]. Available at: http://www.pjm.com/services/training/train-materials.html

17 Power-system state-estimation security: attacks and protection schemes

György Dán, Kin Cheong Sou, and Henrik Sandberg

17.1 Introduction

Supervisory control and data acquisition (SCADA) systems are widely used to monitor and control large-scale transmission power grids. Monitoring traditionally involves the measurement of voltage magnitudes and power flows; these data are collected by meters located in substations. In order to deliver the measured data from the substations to the control centre, the measurement data measured by meters in the same substation are multiplexed by a remote terminal unit (RTU) [1, 2]. Because electric power transmission systems extend over large geographical areas, typically entire countries, wide-area networks (WANs) are used to deliver the multiplexed measurement data from the substations to the control centre.

For large-scale transmission grids it is often not feasible to measure all power flows and voltages of interest. Furthermore, the measurements are often noisy. Therefore the measurement data are usually fed into a model-based state estimator (SE) at the control centre, which is used to estimate the complete physical state (complex bus voltages) of the power grid. The SE is used to identify faulty equipment and corrupted measurement data through the so-called bad-data detection (BDD) system. Apart from BDD, the state estimate is used by the human operators and by the energy-management systems (EMS) found in modern SCADA systems, such as optimal power flow analysis, and contingency analysis (CA), see for example [1]. Future power grids will be even more dependent on accurate state estimators to fulfil their task of optimally and dynamically routing power flows, because clean renewable power generation tends to be less predictable than non-renewable power generation. Consequently, state estimation will need to rely less on historical data and more on measurements, e.g., by incorporating measurement data from phasor measurement units (PMUs) in the state estimator.

Since many SCADA EMS functions rely on the SE, its vulnerability to attacks and the potential to protect it against attacks is an important question. In the following, we describe recent results in this area. In Section 17.2 we present the basics of steady-state power system modelling, we discuss the operation of the SE and the BDD considering various types of meters, and we introduce stealth attacks. In Section 17.3 we survey algorithms to construct minimum-cost stealth attacks against a power system, assuming a point-to-point SCADA network topology. Using these algorithms in Section 17.4 we compare various schemes for deploying protected meters in order to mitigate stealth attacks. In Section 17.5 we build upon the framework developed in Section 17.3 to describe security metrics for a SCADA network in which measurement data are routed through substations. In Section 17.6 we introduce protection mechanisms for the routed SCADA network, which exploit the additional degrees of freedom that routing provides. We summarize the work in Section 17.7.

17.2 Power-system state estimation and stealth attacks

We start with a review of basic steady-state power-network modelling, state-estimation techniques, and bad-data detection. For a more complete treatment of these topics, we refer to [1, 2]. Finally, we discuss the existence of undetectable bad data, referred to as stealth attacks.

17.2.1 Power network and measurement models

We consider transmission power networks with n buses in steady state. A power load or a power supply is often connected to each bus. A typical load on the transmission level is a distribution power grid supplying a town with power. A typical power supply is a power plant. To bus i we associate two state variables: the phase angle δ_i and the voltage level V_i ($V_i e^{j\delta_i}$ is the complex bus voltage). The physical state of the power network is completely determined by its $2n$ state variables. In the following, we consider a simplified setup where only the n phase angles are considered unknown. This is often done in practice and is called *decoupled estimation*, see for example [1]. Also, since there is only weak coupling between the phase angles and the reactive power flows in the network, we will in the following only consider the active power flows. An active power flow from bus i to bus j is denoted P_{ij}. An active power injection at bus i is denoted P_i. The power flows and injections are given by the equations:

$$P_i = V_i \sum_{j \in N_i} V_j \left(G_{ij} \cos \delta_{ij} + B_{ij} \sin \delta_{ij} \right), \tag{17.1}$$

$$P_{ij} = V_i^2 (g_{si} + g_{ij}) - V_i V_j \left(g_{ij} \cos \delta_{ij} + b_{ij} \sin \delta_{ij} \right), \tag{17.2}$$

where $\delta_{ij} = \delta_i - \delta_j$ is the phase angle difference between bus i and bus j, g_{si} and b_{si} are the shunt conductance and susceptance of bus i, $y_{ij} = g_{ij} + \mathrm{j}b_{ij}$ is the admittance of the branch from bus i to bus j, and $Y_{ij} = G_{ij} + \mathrm{j}B_{ij}$ is the ijth entry of the so-called nodal admittance matrix [1]. The neighbourhood set of bus i, which consists of all buses directly connected to this bus, is denoted by N_i.

We denote by M the total number of measurements available to the SE algorithm. M is typically much larger than the number of unknown states n. The general measurement model we use is

$$z = h(x) + e \in \mathbb{R}^M, \tag{17.3}$$

where $x \in \mathbb{R}^n$ is the vector of unknown states (the phase angles here), and $e \in \mathbb{R}^M$ is a vector of independent random variables modelling measurement noise. Each element e_k is assumed to follow a Gaussian distribution of zero mean. Thus $e = \begin{pmatrix} e_1 & \cdots & e_M \end{pmatrix}^T \in \mathcal{N}(0, R)$, where $R := \mathbf{E} ee^T$ is the diagonal measurement covariance matrix. The function $h(x) \in \mathbb{R}^M$ models the measurements' dependence on the state. We consider three types of measurements in this work. The first and most common type is a *power-flow measurement*

$$z_k = h_k(x) = P_{ij} + e_k, \tag{17.4}$$

in this case modelling the measurement of the active power flow from bus i to bus j. The second type is a *power-injection measurement*

$$z_k = h_k(x) = P_i + e_k, \tag{17.5}$$

in this case modelling the measurement of active power injection at bus i. The third, and least common type of measurement is a *PMU measurement* [23]

$$z_k = h_k(x) = \delta_i + \delta_{GPS} + e_k, \tag{17.6}$$

in this case modelling a measurement of the phase angle in bus i. PMUs can measure phase angles directly using a global GPS time stamp. Here δ_{GPS} models a constant phase angle offset, which sets the frame of reference used by all PMU measurements in the system.

Let us consider the simple 4-bus power network in Figure 17.1, with seven measurements. Assuming that the resistance in the transmission lines and all shunt elements are small, which is a common and often accurate assumption, the model for measurements z_1, z_2, and z_7 becomes

$$\begin{pmatrix} z_1 \\ z_2 \\ z_7 \end{pmatrix} = \begin{pmatrix} P_1 \\ P_{12} \\ \delta_2 + \delta_{GPS} \end{pmatrix} + \begin{pmatrix} e_1 \\ e_2 \\ e_7 \end{pmatrix}$$

$$= \begin{pmatrix} b_{12} V_1 V_2 \sin(\delta_1 - \delta_2) + b_{13} V_1 V_3 \sin(\delta_1 - \delta_3) \\ b_{12} V_1 V_2 \sin(\delta_1 - \delta_2) \\ \delta_2 + \delta_{GPS} \end{pmatrix} + \begin{pmatrix} e_1 \\ e_2 \\ e_7 \end{pmatrix}.$$

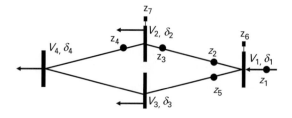

Figure 17.1 A small 4-bus power network. Each bus has a voltage (V_i) and phase angle (δ_i) associated with it. The dots indicate active power-flow/injection measurements (z_1 to z_5). The squares indicate PMU measurements (z_6 and z_7).

Here z_1 is a power-injection measurement, z_2 a power-flow measurement, and z_7 a PMU measurement.

Linearized models and DC approximation. Both in the SE algorithm and in the security analysis to follow, linear approximations are essential. A first-order linear approximation of (17.3) gives

$$z \approx h(x_0) + H_{x_0}(x - x_0) + e, \quad H_{x_0} := \frac{\partial h}{\partial x}(x_0) \in \mathbb{R}^{M \times n}, \quad (17.7)$$

where H_{x_0} is the Jacobian evaluated at $x = x_0$. A common practice is to use the so-called *linear DC approximation*: one assumes that resistances and shunt elements are negligible (as we did in the example above), $x_0 = 0$, and $V_i = 1, \forall i$ (in per unit basis). The corresponding Jacobian is denoted by H, and has the following structure:

$$H := H_0 = \begin{bmatrix} S_1 A D A^T \\ -S_2 D A^T \\ S_3 D A^T \\ S_4 \end{bmatrix}_{M \times n}, \quad (17.8)$$

where A is the arc-to-node incidence matrix of the power network, and D is a diagonal matrix whose diagonal entries are the reciprocals of the reactances of the transmission lines (i.e., b_{ij} corresponding to the arc). S_1, S_2, S_3, S_4 are stacked by the rows of some identity matrices to indicate which injection (block 1), flow (blocks 2–3), or PMU (block 4) measurements are actually measured in the system. For the example in Figure 17.1, we obtain under the assumption $b_{12} = b_{24} = b_{34} = b_{13} = 1$,

$$H = \begin{pmatrix} 2 & -1 & -1 & 0 \\ 1 & -1 & 0 & 0 \\ -1 & 1 & 0 & 0 \\ 0 & 1 & 0 & -1 \\ 1 & 0 & -1 & 0 \\ 1 & 0 & 0 & 0 \\ 0 & 1 & 0 & 0 \end{pmatrix}.$$

As can be seen, the H matrix has a lot of structure. In particular, rows corresponding to injection and flow measurements always sum to zero, and every flow measurement has only two non-zero elements corresponding to the buses at the two ends of the transmission line.

17.2.2 State estimation and bad-data detection

Let us assume that the voltage level V_i of each bus is known, as discussed before. This decoupling assumption is common in the literature, see [1], but can be relaxed to include reactive power-flow measurements and bus-voltage estimates by including more unknown states. Hence, the state-estimation problem we consider consists of estimating the n phase angles δ_i (in x) given the set of M measurements described before.

To obtain a unique estimate, the Jacobian matrix H needs to have full-column rank. If H has full-column rank, the system is called *observable*. This condition is satisfied if there are sufficiently many measurements throughout the system. Nevertheless, if there are no PMU measurements, the system cannot be observable in this sense since all rows sum to zero. Then one chooses one of the buses as reference bus, and assigns the angle zero to it. This removes the corresponding column of the H matrix, and hopefully makes the remaining matrix full-column rank and hence the remaining system observable. In the rest of this work we assume the system is observable, either by means of PMU measurements or by choosing a reference bus and removing the corresponding column.

Given the measurement model (17.3), the state estimate is the weighted least squares estimate of the state. The resulting non-linear equation is usually solved approximately using Gauss–Newton iterations,

$$\hat{x}^{k+1} = \hat{x}^k + (H_{\hat{x}^k}^T R^{-1} H_{\hat{x}^k})^{-1} H_{\hat{x}^k}^T R^{-1}(z - h(\hat{x}^k)), \qquad (17.9)$$

where $\hat{x}^k \in \mathbb{R}^n$ is the estimate of x, k denotes the iteration number, and $H_{\hat{x}^k}$ is the Jacobian defined in (17.7). For the security analysis we use the linear DC approximation. In this case the weighted least squares estimation problem (17.9) can be solved in one step,

$$\hat{x} = (H^T R^{-1} H)^{-1} H^T R^{-1} z. \qquad (17.10)$$

To detect faulty meters or corrupted data, various forms of BDD are used in the control centre [1, 2]. BDD relies on the state estimate $\hat{x}$ to predict what the true power injections, flows, and angles are, and compares them with the measurement data. If enough redundant measurements are available, then the BDD can identify bad data if present. When the DC approximation holds, the prediction of the received measurements is $\hat{z} = H\hat{x} = Kz$, $K = H(H^T R^{-1} H)^{-1} H^T R^{-1} \in \mathbb{R}^{M \times M}$ being the *hat matrix*, and the BDD system calculates the *measurement residual* r, as

$$r := z - \hat{z} = z - H\hat{x} = (I - K)z, \qquad (17.11)$$

where $\hat{x}$ is given by (17.10). If the residual r is larger than expected (measurement errors e will typically make $r \neq 0$), then an alarm is triggered and bad measurements z_i are identified using various techniques, as described in [1, 3, 4].

17.2.3 BDD and stealth attacks

A natural question is whether an *attacker* could corrupt some of the measurement data z without a BDD alarm being triggered. We say that an *attack* a was performed against the measurements if the received measurements at the control centre are $z_a := z + a \in \mathbb{R}^M$, instead of the unattacked measurements z. Note that every non-zero element in a means a corruption of the corresponding measurement.

We define a *stealth attack* to be an attack a such that the BDD system is not triggered (or more accurately, the alarm risk is not significantly increased by the attack a). To corrupt the measurements z into $z + a$ using an arbitrary a will typically trigger a BDD alarm, since the measurement residual r in (17.11) increases. A key observation in [5] is that an attacker that manipulates the measurements from z into $z+a$, where

$$a = Hc \text{ for some } c \in \mathbb{R}^n, \tag{17.12}$$

is undetectable, or stealth, since the residual r is not affected

$$r = z - \hat{z} = z_a - H\hat{x}_a,$$

where $\hat{x}_a = (H^T R^{-1} H)^{-1} H^T R^{-1} z_a = \hat{x} + c$ is the estimate using the attacked measurements z_a. That highly correlated measurement errors, known as 'multiple interacting bad data', are undetectable by residual analysis has been known for a long time in the power systems community, see for example [3, 4]. Intuitively, this is not surprising since z_a corresponds to an actual physical state in the power network (modulo the measurement error e). The BDD system only triggers when the measurements deviate too much from a valid physical state, that is, the received measurements cannot be explained well by the underlying physical model.

The existence of stealth attacks is based on a DC approximation. Despite the simplifying assumptions underlying the DC approximation, stealth attacks based on (17.12) can potentially be made large in real (non-linear) SE software: in the example considered in [6], a power flow measurement was corrupted by 150 MW (57% of the nominal power flow) without triggering BDD alarms.

An attacker can use any stealth attack vector $a, a_k \neq 0$ to perform a stealth attack on measurement k. A rational attacker would, however, be interested in finding an attack vector $a, a_k \neq 0$ with minimum cost. The cost depends on how the attacker can get access to measurement data to manipulate them, i.e., it depends on the communication infrastructure. In the following we consider two different attack cost models. The model presented in Section 17.3 considers that measurement data from certain groups of meters are sent to the control centre over independent communication channels. The model presented in Section 17.5 considers that measurement data from the meters in a substation traverse other substations on the way to the control centre.

17.3 Stealth attacks over a point-to-point SCADA network

We first discuss the case that substation control centre communication is performed over point-to-point links. We consider two attack models.

Under the *stealth meter attack* model the attacker has to gain access to each individual meter it needs to compromise in order to perform a stealth attack against a particular measurement k. The cost of the attacker is the number of meters that have to be compromised, e.g., by tampering with the individual meters in the substations. The protection cost of the operator is the number of meters that are protected, e.g., by investing in tamper-proof meters.

Under the *stealth RTU attack* model an attacker that gains access to an RTU or its communication channel can compromise all measurements multiplexed by the RTU (i.e., measurements from the substation where the RTU is located). The cost of the attacker is the number of RTUs that have to be compromised, e.g., by attacking the point-to-point links that carry the multiplexed measurement data from the RTUs to the control centre (typically the power injections and power flows at the corresponding bus or substation). The protection cost of the operator is the number of RTUs that are protected, e.g., by investing in a tamper-proof RTU and using cryptography to authenticate data.

To capture the cost of the attacker and the system operator, we introduce a partition $\mathcal{M} = \{M_1,\ldots,M_{|\mathcal{M}|}\}$ of the set of measurements $\{1,\ldots,M\}$. The attacker can attack any number of measurements in the same block M_j of the partition at unit cost. For the operator, all measurements belonging to the same block M_j can be protected at unit cost. We denote by S the $|\mathcal{M}| \times M$ matrix whose element $S_{jk} = 1$ if $k \in M_j$, and $S_{jk} = 0$ otherwise. The cost of an attack a for the attacker is then $\|S|a|\|_0$, where $|a|$ denotes the vector of the magnitudes of the elements in a and the symbol $\|\cdot\|_0$ denotes the number of non-zero elements in a vector. We denote the set of protected measurements $\mathcal{P} \subseteq \{1,\ldots,M\}$. When a measurement j is protected, it cannot be attacked (i.e., $a_j \equiv 0$). On the one hand, a measurement can be protected through installing tamper-proof equipment or through cryptography. On the other hand, a measurement might be 'protected' because it is a pseudo measurement whose value is taken from a historical database or is predicted. For example, the injection measurement of a 'transit' bus with neither a load nor a generator connected to it can be modelled as a protected measurement which always takes the value zero.

This notation allows us to consider various attack and protection cost models, including stealth meter attacks and stealth RTU attacks. Stealth meter attacks correspond to a partition $\mathcal{M} = \{\{1\},\ldots,\{M\}\}$, i.e., every measurement is a partition block. Stealth RTU attacks correspond to a partition of size $|\mathcal{M}| = n$ in which the measurements in a bus (alt. substation) form a partition block, and there is an RTU associated with every bus (alt. every substation).

17.3.1 Minimum-cost stealth attacks: problem formulation

In general, the attacker can use any undetectable attack vector $a, a_k \neq 0$ to attack measurement k, but a rational attacker would be interested in finding a stealth attack $a, a_k \neq 0$ with minimum cost. Minimizing the stealth attack cost is equivalent to the number of partition blocks to which the compromised meters belong being minimal, with the constraint that the attacker cannot compromise any protected measurement in $\mathcal{P}$. In order to find a minimum-cost stealth attack on measurement k, the attacker has to solve the problem

$$\alpha_k := \underset{c}{\text{minimize}} \quad \|S|Hc|\|_0$$
$$\text{subject to} \quad (Hc)_k = 1 \quad (17.13)$$
$$(Hc)_j = 0, \quad \forall j \in \mathcal{P},$$

where $(Hc)_k$ denotes the kth element of the vector Hc. In (17.13) the optimization is over all stealth attacks (cf. (17.12)) targeting measurement k, while avoiding those protected by $\mathcal{P}$. A solution c^* to (17.13) can be rescaled such that $a_k = (H\beta c^*)_k = \beta$ for any scalar β. In total, $\alpha_k = \|S|Hc^*|\|_0$ blocks of measurements have to be corrupted to manipulate the measurement value z_k. The problem (17.13) is non-convex and is generally hard to solve for large systems. A simpler problem is to calculate an upper bound for $\min_k \alpha_k$ in the case of meter attacks, which can be done in polynomial time [7].

17.3.2 Exact computation of minimum-cost stealth attacks

We describe two algorithms to compute the minimum-cost stealth attacks in (17.13): a mixed integer linear program (MILP) formulation and an enumerative algorithm that exploits the power system graph topology.

MILP formulation of (17.13). Define the decision vector $c \in \mathbb{R}^n$ as in (17.13), and keep the constraints in (17.13). Two additional binary decision vectors are needed to model the objective function in (17.13). The first vector is $y \in \{0,1\}^M$, describing which measurements can potentially be attacked. This is modelled by the constraint

$$(Hc)_i \leq Ky_i \quad \text{and} \quad -(Hc)_i \leq Ky_i, \quad \forall i \in 1,2,\ldots,M, \tag{17.14}$$

where K is a constant greater than $\|Hc^*\|_\infty$ for at least one optimal solution c^* in (17.13) if (17.13) is feasible. In the power network case, a non-trivial upper bound of K can be obtained from physical insight. To connect the attack on partitions and measurements, another binary decision vector $x \in \{0,1\}^{|\mathcal{M}|}$ is required, with $x_j = 1$ if and only if partition j is attacked. To model the fact that the measurements associated with partition j can be attacked if and only if partition j is attacked, the following constraints have to be satisfied:

$$\sum_{i:S_{ji}=1} y_i \leq Mx_j, \tag{17.15}$$

where M is the number of rows in H. Finally, the objective of the minimum-cost stealth attack problem is to minimize the number of partitions attacked. Hence, the MILP problem can be formulated as

$$\begin{aligned}
\underset{c,x,y}{\text{minimize}} \quad & \sum_j x_j \\
\text{subject to} \quad & (Hc)_k = 1 \\
& (Hc)_j = 0, && \forall j \in \mathcal{P} \\
& (Hc)_i \leq Ky_i, && \forall i \in 1,2,\ldots,M \\
& -(Hc)_i \leq Ky_i, && \forall i \in 1,2,\ldots,M \\
& \sum_{i:S_{ji}=1} y_i \leq Mx_j, && \forall j \in 1,2,\ldots,|\mathcal{M}| \\
& x \in \{0,1\}^{|\mathcal{M}|} \quad \text{and} \quad y \in \{0,1\}^M.
\end{aligned} \tag{17.16}$$

The MILP problem in (17.16) is NP-hard, but moderate instances of the problem are reasonable to solve off-line using solvers such as CPLEX [8] or Gurobi [9].

Graph augmentation algorithm [10]. Assuming H has M rows and rank $n-1$ (i.e., the power system is observable without PMUs), finding α_k is equivalent to finding a set of rows $N \subseteq \{1, \ldots, M\} \setminus \{k\}$ that is maximal in terms of the number of partition blocks M_j it (partially) covers, and for which the following two conditions hold:

$$rank(H_N) = n - 2, \qquad (17.17)$$

$$rank(H_{N \cup \{k\}}) = n - 1, \qquad (17.18)$$

where H_N is the submatrix of H formed by the rows in N. Given N, the attack can be constructed by calculating the nullspace of the submatrix H_N, which is one-dimensional due to the rank-nullity theorem. Since $\forall c \in null(H_N)$, we have $(Hc)_k = 0 \, \forall k \in N$, and N is maximal, it follows that $\alpha_k = \|S|Hc|\|_0$. In general, finding the maximal set N is a combinatorial optimization problem. For sparse power network graphs, however, it is possible to calculate the optimal solutions even for systems with hundreds of state variables and measurement points using the iterative path augmentation algorithm described in the following.

The iteration starts with a not necessarily stealth attack that consists of the partition block to which measurement k belongs, $\mathcal{B}^{(1)} = \{M_j\}, k \in M_j$. In iteration i the algorithm first considers all attacks of cost i. For every attack $B \in \mathcal{B}^{(i)}$ it creates the corresponding attack B' by only keeping the rows l of H for which there is no row j not in attack B that is linearly dependent on row l ($l \sim j$). It then verifies if the set $N = \{1, \ldots, M\} \setminus B'$ satisfies the rank conditions (17.17) and (17.18). If no such attack is found, the algorithm augments every attack $B \in \mathcal{B}^{(i)}$ of cost i with one additional partition block M_k that is unprotected ($M_k \cap \mathcal{P} = \emptyset$) and is neighbouring a partition block already in the attack ($M_k \in \mathcal{N}(M_j)$) for some $M_j \subseteq B$. The pseudo-code of the algorithm is shown in Table 17.1.

17.3.3 Upper bound on the minimum cost

The exact computations described in the previous subsection can be time-consuming for large systems, but efficient algorithms are available for approximating (17.13). In the following we describe two such algorithms.

A simple upper bound [10]. A simple upper bound on α_k can be obtained by inspection of the H matrix. Any column i of H with a non-zero entry in the kth row can be used to construct a stealth attack a that achieves the attack goal, if $H_{ji} = 0 \; \forall j \in \mathcal{P}$. Assume that H_{ki} is non-zero. Then $\frac{1}{H_{ki}} H_{\cdot,i}$, where $H_{\cdot,i}$ denotes the ith column of H, is a stealth attack against measurement k, and is an upper bound for (17.13). By selecting the sparsest vector among all $S|H_{\cdot,i}|$, we obtain an upper bound $\hat{\alpha}_k$ on α_k. Formally, we have

$$\hat{\alpha}_k := \min_{i : H_{ki} \neq 0} \|S|H_{\cdot,i}|\|_0. \qquad (17.19)$$

Since H is typically sparse for power networks, this bound is very fast to compute, and exists whenever $\mathcal{P} = \emptyset$.

Upper bounding via convex relaxation. For ease of presentation, the PMU measurements in (17.8) are not considered in this derivation. However, the PMU case can be handled at the expense of more complicated bookkeeping.

Table 17.1. The iterative path augmentation algorithm used to calculate the attacks with minimal cost for measurement k

1	$\mathcal{B}^{(1)} = \{M_j\}, k \in M_j, \mathcal{B}^* = \emptyset$
2	for $i = 1$ to $|\mathcal{M}| - |\mathcal{P}|$
3	for $B \in \mathcal{B}^{(i)}$
4	$B' = \{l | l \in B, \nexists j \notin B \text{ s.t. } j \sim l\}$
4	if $rank(H_{\{1,...,M\}\setminus B'}) = n - 2$ and $rank(H_{(\{1,...,m\}\setminus B')\cup\{k\}}) = n - 1$ then
4	$\mathcal{B}^* = \mathcal{B}^* \cup B$
5	end if
6	end for
7	if $\mathcal{B}^* \neq \emptyset$ then return $\mathcal{B}^*$
8	for $B \in \mathcal{B}^{(i)}$
9	for $M_j \subseteq B$
10	for $M_k \in \mathcal{N}(M_j), M_k \cap \mathcal{P} = \emptyset, M_k \cap B = \emptyset$
11	$\mathcal{B}^{(i+1)} = \mathcal{B}^{(i+1)} \cup (B \cup M_k)$
12	end for
13	end for
14	end for
15	end for

With some restrictions in (17.13), the structure of the H matrix in (17.8) can be exploited to obtain a more accurate upper bound for α_k than the one in (17.19). The first restriction is that only stealth meter attacks are considered. Therefore, using (17.8) the objective function of (17.13) becomes

$$\|Hc\|_0 = \|S_1 A D A^T c\|_0 + \|S_2 A^T c\|_0 + \|S_3 A^T c\|_0, \quad (17.20)$$

where the matrices A and D are defined in (17.8), and the fact that $\|S_2 D A^T c\|_0 = \|S_2 A^T c\|_0$ and $\|S_3 D A^T c\|_0 = \|S_3 A^T c\|_0$ is used. The second restriction is that the protection set $\mathcal{P}$ cannot include any injection measurement. Hence, there exists an index set $\mathcal{Q}$ such that

$$(Hc)_j = 0, \quad \forall j \in \mathcal{P} \iff (A^T c)_j = 0, \quad \forall j \in \mathcal{Q}. \quad (17.21)$$

Depending on the value of k, the attack can target a power-injection or power-flow measurement. Here only the power-flow case is considered. Hence, there exists an index l such that the target constraint in (17.13) can be written

$$(Hc)_k = 1 \iff (A^T c)_l = 1. \quad (17.22)$$

The injection case is handled indirectly by comparing the power flow α_k bounds available from all measured incident transmission lines.

Putting together (17.20), (17.21), and (17.22) yields an equivalent form of (17.13), which is a difficult combinatorial problem. However, if we *relax* the problem by removing the term $\|S_1 A D A^T c\|_0$ in the objective function, then it can be shown that the relaxed problem becomes a convex minimum cut (min-cut) problem. The solution strategy is then to solve the relaxed problem as a min-cut problem, and use its optimal solution as a suboptimal solution to the original problem in (17.13). For reference, the relaxed problem is

$$\begin{aligned}
\underset{c}{\text{minimize}} \quad & \left\|S_2 A^T c\right\|_0 + \left\|S_3 A^T c\right\|_0 \\
\text{subject to} \quad & (A^T c)_l = 1 \\
& (A^T c)_j = 0, \quad \forall j \in \mathcal{Q}.
\end{aligned} \quad (17.23)$$

A central concept in (17.23) is the 'cutting' of arcs. An arc (i,j) is cut if and only if $c_i \neq c_j$. For example, $\left\|S_2 A^T c\right\|_0$ counts how many arcs representing the measured transmission lines are cut. The concept of cut leads to the following consequence: for any feasible c in (17.23), it is possible to construct another binary feasible solution b such that

$$b_i = \begin{cases} 0, & \text{if } c_i \leq c_{j_2}, \text{ where the target arc } l \text{ is from node } j_1 \text{ to } j_2 \\ 1, & \text{otherwise.} \end{cases}$$

By construction, $(A^T b)_l = 1$. Also, arcs that are not cut in c remain not cut in b. Hence, b is also feasible in (17.23), and its objective function value is no worse. Therefore, the two versions of (17.23), with and without the binary constraint $c_i \in \{0,1\}$, are equivalent. The version of (17.23) with the binary constraint is a *node-partitioning* problem. A node-partitioning problem minimizing the number of arcs cut, subject to some arc-cutting constraints, can be posed as a min-cut problem. The details are as follows.

1. Define a graph $\mathcal{G}$ as the undirected version of the graph of the power network.
2. For each arc j, if $j \in \mathcal{Q}$ then the arc weight is infinity. Otherwise, the weight of arc j is the sum of the nonzero entries in the jth columns of S_2 and S_3. Otherwise, the weight of arc j is zero.
3. Divide the nodes of $\mathcal{G}$ into two partitions. The two nodes of the targeted arc l must be in different partitions. The rest of the nodes are assigned to minimize the sum of the weights of the 'cut' arcs. An arc is cut if and only if the nodes it connects are in different partitions.

The relaxed problem in (17.23) is feasible if and only if the optimal min-cut objective value is finite. The min-cut problem can be solved using efficient polynomial-time algorithms. For a detailed description of the min-cut problem and its solution algorithms, see for instance [11]. Finally, using the results from [12, 13], it is possible to enumerate efficiently all optimal partitions of the min-cut problem. The partition with the best objective value in the original problem in (17.13) is chosen to be the upper bound for α_k.

With a further assumption that S_1, S_2, S_3 are identity matrices and S_4 is empty, it can be shown that the minimum attack cost problem is also a node partitioning problem. For a proof and the ramifications, see [14].

17.3.4 Numerical results

In the following we show numerical results based on the IEEE 14-bus, the IEEE 118-bus, the IEEE 300-bus, and a 2383-bus benchmark system. The H matrices were obtained using MATPOWER [15], and are of the form in (17.8), with $S_1, S_2,$ and S_3 being identity matrices and S_4 being empty (that is, measurements are taken at every power injection and power flow). For each benchmark we calculate the minimal stealth attack cost α_k defined in (17.13) for all k, without any protection (i.e., $\mathcal{P} = \emptyset$).

Figure 17.2 shows the minimum stealth attack cost and the bound $\hat{\alpha}_k$ for the IEEE 14-bus system obtained using the graph-augmentation algorithm. The bound is almost always tight, and the minimal attack cost is quite low in general. Figure 17.3 shows the minimal stealth attack cost α_k for all measurements k in the 118, 300 and 2383-bus benchmark systems in decreasing order of α_k, obtained by solving the MILP problem in (17.16).

To compare the various approximation algorithms we use (17.19), the min-cut relaxation in Section 17.3.3, and a LASSO [16] approximation to solve (17.13) for the 300-bus system. LASSO approximation means that we solve a modified version of (17.13) in which the objective function is $\|S|Hc|\|_1$ (i.e., the 1-norm). The optimal solution of the modified problem is used as a suboptimal solution to the original problem. In terms of accuracy, the min-cut bounds are exact, they coincide with α_k obtained using MILP. On the other hand, the bound in (17.19) results in about 14% of measurements whose α_k are overestimated. The mean and standard deviation of the relative errors, the difference

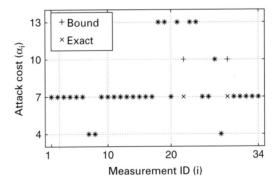

Figure 17.2 The minimum attack costs α_k and their upper bounds $\hat{\alpha}_k$ for the IEEE 14-bus benchmark system.

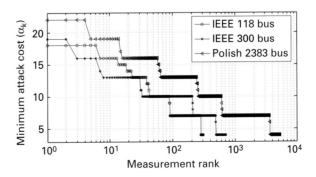

Figure 17.3 Ranked minimum stealth meter attack costs for the 118, 300, and 2383-bus benchmark systems.

between the bound and the exact value of α_k divided by the exact value of α_k, are 66% and 47%, respectively. For LASSO, the attack costs of about 47% of the measurements are overestimated. The corresponding mean and standard deviation of the relative errors are 112% and 121%, respectively. In terms of computation time, MILP takes about 6700 s, while on the same computer it takes 0.044, 0.80, and 42.50 s to obtain the $\hat{\alpha}_k$, min-cut, and LASSO bounds, respectively.

Other approximate schemes for solving (17.13) exist. Notable examples are matching pursuit [17] and basis pursuit [18]. However, similar to LASSO, these generic methods are found to be inferior for the problem in (17.16).

17.4 Protection against attacks in a point-to-point SCADA network

Consider that the operator has a budget π in terms of the number of protected measurement-partition blocks that it can spend. The goal of the operator is to achieve the best possible protection of the state estimator against stealth attacks given its budget π. Given a set $\mathcal{P}$ of protected measurements, let us denote by $C_\mathcal{M}(\mathcal{P})$ the cost of protecting $\mathcal{P}$ considering the partition $\mathcal{M}$. The cost $C_\mathcal{M}(\mathcal{P})$ can be calculated as the number of partition blocks M_j, s.t. $M_j \cap \mathcal{P} \neq \emptyset$, and we have $C_\mathcal{M}(\mathcal{P}) \leq \pi$.

17.4.1 Perfect protection

Ideally, the set of protected measurements $\mathcal{P}$ should be such that no stealth attacks are possible, i.e., $\alpha_k = \infty, \forall k \in \{1, \ldots, M\}$. We refer to such a protection as *perfect protection*.

In the case of the meter attack model, if there are no PMU measurements then in order to achieve perfect protection it is necessary and sufficient for the operator to protect a set $\mathcal{P}$ of measurements chosen such that $rank(H_\mathcal{P}) = n - 1$ [19], where $H_\mathcal{P}$ is the submatrix of H consisting only of the rows in $\mathcal{P}$. The budget required to achieve perfect protection is thus $\pi = n - 1$. If there are PMU measurements then $rank(H_\mathcal{P}) = n$ must hold for perfect protection.

In the case of the RTU attack model the condition $\pi = n - 1$ is not necessary, since we now count the number of protected blocks, which can contain more than one measurement each. Let us call the *RTU-level power-network graph* the graph where each vertex is an RTU in the power system, and every edge is a transmission link between the RTUs. A dominating set of the RTU-level power-network graph is a subset of vertices such that each vertex not in the dominating set is adjacent to at least one member of the dominating set. Consider now that $\mathcal{P}$ is not a dominating set of the RTU-level power-network graph. Then there is an RTU k for which no neighbouring RTU is in the protected set, $j \notin \mathcal{P} \, \forall j \in \mathcal{N}(k)$. A stealth attack can then be constructed based on the column corresponding to the state variable in bus k, as done in (17.19) to obtain a finite $\hat{\alpha}_k \geq \alpha_k$. Hence, a perfect RTU protection $\mathcal{P}$ must be a dominating set of the RTU-level power-network graph [10]. Not all dominating sets constitute a perfect protection, but a dominating set can be used as a starting point to find a perfect protection $\mathcal{P}$.

Dominating-set augmentation algorithm (DSA). Initialize the set of protected measurements $\mathcal{P}$ with a minimal dominating set of the RTU-level power-network graph. Iterate over $k = 1, \ldots, M$ and set $\mathcal{P} = \mathcal{P} \cup M_j$ for $M_j \not\ni k$ if $\alpha_k < \infty$ for some k.

The algorithm terminates after one iteration and provides a perfect protection $\mathcal{P}$. For sparse power-network graphs the budget required to achieve perfect protection is $\pi \ll n$.

17.4.2 Non-perfect protection

If the operator's budget π for protection is insufficient for perfect protection, then the operator might be interested in protecting a set of measurements $\mathcal{P}$ that maximizes the system's protection level according to some metric. Two metrics of particular interest are the *minimum attack cost* and the *average attack cost*. In the following, we describe a greedy algorithm to approximately maximize the minimum attack cost; a similar greedy algorithm can be used to approximately maximize the average attack cost [10].

The goal of the operator to maximize the minimum attack cost among all measurements that are possible to attack can be formulated as

$$\mathcal{P}^{MM} = \arg \max_{\mathcal{P}: C_\mathcal{M}(\mathcal{P}) \leq \pi} \min_k \alpha_k. \qquad (17.24)$$

A simple greedy algorithm that aims to find an optimal set of protected measurements $\mathcal{P}$ in the sense of (17.24) for given budget π is the following.

Most shortest minimal-attack algorithm (MSM). Initially set $\mathcal{P} = \emptyset$. Then in every iteration calculate $\alpha_k, \forall k \in 1, \ldots, m$ and $\min_k \alpha_k$. Pick partition block M_j that appears in most minimal attacks $B \in \mathcal{B}^*$ with least cost, i.e., $C_\mathcal{M}(B) = \min_k \alpha_k$. Set $\mathcal{P} = \mathcal{P} \cup M_j$. Continue until $C_\mathcal{M}(\mathcal{P}) = \pi$.

17.4.3 Numerical results

In the following we show numerical results for the IEEE 14-bus and the IEEE 118-bus benchmark power systems. Figure 17.4 shows the minimum attack cost and the average attack cost ($\sum_{k:\alpha_k < \infty} \alpha_k / |\{k : \alpha_k < \infty\}|$) as a function of the protection budget π for the IEEE 14-bus system. The results were obtained using the greedy MSM algorithm and by selecting measurements to be protected at random [19]. The results for a random selection are obtained from the average of 10 experiments for every protection budget π. Using MSM, both the minimum and the average attack cost increase significantly faster than when selecting at random. For a budget of $\pi = n - 1 = 13$, MSM finds the set of meters that provides perfect protection. Hence, incremental protection of the meters does not lead to extra costs for the operator even if the ultimate goal is perfect protection when using MSM. In the case of random selection this is, however, not the case.

For the case of RTU attacks, perfect protection for the IEEE 14-bus system can be achieved by protecting 4 RTUs, $\mathcal{P} = \{2, 6, 7, 9\}$. MSM finds this set for a budget of $\pi = 4$. This is a minimal dominating set of the power network graph. $\mathcal{P} = \{2, 8, 10, 13\}$, which is also a minimal dominating set of the power network graph, does not provide perfect protection, as RTU 4 can be attacked by tampering with 3 RTUs $(4, 7, 9)$ and

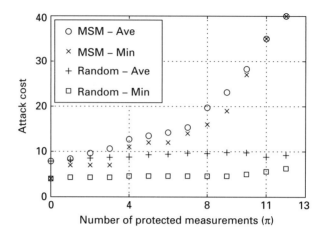

Figure 17.4 The smallest minimum attack cost $\min_k \alpha_k$ and the average minimum attack cost for the IEEE 14-bus system for the case of meter attacks.

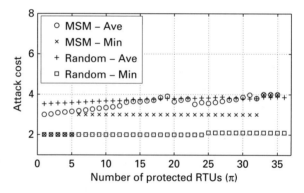

Figure 17.5 The smallest minimum attack cost $\min_k \alpha_k$ and the average minimum attack cost for the IEEE 118-bus system for the case of RTU attacks.

8 measurements ($B = \{4, 7, 9, 22, 23, 29, 42, 43\}$). For the IEEE 118-bus system the minimal dominating set of the power network graph contains 32 RTUs, but it does not provide perfect protection. Using the DSA algorithm we identified 4 RTUs $(5, 23, 69, 77)$ that need to be added to the minimal dominating set in order to achieve perfect protection.

Figure 17.5 shows the minimum and the average attack cost for the case of the RTU attack model for the IEEE 118-bus system. We again compare the MSM algorithm to selecting RTUs to be protected at random; the results for random selection are obtained from the average of 10 experiments for every protection budget π. MSM achieves perfect protection by protecting 36 RTUs, and leads to higher minimal attack cost than random selection for the same protection budget. In terms of average attack cost, the two achieve almost similar performance, however. Compared to the case of measurement attacks (Figure 17.4) we observe, however, that the minimal and average attack costs are rather small even close to perfect protection (i.e., close to $\pi = 36$).

17.5 Stealth attacks over a routed SCADA network

We now turn to the case where the communication between the RTUs and the control centre is performed over a routed network. For modern SCADA systems this scenario is more realistic than the scenario considered in Section 17.3.

Under the *substation attack model* considered in the following, the attacker gets access to the switching equipment located at a subset of the substations. For example, the attacker could get physical access to the equipment in an unmanned substation or could remotely exploit the improper access configuration of the communication equipment. By gaining access to a substation s (i.e., the switching equipment and the RTU), the attacker can potentially manipulate the measurement data that are *measured in* substation s and the data that are *routed through* substation s, unless multi-path routing, physical protection, or data authentication make that impossible. To perform a *stealth attack* on a particular measurement k (its value z_k), the attacker might need to attack several substations simultaneously, which increases the cost of performing the attack.

To protect SCADA communication, the operator can achieve message authentication by installing a secret key in the substation in one of two ways. First, by installing a bump-in-the-wire (BITW) device adjacent to a legacy RTU. Data between the RTU and the BITW device are sent in plain text, hence a BITW does not protect the data if an attacker can gain physical access to the substation. Nevertheless, it protects the data between the BITW device and the control centre. Second, by installing an RTU that supports message authentication. A tamper-proof RTU that supports authentication, though more expensive, ensures data integrity even if the attacker can gain physical access to the substation.

To formulate the problem we introduce a model of the SCADA communication infrastructure. Consider that the n buses of the power system are spread over a set of substations $\mathcal{S}$, $|\mathcal{S}| = S$. We denote the substation at which measurement k is taken by $S(k) \in \mathcal{S}$. We model the SCADA communication system by an undirected graph $\mathcal{G} = (\mathcal{S}, E)$. The vertices of the graph are the substations, and there is an edge between two substations if they are connected by a transmission line. The graph $\mathcal{G}$ is connected but is typically sparse. We consider that the control centre is located near a substation, and denote the substation by $s_c \in \mathcal{S}$. For each substation $s \in \mathcal{S}$ there is a set of established routes $\mathcal{R}_s = \{r_s^1, \ldots, r_s^{R(s)}\}$ from s to s_c through $\mathcal{G}$. $\mathcal{R}$ denotes the collection of all $\mathcal{R}_s$. We represent a route by the set of substations it traverses including s itself and the control centre s_c, i.e., $r_s^i \subseteq \mathcal{S}$. The order in which the substations appear in the route is not relevant to the considered problem. If $R(s) = 1$ then all measurement data from substation s are sent over a single route to the control centre. If $R(s) > 1$ then data is split equally among the routes such that if the data sent over any route gets corrupted, the control centre can detect the data corruption using an error-detection code.

We denote the set of substations that use a BITW device to *authenticate* the data sent to the control centre by $\mathcal{E} \subseteq \mathcal{S}$. For a route r_s^i we denote by $\sigma_{\mathcal{E}}(r_s^i)$ the set of substations in which the data are *susceptible* to attack despite BITW authentication. By definition, $\sigma_{\mathcal{E}}(r_s^i) = \{s\}$ if $s \in \mathcal{E}$ and $\sigma_{\mathcal{E}}(r_s^i) = r_s^i$ otherwise, that is, BITW-authenticated data can only be modified at the substation they originate from, if physical access is possible. To avoid physical access a substation can be protected, e.g., by guards or video surveillance. We denote the set of protected substations by $\mathcal{P} \subseteq \mathcal{S}$. Protected substations are not susceptible to attacks. We assume that the substation where the control centre is located is protected, that is, $s_c \in \mathcal{P}$.

In the following, we define two security metrics to characterize the vulnerability of the system with respect to the vulnerability of individual measurements and with respect to the importance of individual substations. Both metrics depend on the protection measures implemented by the operator, and we use the metrics in Section 17.6 to quantify how various protection measures can decrease the system's vulnerability.

17.5.1 Measurement attack cost

We quantify the vulnerability of measurement k by the minimum number of *substations* that have to be attacked in order to perform a stealth attack against the measurement, and denote it by Γ_k. This metric is analogous to the attack cost α_k defined for the meter and the RTU attack models in Section 17.3, but expresses the attack cost in terms of the number of attacked substations.

If the substation at which the measurement is located is protected and is encrypted ($S(k) \in \mathcal{P} \cap \mathcal{E}$), then the measurement is not vulnerable and we define $\Gamma_k = \infty$. Otherwise, for a measurement k we define Γ_k as the cardinality of the smallest set of substations $\omega \subseteq \mathcal{S}$ such that there is a stealth attack against k involving some measurements k' at substations $S(k')$ such that the unencrypted part of every route of the substations $S(k')$ involved in the stealth attack passes through at least one substation in ω:

$$\Gamma_k = \min_{\omega \subseteq \mathcal{S}; \omega \cap \mathcal{P} = \emptyset} |\omega| \quad \text{s.t.} \quad \exists a, c \text{ s.t. } a = Hc, \ a_k = 1 \text{ and} \quad (17.25)$$
$$a_{k'} \neq 0 \implies \omega \cap \sigma_{\mathcal{E}}(r^i_{S(k')}) \neq \emptyset, \quad \forall r^i_{S(k')} \in \mathcal{R}_{S(k')}.$$

The attack cost of a measurement depends on the routing $\mathcal{R}$, the encrypted substations $\mathcal{E}$, and the protected substations $\mathcal{P}$.

Calculating Γ_k. We can obtain Γ_k by solving a mixed-integer linear programming problem as follows. Define decision vectors $a \in \mathbb{R}^M$ and $c \in \mathbb{R}^n$. a is the attack vector to be determined. We need a to be a stealth attack targeting measurement m and for the solution to be unique we require the attack magnitude on m to be unit

$$a_k = 1 \quad \text{and} \quad a = Hc. \quad (17.26)$$

To describe the connection between the choice of which substations to attack and the set of measurements that can be attacked as a result of the substation attacks, two 0–1 binary decision vectors are needed. One such binary decision vector is $x \in \{0,1\}^n$, with $x_s = 1$ if and only if substation s is attacked. Hence, for protected substations (i.e., $s \in \mathcal{P}$), the following must hold:

$$x_s = 0, \quad \forall s \in \mathcal{P}. \quad (17.27)$$

The other binary decision vector is denoted as $y \in \{0,1\}^M$, with $y_m = 1$ meaning measurement m might be attacked because of attacks on relevant substations. Conversely,

$y_m = 0$ means measurement m cannot be attacked. To apply y as an indicator for which measurements can be attacked, we impose the following constraints:

$$a \leq Ky \quad \text{and} \quad -a \leq Ky, \tag{17.28}$$

where the inequality is entry-wise and K is a scalar which is regarded as 'infinity'. A non-trivial upper bound for K can be obtained from physical insight, as in the case of the MILP formulation in Section 17.3.2. Finally, measurement m can be attacked if and only if the unencrypted part of every route between $S(m)$ and s_c goes through at least one of the attacked substations. This is captured by the following constraints:

$$y_m \leq \sum_{s \in \sigma_{\mathcal{E}}(r^i_{S(k)})} x_s, \quad \forall r^i_{S(k)} \in \mathcal{R}_{S(m)}, \tag{17.29}$$
$$\forall m = 1, 2, \ldots, M.$$

Note that by (17.29) itself it is possible to have $y_m = 0$ for some m, while the sum on the right-hand side can be greater than zero. However, this cannot happen at optimality since the objective is to minimize the sum of all entries of x (i.e., the number of substations to be attacked). In conclusion, Γ_k can be calculated by solving the following problem:

$$\begin{aligned}
\underset{a,c,x,y}{\text{minimize}} \quad & \sum_{s \in \mathcal{S}} x_s \\
\text{subject to} \quad & \text{constraints (17.26) through (17.29)} \\
& x_s \in \{0,1\}, \quad \forall s \\
& y_m \in \{0,1\}, \quad \forall m.
\end{aligned} \tag{17.30}$$

If (17.30) is infeasible, then the measurement attack cost is defined to be $\Gamma_k = \infty$. Otherwise, Γ_k is the optimal objective function value in (17.30).

17.5.2 Substation attack impact

The second metric quantifies the importance of substation s by its *attack impact* I_s, which is the number of measurements against which an attacker can perform a *stealth* attack by getting access to a *single* substation s.

By definition, $I_s = 0$ if the substation is protected ($s \in \mathcal{P}$). Otherwise, we define I_s as follows. A measurement k can be attacked if and only if the unencrypted parts of all routes from $S(k)$ to the control centre contain substation s. Denote by $\mathcal{M}_s \subset \{1, \ldots, M\}$ the index set of all such attackable measurements. Then measurement $k \in \mathcal{M}_s$ can be *stealthily* attacked if and only if the following system of equations has a solution in terms of unknowns $a \in \mathbb{R}^M$ and $c \in \mathbb{R}^n$ (cf. (17.12)):

$$a = Hc, \quad a_{k'} = 0, \ \forall k' \notin \mathcal{M}_s, \quad \text{and} \ a_k = 1. \tag{17.31}$$

The attack impact I_s is then the cardinality of the set of measurements for which (17.31) has a solution. That is,

$$I_s = \left| \{k \mid \exists a \text{ satisfying (17.31)}\} \right|. \tag{17.32}$$

The attack impact of a substation depends on the routing $\mathcal{R}$, the encrypted substations $\mathcal{E}$, the protected substations $\mathcal{P}$, and the power system topology described by H.

Calculating I_s. By a linear algebra fact [20], $a = Hc$ for some c if and only if there exists a matrix N_s such that $N_s a = 0$, where N_s^T is a basis matrix for the null space of H^T. Let us denote by $N_s(:, \mathcal{M}_s)$ the matrix formed by keeping only the columns of N_s in $\mathcal{M}_s$, $a(\mathcal{M}_s)$ as a vector formed by keeping only the entries of a corresponding to $\mathcal{M}_s$. Then (17.31) is solvable if and only if

$$N_s(:, \mathcal{M}_s) a(\mathcal{M}_s) = 0, \quad \text{and} \quad e_i^T a(\mathcal{M}_s) = 1 \qquad (17.33)$$

can be solved, where e_i denotes the ith column of an identity matrix of dimension $|\mathcal{M}_s|$, and the ith entry of $z(\mathcal{M}_s)$ is z_k. Next, let $\tilde{N}_s$ be a basis matrix for the null space of $N_s(:, \mathcal{M}_s)$. Then (17.33) is solvable if and only if there exists a vector $\tilde{c}$ such that $\left(e_i^T \tilde{N}_s \right) \tilde{c} = 1$. This is possible if and only if the ith row of $\tilde{N}_s$ is not identically zero. The above procedure applies to indices other than i. Hence, I_s can be calculated as

$$I_s = \left| \left\{ i \mid \tilde{N}_s(i,:) \neq 0 \right\} \right|.$$

17.5.3 Numerical results

We calculate the attack impact and the measurement attack cost for the IEEE 118 and 300-bus benchmark power systems using the above algorithms. As a baseline we consider that all substations use a single shortest path ($|\mathcal{R}_s| = 1$) to the control centre s_c, which is located at the substation with highest degree. Measurements are taken at every power injection and every power flow, and the system is not protected, i.e., $\mathcal{E} = \mathcal{P} = \emptyset$.

Figure 17.6 shows the attack impact I_s for the substations for which $I_s > 0$ for the two power systems. The results show that there are several substations that would enable an

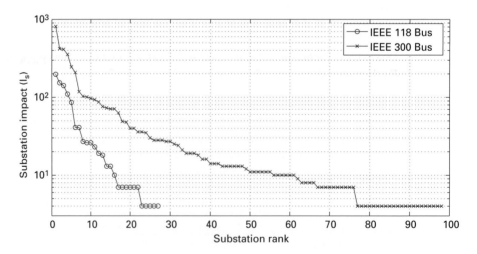

Figure 17.6 Attack impact I_s of the substations in the IEEE 118 and 300-bus systems in decreasing order of attack impact. The case of shortest-path routing.

attacker to perform a *stealth* attack on a significant fraction of the measurements in the power system, e.g., on about 1000 measurements for the 300-bus system (approx. 90% of all measurements). The attack impact decreases slower than exponentially with the rank of the substation, and almost 50% of the substations have non-zero attack impact. The measurement attack costs are low, in the 118-bus (300-bus) system the number of measurements with attack cost 1, 2, 3, and 4 are 374, 78, 11, and 0 (975, 89, 3, and 6), respectively.

17.6 Protection against stealth attacks for a routed SCADA network

Let us consider that the operator wants to improve the system security through, for example, changing single-path routes, using multi-path routing, data authentication, or physical protection. As in the case of point-to-point communication, a natural goal for the operator would be to improve the most vulnerable part of the system, that is, to minimize $\max_{s \in \mathcal{S}} I_s$ or to maximize $\min_{k \in \mathcal{M}} \Gamma_k$. Nevertheless, due to the structure of the graph $\mathcal{G}$ it might happen that $\max_s I_s$ cannot be decreased, but the second-highest attack impact can. Similarly, it might not be possible to increase $\min_k \Gamma_k$, even though the second-lowest attack cost can be increased (cf. Figure 17.9).

An alternative is to formulate the operator's goal as a multi-objective optimization problem. Objective γ is to minimize the number of measurements with attack cost γ, $|\{k | \Gamma_k = \gamma\}|$. The objectives are ordered: objective γ has priority over objective $\gamma' > \gamma$. Formally, we define the vector $w \in \mathbb{N}^{S-1}$ whose γth component is $w_\gamma = |\{k | \Gamma_k = \gamma\}|$. The goal of the operator is then

$$\operatorname*{lexmin}_{\mathcal{P},\mathcal{E},\mathcal{R}} w(\mathcal{P},\mathcal{E},\mathcal{R}), \qquad (17.34)$$

where lexmin stands for lexicographical minimization [21]. w attains its minimum $w_\gamma = 0$ ($1 \leq \gamma \leq S - 1$) when no measurement can be attacked, i.e., $\Gamma_k = \infty$ for all $k \in \mathcal{M}$. Due to the definition of w, the solution to (17.34) is a solution to $\max_{\mathcal{P},\mathcal{E},\mathcal{R}} \min_{k \in \mathcal{M}} \Gamma_k$. Furthermore, since I_s and Γ_k are related, $I_s = 0 \,\forall s \in \mathcal{S} \iff \min_k \Gamma_k > 1$, it is also a solution to $\min_{\mathcal{P},\mathcal{E},\mathcal{R}} \max_{s \in \mathcal{S}} I_s$ if $\max_{\mathcal{P},\mathcal{E},\mathcal{R}} \min_{k \in \mathcal{M}} \Gamma_k > 1$. A simple way to solve the lexicographical minimization in (17.34) is the following iterative algorithm [21].

Consider given $\mathcal{P},\mathcal{E},\mathcal{R}$ and let $\gamma^* = \min\{\gamma | w_\gamma > 0\}$. If $\gamma^* = \infty$, the system is not vulnerable. Otherwise, use the following algorithm to decrease w_γ for some $\gamma \geq \gamma^*$.

Critical first algorithm

(i) For every measurement k with $\Gamma_k = \gamma^*$, calculate the set of RTU stealth attacks using one of the algorithms in Section 17.3.2. Every such stealth attack is a set of substations including $S(k)$. Consider only the stealth attacks that can be performed by attacking some set of substations ω with cardinality γ^* as defined in (17.25). For every measurement k with $\Gamma_k = \gamma^*$ there is at least one such stealth attack.

(ii) For every measurement k find the substations that appear in all of the corresponding stealth attacks. Call these *critical* substations for measurement k. There is always at least one *critical* substation $\hat{s}$ for every measurement k, $S(k)$.
(iii) For every *critical* substation $\hat{s}$ create alternate protection schemes $\mathcal{P}'$, $\mathcal{E}'$, and $\mathcal{R}'$ as described in the following subsections.
(iv) Calculate Γ'_k by solving (17.30) for every measurement assuming $\mathcal{P}'$, $\mathcal{E}'$, and $\mathcal{R}'$. Among all $\mathcal{P}'$, $\mathcal{E}'$, and $\mathcal{R}'$ discard the ones that result in $w'_\gamma > w_\gamma$ for some $\gamma \leq \gamma^*$. If an alternate protection scheme remains, then pick the alternate protection scheme for which w'_{γ^*} is minimal. Otherwise, if $\gamma^* \geq \max\{\gamma | w_\gamma > 0\}$, terminate. If not, set $\gamma^* = \gamma^* + 1$ and restart from (i).

This fairly general algorithm can be used to deploy different forms of protection in the system through the calculation of the alternate protection schemes $\mathcal{P}'$, $\mathcal{E}'$, or $\mathcal{R}'$. In the following we show examples for routing and for authentication.

17.6.1 Single-path and multi-path routing

We first consider single-path and multi-path routing, because modifying routes has smaller complexity than deploying authentication or protection. In both cases the alternate protection schemes differ only in terms of routing, and consequently $\mathcal{P}' = \mathcal{P}$ and $\mathcal{E}' = \mathcal{E}$.

For the case of single-path routing to obtain $\mathcal{R}'$ from $\mathcal{R}$ for a critical substation $\hat{s}$, we modify the only route $r_1^{\hat{s}}$ in $\mathcal{R}_{\hat{s}}$. For a route $r_1^{\hat{s}}$ we create the shortest alternate route $r_1^{\hat{s}'}$ that avoids the substation $s \in r_1^{\hat{s}}$ that appears in most substation attacks ω with cardinality γ^*.

For the case of multi-path routing to obtain $\mathcal{R}'$ from $\mathcal{R}$ for a critical substation $\hat{s}$, we consider the single route $r_1^{\hat{s}}$ in $\mathcal{R}_{\hat{s}}$, and construct the shortest route $r_2^{\hat{s}'}$ such that $r_2^{\hat{s}'}$ and $r_1^{\hat{s}}$ are node-disjoint. The routes in $\mathcal{R}_{\hat{s}}'$ are then $r_1^{\hat{s}'} = r_1^{\hat{s}}$ and $r_2^{\hat{s}'}$.

Figure 17.7 shows the maximum normalized substation attack impact, i.e., $\max_s I_s / M$, as a function of the number of single-path routes changed in the 118-bus system. The maximum attack impact shows a very fast decay, and decreases by almost a factor of two. At the same time the average path length to the control centre increases by only 10%. Figure 17.8 shows the number of measurements that have attack cost 1, 2, and 3 (i.e., w_1, w_2, and w_3) as a function of the number of routes changed in the 118-bus system. By changing single-path routes, the algorithm could increase the attack cost for about 200 measurements from $\Gamma_k = 1$ to $\Gamma_k = 2$, and for some measurements to $\Gamma_k = 3$ (e.g., at iteration 5).

Figure 17.9 shows the maximum normalized substation attack impact and the number of measurements with attack costs 1 to 4 as a function of the number of multi-path routes in the system. Multi-path routing introduces complexity in the management of the communication infrastructure. In the case of SDH at the link layer several virtual circuits have to be configured and maintained. In the case of Ethernet some form of traffic engineering is required (e.g., MPLS). Since the cost of establishing a multi-path route from a substation to the control centre has a higher cost than changing a single-path

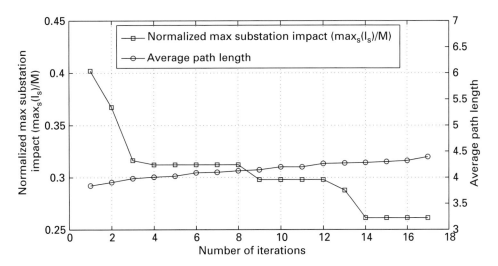

Figure 17.7 Maximum normalized attack impact and average path length vs. number of single-path routes changed in the IEEE 118-bus system.

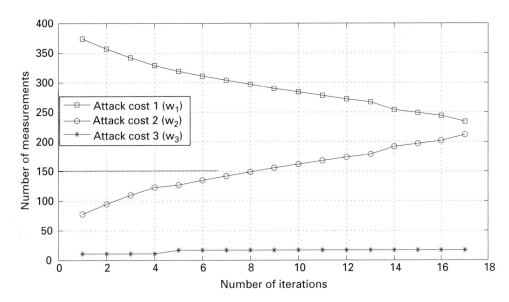

Figure 17.8 Number of measurements for various attack costs vs. number of routes changed in the IEEE 118-bus system.

route, we took the set of single-path routes $\mathcal{R}$ obtained in the last iteration shown in Figures 17.7 and 17.8 as the starting point for deploying multi-path routing. Compared to single-path routing, multi-path routing could decrease the maximum attack impact by 50% through increasing the number of measurements with attack cost $\Gamma_k = 2$ and $\Gamma_k = 3$. About 80 measurements have attack cost 1 when the algorithm terminates.

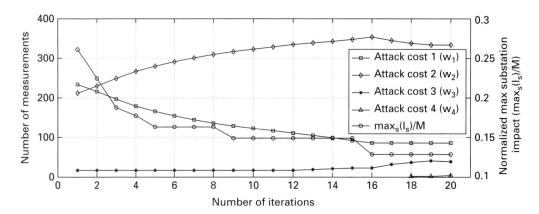

Figure 17.9 Maximum attack impact and number of measurements for various attack costs vs. number of multi-path routes. IEEE 118-bus system.

17.6.2 Data authentication and protection

In the case of authentication, the alternate protection schemes differ in terms of the set of authenticated substations $\mathcal{E}$. Consequently, $\mathcal{P}' = \mathcal{P}$ and $\mathcal{R}' = \mathcal{R}$. To obtain $\mathcal{E}'$ from $\mathcal{E}$ for a critical substation $\hat{s}$, we add substation $\hat{s}$ to the set of substations using authentication, i.e., $\mathcal{E}' = \mathcal{E} \cup \hat{s}$.

Figure 17.10 shows the maximum normalized substation attack impact and the number of measurements with attack cost 1 to 5 as a function of the number of authenticated RTUs in the system. Again, we took the set of routes $\mathcal{R}$ obtained in the last iteration of the algorithm for single-path routing as the starting point for deploying authentication. Authentication eliminates measurements with attack cost $\Gamma_k = 1$ after 25 substations are authenticated. Upon termination, more measurements have attacks cost $\Gamma_k \geq 3$ than using multi-path routing.

Protection can be considered in a similar way, i.e., $\mathcal{P}' = \mathcal{P} \cup \hat{s}$. In the case $\mathcal{E} = \mathcal{P}$ (e.g., tamper-proof RTU equipped with secret key), $\sigma_{\mathcal{E}}(r_i(s)) = \emptyset \ \forall s \in \mathcal{E}$, that is, the measurements originating from substations $s \in \mathcal{E}$ are not susceptible to attacks [22]. This is equivalent to the RTU attack and protection model considered in Section 17.3. Hence, it is possible to achieve $\Gamma_k = \infty \ \forall k$ by letting $\mathcal{E} = \mathcal{P}$ be an appropriate dominating set of $\mathcal{G}$.

17.7 Conclusion

In this chapter, we have given an overview of recent work on the vulnerability and protection of the state estimator in SCADA EMS systems. We have briefly presented the principles of power system modelling, power-system state estimation, and bad-data detection. Based on these models, we have discussed the possibility of stealth attacks against the state estimator and the bad-data detection algorithms. We have then considered two models of the substation to control centre SCADA communication

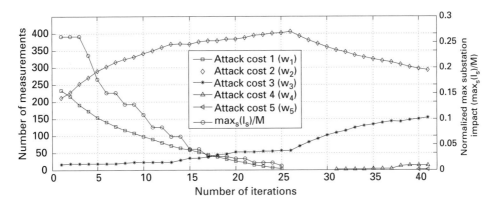

Figure 17.10 Maximum attack impact and number of measurements for various attack costs vs. number of authenticated RTUs ($|\mathcal{E}|$). IEEE 118-bus system.

infrastructure and developed security metrics and algorithms to calculate the security metrics. We used these metrics and algorithms to investigate how stealth attacks can be mitigated. For the case of point-to-point substation to control centre communication channels we considered data authentication as the primary means of mitigation. For the case of substation to control centre communication over a routed network we also considered routing as a potential candidate for mitigation. We have used IEEE benchmark power systems to illustrate that, especially in the case of a routed communication network, the system can be attacked at a rather low cost. Nevertheless, it is also clear from our results that by securing strategically chosen components of the communication infrastructure, stealth attacks can be very efficiently mitigated.

References

[1] A. Abur and A. G. Exposito, *Power System State Estimation: Theory and Implementation*. Marcel Dekker, Inc., 2004.

[2] A. Monticelli, 'Electric power system state estimation', in *Proceedings of the IEEE*, 2000.

[3] L. Mili, T. V. Cutsem, and M. Ribbens-Pavella, 'Bad data identification methods in power system state estimation – a comparative study', *IEEE Transactions on Power Apparatus and Systems*, vol. 104, no. 11, pp. 3037–3049, November 1985.

[4] F. F. Wu and W.-H. E. Liu, 'Detection of topology errors by state estimation', *IEEE Transactions on Power Systems*, vol. 4, no. 1, pp. 176–183, February 1989.

[5] Y. Liu, P. Ning, and M. Reiter, 'False data injection attacks against state estimation in electric power grids', in *Proceedings of ACM Conference on Computer and Communications Security*, pp. 21–32, November 2009.

[6] A. Teixeira, G. Dán, H. Sandberg, and K. H. Johansson, 'A cyber security study of a SCADA energy management system: Stealthy deception attacks on the state estimator', in *Proceedings of IFAC World Congress*, August 2011.

[7] O. Kosut, L. Jia, R. Thomas, and L. Tong, 'Malicious data attacks on smart grid state estimation: attack strategies and countermeasures', in *Proceedings of IEEE SmartGridComm*, October 2010.
[8] Cplex [online]. Available at: http://www-01.ibm.com/software/integration/optimization/cplex-optimizer/
[9] Gurobi. http://www.gurobi.com/
[10] G. Dán and H. Sandberg, 'Stealth attacks and protection schemes for state estimators in power systems', in *Proceedings of IEEE SmartGridComm*, October 2010.
[11] J. Tsitsiklis and D. Bertsimas, *Introduction to Linear Optimization*. Athena Scientific, 1997.
[12] J.-C. Picard and M. Queyranne, 'On the structure of all minimum cuts in a network and applications', in *Combinatorial Optimization II*, ser. Mathematical Programming Studies, 1980, vol. 13, pp. 8–16.
[13] L. Schrage and K. R. Baker, 'Dynamic programming solution of sequencing problems with precedence constraints', *Operations Research*, vol. 26, no. 3, pp. 444–449, 1978.
[14] K. Sou, H. Sandberg, and K. Johansson, 'Electric power network security analysis via minimum cut relaxation', in *Proceedings of IEEE Conference on Decision and Control (CDC)*, December 2011.
[15] R. Zimmerman, C. Murillo-Sánchez, and R. Thomas, 'MATPOWER steady-state operations, planning and analysis tools for power systems research and education', *IEEE Transactions on Power Systems*, vol. 26, no. 1, pp. 12–19, 2011.
[16] R. Tibshirani, 'Regression shrinkage and selection via the lasso', *Journal of the Royal Statistical Society, Series B*, vol. 58, no. 1, pp. 267–288, 1996.
[17] S. Mallat and Z. Zhang, 'Matching pursuit with time-frequency dictionaries', *IEEE Transactions on Signal Processing*, vol. 41, pp. 3397–3415, 1993.
[18] S. Chen, D. Donoho, and M. Saunders, 'Atomic decomposition by basis pursuit', *SIAM Journal on Scientific Computing*, vol. 20, no. 1, pp. 33–61, 1999.
[19] R. B. Bobba, K. M. Rogers, Q. Wang, H. Khurana, K. Nahrstedt, and T. J. Overbye, 'Detecting false data injection attacks on dc state estimation', in *Preprints of the First Workshop on Secure Control Systems, CPSWEEK 2010*, 2010.
[20] G. Strang, *Introduction to Applied Mathematics*. Wellesley-Cambridge Press, 1986.
[21] J. Ignizio and T. Cavalier, *Linear Programming*. Prentice Hall, 1994.
[22] O. Vuković, K. Sou, G. Dán, and H. Sandberg, 'Network-layer protection schemes against stealth attacks on state estimators in power systems', in *Proceedings of IEEE SmartGridComm*, October 2011.
[23] D. Atanackovic, J. H. Clapauch, G. Dwernychuk, J. Gurney, and H. Lee, 'First steps to wide area control', *IEEE Power and Energy Magazine*, vol. 6, no. 1, pp. 61–68, 2008.

18 A hierarchical security architecture for smart grid

Quanyan Zhu and Tamer Başar

18.1 Introduction

The smart grid aims to provide reliable, efficient, secure, and quality energy generation/distribution/consumption using modern information, communications, and electronics technologies. The integration with modern IT technology moves the power grid from an outdated, proprietary technology to more common ones such as personal computers, Microsoft Windows, TCP/IP/Ethernet, etc. It can provide the power grid with the capability of supporting two-way energy and information flow, isolate and restore power outages more quickly, facilitate the integration of renewable energy resources into the grid, and empower the consumer with tools for optimizing energy consumption. However, in the meantime, it poses security challenges on power systems as the integration exposes the system to public networks.

Many power grid incidents in the past have been related to software vulnerabilities. In [1], it is reported that hackers have inserted software into the US power grid, potentially allowing the grid to be disrupted at a later date from a remote location. As reported in [2], it is believed that an inappropriate software update has led to a recent emergency shutdown of a nuclear power plant in Georgia, which lasted for 48 hours. In [3], it has been reported that a computer worm, Stuxnet, has been spread to target Siemens SCADA systems that are configured to control and monitor specific industrial processes. On 29 November 2010, Iran confirmed that its nuclear programme had indeed been damaged by Stuxnet [4, 5]. The infestation by this worm may have damaged Iran's nuclear facilities in Natanz and eventually delayed the start-up of Iran's nuclear power plant.

Modern power systems do not have built-in security functionalities and the security solutions in regular IT systems may not always apply to systems in critical infrastructures. This is because critical infrastructures have different goals and assumptions concerning what needs to be protected, and have specific applications that are not originally designed for a general IT environment. Hence, it is necessary to develop unique security solutions to fill the gap where IT solutions do not apply.

In this chapter, we describe a layered architecture to address the security issues in power grids, which facilitates identifying research problems and challenges at each layer and building models for designing security measures for control systems in critical infrastructures. We also emphasize a cross-layer viewpoint towards security issues in power grids in that each layer can have security dependence on the other layers. We need to understand the tradeoff between the information assurance and the physical layer system performance before designing defence strategies against potential cyber threats and attacks. As examples, we address three security issues of smart grid at different layers, namely, the resilient control-design problem at the physical power plant, the data-routing problem at the network and communication layer, and the information security management at the application layers.

The concept of hierarchical structures has been adopted as a solution for the Internet and manufacturing operations. The well-known layered structure of OSI model for the Internet has influenced the integration between software and hardware [6]. The upper layers of the OSI model represent software that implements network services like encryption and management. The lower layers implement more primitive, hardware-oriented functions like routing, addressing, and flow control. The layered structure introduces a practical framework for network technology development at individual layers and also allows cross-layer methods to investigate issues across these virtual boundaries between layers [7].

The integration between enterprise and industrial control systems is guided by ISA95 standards for information exchange between enterprise and manufacturing control activities and their supporting IT systems. It defines levels within a manufacturing operation based on the Purdue reference model (PRM) for computer integrated manufacturing (CIM) [8]. PRM has formed the basis for ISA95 standards today, providing the openness necessary to unify plant resource requirements.

The hierarchical model in this chapter extends the notions from OSI and PRM models and integrates them for smart grids. Our model is related to the notion of resilient control systems proposed in [9–11]. The goal of a resilient control system is to maintain state awareness and an accepted level of operational normalcy in response to disturbances, including threats of an unexpected and malicious nature [9]. To achieve resilience, the control system design is divided into four parallel areas: human systems, complex networks, cyber awareness, and data fusion. The hierarchical perspective shares the similar divide-and-conquer philosophy but views the system differently in a hierarchically structured way.

The model is also related to that in [12], where a novel framework of security solution for power-grid automation has been proposed. The integrated security framework has three layers, namely, power, automation and control, and security management. The automation and control system layer monitors and controls power transmission and distribution processes, while the security layer provides security features.

The rest of the chapter is organized as follows. In Section 18.2, we first describe the general hierarchical architecture of cyber–physical systems and the related security issues associated with each layer. In Section 18.3, we focus on the cyber and physical layers of the smart grid and propose a general cross-layer framework for robust and resilient controller design. We use a single-machine infinite-bus system to illustrate as an example the design methodology for the generator voltage regulation problem. In Section 18.4, we study a secure network-routing problem at the data communication and networking layers of the smart grid. In addition, we compare the centralized vs. decentralized routing protocols and propose a hybrid architecture as a result of the tradeoff between robustness and resilience in the smart grid. In Section 18.5, we study the information security at the management layer of the grid. More specifically, we propose a model to derive optimal patch scheduling for the control systems in the smart grid. We conclude finally in Section 18.6 and discuss future work that can follow from the hierarchical model.

18.2 Hierarchical architecture

Smart grid comprises physical power systems and cyber information systems. The integration of the physical systems with the cyber space allows new degrees of automation and human–machine interactions. The uncertainties and hostilities existing in the cyber environment have brought emerging concerns for modern power systems. It is of supreme importance to have a system that maintains state awareness and an accepted level of operational normalcy in response to disturbances, including threats of an unexpected and malicious nature [9, 13].

The physical systems of the power grid can be made resilient by incorporating features such as robustness and reliability [14], while the cyber components can be enhanced by many cyber-security measures to ensure dependability, security, and privacy. However, the integration of cyber and physical components does not necessarily ensure overall reliability, robustness, security, and resilience of the power system. The interaction between the two environments can create new challenges in addition to the existing ones. To address these challenges, we first need to understand the architecture of smart grids.

A smart grid can be organized hierarchically into six layers, namely, physical layer, control layer, data communication layer, network layer, supervisory layer, and management layer. This hierarchical structure is depicted in Figure 18.1. The first two layers, physical layer, and control layer, can be seen jointly as the physical environment

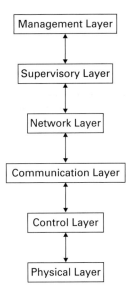

Figure 18.1 The hierarchical structure of ICSs composed of six layers. The physical layer resides at the bottom level and the management layer at the highest echelon.

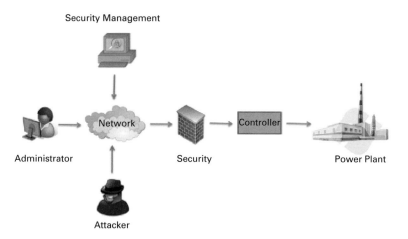

Figure 18.2 The hierarchical view towards security of a power plant.

of the system. The data communication layer and network layer comprise the cyber environment of the power grid. The supervisory layer together with the management layer constitute the higher-level application layer where services and human–machine interactions take place.

A simplified structure of power grid is depicted in Figure 18.2. The power plant is at the physical layer and the communication network and security devices are at the network and communication layers. The controller interacts with the communication layer and the physical layer. An administrator is at the supervisory layer to monitor and control the network and the system. Security management is at the highest layer where security policies are made against potential threats from attackers. SCADA is the fundamental monitoring and control architecture at the control area level. The control centres of all major US utilities have implemented a supporting SCADA for processing data and coordinating commands to manage power generation and delivery within the EHV and HV (bulk) portion of their own electric power system [15].

The layered structure is also commonly seen in SCADA systems [16]. In large SCADA systems, there is usually a communication network connecting individual PLCs to the operator interface equipment at the central control room. There are communication networks used at a lower level in the control-system architecture for communication between different PLCs in the same subsystems or facility, as well as for communication between field devices and individual PLCs. Figure 18.3 describes typical SCADA network levels, where four layers are depicted, namely, supervisory level, communication level, control level, and device/physical level.

The information structure of SCADA systems in today's power grids is highly hierarchical. Each primary controller utilizes its own local measurement only, and each control area utilizes measurements in its own utility only and has its own SCADA system. Protection mechanisms are preprogrammed to protect individual pieces of equipment and rarely require communication [17, 18].

A hierarchical security architecture for smart grid

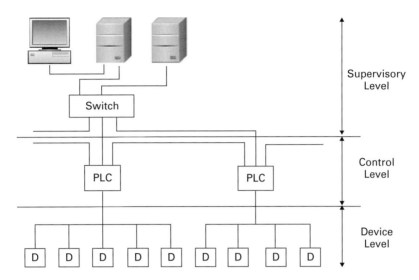

Figure 18.3 Typical SCADA network levels.

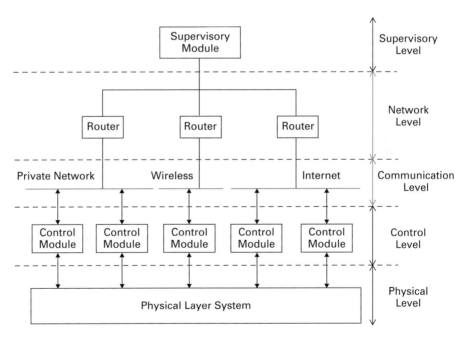

Figure 18.4 A conceptual control system with layering.

To further describe the functions at each layer, we resort to Figure 18.4, which describes conceptually a smart grid system with a layering architecture. The lowest level is the physical layer where the physical/chemical processes we need to control or monitor reside. The control layer includes control devices (Figure 18.5) that are

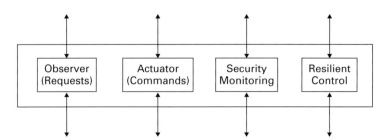

Figure 18.5   Control module.

encoded with control algorithms that have robust, reliable, secure, fault-tolerant features. The communication layer passes data between devices and different layers. The network layer includes the data packet routing and topological features of control systems. The supervisory layer offers human–machine interactions and capability of centralized decision-making. The management layer makes economic and high-level operational decisions.

In subsequent subsections, we identify problems and challenges at each layer and propose problems whose resolution requires a cross-layer viewpoint.

18.2.1   Physical layer

The physical layer comprises the physical plant to be controlled. It is often described by an ordinary differential equation (ODE) model from physical or chemical laws. It can also be described by difference equations, Markov models, or model-free statistics. We have the following challenges that pertain to the security and reliability of the physical infrastructure. First, it is important to find appropriate measures to protect the physical infrastructure against vandalism, environmental change, unexpected events, etc. Such measures often need a cost-and-benefit analysis involving the value assessment of a particular infrastructure. Second, it is also essential for engineers to build the physical systems with more dependable components and more reliable architecture. This raises concerns over the physical maintenance of the control system infrastructure that demands a cross-layer decision-making between the management and physical layers [19].

18.2.2   Control layer

The control layer consists of multiple control components, including observers/sensors, intrusion-detection systems (IDSs), actuators, and other intelligent control components. An observer has the sensing capability that collects data from the physical layer and may estimate the physical state of the current system. Sensors may need to have redundancies to ensure correct reading of the states. The sensor data can be fused locally or sent to the supervisor level for global fusion. A reliable architecture of sensor data fusion will be a critical concern.

An IDS protects the physical layer as well as the communication layer by performing anomaly-based or signature-based intrusion detection. An anomaly-based ID is more common for the physical layer, whereas a signature-based ID is more common for the packets or traffic at the communication layer. If an intrusion or an anomaly occurs, an IDS raises an alert to the supervisor or works hand-in-hand with built-in intrusion-prevention systems (related to emergency responses, e.g., control reconfiguration) to take immediate action. There lies a fundamental tradeoff between local decisions versus a centralized decision when intrusions are detected. A local decision, for example, made by a prevention system, can react in time to unanticipated events; however, it may incur a high packet drop rate if the local decision suffers high false negative rates due to incomplete information. Hence, it is an important architectural concern whether the diagnosis and control modules need to operate locally with IDS or globally with a supervisor.

18.2.3 Communication layer

The communication layer is where we have a communication channel between control-layer components and network-layer routers. The communication channel can take multiple forms: wireless, physical cable, bluetooth, etc. The communication layer handles the data communication between devices and layers. It is an important vehicle that runs between different layers and devices. It can often be vulnerable to attacks, such as jamming and eavesdropping. There are also privacy concerns over the data at this layer. Such problems have been studied within the context of wireless communication networks [20]. However, the goal of a critical infrastructure may require different types of study than the conventional studies of these issues.

18.2.4 Network layer

The network layer concerns the topology of the architecture. It comprises two major components: one is network formation and the other one is routing. We can randomize the use of routes to disguise or confuse the attackers so as to achieve certain security or secrecy or minimum delay. Moreover, once a route is chosen, how much data should be sent on that route has long been a concern for researchers in communications [21–23]. In control systems, many specifics to the data form and rates may allow us to reconsider this problem in a control domain.

18.2.5 Supervisory layer

The supervisory layer coordinates all layers by designing and sending appropriate commands. It can be viewed as the brain of the system. Its main function is to perform critical data analysis or fusion to provide immediate and precise assessment of the situation. It is also a holistic policy-maker that distributes resources in an efficient way. The resources include communication resources and maintenance budget, as well as control efforts. In centralized control, we have one supervisory module that collects and stores all historical data and serves as a powerful data-fusion and signal-processing centre [24, 25].

18.2.6 Management layer

The management layer is a higher-level decision-making engine, where the decision-makers take an economic perspective towards the resource allocation problems in control systems. At this layer, we deal with problems such as: (i) how to budget resources to different systems to accomplish a goal; and (ii) how to manage patches for control systems, e.g., disclosure of vulnerabilities to vendors, development and release of patches [26].

18.3 Robust and resilient control

The layered architecture in Figure 18.4 can facilitate an understanding of the cross-layer interactions between the physical world and the cyber world. In this section, we aim to establish a framework for designing a resilient controller for the physical power systems. In Figure 18.6, we describe a hybrid system model that interconnects the cyber and physical environments. We use $x(t)$ and $\theta(t)$ to denote the continuous physical state and the discrete cyber state of the system, which are governed by the laws f and Λ, respectively. The physical state $x(t)$ is subject to disturbances w and can be controlled by u. The cyber state $\theta(t)$ is controlled by the defence mechanism l used by the network administrator as well as the attacker's action a.

We view resilient control as a cross-layer control design, which takes into account the known range of unknown deterministic uncertainties at each state as well as the random unanticipated events that trigger the transition from one system state to another. Hence, it has the property of disturbance attenuation or rejection of physical uncertainties as well as damage mitigation or resilience to sudden cyber attacks. It would be possible to derive resilient control for the closed-loop perfect-state measurement information structure in a general setting with the transition law depending on the control action, which can further be simplified to the special case of the linear quadratic problem.

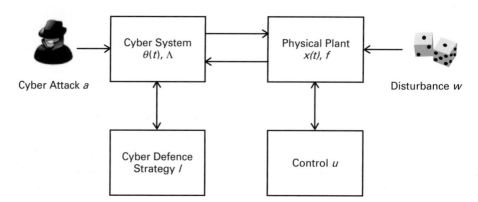

Figure 18.6 The interactions between the cyber and physical systems are captured by their dynamics governed by the transition law Λ and the dynamical system f. The physical system state $x(t)$ is controlled by u with the presence of disturbances and noises. The cyber state $\theta(t)$ is controlled by the defence mechanism l used by the network administrator as well as the attacker's action a.

The framework depicted in Figure 18.6 can be used to describe the voltage regulation problem of a power generator subject to sudden faults or attacks. A power system has multiple generators interconnected through a large dynamic network. A common approach to designing control systems for generators is to model the dynamics of a single generator and to approximate everything else as an infinite bus, i.e., the voltage and phase of the entire network are not affected by the input power or field excitation of the generator. Shown in Figure 18.7, a single generator is connected to an infinite bus through parallel transmission lines. It is possible to design a stabilizing control, called the power system stabilizer (PSS), to damp out the low-frequency oscillations for a single-machine infinite-bus (SMIB) system [27–29]. A fault can occur as a result of an unanticipated cyber attack. For example, an attacker can break into the IT system and damage the circuit breakers in a power grid, leading to operation under a faulty state. It is important that one designs a controller to regulate the system to equilibrium as quickly as possible if such a failure occurs [30], and at the same time a defence mechanism to protect the systems from possible attacks. In Figure 18.8, we describe a two-state operation. One is under the normal state ($\theta = 1$) and the other is the post-attack state ($\theta = 2$). We now consider a specific mathematical formulation.

Let δ be the power angle, ω the relative speed, P_e the active electrical power delivered by the generator, and let u_f be the input of the amplifier of the generator, taken as the

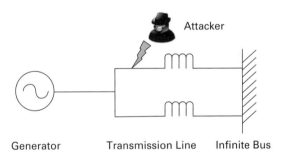

Figure 18.7 An illustration of a synchronous-machine infinite-bus system. A cyber attack can lead to the breaker failure and result in power loss. A control scheme is needed to regulate the system to equilibrium as quickly as possible after the attack.

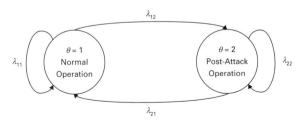

Figure 18.8 A two-state operation model for a power system with one normal operation state ($\theta = 1$) and one post-attack state ($\theta = 2$); $\lambda_{ij}, i, j = 1, 2$, are the transition rates between the two states.

Table 18.1. Table of parameters

Symbol and value	Meaning
$k_c = 1$	Gain of the excitation amplifier
$D = 5.0$	Per-unit damping constant
$H = 4.0$	Per-unit inertia constant
$\omega_0 = 100\pi$	Synchronous machine speed
$P_m = 0.9$	Mechanical input power
$T_{d0} = 6.9$	Direct-axis transient short-circuit time constant
$V_s = 0.91$	Infinite bus voltage
$x_d = 1.863$	Direct-axis reactance of the generator
$x'_d = 0.257$	Direct-axis transient reactance of the generator
$x_T = 0.127$	Reactance of the transformer
$x_L = 0.4853$	Reactance of the transmission line

control variable. The system equations to model SMIB are then:

$$\dot{\delta}(t) = \omega(t)$$
$$\dot{\omega}(t) = -\frac{D}{2H}\omega(t) + \frac{\omega_0}{2H}(P_m(t) - P_e(t))$$
$$\dot{P}_e(t) = -\frac{1}{T'_{d0}}P_e(t) + \frac{1}{T'_{d0}}\left\{\frac{V_s}{x_{ds}}\sin(\delta(t))\left[k_c u_f + T'_{d0}(x_d - x'_d)\frac{V_s}{x'_{ds}}\omega(t)\sin(\delta(t))\right]\right.$$
$$\left. + T'_{d0}\omega(t)\cot(\delta(t))\right\} + w, \tag{18.1}$$

where w is the disturbance, $T'_{d0} = \frac{x'_{ds}}{x_{ds}}T_{d0}$, $x_{ds} = x_T + x_L + x_d$, and $x'_{ds} = x_T + x_L + x'_d$; the main parameters are listed in Table 18.1 and their values are chosen based on [28].

Under normal operation ($\theta = 1$), the control objective is to regulate the synchronous machine states (δ, ω, P_e) to the level of $(\delta_0, 0, P_m)$; here one can linearize system model (18.1) at $(\delta_0, 0, P_m)$. An unanticipated fault caused by a cyber attack can happen at the rate λ_{12}. When a circuit breaker is compromised, the total reactance on the transmission line will change accordingly.

The transition rates $\lambda_{ij}, i, j = 1, 2$, take the following parametrized form: $\lambda_{12} = p$, $\lambda_{11} = -p, \lambda_{21} = \lambda_{22} = 0$, where we have assumed that the operation cannot immediately be restored following an attack. At the cyber layer, the administrator can take two actions, namely, to defend ($l_1 = $ D) or not to defend ($l_2 = $ ND). The attacker can also take two actions, namely, to attack ($a_1 = $ A) or not to ($a_2 = $ NA). The parameter p determines the transition law with respect to pure strategies, and its values are tabulated in Table 18.2.

In Table 18.2, we have assumed a higher probability of transition to a failure state if the attacker launches an attack while the cyber system does not have proper measures to defend itself. On the other hand, the probability is lower if the cyber system can defend itself from attacks. In this table, we have assumed a base transition rate of 0.05 to denote the inherent reliability of the physical system without exogenous attacks.

Let $x := [x_1, x_2, x_3]' = [\delta, \omega, P_e]'$ be the aggregate state. The goal of robust control is to find an optimal state-feedback control $u(t) = \mu^*(t, x, \theta)$ so that the infinite-horizon

Table 18.2. Defender vs. attacker

	A	NA
D	0.1	0.05
ND	0.95	0.05

cost function L, defined below, is minimized for the worst-case disturbance $w(t)$ to achieve the best noise attenuation [31]:

$$L(x,u,w;\theta,t_0) = \mathbb{E}\int_{t_0}^{\infty}(|x(t)|_{Q^i}^2 + |u(t)|_{R^i}^2 - \gamma^2|w(t)|^2)dt, \qquad (18.2)$$

where $\gamma > \gamma^*$ is a noise-attenuation level greater than the best achievable attenuation level γ^*, $|\cdot|$ denotes the Euclidean norm with appropriate weighting, $A^i, B^i, D^i, Q^i, R^i, i = 1,2$, are matrices of appropriate dimensions, whose entries are continuous functions of time t, with $Q^i(\cdot) \geq 0, R^i(\cdot) > 0$, which are actually taken to be constant matrices. One can choose these weighting matrices as

$$Q^1 = Q^2 = \begin{bmatrix} 1000 & 0 & 0 \\ 0 & 1 & 0 \\ 0 & 0 & 10 \end{bmatrix}, \quad R^1 = 10, R^2 = 1,$$

which emphasizes the regulation of power angle and the willingness to use more control in a post-attack state.

The optimal control μ^* depends on the interactions between the administrator and the attacker. Let **f** and **g** be the cyber strategies of the administrator and the attacker, respectively. The value function $V^i, i = 1,2$, achieved under the optimal control μ^*, is hence dependent on **f** and **g**.

The goal of resilient control is to find an optimal cyber-policy **f*** to defend against an attacker's optimal strategy **g*** at state $i = 1$. The defence against attacks happens on a longer time scale and involves decision-making at the human and cyber levels of the system. Using time-scale separation, the optimal defence mechanism can be designed by viewing the physical control system at its steady state at each cyber state θ at a given time k. The interaction between an attacker and a defending administrator can be captured by a zero-sum stochastic game with the defender aiming to maximize the long-term system performance or payoff function V_β, defined below, whereas the attacker aims to minimize it [32]:

$$V_\beta(i,\mathbf{f}(k),\mathbf{g}(k)) := \int_0^\infty e^{-\beta k}\mathbb{E}_i^{\mathbf{f}(k),\mathbf{g}(k)}V^i(k,\mathbf{f}(k),\mathbf{g}(k))dk,$$

where β is the discount factor. The operator $\mathbb{E}_i^{\mathbf{f},\mathbf{g}}, i = 1,2$, is the expectation operator and $V^i(k,\mathbf{f},\mathbf{g})$ is the optimal value function at state i with starting time at k, i.e.,

$$V^i(k,\mathbf{f},\mathbf{g}) = \min_u \max_w L(x,u,w;i,k). \qquad (18.3)$$

We can consider a class of mixed stationary strategies $\mathbf{f}$ that randomizes between $l_1 = $ D (to Defend) and $l_2 = $ ND (Not to Defend), and $\mathbf{g}$ that randomizes between $a_1 = $ A (to Attack) and $a_2 = $ NA (Not to Attack). It can be seen that the problem of finding $\mathbf{f}^*$ and $\mathbf{g}^*$ depends on the optimal control μ^* at the physical layer. Hence, in order to find the optimal control μ^* and optimal defence $\mathbf{f}^*$, one needs to solve the optimization problems (18.2) and (18.3) jointly.

Using jointly the optimality criteria for (18.2) and (18.3) [31, 32], one can obtain stationary saddle-point strategies $\mathbf{f}^* = [1, 0]'$, $\mathbf{g}^* = [0, 1]'$, and

$$\mu^*(t, x, 1) = -(R^1)^{-1} B^{1'} Z^1 x, \quad \mu^*(t, x, 2) = -(R^2)^{-1} B^{2'} Z^2 x,$$

where

$$Z^1 = \begin{bmatrix} 399.3266 & 31.8581 & -162.2334 \\ 31.8581 & 5.7083 & -15.2963 \\ -162.2334 & -15.2963 & 149.7459 \end{bmatrix}$$

and

$$Z^2 = \begin{bmatrix} 2.8512 & 0.1066 & -2.8575 \\ 0.1066 & 0.0345 & -0.1041 \\ -2.8575 & -0.1041 & 4.1506 \end{bmatrix}.$$

The pair of saddle-point strategies $\mathbf{f}^*$ and $\mathbf{g}^*$ indicates that the defender should always be defending and the attacker has no intent of launching an attack. The attenuation level under μ^* for the worst-case disturbance is $\gamma^* = 8.5$. Figure 18.9 shows the evolution of state ω (or x_2) with failure occuring at time $t = 10$. The optimal control design allows the state ω to be stabilized after the fault occurs.

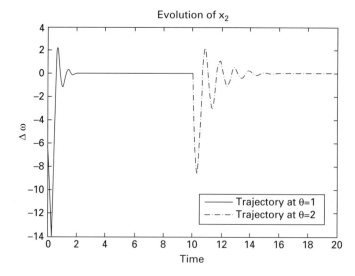

Figure 18.9 Evolution of state ω (x_2) with failure occuring at $t = 10$.

18.4 Secure network routing

One of the challenging issues at the data communication and networking layers of the smart grid in Figure 18.4 is the assurance of secure routing of phasor measurement unit (PMU) and smart meter (SM) data in the open network, which is enabled by the adoption of IP-based network technologies.

It is forecast that 276 million smart grid communication nodes will be shipped worldwide during the period from 2010 to 2016, with annual shipments increasing dramatically from 15 million in 2009 to 55 million by 2016 [33]. The current dedicated network or leased-line communication methods are not cost-effective to connect large numbers of PMUs and SMs. Thus, it is foreseen that IP-based network technologies will be widely adopted since they enable data to be exchanged in a routable fashion over an open network, such as the Internet [34–38]. This will bring benefits such as efficiency and reliability, and risks of cyber attacks as well. Without doubt, smart grid applications based on PMUs and SMs will change the current fundamental architecture of communication network of the power grid, and bring new requirements for communication security. Delay, incompleteness, and loss of PMU and SM data will adversely impact smart grid operation in terms of efficiency and reliability. Therefore, it is important to guarantee integrity and availability of those PMUs and SMs data. To meet the QoS requirements in terms of delay, bandwidth, and packet loss rate, QoS-based routing technologies have been studied in both academia and the telecommunications industry [39–42]. Unlike video and voice, data communications of PMUs and SMs have different meanings of real time and security, especially in terms of timely availability [34, 43–47]. Therefore, QoS-based and security-based routing schemes for smart grid communications should be studied and developed to meet smart grid application requirements in terms of delay, bandwidth, packet loss, and data integrity.

In this section, we leverage the hierarchical structure of power grids and propose a routing protocol that maximizes the QoS along the routing path to the control room. In addition, we optimize the data communication rates between the super data concentrator at the penultimate level with the control centre. We propose a hybrid structure of routing architecture to enable the resilience, robustness, and efficiency of the smart grid.

18.4.1 Hierarchical routing

Smart grid has a hierarchical structure that is built upon the current hierarchical power grid architecture. The end-users, such as households, communicate their power usage and pricing data with a local area substation which collects and processes data from SMs and PMUs. In the smart grid, the path for the measurement data may not be predetermined. The data can be relayed from smaller-scale data concentrators (DCs) to some super data concentrators (SDCs) and then to the control room. With the widely adopted IP-based network technologies, the communications between households and DCs can be in a multi-hop fashion through routers and relay devices. The goal of each household is to find a path with minimum delay and maximum security to reach DCs and then substations. This optimal decision can be enabled by the automated energy-management systems built

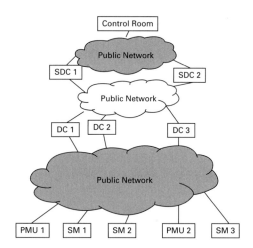

Figure 18.10 An example of the physical structure of the hierarchical smart grid communication network.

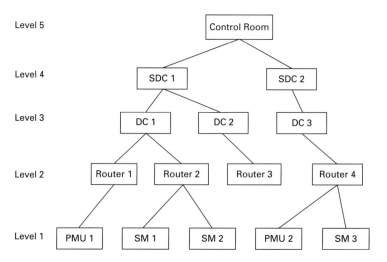

Figure 18.11 A snapshot of routing paths in the hierarchical smart grid.

in SMs. Figure 18.10 illustrates the physical structure of the smart grid communication network. The PMUs and SMs send data to DCs through a public network. DCs process the collected data and send the processed data to SDCs through (possibly) another public network.

The physical communication structure from local meters to the control room can be divided logically into several levels. Figure 18.11 illustrates a snapshot of routing paths in a simplified example of the hierarchical structure of the physical communication architecture, which is divided logically into five levels. For simplicity, we depict only the level of routers in Figure 18.11. In practice, there can be multiple levels of routers and they can be found in the public network between DCs and SDCs as well as SDCs and the control room.

In the depicted smart grid, the data from a PMU or SM has to make four hops to reach the control room. The decision for a meter to choose a router depends on the communication delay, security enhancement level, and packet loss rate. In addition, the decisions for a DC to choose a SDC also depend on the same criteria. The communication security at a node is measured by the number of security devices such as firewalls, intrusion-detection systems (IDSs), and intrusion-prevention systems (IPSs) deployed to reinforce the security level at that node. We assign higher utility to network routers and DCs that are protected by a larger number of firewalls, IDSs/IPSs, and dedicated private networks in contrast to public networks. This relatively simple metric only considers one aspect of the control system cyber security. It can be extended further to include more security aspects by considering the authorization mechanisms, the number of exploitable vulnerabilities, potential damages, as well as recovery time after successful attacks. The readers can refer to [13, 49–51] for more comprehensive metrics.

A tradeoff with higher security is the latency and packet loss rate incurred in data transmission. A secure network inevitably incurs delays in terms of processing (encrypting/decrypting) and examining data packets. We can model the process of security inspection by a tandem queueing network. Each security device corresponds to an M/M/1 queue whose external arrival rate follows a Poisson process and the service time follows an exponential distribution. The latency caused by the security devices such as IDSs/IPSs is due to the number of pre-defined attack signatures and patterns to be examined [19, 52, 53]. In addition, devices such as IPSs can also lead to high packet loss due to their false negative rates in the detection.

Furthermore, a node with higher level of security may be preferred by many meters or routers, eventually leading to a high volume of received data and hence a higher level of congestion delay. This fact enables a game-theoretical approach to analyse the distributed routing decisions in the smart grid. The solution concept of mixed-strategy Nash equilibrium [48] as a solution outcome is desirable for two reasons. First, in theory, a mixed-strategy Nash equilibrium always exists for a finite matrix game [48] and many learning algorithms such as fictitious play and replicator dynamics can lead to it [54–56]. Second, the randomness in the choice of routes makes it harder for an attacker to map out the routes in the smart grid.

18.4.2 Centralized vs. decentralized architectures

Section 18.4.1 introduced a hierarchical routing problem for the smart grid. The decision at the penultimate level is trivial since all the data should route to the control station. In reality, a local control station pulls data from its communicating SDCs. SDCs cannot send data to the control station at an arbitrary rate. The bandwidth and communication resources of a control station are often constrained. Hence, it becomes important to consider an appropriate resource allocation at the data communication between SDCs and the control station. The resource allocation scheme can be either centralized, i.e., determined by the control station, or decentralized, i.e., determined by individual SDCs.

A centralized routing architecture ensures global efficiency and is robust to small disturbances from SMs and individual DCs or SDCs. However, it is costly to implement centralized planning on a daily basis for a large-scale smart grid. In addition, global solutions can be less resilient to unexpected failures and attacks as they are less nimble for changes in routes and it takes time for the centralized planner to respond in a timely manner.

On the other hand, decentralized decision-making can be more computationally friendly based on local information and hence the response time to emergency is relatively fast. The entire system becomes more resilient to local faults and failures, thanks to the independence of the players and the reduced overhead on the response to unanticipated uncertainties. However, the decentralized solution can suffer from high loss due to inefficiency [24, 25]. Hence, we need to assess the tradeoff among efficiency, reliability, and resilience to design the communication protocol between the control stations and the SDCs.

Illustrated in Figure 18.12, a hybrid architecture of the communication infrastructure of L levels allows us to incorporate desirable features of the two architectures. One can adopt a centralized planning at the top levels $L-1$ and L while building a decentralized routing protocol between levels 1 and $L-1$. Such an architecture can enable robustness at the critical data centres and resilience at the lower user level.

Hence, the last-stage utility is determined in a centralized manner by the control room based on priority and load. The resource allocation at the last level is robust to small parametric disturbances and is independent of routing decisions between level 1 and level $L-1$. The routing decisions of the meters are resilient to router failures in a public

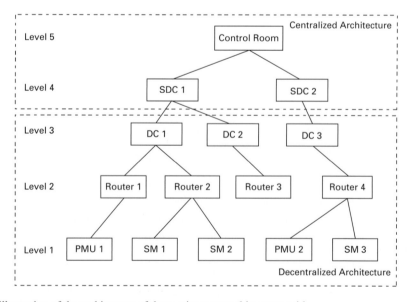

Figure 18.12 An illustration of the architecture of the routing protocol in smart grid.

network. The learning algorithms in [54–56] can be used to respond to a dysfunctional router by selecting a new router once the one in use is compromised.

18.5 Management of information security

The use of technologies with known vulnerabilities exposes power systems to potential exploits. In this section, we discuss information security management which is a crucial issue for power systems at the management layer in Figure 18.4.

18.5.1 Vulnerability management

The timing between the discovery of new vulnerabilities and their patch availabilities is crucial for the assessment of the security risk exposure of software users [57, 59–61]. The security focus in power systems is different from that in computer or communication networks. The application of patches for control systems needs to take into account the system functionality, avoiding the loss of service due to unexpected interruptions. The disclosure of software vulnerabilities for control systems is also a critical responsibility. Disclosure policy indirectly affects the speed and quality of the patch development. Government agencies such as CERT/CC (Computer Emergency Response Team/Coordination Centre) currently act as a third party in the public interest to set an optimal disclosure policy to influence the behaviour of vendors [62].

The decisions involving vulnerability disclosure, patch development, and patching are intricately interdependent. In Figure 18.13, we illustrate the relationship between these decision processes. A control system vulnerability starts with its discovery. It can be discovered by multiple parties, for example, individual users, government agencies, software vendors or attackers, and hence can incur different responses. The discoverer may choose not to disclose it to anyone, may choose to fully disclose through a forum

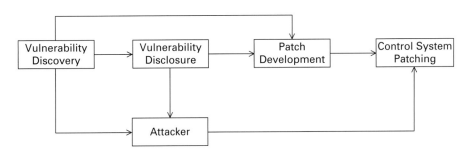

Figure 18.13 A holistic viewpoint towards vulnerability discovery, disclosure, development, and patching. An attacker can discover a vulnerability or learn it from a disclosure process, eventually influencing the speed of patch application. A discoverer can choose to fully disclose through a forum or report to the vendor or may provide to an attacker. A vulnerability can be disclosed to a vendor for patch development or leaked to the attacker.

such as bugtraq [58], may report to the vendor, or may provide to an attacker. Vulnerability disclosure is a decision process that can be initiated by those who have discovered the vulnerability. Patch development starts when the disclosure process reaches the vendor and finally a control system user decides on the application of the patches once they become available. An attacker can launch a successful attack once it acquires the knowledge of vulnerability before a control system patches its corresponding vulnerabilities. The entire process illustrated in Figure 18.13 involves many agents or players, for example, system users, software vendors, government agencies, attackers. Their state of knowledge has a direct impact on the state of vulnerability management.

We can compartmentalize the task of vulnerability management into different submodules: discovery, disclosure, development, and patching. The last two submodules are relatively convenient to deal with since the agents involved in the decision-making are very specific to the process. The models for discovery and disclosure can be more intricate in that these processes can be performed by many agents and hence specific models should be used for different agents to capture their incentives, utility, resources, and budgets.

18.5.2 User patching

In this subsection, we establish a model for users to determine the optimal time to patch their control systems. In control systems, it is known that the attack rate is low and the patching rate is low as well. It often occurs that users do not patch until there is a security alert, an available patch announcement, or an experienced security breach. The operation of the control system is separated into several operating periods. A control system cannot halt its operation until the end of an operating period. Let $B_k, k = 1, 2, \ldots$, denote the kth operating period since the last patching and $T_k = T_{k-1} + B_k = \sum_{i=1}^{k} B_i$, where T_0 is the beginning of the first operation period. Let τ be the time length between the start of the first operation period and a security alert or an attack. Let $f_\tau(t), t \geq 0$, be its probability density function, which is taken to be hyper-exponential with n phases and parameters $\lambda = (\lambda_1, \ldots, \lambda_n)$ and normalized weighting factor $\mathbf{q} = (q_1, \ldots, q_n)$, i.e.,

$$f_\tau(t) = \sum_{i=1}^{n} q_i g_i(t) = \sum_{i=1}^{n} q_i \lambda_i \exp(-\lambda_i t), \quad \sum_{i=1}^{n} q_i = 1. \quad (18.4)$$

Each phase i can be interpreted differently. For example, let $g_1(t)$ be the distribution of the arrival rate of alerts; $g_2(t)$ be the arrival rate of an attack; $g_3(t)$ be the arrival rate of an announcement of an available patch. We illustrate this model in Figure 18.14. At every $T_{k-1}, k \geq 1$, a control system starts to operate for a period B_k and then stops for monitoring and patching.

The decision of an administrator is to determine B_k so that the risk of an unpatched control system subject to potential attacks is minimized. The decision-making process can be viewed as a black box as in Figure 18.15, where the decision input is the knowledge of the arrival rates of an attack; the intrinsic system parameters are the monitoring cost c_m and the production cost parameters c_0 and c_1; and the decision output is the operation

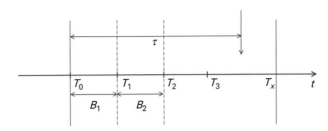

Figure 18.14 An illustration of the control system patching model.

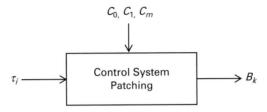

Figure 18.15 A system viewpoint towards control system patching decision.

period B_k. The input and output characteristics of the decision process assist us in integrating it into the system model in Figure 18.13.

Let $c_m \in \mathbb{R}_+$ be the monitoring cost at the end of each operation period and $c_p(t_k, b_{k+1}) : \mathbb{R}_+^2 \to \mathbb{R}_+$ be the operation cost of running the plant for b_{k+1} starting from t_k. We have the following dynamic programming (DP) equation to find the optimal decision policy to be taken at each period, taking into account the whole lifetime of the system:

$$V_k^*(t_k) = \min_{b_{k+1} \geq 0} \{\mathbb{E}[c(t_k, b_{k+1}) + \mathbb{P}(\tau_{t_k} > b_{k+1})V_{k+1}^*(t_{k+1})]\}, \qquad (18.5)$$

where b_k is the decision variable of operating time; $V_k^*(t_k)$ represents the optimal cost at time t_k. The term $\mathbb{P}(\tau_{t_k} > b_{k+1})$ represents the transition probability given by the conditional probability

$$\mathbb{P}(\tau_{t_k} > b_{k+1}) := P(\tau > t_k + b_{k+1} | \tau > t_k). \qquad (18.6)$$

The term $\mathbb{E}[c(t_k, b_{k+1})]$ is the stage cost at t_k given by

$$\mathbb{E}[c(t_k, b_{k+1})] = \hat{\gamma}\mathbb{E}[(b_{k+1} - \tau_{t_k})\mathbf{1}_{(\tau_{t_k} \leq b_{k+1})}] + c_m + c_p(t_k, b_{k+1}), \qquad (18.7)$$

where $\hat{\gamma}$ denotes the unit cost of untimely patching; τ_{t_k} is the conditional residual time counting from t_k given that $\tau_{t_k} > t_k$. By solving the DP equation, we can find a dynamic policy for the operation of the plants at each starting time t_k.

We can simplify the general model by assuming that the arrival of security alerts or breaches forms a single Poisson process with rate λ and the arrival time τ and the conditional residual time τ_t are exponentially distributed with parameter λ. Let $C_k^0 = \frac{1}{2}c_0 b_k^2$ be the cost for operating a plant non-stop for a period of time b_k. Denote by $C_k^1 = c_1 b_k$ the linear gain or profit from running the plant. Hence, we can assume that the cost c_p is given by

$$c_p(t_k, b_{k+1}) := C_k^0 - C_k^1 = \frac{1}{2}c_0 b_k^2 - c_1 b_k. \qquad (18.8)$$

Due to the memoryless property of exponential distribution, the DP equation in (18.5) can be simplified to

$$V^*(\lambda) = \min_{b \geq 0} \left\{ \hat{\gamma} \mathbb{E}[(b - \tau(\lambda))\mathbf{1}_{(\tau(\lambda) \leq b)}] + \frac{1}{2}c_0 b^2 - c_1 b + P(\tau(\lambda) > b)V^*(\lambda) \right\}.$$

For each fixed $b \geq 0$, $V(\lambda)$ can be solved from (18.6) (without the minimum) to yield

$$V(\lambda, b) = \frac{\hat{\gamma} \mathbb{E}[(b - \tau(\lambda))\mathbf{1}_{(\tau(\lambda) \leq b)}] + \frac{1}{2}c_0 b^2 - c_1 b}{1 - P(\tau(\lambda) > b)}. \qquad (18.9)$$

Note that in the above,

$$\mathbb{E}[(b - \tau(\lambda))\mathbf{1}_{(\tau(\lambda) \leq b)}] = \frac{\lambda b - 1 + \exp(-\lambda b)}{\lambda}, \qquad (18.10)$$

and the term in the denominator of (18.9) is as follows:

$$P(\tau(\lambda) > b) = \exp(-\lambda b). \qquad (18.11)$$

Hence, solving the DP is equivalent to finding a solution to the optimization problem (R-OPT) that takes into account the risk factor of potential threats and attacks:[1]

$$\text{(R-OPT)} \quad V^*(\lambda) = \min_{b \geq 0} V(\lambda, b). \qquad (18.12)$$

Note that a simple solution for operation time b without security risk consideration, i.e., ignoring the potential costs that can be incurred by attacks in (R-OPT), is based on a cost and benefit analysis solving the risk-free optimization problem (NR-OPT):[2]

$$\text{(NR-OPT)} \quad \min \frac{1}{2}c_0 b_k^2 - c_1 b_k + c_m, \qquad (18.13)$$

which yields a benchmark solution $b^* = c_1/c_0$.

---

[1] 'R-OPT' stands for risk-based optimization.
[2] 'NR-OPT' stands for no-risk-based optimization.

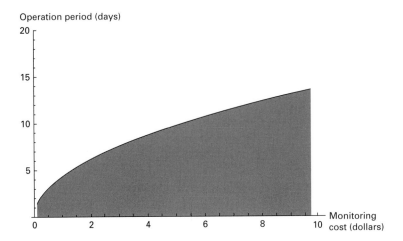

Figure 18.16 Operation period vs. monitoring cost.

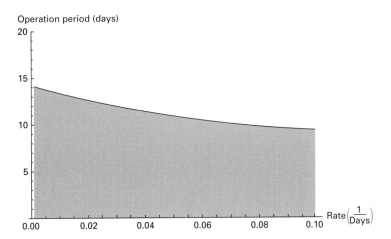

Figure 18.17 Operation period vs. attack rate.

We can solve the DP equations numerically for a given scenario of control systems. Set the parameters $c_1 = 10, c_0 = 0.1, \hat{\gamma} = 10, c_m = 10, \lambda = 0.001$ as the nominal case. In Figure 18.16, we show the optimal operation period versus the monitoring cost. It can be seen that when the cost becomes higher, the control system cannot afford a frequent checking and monitoring and hence the operation period increases. In Figure 18.17, we show the optimal operation period versus the attack rate. We observe that when the attack rate is high, the control system needs to decrease its operation period to monitor and update its system more often. From the simulation results, we notice that an optimal operation period of control systems given the currently estimated attack rate is roughly around half a month.

18.6 Conclusion

Security issues that arise in the smart grid constitute a pivotal concern in modern power-system infrastructures. In this chapter, we have discussed a six-layer security architecture for smart grid, motivated by the OSI for the Internet and PRM models for enterprise and control systems. We have identified the security challenges present at each layer and pinpointed a holistic viewpoint for security solutions in smart grid. The layered architecture facilitates the understanding of the tradeoff between the information assurance at the cyber-related layers and the physical-layer system performance.

We have addressed security issues at three different layers. The resilient control design at the physical system is pivotal for modern power systems. We have discussed a hybrid framework in which the occurrence of unanticipated events is modelled by a stochastic switching, and deterministic uncertainties are represented by the known range of disturbances. This framework has taken the resilience of physical systems into consideration and has enabled a cross-layer control design for modern power grids.

At the data communication and network layers, we have investigated the secure routing problem in smart grid, which arises from the adoption of IP-based network technologies due to the wide use of PMUs and smart meters. We have leveraged the hierarchical structure of power grids and discussed a routing protocol that is based on distributed optimization of the quality-of-service along individual routing paths. We have also illustrated the hybrid structure of the routing protocol to incorporate the desirable features of the centralized and decentralized architectures.

The use of information technologies in power systems poses additional potential threats due to the frequent disclosure of software vulnerabilities. At the higher level of the information security management layer, we have discussed a series of policy-making decisions on vulnerability discovery, disclosure, patch development, and patching. We have adopted a system approach to understand the interdependencies of these decision processes.

The future work that follows the hierarchical model involves the integration of learning algorithms for detection, automation, and reconfiguration in smart grid. In addition to the security problems illustrated in this chapter, there are other security and privacy issues existing at each layer, for example, the jamming and eavesdropping problems at the data communication layer, the user data privacy problem at the management layer, and the system reliability problem at the network layer. Furthermore, the hierarchical framework can be extended to study multi-agent systems. The interactions between subsystems in the smart grid can reside at the network, communication, and physical layers. It will be interesting to investigate the competition and cooperation for resources at multiple layers.

References

[1] S. Gorman, 'Electricity grid in U.S. penetrated by spies', *Wall Street Journal*, 8 April 2009 [online]. Available at: http://online.wsj.com/article/SB123914805204099085.html

[2] B. Krebs, 'Cyber incident blamed for nuclear power plant shutdown', *Washington Post*, 5 June 2008 [online]. Available at: http://www.washingtonpost.com/wp-dyn/content/article/2008/06/05/AR2008060501958.html

[3] R. McMillan, 'Siemens: Stuxnet worm hit industrial systems', *Computerworld*, 16 September 2010 [online]. Available at: http://www.computerworld.com/s/article/print/9185419 29.

[4] 'Iran confirms Stuxnet worm halted centrifuges', *CBS News*, November 2010 [online]. Available at: http://www.cbsnews.com/stories/2010/11/29/world/main7100197.shtml

[5] S. Greengard, 'The new face of war', *Communications of the ACM*, vol. 53, no. 12, pp. 20–22, December 2010.

[6] J. F. Kurose and K. W. Ross, *Computer Networking: A Top-Down Approach*. Addison Wesley, 5th edition, 2010.

[7] J. Wang, L. Li, S. H. Low, and J. C. Doyle, 'Cross-layer optimization in TCP/IP networks', *IEEE/ACM Transactions on Networking*, vol. 13, no. 3, pp. 582–568, June 2005.

[8] T. J. Williams (ed.), 'A reference model for computer integrated manufacturing (CIM): a description from the viewpoint of industrial automation', *International Purdue Workshop on Industrial Computer Systems*.

[9] C. G. Rieger, D. I. Gertman, and M. A. McQueen, 'Resilient control systems: next generation design research', in *Proceedings of 2nd Conference on Human System Interactions* (Catania, Italy, 21–23 May 2009). IEEE Press, pp. 629–633.

[10] C. Rieger, 'Notional examples and benchmark aspects of a resilient control systems', in *Proceedings of International Symposium on Resilient Control Systems (ISRCS)*, Idaho Falls, ID, August 2010.

[11] K. Villez, V. Venkatasubramanian, T. Spinner, R. Rengaswamy, H. Garcia, and C. Rieger, 'Achieving resilience in critical infrastructures: a case study for a nuclear power plant cooling loop', in *Proceedings of International Symposium on Resilient Control Systems (ISRCS)*, Idaho Falls, ID, August 2010.

[12] W. Dong, L. Yan, M. Jafari, P. Skare, and K. Rohde, 'An integrated security system of protecting smart grid against cyber attacks', in *Proceedings of Innovative Smart Grid Technologies (ISGT)*, pp. 1–7, January 2010.

[13] D. Wei and K. Ji, 'Resilient industrial control system (RICS): concepts, formulation, metrics, and insights', in *Proceedings of International Symposium on Resilient Control Systems (ISRCS)*, Idaho Falls, ID, August 2010.

[14] K. Moslehi and R. Kumar, 'Smart grid – a reliability perspective', in *Proceedings of Innovative Smart Grid Technologies (ISGT)*, pp. 1–8, January 2010.

[15] F. F. Wu, K. Moslehi, and A. Bose, 'Power system control centers: past, present, and future', *Proceedings of the IEEE*, vol. 93, no. 11, pp. 1890–1909, November 2005.

[16] Technical Manual, 'Supervisory control and data acquisition (SCADA) systems for command, control, communications, computer, intelligence, surveillance, and reconnaissance (C4ISR) facilities', Department of the Army [online]. Available at: http://www.army.mil/usapa/eng/DR_pubs/dr_a/pdf/tm5_601.pdf

[17] A. G. Phadke and J. S. Thorp, *Computer Relaying for Power Systems*. John Wiley, 1988.

[18] M. Ilic, 'From hierarchical to open access electric power systems', *Proceedings of the IEEE*, vol. 95, no. 5, pp. 1060–1084, May 2007.

[19] Q. Zhu and T. Başar, 'Towards a unifying security framework for cyber-physical systems', in *Proceedings of Workshop on the Foundations of Dependable and Secure Cyber-Physical Systems (FDSCPS-11)*, CPSWeek 2011, Chicago.

[20] M. Manshaei, Q. Zhu, T. Alpcan, T. Başar, and J.-P. Hubaux, 'Game theory meets network security and privacy', Technical Report EPFL-REPORT-151965, EPFL, 2010.

[21] W. Saad, Q. Zhu, T. Başar, Z. Han, and A. Hjorungnes, 'Hierarchical network formation games in the uplink of multi-hop wireless networks', in *Proceedings of 28th IEEE Conference on Global Telecommunications (GLOBECOM)*, Honolulu, HI, pp. 2390–2395, December 2009.

[22] Q. Zhu, D. Wei, and T. Başar, 'Secure routing in smart grids', in *Proceedings of Workshop on the Foundations of Dependable and Secure Cyber-Physical Systems (FDSCPS-11)*, CPSWeek 2011, Chicago.

[23] Q. Zhu, Z. Yuan, J. B. Song, Z. Han, and T. Başar, 'Dynamic interference minimization routing game for on-demand cognitive pilot channel', in *Proceedings of 28th IEEE Conference on Global Telecommunications (GLOBECOM)*, Miami, FL, 2010.

[24] T. Başar and Q. Zhu, 'Prices of anarchy, information, and cooperation in differential games', *Dynamic Games and Applications*, vol. 1, no. 1, pp. 50–73, 2010.

[25] Q. Zhu and T. Başar, 'Price of anarchy and price of information in linear quadratic differential games', in *Proceedings of American Control Conference (ACC)*, Baltimore, MD, 2010.

[26] Q. Zhu, M. McQueen, C. Rieger, and T. Başar, 'Management of control system information security: control system patch management', in *Proceedings of Workshop on the Foundations of Dependable and Secure Cyber-Physical Systems (FDSCPS-11)*, CPSWeek 2011, Chicago.

[27] Y.-N. Yu, K. Vongsuriya, and L. N. Wedman, 'Application of an optimal control theory to a power system', *IEEE Transactions on Power Apparatus and Systems*, vol. PAS-89, no. 1, January 1970.

[28] Y. Wang, D. J. Hill, R. H. Middleton, and L. Gao, 'Transient stability enhancement and voltage regulation of power systems', *IEEE Transactions on Power Systems*, vol. 8, no. 2, pp. 620–627, May 1993.

[29] E. J. Davison and N. S. Rau, 'The optimal output feedback control of a synchronous machine', *IEEE Transactions on Power Apparatus and Systems*, vol. 90, no. 5, pp. 2123–2134.

[30] P. W. Sauer and M. A. Pai, *Power System Dynamics and Stability*. Prentice Hall, 1st edition, 1997.

[31] T. Başar and P. Bernhard, *H-infinity Optimal Control and Related Minimax Design Problems: A Dynamic Game Approach*. Birkhäuser, 1995.

[32] L. S. Shapley, 'Stochastic games', *Proceedings of the National Academy of Science, USA* no. 39, pp. 1095–1100, 1953.

[33] Pike Research's Report, 'Smart grid networking and communications', 2010.

[34] K. Tomsovic, D. Bakken, V. Venkatasubramanian, and A. Bose, 'Designing the next generation of real-time control, communication, and computations for large power systems', *Proceedings of the IEEE*, vol. 93, no. 5, pp. 965–979, May 2005.

[35] G. N. Ericsson, 'On requirements specifications for a power system communications system', *IEEE Transactions on Power Delivery in IEEE Transactions on Power Delivery*, vol. 20, no. 2, pp. 1357–1362, April 2005.

[36] M. S. Amin and B. F. Wollenberg, 'Toward a smart grid: power delivery for the 21st century', *IEEE Power and Energy Magazine*, vol. 3, no. 5, pp. 34–41, October 2005.

[37] E. Santacana, G. Rackliffe, L. Tang, and X. Feng, 'Getting smart', *IEEE Power and Energy Magazine*, vol. 8, no. 2, pp. 41–48, March 2010.

[38] US Department of Energy, 'Grid 2030 – a national vision for electricity's second 100 years', Technical Report, July 2003.

[39] D. H. Lorenz and A. Orda, 'QoS routing in networks with uncertain parameters', *IEEE/ACM Transactions on Networking*, vol. 6, no. 6, pp. 768–778, 1998.

[40] P. Van Mieghem and F. A. Kuipers, 'Concepts of exact QoS routing algorithms', *IEEE/ACM Transactions on Networking*, vol. 12, no. 5, pp. 851–864, 2004.

[41] G. Xue and S. K. Makki, 'Multi-constrained QoS routing: a norm approach', *IEEE Transactions on Computers*, vol. 56, no. 6, pp. 859–863, 2007.

[42] F. Kuipers, P. Van Mieghem, T. Korkmaz, and M. Krunz, 'An overview of constraint-based path selection algorithms for QoS routing', *IEEE Communications Magazine*, vol. 40, no. 12, pp. 50–55, 2002.

[43] US Department of Energy, 'Communications requirement of smart grid technologies', 5 October, 2010.

[44] G. N. Ericsson, 'On requirements specifications for a power system communications system', *IEEE Transactions on Power Delivery*, vol. 20, no. 2, pp. 1357–1362, April 2005.

[45] P. McDaniel and S. McLaughlin, 'Security and privacy challenges in the smart grid', *IEEE Security & Privacy*, vol. 7, no. 3, pp. 75–77, June 2009.

[46] National Institute of Standards and Technology, 'NIST framework and roadmap for smart grid interoperability standards', Release 1.0, January 2010.

[47] C. H. Hauser, D. E. Bakken, and A. Bose, 'A failure to communicate: next generation communication requirements, technologies, and architecture for the electric power grid,' *IEEE Power and Energy Magazine*, vol. 3, no. 214, pp. 47–55, March 2005.

[48] T. Başar and G. J. Olsder, *Dynamic Noncooperative Game Theory*. Classics in Applied Mathematics, SIAM, 2nd edition, 1999.

[49] Department of Homeland Security, 'Primer control systems cyber security framework and technical metrics', Technical Report, June 2009.

[50] W. Boyer and M. McQueen, 'Ideal based cyber security technical metrics for control systems', in *Proceedings of 2nd International Workshop on Critical Information Infrastructures Security*, October 2007.

[51] A. McIntyre, B. Becker, and R Halbgewachs, 'Security metrics for process control systems', SANDIA Report, SAND2007-2070P, September 2007.

[52] Q. Zhu and T. Başar, 'Dynamic policy-based IDS configuration', in *Proceedings of 48th IEEE Conference on Decision and Control (CDC)*, Shanghai, December 2009.

[53] Q. Zhu, H. Tembine, and T. Başar, 'Network security configuration: a nonzero-sum stochastic game approach', in *Proceedings of American Control Conference (ACC)*, Baltimore, MD, 2010.

[54] D. Fudenberg and D. K. Levine, *The Theory of Learning in Games*. MIT Press, 1998.

[55] Q. Zhu, H. Tembine, and T. Başar, 'Distributed strategic learning with application to network security', in *Proceedings of American Control Conference (ACC)*, San Francisco, CA, 2011.

[56] Q. Zhu, H. Tembine, and T. Başar, 'Heterogeneous learning in zero-sum stochastic games with incomplete information', in *Proceedings of 49th IEEE Conference on Decision and Control (CDC)*, Atlanta, GA, 2010.

[57] S. Frei, B. Tellenbach, and B. Plattner, '0-day patch: exposing vendors (in)security performance', *BlackHat*, 2008.

[58] Bugtraq. http://www.securityfocus.com/

[59] M. A. McQueen, W. F. Boyer, T. A. McQueen, and S. McBride, 'Empirical estimates of 0 day vulnerabilities in control systems', *SCADA Security Scientific Symposium*, 21–22 January 2009.

[60] S. Ragan, 'The new era of vulnerability disclosure – a brief chat with HD Moore', August 16, 2010 [online]. Available at: http://www.thetechherald.com/article.php/201033/6025/The-new-era-of-vulnerability-disclosure-a-briefchat-with-HD-Moore

[61] The H Security, 'Pressure mounts for a swifter response to vulnerabilities' [online]. Available at: http://www.h-online.com/security/news/item/Pressure-mounts-for-a-swifter-response-to-vulnerabilities-1050624.html

[62] Symantec Inc. 'Symantec Internet security threat report'. http://www.symantec.com, 2003.

19 Application-driven design for a secured smart grid

Robin Berthier, Rakesh B. Bobba, Erich Heine, Himanshu Khurana, William H. Sanders, and Tim Yardley

19.1 Introduction

As a core critical infrastructure, the national electric grid is at a crossroads, with modernization efforts driven by advanced cyber-system capabilities on the one hand and risks from cyber attack on the other. All stakeholders are concerned about these risks and view the need to incorporate resilience and adequate cyber security measures into the grid as crucial to all modernization efforts. The Federal Energy Regulatory Commission (FERC) recently released its policy statement [1] on smart grid technologies, which identified cyber security as one of two key priority areas.

As tasked by the Energy Independence and Security Act of 2007, the National Institute of Standards and Technology (NIST) is leading a major effort to develop a comprehensive framework for interoperability in smart grid. In their preliminary Roadmap for Interoperability, the development of a cyber security risk-management framework was identified as a major challenge. The ongoing roadmap to secure energy delivery systems [2] is another example of an important public/private dialogue that has identified milestones and goals for achieving resilience, such as designing, installing, operating, and maintaining control systems by 2015 that can survive an intentional cyber attack without loss of critical function. In addition, over $9 billion has been committed by the electric sector and the Department of Energy as part of ARRA (American Recovery and Reinvestment Act) recovery investment efforts on modernization of the grid, with cyber security being an important focus. This investment offers opportunities and challenges in realizing a resilient electric grid for the future.

Risk in electric grid systems is determined by the threats against the system (i.e., adversaries intent on attacking the grid), vulnerabilities present in the software and hardware components, and consequences of a successful attack. The recent Stuxnet [3] attack demonstrated that the risks are real. In that attack, malware targeting industrial control systems was able to start its attack path on mobile assets, proceed to penetrate air-gapped networks via physical means, and eventually modify code on programmable logic controllers. Vulnerabilities in electric grid systems are continually being discovered, further emphasizing the concern that systems need to be developed with stronger security requirements in mind. As a case in point, security flaws have been found in multiple metering devices [4], and it has been demonstrated that malware could be designed specifically to infect smart meters and propagate [5].

Table 19.1. Electric grid and cyber security domains and properties

Smart grid and cyber security	Category/type	Description
Smart grid domains	Customer	Meters, home-area networks, electric vehicles, distributed generation, storage, thermostat, smart appliances, energy management system
	Distribution	Field area devices and networks, distributed generation
	Transmission	Substation devices and networks, wide-area networks
	Bulk generation	Generators, plant control systems
	Markets	ISO/RTO, aggregator, retailer/wholeseller, clearninghouse
	Operations	RTO/ISO, EMS, SCADA, WAMS, demand response, meter data-management system, utility networks
	Service providers	Billing, energy management, aggregation
Computer and network security	Properties	Confidentiality, integrity, availability
	Functions	Authentication, authorization, audit
	Objectives	Protect, detect, respond
	Components	Key management, anti-tamper, authentication and access control, intrusion detection and response
	Broader efforts	People, processes

As smart grid systems are designed, developed, and deployed, they create enhanced and modified control loops across the grid. These extended loops face new risks from attacks and failures, and therefore need to be secure and resilient. Achieving this security and resilience requires that adequate protection, detection, and response capabilities be integrated into these systems. Given that the risks vary for the various components and systems, investment in security needs to match those risks for cost effectiveness. Ideally, we would like a scientific set of cyber security tools that provide quantitative risk-driven solutions for the grid. While a plethora of capable cyber security methods and tools exist today, a complete scientific foundation is still under development [6].

In this chapter, we discuss an application-driven design approach that builds on the large cyber security toolset. In this approach, we focus on specific applications and systems, first to understand the cyber security requirements and then to develop a solution reflecting those requirements. A key element of this approach is careful enumeration of the control-system-specific aspects of each system and an integrated study of these aspects, cyber security properties, and solutions.

Successful adoption of this application-driven design approach requires careful identification of the applications and systems that need to be studied, the cyber security properties that must be addressed, and the underlying grid properties that must be maintained. In Table 19.1, we enumerate some of the common grid domains and functions, and key cyber security properties. The list is meant to be representative and not exhaustive.

The idea is that the intersection of these aspects gives rise to potentially interesting and challenging research problems. Based on this approach and an understanding of some of the key emerging smart grid systems, we have identified and worked on the following three specific problems in the past and we build on that work here [7–9].

1. *Intrusion detection for advanced metering infrastructure (AMI)* [7]. AMI networks will carry monitoring and control information for many smart meters. We address the challenge of developing effective intrusion detection and response solutions for these networks to help ensure their security in the face of cyber attacks. Key constraints include the limited capability and capacity of the devices and networks in AMI systems.
2. *Converged networks for supervisory control and data acquisition (SCADA) systems* [8]. SCADA systems are increasingly adopting IP-based common off-the-shelf components to realize advanced applications. We address the challenge of network convergence for SCADA systems to build a strong foundation for secure and timely operations that are needed for a range of SCADA applications.
3. *Design principles for authentication of SCADA protocols* [9]. SCADA protocols are being enhanced to support new applications and to be secure in the face of attacks. We address the problem of developing effective design principles that result in effective authentication protocols while addressing grid-specific needs of efficiency, availability, and evolvability.

The remaining sections of this chapter discuss each of the above problems and our approach to them. Specifically, we discuss intrusion detection for AMI in Section 19.2, converged networks for SCADA in Section 19.3, design principles for authentication in SCADA protocols in Section 19.4, and conclude in Section 19.5.

19.2 Intrusion detection for advanced metering infrastructures

Advanced metering infrastructures (AMIs) are starting to be deployed throughout the world. The rapid growth is accompanied by serious security concerns about the potential vulnerabilities of the new technologies that are being introduced. These concerns have been fueled by recent discoveries of security flaws in multiple metering devices.

To build a secure AMI architecture we need adequate technologies in three areas: (i) prevention, (ii) detection, and (iii) mitigation or resilience. In our current work, we are focusing on detection, specifically on monitoring of the different communication networks of an AMI. Our objective is to define the requirements for an efficient and complete network and host an intrusion detection system that can identify any attempt to violate the security policy. To reach this objective, we address the following questions:

- What is the threat model of an AMI, and how would attacks manifest in the different communication networks?
- Which components need to be monitored, and which monitoring architecture should be deployed?

- What are the unique constraints of an AMI, and which detection technology can best identify malicious activity?

We first review the potential security issues introduced by smart meters, and then present a comprehensive monitoring architecture. In the last part of this section, we focus on an innovative approach called *specification-based* intrusion detection, which enforces a security policy for AMIs through the precise definition of legitimate use cases and security requirements.

19.2.1 Smart meters and security issues

An AMI includes several communication networks, identified according to their spatial scope:

- The wide-area network (WAN) serves as a communication link between headends in the local utility network and either data concentrators or smart meters. This network uses long-range and high-bandwidth communication technologies, which may be WiMAX, cellular, satellite, or power-line communication (PLC).
- Neighbourhood-area networks (NANs) ensure communication between data concentrators or access points and smart meters that play the role of interfaces with home-area networks (HANs). Utilities often deploy NANs with either PLCs or radio frequency (RF) mesh networks. The scale of this NAN ranges from a few hundred to tens of thousands of nodes.
- Field-area networks (FANs) allow the utility workforce to connect to equipment in the field.

A major challenge in protecting an AMI against malicious activities is the need to create a monitoring solution that covers the heterogeneity of communication technologies through both their requirements (e.g., encryption and real time) and their constraints (e.g., topology and bandwidth). It is critical to identify these elements for two reasons: (i) they can help to define the potential impact of malicious activities targeting an AMI, and (ii) they can impose limits on the functionality and security of a monitoring solution. Because large portions of an AMI network are wireless and use a mesh network topology, which facilitates network-related attacks such as traffic interception, the design of the monitoring architecture is more challenging than it would be in a wired network. Moreover, a large number of nodes are deployed in the field or in consumer facilities, which means that attacks requiring physical access are easier to conduct than they would be for traditional substation or pole-based nodes.

The threats targeting an AMI can be viewed in three different ways: (i) by type of attacker, (ii) by motivation, and (iii) by attack technique. We now explore the different perspectives to understand how the different attacks will likely manifest themselves in an AMI, and, as a result, how we can automatically detect them. To do so, we use the terminology from [10] to list the types of attackers and their motivations. Six types of attackers are considered:

- *Curious eavesdroppers*, who are motivated to learn about the activity of their neighbours by listening in on the traffic of the surrounding meters or HANs.
- *Motivated eavesdroppers*, who wish to gather information about potential victims as part of an organized theft.
- *Unethical customers*, who are motivated to steal electricity by tampering with the metering equipment installed inside their homes.
- *Overly intrusive meter data-management agencies*, which are motivated to gain high-resolution energy and behaviour profiles about their users, which can violate customer privacy. This type of attacker also includes employees who could attempt to spy illegitimately on customers.
- *Active attackers*, who are motivated by financial gain or terrorist goals. The objective of a terrorist could be to create large-scale disruption of the grid, either by remotely cutting off many customers or by creating instability in the distribution or transmission networks. Active attackers attracted by financial gain could also use disruptive actions, such as denial-of-service (DoS) attacks, or they could develop self-propagating malware in order to create revenue-making botnets.
- *Publicity seekers*, who use techniques similar to those of other types of attackers, but in a potentially less harmful way, because they are more interested in fame and usually have limited financial resources.

Attackers may use a variety of attack techniques to reach their objectives (e.g., interception, modification, injection, bypassing authentication, spoofing, compromise, or denial-of-service), and the list above offers only a high-level classification of possible malicious activities. At a lower level, vulnerabilities can be found and exploited in flaws or misuses of (i) routing, (ii) configuration, (iii) name resolution, (iv) encryption, or (v) authentication protocols for network compromise. Vulnerabilities at a higher level include: (i) software and firmware vulnerabilities, (ii) hardware vulnerabilities, (iii) read and write access to data storage, or (iv) access to encryption keys. Figure 19.1 illustrates the potential impact of the exploitation of these vulnerabilities. A compromised meter or a computer spoofing a meter could be used as an entry point to send harmful remote disconnect commands to other meters, or to launch a denial-of-service attack against network bottlenecks. More information about low-level system vulnerabilities and attacks can be found in [11, 12].

19.2.2 Architecture for situational awareness and monitoring solution

Informed by the different threats and attack techniques reviewed in Section 19.2.1, we now discuss the design of an IDS architecture and explore the different monitoring operations that could be used to ensure detectability of both known and unknown malicious activity.

A traditional IDS architecture is made of a collection of lightweight sensors that report to a centralized management server. The core data-processing and detection intelligence of this architecture resides on the central component, which makes scalability difficult. In the case of an AMI that could have millions of nodes, this type of architecture does

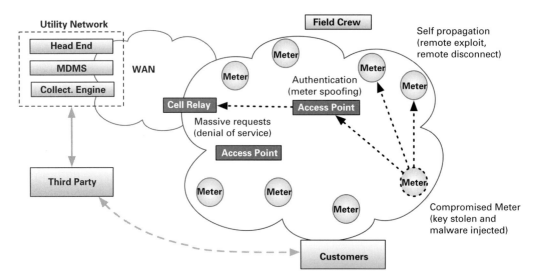

Figure 19.1 Attack scenarios on AMI components.

not appear to provide an optimal way to handle the load. An alternative solution is to use a distributed architecture in which most of the data processing is handled by the sensors. This type of architecture also includes a central component, but it only has the tasks of coordinating sensors and collecting high-level intrusion information. We can realize this by taking advantage of the distributed topology of AMIs to deploy sensors that can analyse network traffic throughout an AMI. Locations in which it would be possible to deploy sensors that can host distributed monitoring operations include: (i) smart meters, (ii) dedicated intrusion-detection devices in the field, and (iii) access points. A dedicated device can analyse the traffic of a given wireless network without having to instrument every network node. This type of sensor is more expensive to deploy and maintain, but it offers higher computation resources and better protection against security compromises than typical network nodes, such as smart meters.

We note that distribution of monitoring operations is also important in enhancing the reliability of the IDS against accidental failures and malicious attacks. The system needs to operate even if a subset of sensors or the management server is unavailable or compromised. A distributed topology offering redundancy helps to eliminate single points of failure. We also emphasize that attacks against sensors are greatly limited if the sensors are isolated either through virtualization or by using their own hardware rather than being embedded in the components they monitor. Techniques to detect compromised nodes include the use of a reputation system to evaluate the level of trust in alert reports or the use of a distributed proof system to prevent a single node from having too much visibility. Finally, the reliability of communications between sensors and management servers can be increased through the use of a separate communication network. However, the cost of this solution is often prohibitive.

An important concern with any large-scale monitoring solution is the generated volume of alerts. The lack of a comprehensive alert management process will undermine the usefulness and practicality of an IDS. An alert management process consists of two steps: (i) data reduction (including data aggregation), and (ii) alert correlation. The first step groups alerts that share similar attributes, while the second step provides high-level information about the intrusions detected. The second step uses correlation rules to map alerts over time, space, and logical sequence. That allows an IDS to greatly reduce the number of alerts that are communicated to a human operator. Moreover, it can calculate a criticality value for each alert, enabling escalation and prioritization procedures.

Based on the threats reviewed in Section 19.2.1 and the distributed monitoring architecture, we identified monitoring operations which are described in more detail in [7]. They include stateful checking of network protocols (e.g., configuration, routing, and application protocols), stateless analysis of traffic information (e.g., firewall), device log analysis through periodic health reports, firmware integrity checking, and monitoring of network characteristics (e.g., wireless signal strength). These monitoring operations are distributed to the three different sensor locations based on computational power and activity visibility. Access points and dedicated sensors in the field have sufficient computing resources and memory to host the core stateful detection technologies. They can also aggregate logs and alerts sent by the metering devices. The embedded hardware of meters cannot support sophisticated monitoring functions, but it can periodically check the integrity of the software and send health reports to its access points. We explain in Section 19.2.3 how specification-based detection technology can leverage a security policy to check meter behaviour and ensure security compliance.

19.2.3 Enforcing security policies with specification-based IDS

The process of detecting malicious activity can be based on any of three distinct approaches. Signature-based detection, also known as *misuse detection*, consists of looking for patterns of malicious behaviour using a database of predefined attack signatures. Anomaly-based detection consists of identifying deviations from a normal behaviour profile predefined with statistical measures. Specification-based detection consists of identifying deviations from a correct behaviour profile predefined with logical specifications.

The detection techniques differ in two fundamental ways. First, signature-based IDS uses a blacklist approach, while anomaly-based IDS uses a whitelist approach and specification-based IDS uses a combination of both. A blacklist approach requires creation of a knowledge base of malicious activity, while a whitelist approach requires training of the system and identification of its normal or correct behaviour. Obviously, these two approaches have complementary limitations and advantages: a blacklist-based IDS will not be able to detect unknown attacks and will require frequent updates, while a whitelist-based IDS will often be more expensive to train and to tune. Another limitation of whitelist approaches is that they provide little information about the root causes of attacks.

The second fundamental difference lies in the level of understanding required by each approach. Signature and anomaly-based IDS monitor activity at a low level, while specification-based IDS requires a high-level and stateful understanding of the activity monitored. For example, in the context of a network IDS (NIDS), signature and anomaly-based IDS will monitor network activity at layers 2, 3, and 4 of the OSI model. A signature-based IDS, such as Snort, will also analyse payload information at layers 5 and 7, but in a stateless manner, without tracking the behaviour of the underlying applications over time. On the other hand, a specification-based IDS will work by building a state machine of a process, and then monitoring activity to check whether anything escapes from the system boundaries specified previously. This second approach is typically more expensive to develop and less scalable than the first approach. However, it has the strong advantage of being more accurate.

In the context of an AMI, we believe that the best approach is to develop critical monitoring operations using a specification-based IDS. This choice is founded on three arguments. First, the potentially higher accuracy of specification-based IDS compared with signature-based IDS is better suited for the criticality of an AMI. Second, the lack of empirical attack data for the AMI makes the construction of a blacklist of signatures difficult. Third, the limited number of protocols and applications that must be monitored in an AMI reduces the development cost of specification-based sensors. This last argument explains why specification-based IDS have been applied to specific problems such as mobile ad hoc networks, spanning tree protocols (STP), and voice-over-IP (VoIP) protocols. It would be very challenging to specify state machines for a large system made of thousands of different protocols, such as the Internet. But an AMI offers a controlled environment in which the task of developing specifications would be cost-efficient.

The monitoring operations identified in Section 19.2.2 were divided into three categories (stateful specification-based monitoring, stateless specification-based monitoring, and anomaly-based monitoring). The three categories provide information about the computation resources needed by the different monitoring operations, arranged from high to low. Some monitoring operations work by checking system and network behaviour against configurations, policies, and protocol specifications. These resources have to be defined as part of the IDS design process, in close collaboration with the developers and users of the AMI systems and networks.

For most AMI detection operations, the different resources required are: (i) a network configuration to provide information about network topology and access control rights, (ii) protocol specifications to determine correct header and payload formats, as well as the behavioural state machines that can capture coherent sequences of request and reply operations, (iii) system and network security policies to specify the allowed behaviours of applications and processes, and (iv) statistical profiles to determine boundaries of normal network traffic characteristics.

To illustrate how a specification-based approach would be implemented, we developed a specification-based monitoring prototype for the C12.22 protocol, which is a standard widely used in the USA to transport information between meters at the application layer. The goal is to show how we can leverage a formal verification framework introduced by [13] to ensure that the behaviour of the smart meters never violates the security policy. The key idea is to build a model of our system and formally verify that no network trace could violate the security policy without being detected. This process ensures that the monitoring operations defined and run locally between sets of a few meters are sufficient to cover a security policy defined globally at the AMI level.

The framework is built hierarchically in five layers: (i) a model of the network, (ii) a set of monitoring operations and a set of assumptions (to cover unmonitored items), (iii) the protocol specifications, (iv) the security policy and the security requirements, and (v) a verification theorem. The theorem determines whether all possible network traces that respect the first three layers of the framework will also respect the fourth layer. This means that no network trace can escape the monitoring operations while violating the security policy.

The five layers are implemented as functions and data structures in ACL2, which is a software tool that combines a programming language, a logic, and a theorem prover based on Common Lisp. ACL2 automates most of the proof effort using techniques such as rewriting and mathematical induction, but the user often needs to guide the proof by adding lemmas that the mechanical prover cannot deduce by itself. We note that the manual operation is very useful for discovering hidden assumptions and incomplete functions.

We have identified the following four key rules for constraining meters to an expected behaviour that prevents malicious nodes from compromising the infrastructure:

- *Only well-formed C12.22 requests and responses are authorized at the application layer.*
- *Only the meter collection engine (CE) can initiate C12.22 requests to meters.*
- *Requests labelled as sensitive (e.g., remote disconnect) cannot target a range of more than Y meters per hour.*
- *Correctly configured and working meters respond to requests in less than Z seconds.*

The first rule prevents malformed connection attempts or bug format exploitations from being undetected. The second rule ensures that a compromised meter trying to initiate malicious connections to its neighbours would be identified. The third rule covers compromised meters or rogue hosts spoofing the CE and violating the usage boundaries of legitimate use cases. Finally, the fourth rule is designed to detect denial-of-service attacks. Of course, other rules could be added to this security policy based on the needs of utilities, to best use their AMIs. But the important concept of this framework is that these rules define a set of *authorized traces*, and we can now implement local monitoring operations to identify *valid traces*. The goal of the formal proof is to show that the set of valid traces is a subset of the set of authorized traces [13]. If we add a new rule to the policy and the proof no longer holds, then it means that a possible trace was identified violating the security requirement without being detected. In such a case, we need to modify the set of monitoring operations to cover this type of trace.

Once the proof is successfully carried out, the final step is to implement the monitoring operations defined in the formal model. We note that vulnerabilities can be introduced when translating a theoretical model into a practical implementation, so special care must be taken to test the prototype extensively at each stage of the development in order to validate it.

Specification-based security is a powerful approach for addressing security challenges in critical environments. This section illustrated the concept of application-driven design by describing how a security policy that captures legitimate use cases can be enforced through formal specification-based monitoring operations. This approach could be leveraged and applied successfully to other smart grid systems, including home-area networks (HANs) or SCADA environments.

19.3 Converged networks for SCADA systems

In this section, we consider SCADA systems (specifically those in power systems) and the digital communication between control centres and field devices (e.g., PLCs and relays) that provide protection, monitoring, and maintenance. SCADA systems provide centralized monitoring and control of field devices spread over large geographic areas. The exchange of data and commands over SCADA networks also provides support for several classes of applications. These applications require isolation from each other and timely delivery of data and commands to support the overall real-time needs of grid applications.

Many architectures deployed in the field today use multiple physical networks to carry different types of data and ensure isolation and timeliness properties. This approach is based on the accepted use of dedicated hosts and networks for such critical applications. In contrast, use of quality-of-service (QoS) managed solutions [14] offers the potential for integration of all the above-mentioned SCADA services and applications over a single network. Such an approach would greatly reduce management overheads, as control centres would have to manage only a single network and its redundant backup, and it would be easy to deploy new applications that can safely share existing resources. Extensive relevant research already exists in the areas of QoS, real-time systems, and resource management, such as [15–18].

A primary objective in our work was to design, implement, and experiment with a real-time middleware system that allows SCADA networks to converge, which in turn allows existing and emerging SCADA applications and services to benefit from high-speed networking capabilities. A key observation supporting this work is the increasing use of commercial operating systems (e.g., Windows and Linux) in SCADA and other modern smart grid environments. Such systems motivate development of a real-time middleware solution, because they typically lack effective resource management. At the same time, they allow for development of the middleware, because of the inherent system flexibility.

We first conducted a detailed study of SCADA application requirements, and demonstrated the need for strong host-level CPU and network resource management capabilities that support the ability to meet millisecond-level deadlines. Based on those requirements, we designed, implemented, and experimented with our real-time middleware solution, converged networks for SCADA (CONES). The middleware provides (i) host-based process management, (ii) network resource management, and (iii) host-based network resource management.

The primary contributions of our work are (i) a study of SCADA application requirements to specify resource management needs for convergence of SCADA networks, (ii) a practical implementation on the Linux platform that provides fine-grained host-level CPU and network resource management, and (iii) extensive experimentation on an integrated system in a testbed environment to demonstrate the feasibility of the solution. The CONES implementation leverages the iDSRT software implementation [16] and complements ongoing work in the GridStat project, which looks at QoS over wide-area electric grid networks, such as those that support sharing of synchrophasor data [19].

19.3.1 Requirements and challenges for convergence

The vision for our work was to develop a communication system that supports convergence of SCADA applications while ensuring isolation and timeliness properties of individual applications. In keeping with product development and deployment trends in the electric sector, we aimed to implement such a system by using commercial platforms and operating systems. This section enumerates and explores the technical requirements that need to be met by the desired communication system.

As a first step in the process, we studied some common SCADA applications and their QoS requirements, including bandwidth, latency, and priority, among others. We consulted with several electric sector industry experts and identified a few key applications. The applications crossed several classes of traffic (e.g., SCADA, engineering, and monitoring) and examined their properties, particularly bandwidth and latency requirements.

In addition, IEEE Standard 1646 [20] provides a detailed study of communication delivery performance requirements for substations. The standard details data delivery times ranging from 4 ms to minutes to support a variety of applications. The standard includes detailed timing information of several applications. From the standard, we identified two key applications that drive the technical requirements below:

- *Peer-to-peer substation protection.* This application requires occasional communication between IEDs in a substation with high priority and a data delivery duration of 4 ms.
- *System protection with synchrophasor data.* This application requires a high rate of communication (30 to 120 samples per second) to control centres from substations with medium priority and a data delivery duration of 20 to 30 ms.

After studying the applications and general QoS requirements, we identified the following three technical requirements that must be met to achieve our goals for a converged network for SCADA. We envision that these requirements will be met using commercial platforms and operating systems. That will allow for widespread deployment of cost-effective solutions.

- *Host-based process management.* In order to support real-time applications, CPU resources at hosts must be managed. On commercial platforms, management is achieved by scheduling processes based on priorities and deadlines [21, 22]. Additionally, in practice, one can achieve soft real-time guarantees that match requirements for many applications [15]. Our objective was to meet the end-to-end real-time requirements for SCADA systems outlined above and exemplified by the two key applications of substation and system protection.

- *Network resource management.* In order to support real-time applications that involve data delivery over routed networks, there is a need to engineer the network properly and provide QoS assurances. Our objectives were to leverage existing technologies and integrate them into an overall architecture to provide end-to-end real-time guarantees for SCADA applications on IP networks.
- *Host-based network resource management.* As part of the overall QoS objectives, it is important to manage the network resources at the host level as well. The scheduling of *outgoing* packets at each host is a key element to successful network resource management. That has recently been discussed in the context of wireless networks in which such scheduling affects real-time delivery [23, 24]. However, it has not been considered an issue in wired networks in which gigabit capacity networks seemingly offer sufficient resources. We demonstrated that such network resources must be managed to meet strict deadlines in typical SCADA environments. Our requirements are to provide host-based network resource management, and to integrate it with an architecture to provide end-to-end real-time guarantees for SCADA applications.

19.3.2 Architecture with time-critical constraints

In this section, we discuss our approach to building a real-time middleware system that addresses the requirements identified earlier. We call this middleware converged networks for SCADA (CONES) to reflect its primary objective of enabling the use of a single physical network to carry data for all current and future SCADA applications.

The use of middleware to address the needs of flexible and cost-effective management for distributed real-time and embedded systems is considered to be an effective approach [15]. We adopted this approach in our work, with an emphasis on low-level properties and coordination, to achieve the millisecond-level deadlines identified earlier. Recently, iDSRT was proposed as an example middleware solution that explores the use of wireless networks for power grid environments [16]. We built on iDSRT and leveraged many of its existing features while significantly enhancing others to form one of the core components of the CONES system.

Figure 19.2 shows the CONES architecture in the context of SCADA networks. We envision SCADA host machines (e.g., IEDs or aggregators) to be connected via IP LAN technologies in a substation environment. A subset of these host machines would also be connected to control centre host machines (e.g., a SCADA master or engineering workstation) via IP WAN technologies. All host machines use an enhanced Linux kernel in which the CPU scheduler and networking stacks are modified; such enhanced host machines are called 'smart' hosts. Furthermore, by utilizing our developed CONES API, applications leverage these modified kernel capabilities to manage CPU and networking

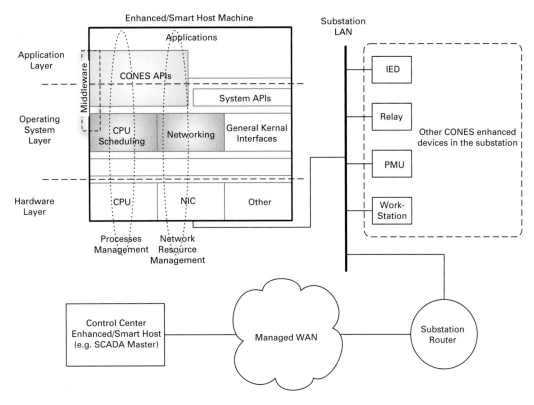

Figure 19.2 CONES architecture and approach [8].

resources on these hosts. On the WAN side, we envision use of available tools for network provisioning and QoS assurance, such as MPLS, traffic engineering, and DiffServ.

Design, implementation, and experimentation

To meet the goals laid out in Section 19.3.1, we chose a middleware approach. It allows for simple application integration, while also allowing for CPU and network resource management across distributed SCADA hosts and providing soft real-time guarantees. To build the middleware, we had to integrate several components across various system layers, including the operating system at both the kernel and system library levels, as well as application-layer programming interfaces. Further, all of the components needed to be readily available, off-the-shelf systems, or easily integrable with such systems.

For real-time scheduling we chose iDSRT for two reasons: it was a Linux extension, meaning it worked well in our environment, and it also included explicit network resource scheduling. The architecture of iDSRT is fully described in [16]. In brief, the software uses the earliest deadline first (EDF) algorithm to schedule network traffic and CPU time. It uses both network and CPU deadlines as factors in its scheduling, allowing for finer-grained control of resources.

For CPU scheduling, iDSRT uses EDF to select which process to run based on three factors: (i) a known deadline, (ii) a known worst-case run-time, and (iii) a period, allowing for periodic time-bounded tasks. To implement the EDF algorithm on Linux, the iDSRT team created a scheduler module that allows processes that overrun their allotted time, as well as non-real-time processes, to use traditional 'best effort' scheduling. Similarly, the networking scheduler in iDSRT, called *implicit EDF (iEDF)*, implements the Implicit EDF algorithm which allows the scheduling of network resources for non-existent tasks. To do so, iEDF hooks into the packet queuing subsystem of the Linux networking stack. That effectively gives the scheduler control of the networking resources during packet selection, allowing for a deadline-based selection prior to standard best-effort queues. This functionality is used to coordinate transmission over a shared medium, thus reducing collision. The implicit capability is not currently used in CONES; however, future work may change that.

The CPU EDF scheduler was already adequate for CONES but iDSRT's iEDF module required three major modifications to work in the CONES environment.

- *Process management.* We changed process tracking for iEDF from a simple reuse of the TOS bits field in the IP header to an in-kernel data structure that tracks real-time designation per process. That allows an arbitrary number of processes to be real time, and allows an arbitrary process start and stop ordering.
- *Standards compliance.* By releasing the TOS from process identification, we made the associated fields available for use by any process in CONES-enhanced systems, allowing the IP stack to be standards-compliant.
- *Real-time parameter control.* We modified the networking stack by adding the SOCK_REALTIME socket option. As a result, sockets can identify themselves as real time and any given packet can be associated with a real-time or best-effort stream on the same socket. In addition, a process can now have multiple sockets, some real time and some best effort.

Experimental analysis of the CONES architecture showed its viability in several situations, which are detailed in [8]. The experiments compared CONES-enhanced hosts with stock-kernel-based hosts, and created several resource contention scenarios. These experimental results show that CONES can be used to build a solution using commercial hardware/software systems that meets the requirements identified at the beginning of this section, as explained below.

- *Peer-to-peer substation protection.* This application requires occasional communication between IEDs in a substation with high priority and a data delivery duration of 4 ms.

For this requirement, we looked at host-based CPU and network management. In looking at how a host behaves in the presence of CPU contention, we realized that a host machine requires real-time capabilities like those provided by CONES, to meet the 4 ms requirement. Without real-time support, we saw delays of several seconds. On the other hand, with CONES, even under extreme CPU or network contention, delays were limited to 1 ms, allowing sufficient time to meet the 4 ms deadline.

- *System protection with synchrophasor data.* This application requires high-rate communication (30 to 120 samples per second) to control centres from substations with medium priority and a data delivery duration of 20 to 30 ms.

For this requirement, we again looked at host-based resource management, as well as wide-area network resource management. In addition to the host-based contention issues, we showed that without network resource management, delays of 15 to 20 ms one way were observed in a WAN environment. However, with appropriate network management, these delays come down to 2 ms, making it easy to achieve the targeted deadlines for this requirement. Collectively, it can be argued that a CONES-based solution (or a similar one) is necessary to satisfy this set of requirements for commercial platforms.

Real-time and secure platforms play an important role in realizing trustworthy smart grid systems. The techniques developed in this work are generalizable across many control and monitoring applications in the power grid. For example, these tools can be used to build gateway systems that are needed for protocol translation, secure network transport, and publish/subscribe communications, all while maintaining the timeliness constraints on information delivery.

19.4 Design principles for authentication

Given the threats against smart grid systems, an important goal is the development of effective cyber security protocols and standards. Authentication is recognized as an important security requirement for smart grid communication protocols. In fact, academic researchers and industry practitioners have recently focused on developing authentication protocols and standards for the electric grid; for example, the DNP Users Group developed a standard for DNP3 authentication [25] and the IEC developed the 62351-5 standards [26] for securing transmission protocols in telecontrol equipment and systems.

Intuitively, it seems natural to leverage existing authentication protocols and standards for Internet systems while designing such authentication protocols for the grid. Development of authentication protocols for Internet systems using cryptographic tools and techniques has been a practice for more than two decades. However, prior work has shown that if adequate care is not taken, the process is prone to significant errors [27–29]. To help avoid such errors, researchers have developed design principles that have been quite effective for new protocols. While those principles are relevant to authentication protocols for the electric grid, they do not address some of the unique constraints and properties of the grid that distinguish it from more well-known Internet systems. To address the gaps, we build upon prior work in design principles for authentication protocols, but focus the discussion on grid-specific issues.

19.4.1 Requirements and challenges in designing secure authentication protocols for smart grid

For authentication standards that are going to be deployed across the grid, one can characterize common constraints of the grid today as follows: (i) devices have limited computation capabilities, (ii) networks have limited bandwidth, (iii) there is a need to integrate legacy protocols and systems, and (iv) there is no system-wide cyber security infrastructure. That said, the entire cyber infrastructure of the grid is being modernized, and this modernization is expected to eliminate some of those common constraints in the next 5 to 15 years.

Whether the grid comprises an advanced or an outdated cyber infrastructure (or perhaps a combination), it is important to realize that the grid will always be different from Internet systems in fundamental ways [30, 31]. These differences arise largely from three factors: (i) the critical and real-time nature of electric grid systems, (ii) their impact on the safety of equipment and personnel, and (iii) their long component life. These differences lead to constraints and properties that impact cyber security protocols in general, and authentication protocols in particular. Specifically, grid applications require: (i) high performance in terms of latency and jitter in message exchange, (ii) high availability in terms of tolerating faults and failures (i.e., they need graceful degradation of services), (iii) the ability of the computation and communications subsystem to meet the real-time requirements of applications, (iv) comprehensive security design (as grid applications are likely targets for sophisticated cyber attacks), and (v) adaptable and evolvable designs (since components typically have a lifetime of 15 or more years once deployed). We argue that protocols must be developed with those fundamental constraints and properties in mind, and not significantly influenced by today's constraints that are likely to be eliminated with modernization.

19.4.2 Design principles for authentication protocols

As has been observed in practice, authentication protocols, and security protocols in general, can be surprisingly hard to design correctly, e.g. [32, 33]. This is true even when an existing protocol is used in a new environment. Many protocols that were once considered secure were later found to be flawed. Learning from experience, security researchers have developed several principles that can be very helpful in avoiding pitfalls and common errors, both when designing new and adapting existing security protocols.

While we note that design principles are neither necessary nor sufficient for protocol correctness, we believe that studying them for control systems such as the electric grid can significantly help with both designing new and customizing existing protocols for smart grid. A few key design principles gathered from the literature are shown in Table 19.2. A larger set of design principles is identified and discussed in [9]. We will illustrate the value of design principles in the next section by applying the design principles from Table 19.2 to a draft DNP3 Secure Authentication Supplement [25].

In addition to the principles identified in Table 19.2, there are two overarching principles that are important to keep in mind. First, as discussed in [34], when designing

Table 19.2. Selected design principles for security protocols

1. Explicit names. If the identity of a principal is essential to the meaning of a message, it is prudent to mention the principal's name explicitly in the message [32].

2. Explicit trust assumptions. The protocol designer should know the trust relations upon which his protocol depends, and why the dependence is necessary. The reasons why particular trust relations are acceptable should be explicit, though they will be founded on judgment and policy rather than on logic [32].

3. Protocol boundaries. Often the specification of a protocol and its verification focus on the core of the protocol and neglect its boundaries. However, these boundaries are far from trivial; making them explicit and analysing them are important parts of understanding the protocol in context. These boundaries include: (i) interfaces and rules for proper use of the protocol, (ii) interfaces and assumptions for auxiliary functions and participants, such as cryptographic algorithms and network services, (iii) traversals of machine and network boundaries, (iv) preliminary protocol negotiations, and (v) error handling [29].

4. Explicit security parameters. Be explicit about the security parameters of crypto primitives. A key generation routine should be claimed as good for so many keys; a threshold scheme for resistance to conspirators; a block cipher for blocks; and so on [27, 32].

or adapting a protocol to be secure, it is important to know the threat environment in which the protocol is expected to operate and design accordingly. A common mistake is to adopt an existing protocol into a threat environment that is different from what the protocol was designed to be secure against, without re-evaluating the security of the protocol in the new environment. The second overarching principle is that of explicitness of communication, stated as follows in [32]: *every message should say what it means: the interpretation of the messages should depend only on its content. It should be possible to write down a straightforward English sentence describing the content – though if there is a suitable formalism that is good too*. Similar notions are also found in [35, 36]. Some of the principles found in the literature, e.g., Principle 1 in Table 19.2, are specific instances of this overarching principle.

Apart from security, a few key design issues to consider when developing secure authentication protocols for the smart grid environment are: (i) efficiency of the protocol, (ii) availability of both the authentication mechanism and the system incorporating it, and (iii) evolvability of the protocol. Efficiency and availability are important because real-time critical applications need to be supported by the grid, while evolvability is important as devices deployed in the grid will likely be in operation for a long time, typically 1 to 15 years. In the following section, we discuss efficiency and illustrate some tradeoffs, using the draft DNP3 Secure Authentication Supplement as an example. A discussion on availability and evolvability can be found in [9].

19.4.3 Use case: secure authentication supplement to DNP3

Under contract from the Electric Power Research Institute (EPRI), we reviewed an initial draft of the DNP3 Secure Authentication Supplement [25]. In this section, we illustrate the value of design principles for smart grid protocols by applying the principles from Table 19.2 to the draft DNP3 Secure Authentication Supplement. We then discuss the design choices for efficiency in the DNP3 Secure Authentication Supplement.

Principle 1 (explicit names). This principle was proposed in [32] to prevent impersonation attacks and implies that every entity with which a principal interacts must have a unique name. Currently, there is no global naming scheme for principals in the power grid; in fact, it is common to find multiple entities with the same name. However, a global naming scheme is not the only possible approach. For example, one can encode pairwise relations in keys by using unidirectional, pairwise symmetric keys as recommended by the ISO/IEC 9798-4 standard [37]. Indeed, the draft DNP3 Secure Authentication Supplement uses unidirectional pairwise symmetric keys to address this global naming problem [25].

Principle 2 (explicit trust assumptions). Abadi and Needham [32] argue for caution and clarity when deciding which entities are trusted for the correct execution of the security protocol and the extent to which they are trusted. In the context of the power grid, this principle argues the need to clearly state and analyse all trusted entities and the extent of trust in them. Some of those entities may include users who manage keys and passwords, servers that manage keys or time, and software that generates nonces and sequence numbers. As an example of this principle, the draft DNP3 Secure Authentication Supplement that we reviewed [25] mandated maintenance of per-user unidirectional, pair-wise keys for authenticating individual users and distinguishing them from the master device. However, unless it is ensured that the pairwise keys are not openly available to the master, which acts on behalf of the users, achieving the desired property is not possible unless the master is trusted.

One way to ensure that the user keys are not available to the master is by storing them on user smart cards, but in practice, the use of smart cards or other such methods is not common in control systems. Therefore, the master often needs to be a trusted entity, in which case user-level authentication can be achieved through other simpler, and less expensive, ways. Trusting the master to add appropriate user-ids to transactions might be one way.

Principle 3 (protocol boundaries). Abadi [29] highlights the fact that security protocols do not execute in isolation, but rather in an environment. Therefore, it is crucial to understand this environment and ensure that the protocol can function correctly in it. For the power grid, this principle has far-reaching consequences, and to the extent possible, a thorough analysis of the environment is essential. It could include the analysis of the underlying messaging protocols, the real-time nature of control systems, legacy integration issues, choice of cryptographic primitives, networking constraints, and error handling.

Principle 4 (explicit security parameters). Anderson and Needham observed in [27] that most cryptographic primitives have limitations. Such limitations should be made explicit and taken into account, e.g., when specifying key refresh times. Many remote devices in the power grid operate in deploy-and-forget mode. Making the security parameters explicit and taking them into account correctly will help balance security needs with the desire for low management overheads. To illustrate this, the draft DNP3

Secure Authentication Supplement [25] prescribed minimum and maximum session or key-refresh intervals, however, it did not tie those prescriptions to the strengths of the constituent cryptographic mechanisms, and thus left the door open for misconfiguration of the protocol.

For smart grid systems, it is also important to study the intersection of security and efficiency. Authentication efficiency is gauged in terms of the computation and communication overhead imposed by the authentication protocol. Computation overhead arises from expensive operations such as cryptographic operations (e.g., HMAC, symmetric or asymmetric encryption, and digital signatures). Communication overhead arises from additional rounds of messages and message expansion due to the authentication protocol (e.g., added sequence numbers, nonces, and digital signatures).

The communication overhead of adding authentication can be reduced in several ways: (i) by minimizing the additional number of messages needed, (ii) by minimizing the size of those additional messages, and (iii) by minimizing the size of additional fields added to existing messages. For example, one could opt for one-pass (one message), instead of two-pass (two message) authentication mechanisms. Similarly, one could use symmetric key-based keyed message authentication codes (MACs) instead of asymmetric key-based digital signatures to reduce byte overhead. Another option is to use smaller digital signatures, such as 326-bit elliptic curve digital signature algorithm (ECDSA) signatures versus 1024-bit RSA signatures.

The computation overhead can be reduced by minimizing the number of expensive operations (e.g., cryptographic ones) in an authentication mechanism. Cryptographic operations such as exponentiations in finite groups are orders of magnitude more expensive than additions or multiplications. Typically, symmetric key-based cryptosystems are considered to be more efficient than their asymmetric counterparts. For example, digital signatures can be two to three orders of magnitude more expensive than HMACs in terms of computation. Therefore, for authentication in networks with frequent messaging (e.g., SCADA), digital signatures must only be used after careful analysis. However, they may have distinct benefits over symmetric key-based cryptosystems, such as their ability to provide non-repudiation. Similarly, key distribution and management can be simpler when asymmetric cryptosystems are used.

There are tradeoffs for each of the design choices for efficiency, and there is some tension between optimizing an authentication protocol for efficiency and its security. This is also implied by the more general and overarching principle that all communication must be explicit. Thus, optimization of a security protocol should be undertaken with care.

The draft DNP3 Secure Authentication Supplement we reviewed used a two-pass (challenge–response) unilateral authentication as the default to allow for authentication only when necessary (challenged), although it does support an optional mode with one-pass unilateral authentication. However, it used symmetric key-based MACs for authentication instead of asymmetric primitives, saving both bandwidth and computation. It further reduced the communication overhead by using truncation. It truncated the MAC to 4 bytes in the case of HMAC-SHA1 over serial links. Interestingly, there are no formally established results regarding the security of a truncated HMAC; however, standards such as NIST Special Publication 800-107 [38] currently recommend 64 to 96-bit output to provide sufficient security for most applications. Thus, truncating the MAC to 4 bytes in a protocol might be a security risk. An alternative way to save bandwidth might have been to use only sequence numbers instead of using both sequence numbers and *nonces*. Note, the ISO 9897-4 standard states that a HMAC can be used *either* with nonces *or* with sequence numbers to achieve secure authentication. This would have allowed the specification to increase the HMAC size to the NIST recommended standard.

Design of secure protocols is a major challenge being addressed by researchers and standards development organizations today. For example, the cyber-security working group (CSWG), part of the smart grid interoperability panel (SGIP) effort led by the NIST, has a group devoted to analysing the security of smart grid standards[1] and identifying relevant gaps. Principles and common failure modes studied in this work indicate the level of care and effort that would be required to ensure that (i) all of the computing and communication-related smart grid protocols are designed to minimize cyber risks, and (ii) a suite of core security protocols that are well-designed and validated exist and can easily be integrated into smart grid protocols. The initial NIST framework for smart grid interoperability standards has identified 19 sets of smart grid protocols and seven sets of cyber security protocols. Clearly, a significant effort is needed to address protocol security, and having principles to guide the process will be very valuable.

19.5 Conclusion

As smart grids are designed, developed, and deployed, they create enhanced and modified control loops across the grid. These extended loops face new risks from attacks and failures, and therefore, need to be secure and resilient. In this chapter, we discussed an application-driven design approach for specific applications and systems according to which we first understand the cyber security requirements and then develop a solution reflecting those requirements. Based on that approach and an understanding of some of the key emerging smart grid systems, we identified and worked on three specific problems.

We have reviewed threats against AMI and described how a comprehensive intrusion detection system could be designed and deployed to identify compromised smart meters and to gain situational awareness over security issues. We have also explained how the concept of specification-based intrusion detection can be applied to the AMI environment in order to capture legitimate system and network behaviour and then trigger alerts or automated responses when suspicious activity occurs. We have discussed how such techniques can be developed within a formal verification framework to guarantee that the global AMI security policy is enforced through the distributed monitoring operations.

We have looked at the design, development, and experimentation of CONES, which is a real-time middleware toolkit aimed at supporting converged SCADA networks. CONES provides (i) host-based process management, (ii) network resource

---

[1] http://www.nist.gov/smartgrid/

management, and (iii) host-based network resource management. These capabilities collectively allow convergence of SCADA networks while ensuring that isolation and timeliness properties of all SCADA applications (e.g., protection, monitoring, and maintenance) are supported as per the requirements of those applications. We have discussed the architecture, an advanced prototype implementation, and extensive experimentation results to demonstrate the feasibility of CONES.

Finally, we have looked at design principles, best practices, and guidelines for the development of smart grid authentication protocols. We have summarized key design principles that have been proposed in other areas to help improve protocol security. While our analysis is focused on authentication protocols, it applies to most cyber security protocols such as those for key exchange, encryption, and non-repudiation. However, to develop a comprehensive set of key design principles, other protocols deserve a similar analysis. Use of the design principles in developing protocols is very helpful but not sufficient for ensuring security of protocols. As recommended by Syverson [28], design principles should be used throughout the beginning, middle, and end of the protocol design process. In addition, protocol security can be enhanced greatly by having external expert reviewers use a variety of tools (including formal verification) to ensure that the security goals of the protocols are met and to potentially provide enhancements to mitigate vulnerabilities.

Acknowledgements

This material is based upon work supported by the National Science Foundation under grant no. CNS-0524695 and the Department of Energy under award no. DE-OE0000097. Additional support for the Design Principles section was provided by the Electric Power Research Institute under contract number EP-P30283/C14161. Part of Himanshu Khurana's contributions to this work were made while he was at the University of Illinois, working on the acknowledged grants and contracts.

References

[1] Federal Energy Regulatory Commission, 'Smart grid policy', http://www.ferc.gov/whats-new/comm-meet/2009/071609/E-3.pdf, 2009.
[2] Department of Energy, 'Interactive energy roadmap to secure control systems', http://www.controlsystemsroadmap.net/, 2011.
[3] N. Falliere, L. Murchu, and E. Chien, 'W32. stuxnet dossier', Symantec Technical Report, Version 1.4, February 2011.
[4] J. Wright, 'Smart meters have security holes', http://www.msnbc.msn.com/id/36055667, 2010.
[5] M. Davis, 'Smart grid device security', http://www.ioactive.com/news-events/DavisSmart-GridBlackHatPR.php, 2009.
[6] JASON Group, 'Science of cyber security', Technical Report JSR-10-102, MITRE, 2010.

[7] R. Berthier, W. Sanders, and H. Khurana, 'Intrusion detection for advanced metering infrastructures: requirements and architectural directions', in *Proceedings of 1st IEEE International Conference on Smart Grid Communications (SmartGridComm)*, pp. 350–355, 2010.

[8] E. Heine, H. Khurana, and T. Yardley, 'Exploring convergence for SCADA networks', in *Proceedings of 2nd IEEE PES Innovative Smart Grid Technologies (ISGT 2011)*, 2011.

[9] H. Khurana, R. Bobba, T. Yardley, P. Agarwal, and E. Heine, 'Design principles for power grid cyber-infrastructure authentication protocols', in *Proceedings of 43rd Annual Hawaii International Conference on System Science (HICSS)*, 2010.

[10] M. LeMay, G. Gross, C. Gunter, and S. Garg, 'Unified architecture for large-scale attested metering', in *Proceedings of 40th Annual Hawaii International Conference on System Science (HICSS)*, 2007.

[11] T. Goodspeed, D. Highfill, and B. Singletary, 'Low-level design vulnerabilities in wireless control system hardware', in *Proceedings of Scada Security Science Symposium (S4)*, pp. 3–1–3–26, 2009.

[12] M. Carpenter, T. Goodspeed, B. Singletary, E. Skoudis, and J. Wright, 'Advanced metering infrastructure attack methodology', http://inguardians.com/pubs/AMI_Attack_Methodology.pdf, 5 January, 2009.

[13] T. Song, C. Ko, C. Tseng, P. Balasubramanyam, A. Chaudhary, and K. Levitt, 'Formal reasoning about a specification-based intrusion detection for dynamic auto-configuration protocols in ad hoc networks', in *Proceedings of 3rd International Workshop on Formal Aspects in Security and Trust*, pp. 16–33, 2005.

[14] S. Chen and K. Nahrstedt, 'An overview of quality of service routing for next-generation high-speed networks: problems and solutions', *IEEE Network*, vol. 12, no. 12, pp. 64–79, 1998.

[15] R. Schantz, J. Loyall, C. Rodrigues, D. Schmidt, Y. Krishnamurthy, and I. Pyarali, 'Flexible and adaptive QoS control for distributed real-time and embedded middleware', in *Proceedings of ACM/IFIP/USENIX 2003 International Conference on Middleware (Middleware)*, pp. 374–393, 2003.

[16] H. Nguyen, R. Rivas, and K. Nahrstedt, 'iDSRT: integrated dynamic soft real-time architecture for critical infrastructure data delivery over WLAN', in *Proceedings of Quality of Service in Heterogeneous Networks (QSHINE)*, pp. 185–202, 2009.

[17] R. Johnston, C. Hauser, K. Gjermundrod, and D. Bakken, 'Distributing time-synchronous phasor measurement data using the GridStat communication infrastructure', in *Proceedings of 39th Annual Hawaii International Conference System Science (HICSS)*, 2006.

[18] D. Bakken, C. Hauser, H. Gjermundrod, and A. Bose, 'Towards more flexible and robust data delivery for monitoring and control of the electric power grid', Technical Report EECS-GS-009, EECS, Washington State University, 2007.

[19] D. Bakken, D. Whitehead, and G. Zweigle, 'Smart generation and transmission with coherent, real-time data', Technical Report, TR-GS-015, EECS, Washington State University, 2010.

[20] 'IEEE standard communication delivery time performance requirements for electric power substation automation', IEEE Std 1646-2004, 2005.

[21] D. Schmidt, M. Deshpande, and C. O'Ryan, 'Operating system performance in support of real-time middleware', in *Proceedings of 7th IEEE International Workshop on Object-Oriented Real-Time Dependable Systems (WORDS)*, pp. 199–206, 2002.

[22] S. Wang, S. Kodase, K. G. Shin, and D. L. Kiskis, 'Measurement of OS services and its application to performance modeling and analysis of integrated embedded software', in *Proceedings of 8th IEEE Real-Time and Embedded Technology and Applications Symposium*, pp. 113–122, 2002.

[23] M. Caccamo, L. Zhang, L. Sha, and G. Buttazzo, 'An implicit prioritized access protocol for wireless sensor networks', in *Proceedings of 23rd IEEE Real-Time Systems Symposium*, pp. 39–48, 2002.

[24] T. Crenshaw, S. Hoke, A. Tirumala, and M. Caccamo, 'Robust implicit EDF: a wireless MAC protocol for collaborative real-time systems', *ACM Transactions on Embedded Computing Systems*, vol. 6, no. 4, 2008.

[25] DNP3 Users Group Technical Committee, 'DNP3 secure authentication specification version 2.0, DNP users group documentation as a supplement to volume 2 of DNP3', 2008.

[26] 'IEC 62351-5 power systems management and associated information exchange – data and communication security – part 5: security for IEC 60870-5 and derivatives', *International Electrotechnical Commission*, 2009.

[27] R. Anderson and R. Needham, 'Robustness principles for public key protocols', in *Proceedings of 15th Annual International Conference on Advances in Cryptology (CRYPTO)*, pp. 236–247, 1995.

[28] P. Syverson, 'Limitations on design principles for public key protocols', in *Proceedings 1996 IEEE Symposium on Security and Privacy (SP)*, pp. 62–72, 1996.

[29] M. Abadi, 'Security protocols and specifications', in *Proceedings of 2nd International Conference on Foundations of Software Science and Computation Structure (FoSSaCS)*, pp. 1–13, 1999.

[30] T. Fleury, H. Khurana, and V. Welch, 'Towards a taxonomy of attacks against energy control systems', *Critical Infrastructure Protection II*, IFIP, vol. 290, 2009.

[31] K. Stouffer, J. Falco, and K. Kent, 'Guide to supervisory control and data acquisition (SCADA) and industrial control systems security', NIST Special Publication 800-82, June 2011.

[32] M. Abadi and R. Needham, 'Prudent engineering practice for cryptographic protocols', *IEEE Transactions on Software Engineering*, vol. 22, pp. 6–15, 1996.

[33] G. Lowe, 'Some new attacks upon security protocols', in *Proceedings of 9th IEEE Computer Security Foundations Workshop*, pp. 162–169, 1996.

[34] R. Canetti, C. Meadows, and P. Syverson, 'Environmental requirements for authentication protocols', *Software Security: Theories and Systems,* LNCS Vol. 2609, pp. 339–355. Springer, 2003.

[35] C. Boyd and W. Mao, 'On a limitation of BAN logic', *EUROCRYPT*, pp. 240–247, 1993

[36] T. Woo and S. Lam, 'A lesson on authentication protocol design', *Operating Systems Review*, vol. 28, no. 3, pp. 24–37, 1994.

[37] International Standards Organization and International Electrotechnical Commission, 'ISO/IEC 9798-4:1999 information technology – security techniques – entity authentication – part 4: mechanisms using a cryptographic check function', Standard, 1999.

[38] Q. Dang, 'Recommendation for applications using approved hash algorithms', NIST special publication 800-107, February 2009.

Part VI

Field trials and deployments

20 Case studies and lessons learned from recent smart grid field trials

Rose Qingyang Hu and Yi Qian

20.1 Introduction

The power industry has recently undergone a significant transformation of their information and communication technology (ICT) systems to support both current and future business models of smart grid operations. Moreover, the power industry is transforming from the traditional models of business to embrace a number of new and enhanced technologies that support future smart grid operations. This chapter discusses several smart grid field trials of the last few years. We summarize these field trials in three categories: (i) smart power grids, which include the Jeju smart grid testbed, ADS programme for Hydro One, and the SmartHouse project; (ii) smart electricity systems, which include intelligent protection relay; and (iii) smart consumers, which include several dynamic pricing schemes tested by PEPCO, Commonwealth Edison, Connecticut light and power, and California statewide pricing pilot. At the end of the chapter, we briefly discuss the lessons learned from these smart grid field trials.

20.2 Smart power grids

A smart power grid is a power grid that allows various kinds of interconnections between areas for energy consumers and energy supply sources. The rolling out of such networks will pave the way for new business models and for the building of a power grid self-detecting and automatic recovery system that will ensure a reliable and high-quality power supply. In this section, we give an overview of three recent smart power grid pilot projects, the Jeju smart grid testbed, ADS programme for Hydro One, and the SmartHouse project.

20.2.1 The Jeju smart grid testbed

The Jeju smart grid testbed includes all areas of the smart grid implementations – a smart power grid, smart buildings and homes, smart transportations, smart renewable energy, and smart electricity services [1]. For the smart power grid testbed, Korea Electric Power Corporation (KEPCO) and several other participating companies have been developing smart transmitter and automated protection and recovery technologies. In the smart building and home areas, SK Telecom, KT, LG, KEPCO, and many other participating companies have developed and standardized advanced metering infrastructure technologies and smart home and smart building technologies. KEPCO, SK Energy, GS Caltex, and some other companies that participated in smart transportations have developed electric vehicle parts and charging stations, vehicle-to-grid and ICT systems, and additional service models. In the smart renewable energy area, KEPCO, Hyundai, Posco, and other companies have worked on coordination and stabilization technologies for renewable power with efficient implementation across the power grid. Finally, Korea Power Exchange, KEPCO, and several other companies will develop real-time pricing, demand response, and online consumer power-trading systems for smart electricity services. Figure 20.1 shows an overview of the Jeju smart grid testbed.

The construction of a total operation centre (TOC) for the Jeju smart grid testbed was completed in 2010. At the same time, legislation – including the special act on support for the smart grid – was proposed in 2010 to begin establishing the basic laws and regulations needed to secure long-term investment funding and the provision of incentives. A variety of tests and implementations of smart grid technologies in the Jeju smart grid testbed will be carried out through 2013, when the project is scheduled to be completed. As follow-ups to the Jeju project, smart grid pilot city projects and smart grid implementation projects in metropolitan areas in Korea will be developed by 2020, followed by the nationwide smart grid implementation that will be completed by 2030 [1].

The Jeju smart grid testbed is one of the world's first all-inclusive smart grid testbeds. Spearheaded by Korean conglomerates including KEPCO, LG, KT, and POSCO, the Jeju smart grid testbed gives an opportunity to fully experience the wide array of prototypes available, including smart meters, in-home displays, smart appliances, electric vehicle (EV) charging facilities, wind turbines, and photovoltaic

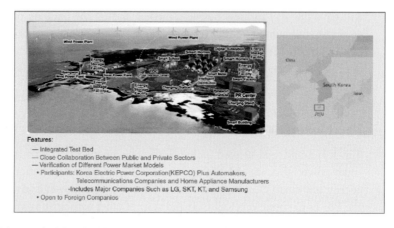

Figure 20.1   The Jeju smart grid testbed [1].

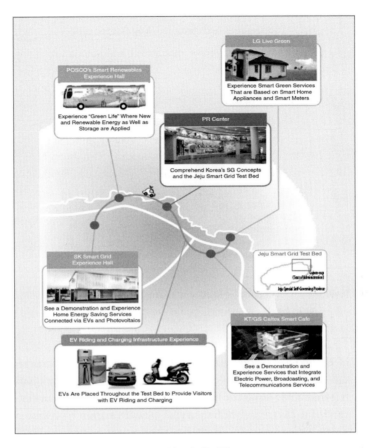

Figure 20.2 The Jeju smart grid testbed centre and exhibition halls [1].

equipment. Figure 20.2 shows the various installations that make up the Jeju smart grid testbed.

20.2.2 ADS program for Hydro One

Hydro One established a long-term vision in 2005 to increase innovation and continue its leading role in providing safe, reliable, and cost-effective transmission and distribution of electricity from various supply sources to Ontario electricity users in Canada [2]. Hydro One believes that transmission and distribution companies must transform themselves and change how they do business. Hydro One is focused on innovating and establishing an electricity grid that is modern, flexible, and smart – one that will not only support and drive consumer choices about electricity, but also set Hydro One on the path to becoming the leading electricity delivery company in North America [2].

The advanced distribution system (ADS) programme for Hydro One includes the implementation of a distribution management system to meet the four business objectives

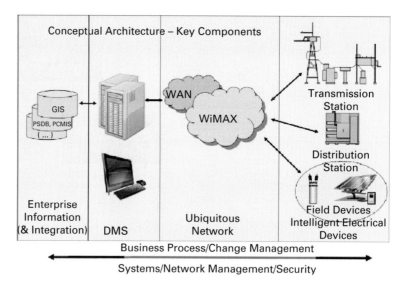

Figure 20.3   Hydro One's ADS smart grid conceptual architecture [2].

of Hydro One: (i) optimize connection of distributed generation (DG) on the distribution network; (ii) improve distribution reliability and operations; (iii) optimize outage restoration; (iv) optimize network asset planning. The system that is used to automate and control an electrical distribution grid includes a central software component that is called a distribution management system. This software system is a powerful network planning, analysis, and operations tool that uses a detailed model of the power grid, telemetry, and information about power flow patterns to help manage the system in real time. Several core components of Hydro One's ADS programme are represented in Figure 20.3 [2], which includes the distribution management system that acts as the "brain" of the Hydro One ADS programme. A distribution management system is a suite of decision-support software applications which will assist Hydro One's control rooms and field operating personnel in monitoring and controlling the distribution system. It is essential for close integration of the distribution management system with other enterprise applications and data stores.

Hydro One uses the concept of 'domains' to classify the possible implications for privacy in the smart grid, and to impose certain architectural decisions that will meet privacy requirements, while delivering the necessary functionality. As illustrated in Figure 20.4, there are three domain types identified: 'customer domain', 'services domain', and 'grid domain' [2].

ADS is designed to function with integration of the grid domain from the services domain via the implementation of a service-oriented architecture. This design will deliver services using transaction and message-management tools known in the industry as an enterprise services bus. In this way, separation via the services domain is controlled between the customer domain and grid domain [2]. The ADS uses aggregated average

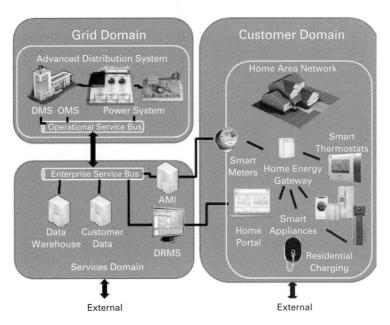

Figure 20.4 Grid domain, services domain, and customer domain for Hydro One [2].

consumption profiles from a meter data system that is segregated from the commercial and personal information of the consumer. For the purposes of analysing the network, the location of the meter's connection to the grid is important, allowing the system to represent load along the length of a particular circuit aggregating load to transformers, substations, and the entire system [3].

20.2.3 The SmartHouse project

The SmartHouse project sets out to validate and test how ICT-enabled collaborative technical–commercial aggregations of smart houses provide an essential step to achieve the needed radically higher levels of energy efficiency in Europe. Figure 20.5 shows the key concept of the SmartHouse project [4].

There are three goals that the SmartHouse project is heading towards: (i) improving energy efficiency; (ii) increasing the penetration of renewable energies; and (iii) diversifying and decentralizing the energy mix of Europe. ICT will play a key role in the transformation of the electricity sector, enabling it to cope with more decentralized and renewable generation efficiently. Intelligent networked ICT technology for collaborative technical–commercial aggregations enables smart houses to communicate, interact, and negotiate with both customers and energy devices in the local energy grid so as to achieve maximum energy efficiency [5].

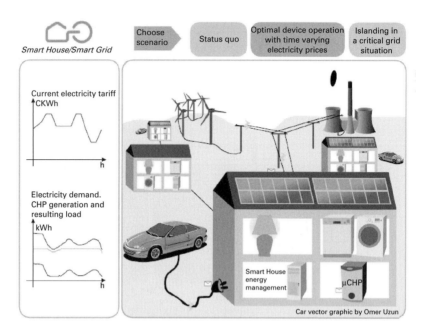

Figure 20.5 The SmartHouse project [4].

20.3 Smart electricity systems

In this section we give an overview on a smart electricity system, an intelligent protection relay system for smart grid [6]. In recent years, trends towards electricity liberalization and the increased importance of environmental issues have led to diversification of power system configurations and characteristics, through the installation of distributed power sources based on photovoltaic and wind power generation technologies. Under these circumstances, there is a concern regarding the effect on the reliability of power networks with respect to aspects such as overload, frequency variation, and harmonic content. In [6], the concepts of protection relay systems for operation within a smart grid are proposed, and a prototype development with experimental results is described, which will be summarized in the rest of this section.

A field trial was established in the 66-kV system operated by Shikoku Electric Power Co., Inc. in Japan from June 2008 to February 2009. The system configuration for the field trial is shown in Figure 20.6 [6]. There was an information management server and a client located within the head office, there were three protection relays (RYI, RY2, and RY3) situated within two substations, and a dedicated network provided for maintenance. Although no internal faults occurred during the operation of the trial system, the results of the evaluation of power transient fluctuations utilizing wide-area data during system operations (e.g., disconnecting one circuit, 1L shown in Figure 20.6, of a double-circuit parallel line) will be introduced below.

According to [6], the 1L and 2L current waveforms of RY1 were collected in the information management server on disconnection of 1L of the double-circuit parallel

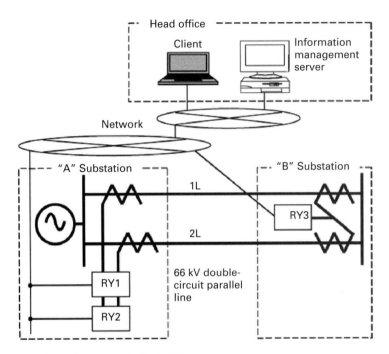

Figure 20.6 The system configuration for the field trial [6].

line. These waveforms conform to the system phenomenon because the 1L currents became '0' and the 2L currents doubled just after this system operation. In addition, it could be confirmed simultaneously that the disturbance record data stored by the trigger of the OCIEF relay operation were properly sent to the information management server based on the evaluation executed in RY1. The loci on the relay characteristics of the segregated and zero-sequence transverse differential currents based on the disturbance record data. Although some transverse currents for each phase are generated, it can easily be recognized that none of the loci exceed the threshold. On the other hand, in accordance with the evaluation algorithm, the change in system characteristics can be detected correctly by means of evaluating the zero-sequence transverse currents, which are generated transiently just after the 1L disconnection. The results of the field trial demonstrate that the proposed intelligent protection relay system has the capability to evaluate power system characteristics in detail on a real-time basis and contribute towards supervision, coordination check, and optimization of setting values.

20.4 Smart consumers

In this section we survey several smart grid field trials in terms of smart consumers. Various forms of dynamic pricing are being implemented or considered around the United States for smart consumers, including time-of-use (ToU), real-time pricing (RTP), critical peak pricing (CPP), and critical peak rebates (CPR) [7]. The ToU rate is a traditional two-part rate, with on-peak and off-peak periods. ToU is the only rate mentioned which does not require the use of enabling technology, given that prices are pre-established, known in advance, and typically do not change. The CPP rate is similar to ToU pricing, with the addition of much steeper rates during 'critical peak' hours and days. The CPR rate subjects customers to the same critical peak periods as CPP, but customers have the option of earning rebates for the amount of energy reduced during peak periods. A major difference from ToU is that CPP and CPR are only implemented for a certain number of event days in a year. RTP pricing, also known as 'hourly pricing' or HP in certain cases, offers electricity prices which reflect the cost of energy at the wholesale level, and can change as often as an hourly basis [7]. In the rest of this section, we summarize several case studies for smart consumers using dynamic pricing programmes offered by electric utilities.

20.4.1 PEPCO

PEPCO began a pilot programme in 2008, involving nearly 900 customers in Washington, DC. The pilot programme, PowerCentsDC, was designed to assess the impacts of smart meters, enabling technology, and various pricing structures on consumer behaviour. Each customer received a free smart meter, and those with central air conditioning were provided with the option of a free smart thermostat. Figure 20.7 depicts the peak reductions achieved according to price structure and season. The greatest reduction shown in energy usage resulted from the CPP rate structure [7]. The thermostats could also be programmed, automatically reducing consumption of air conditioning and central heating during preset hours. About a third of the participants, had smart thermostats that were used to reduce central AC compressor use upon receiving a radio-controlled signal during critical peak periods. CPP customers with smart thermostats reduced their summer peak demand usage by 49%, as opposed to 29% for those without the technology. CPR and HP customers were also found to have larger reductions in energy usage with a smart

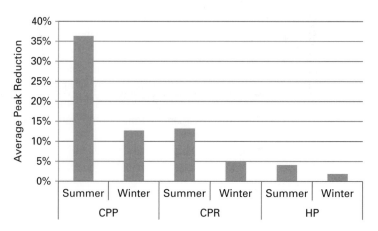

Figure 20.7 Reductions in peak-energy usage by residential customers according to season and pricing structure CPP, CPR, and HP [7].

thermostat. The largest peak reductions were achieved by CPP participants, whereas the smallest reductions occurred in the hourly pricing group. These results are partially due to prices in the HP programme not being as high as they were in the CPP and CPR programmes. Both the customer surveys and actual pilot results indicate that larger financial incentives result in larger reductions in energy consumption. Combining an effective pricing programme such as CPP with an enabling technology such as a smart thermostat resulted in a further 20% reduction in energy usage [8].

20.4.2 Commonwealth Edison

Commonwealth Edison (ComEd) expanded its pilot residential real-time pricing (RRTP) programme in 2007 to encompass 110,000 participants in Illinois in a span of 4 years. ComEd is providing this programme only to select customers who are regular Internet users and who understand and accept real-time pricing. Customers also need to be a part of the programme for at least one year, shift their heavy appliance usage to non-peak hours, and pay an additional lease fee per month for the new smart meter [7, 9]. The RRTP programme works via multiple interfaces. The customer can visit a website to view the day-ahead prices of electricity to plan usage in advance. These generally provide an accurate picture of the next-day hourly prices. The customer can also receive a notification, via email, text message, or automated phone call, on days when the electricity rate rises higher than 14 cents/kWh, as shown in Figure 20.8 [7]. These signals allow the

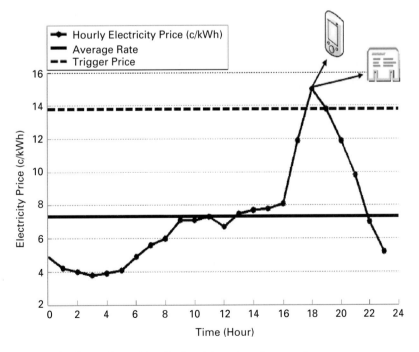

Figure 20.8 Chart showing the fluctuations of hourly electricity prices that can trigger a customer notification [7].

customer to plan ahead, using more energy during the lower range of prices instead of during high-rate hours. The price shown and charged to the customer is the market rate, and ComEd does not mark up the price.

20.4.3 Connecticut light and power

Connecticut light and power conducted a pilot study in 2009, 'plan-it wise', involving approximately 1100 residential customers to gauge their interest and responsiveness to smart grid technologies, and dynamic pricing rates. Participation was voluntary and the pilot lasted for the duration of the summer of 2009. Randomly selected customers were provided with one of four types of enabling technologies, along with information online: an A/C control switch, a smart thermostat, an energy monitor or an in-home display. The A/C control switch was designed to allow the utility to regulate air conditioning during predetermined times, while the energy monitor and in-home display show price changes and real-time feedback, respectively [7, 10]. Figure 20.9 shows the results from high price differentials among different rate structures, with and without enabling technology. The rates that were in effect for the fewest number of hours with the highest price differentials achieved the greatest peak reductions [7].

20.4.4 California statewide pricing pilot

California's three investor-owned utilities with two regulatory bodies conducted a pilot involving roughly 2500 customers for the duration of 1.5 years started in July 2003. Customers were recruited via an enrolment package. The objective of this pilot was to determine whether and to what extent customers would reduce peak usage in the presence of dynamic pricing signals. This study was designed to address concerns following the

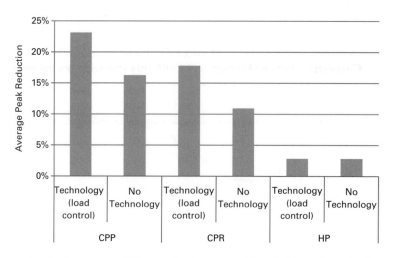

Figure 20.9 Average peak reductions among different rate structures, with and without the technologies [7].

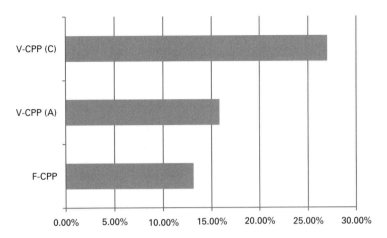

Figure 20.10 Peak reduction among different rates offered in the California statewide pricing pilot [7].

energy crisis experienced by California during 2000–2001 [7, 11]. Figure 20.10 shows the breakdown of peak demand reduction on critical days during the summer for all rate schemes tested. The results varied across climate zones as expected. The F-CPP rate was able to achieve an estimated peak reduction of 13% on critical days [7].

20.5 Lessons learned

In this chapter we have surveyed several smart grid field trials in three categories, smart power grids, smart electricity systems, and smart consumers. In power smart grids, we summarized the Jeju smart grid testbed in Korea, the ADS programme for Hydro One in Canada, and the SmartHouse project in Europe. In smart electricity systems, we revisited the intelligent protection relay field trial in Japan. In the category of smart consumers, we examined several dynamic pricing schemes tested by PEPCO, Commonwealth Edison, Connecticut light and power, and California statewide pricing pilot in the USA. From the experience of these field trials over the globe, we can see that for smart grid with communication infrastructures, we expect that the traditional carbon fuel-based power plants can cooperate with emerging distributed renewable energy such as wind, solar, etc., to reduce the carbon fuel consumption and consequent greenhouse gas such as carbon dioxide emission. Consumers can minimize their energy expenses by adjusting their intelligent home appliance operations to avoid peak hours and utilize renewable energy instead. We should explore further the challenges for a communication infrastructure as part of a complex smart grid system. Since a smart grid system might have millions of consumers and devices, the demand for its reliability and security is extremely critical. Through a communication infrastructure, a smart grid will improve power reliability and quality to eliminate electricity blackout.

20.6 Conclusion

In this chapter we have surveyed several smart grid field trials in three categories, smart power grids, smart electricity systems, and smart consumers. In smart power grids, we summarized the Jeju smart grid testbed in Korea, the ADS programme for Hydro One in Canada, and the SmartHouse project in Europe. In smart electricity systems, we revisited the intelligent protection relay field trial in Japan. In the category of smart consumers, we examined several dynamic pricing schemes tested by PEPCO, Commonwealth Edison, Connecticut light and power, and California statewide pricing pilot in the USA. On the basis of the results from field experiments, we can focus on those challenges to smart grid infrastructures in system design and operations as well as consumer experience, and define a roadmap towards mass-market smart grid deployments.

References

[1] J. Kim and H. I. Park, 'Policy directions for the smart grid in Korea', *IEEE Power and Energy Magazine*, vol. 9, no. 1, pp. 40–49, January–February 2011.

[2] Information and Privacy Commissioner, Ontario, Canada, 'Operationalizing privacy by design: The Ontario smart grid case study', February 2011 [online]. Available at: http://www.ipc.on.ca/images/Resources/pbd-ont-smartgrid-casestudy.pdf

[3] S. A. Dougherty, 'Privacy by design: from resolution to reality – a Hydro One case study through privacy by domains', IBM Corporation, 2011 [online]. Available at: http://www.cio.gov.bc.ca/local/cio/informationsecurity/documents/PS_2011_PDFs/Dougherty_Steven-WorkshopF.pdf

[4] SmartHouse/SmartGrid, http://www.smarthouse-smartgrid.eu/

[5] K. Kok, S. Karnouskos, J. Ringelstein, A. Dimeas, A. Weidlich, C. Warmer, S. Drenkard, N. Hatziargyriou, and V. Lioliou, 'Field-testing smart houses for a smart grid', in *Proceedings of 21st International Conference on Electricity Distribution CIRED*, Frankfurt, 2011.

[6] F. Kawano, G. P. Baber, P. G. Beaumont, K. Fukushima, T. Miyoshi, T. Shono, M. Ookubo, T. Tanaka, K. Abe, and S. Umeda, 'Intelligent protection relay system for smart grid', in *Proceedings of 10th IET International Conference on Developments in Power System Protection (DPSP 2010)*, Manchester, UK, 29 March – 1 April 2010.

[7] J. Wang, M. A. Biviji, and W. M. Wang, 'Lessons learned from smart grid enabled pricing programmes', in *Proceedings of 2011 IEEE Power and Energy Conference at Illinois (PECI)*, pp. 1–7, 25–26 February 2011.

[8] PEPCO, 'PowerCentsDC programme final report', September 2010 [online]. Available at: http://www.powercentsdc.org/ESC%2010-09-08%20PCDC%20Final%20Report%20-%20FINAL.pdf

[9] The ComEd Residential Real-Time Pricing Program, 'Guide to real-time pricing', 2010 [online]. Available at: http://www.thewattspot.com/pdf/RRTPGuide200903.pdf

[10] A. Faruqui and S. Sergici, 'Impact evaluation of CL&P's plan-it wise energy programme: final results', The Brattle Group, November 2009 [online]. Available at:

http://nuwnotes1.nu.com/apps/clp/clpwebcontent.nsf/AR/PlanItWiseAppendix/$File/Plan-it%20Wise%20Pilot%20Results%20Appendix.pdf

[11] S. George and A. Faruqui, 'California's statewide pricing pilot: overview of key findings', Charles River Associates, May 2005 [online]. Available at: http://sites.energetics.com/MADRI/pdfs/california_050405.pdf

Index

3GPP LTE, 134, 158
3GPP LTE-advanced, 158
3GPP MTC, 160
3GPP-LTE, 123
3rd Generation Partnership Project (3GPP), 158
6LoWPAN, 129, 138, 153, 295, 327, 333

actuator, 310
advanced encryption standard (AES), 25
advanced metering infrastructure (AMI), 6, 45, 168, 283, 309, 441
ancillary services, 92, 167
ANR+, 154
ARRA (American Recovery and Reinvestment Act), 439
automatic generation control (AGC), 35, 41
autonomous operation, 148
average revenue per unit (ARPU), 156

BACnet protocol, 8
billing, 170
Bluetooth, 141
broadband over power line (BPL), 18
broadband PON (BPON), 20
broadband power-line communications, 126
building-area network (BAN), 8

CDMA2000, 155
cellular systems, 150
CEN, 160
CENELEC, 160
Chi-square test, 177
circuit breaker, 284
cognitive radio (CR), 135
compressive sensing (CS), 316
constrained application protocol (CoAP), 331, 336
contention-free scheduling TDMA (CF-TDMA) protocol, 288
context-aware intelligent control, 316
critical peak pricing (CPP), 55, 472
critical peak rebates (CPR), 472
cryptosystem, 457

customer information/messaging (CMSG), 239
cyber-attack, 356
cyber-security working group (CSWG), 458

DASH7, 154
data and service centre (DSC), 320
data-aggregation point (DAP), 256
day-ahead market, 51, 52
demand response (DR), 45, 169, 309
demand response–direct load control (DRDLC), 238
demand-side management (DSM), 8, 45, 69, 169, 309
denial-of-service (DoS) attack, 443
DER, 166
destination-oriented directed acyclic graph (DODAG), 329
DiffServ, 17
direct load control (DLC), 59, 169
directed diffusion (DD) protocol, 291
dispatch distributed customer storage (DDCS), 238
distributed automation, 167
distributed electricity storage, 91
distributed energy resource (DER), 6, 166
distributed generation (DG), 6, 166
distribution management system (DMS), 238
distribution network, 163, 165
distribution system demand response–centralized control (DSDRC), 238

electric vehicle (EV), 91
Electrical Power Research Institute (EPRI), 5
elliptic curve digital signature algorithm (ECDSA), 457
energy flow, 305
energy service interface (ESI), 8
energy storage, 166
energy systems interface (ESI), 244
energy-aware plug-and-play (EPnP), 14
energy-consumption scheduler (ECS), 71
energy-efficiency, 149
energy-management system (EMS), 318
enhanced data rates for GSM evolution (EDGE), 155

ESRT (event-to-sink reliable transport) protocol, 296
Ethernet PON (EPON), 20
ETSI, 160
ETSI M2M, 158
European Commission Research, 5
European Cooperation in Science and Technology (COST), 249
European Telecommunications Standard Institute (ETSI), 158

fault clear, isolation, and reconfigure (FCIR), 238
Federal Energy Regulatory Commission (FERC), 439
fibre Bragg grating (FBG) strain sensors, 279
field distribution automation maintenance–centralized control (FDAMC), 238
field-area network (FAN), 442
flat-routing, 291
flexible AC transmission systems (FACTS), 42
free-space optical (FSO) communications, 121
frequency regulation, 167

G3-PLC, 129
GAF (geographic adaptive fidelity) protocol, 291
GEAR (geographic and energy-aware routing) protocol, 292
general packet radio service (GPRS), 155
generation, 163
generation dispatch (GD), 308
gigabit PON (GPON), 20
global positioning system (GPS), 165
global system for mobile communications (GSM), 154
grid monitoring and control (GMC), 307
grid-connected mode, 91
GridStat, 18, 46, 50
GridWise Architecture Council (GWAC), 10

heterogeneous network platform (HNP), 314
hierarchical routing, 291
high-speed downlink packet access (HSDPA), 155
high-voltage DC (HVDC), 42
home energy-management system (HEMS), 8, 13
home sensor and actuator network (HSANET), 320
home-area network (HAN), 7
HomePlug, 99, 127, 128, 136
hybrid wireless-broadband power line network (W-BPL), 19

IEEE
  802.15.4, 150
IEEE 1901, 126
IEEE 802.11, 150, 153, 158

IEEE 802.11s, 124
IEEE 802.15.1, 141, 158
IEEE 802.15.3a, 141
IEEE 802.15.4, 153, 158, 326, 333
IEEE 802.15.4e/g/k, 153
IEEE 802.22, 135
IETF, 152
IETF CoRE, 153
IETF ROLL, 153
information flow, 306
intelligent electronic device (IED), 38, 167
intelligent islanding, 220
Internet of things, 60, 138, 324, 326
Internet protocol (IP), 168
Internet VPN, 16
intrusion-detection system (IDS), 427
intrusion-prevention system (IPS), 427
IntServ, 17
IP over foo, 327
IP security (IPSec), 25
IP-based networking, 137
IP-based networks, 16
IPSec, 138
IPSec (IP Security), 17
IPv6, 153, 328
islanded distributed customer storage (IDCS), 238
islanding, 167
ITU-T G.9960/61, 127

jamming-based attack, 375

LEACH (low energy adaptive clustering hierarchy) protocol, 291
load frequency control (LFC), 41
load management system (LMS), 239
load shedding, 219
location-based routing, 291
locational marginal price (LMP), 56, 376
locational marginal pricing (LMP) algorithm, 44

M2M, 140, 147
  applications, 156
  area network, 141
  capillary, 150, 152, 158, 163
  cellular, 150, 154, 158, 163, 171
  ETSI, 158
  gateway, 141
  hybrid solutions, 152
  security, 149
  server, 141
  wired, 150
machine-to-machine, 147
machine-type-communications (MTC), 160
meter data-management system (MDMS), 7, 235, 447

meter reading, 239
microgrid, 166
mobility and remote operation, 148
MPLS, 213
multiprotocol label switching (MPLS), 17, 138, 214

NASPI, 165
   NASPInet, 165
National Institute of Standards and Technology (NIST), 5, 235, 439, 458
neighborhood-area network (NAN), 7, 442
network address translation (NAT), 153
New York State Electric and Gas Company (NYSEG), 7
North American Reliability Corporation (NERC), 353

Open Smart Grid (OpenSG) working group, 235
OpenSG matrix, 235
orthogonal frequency-division multiple access (OFDMA), 133

peak shaving, 93, 167
peak-load pricing (PLP), 86
peak-to-average ratio (PAR), 93
pervasive service-oriented network (PERSON), 314
phasor data collector (PDC), 9
phasor data concentrator (PDC), 165, 206
phasor gateway (PGW), 165
phasor measurement unit (PMU), 9, 165, 205, 425
PLC coexistence, 130
plug-in hybrid electric vehicle (PHEV), 69, 170, 236, 239
power market, 375
power-line communication (PLC), 18, 99, 126, 212
power-system stabilizer (PSS), 40
PoweRline Intelligent Metering Evolution (PRIME), 129
premise-area network (PAN), 8
PSFQ (pump slowly fetch quickly) protocol, 296

quality-of-service (QoS), 155, 205

real-time market, 53
real-time operating system (RTOS), 209
real-time pricing (RTP), 55, 471
remote terminal unit (RTU), 388
renewable energy integration, 93
renewable energy source (RES), 45
renewable integration, 167
representational state transfer (REST), 331
retail market, 54

RFC768, 334
RMST (reliable multi-segment transport) protocol, 296
roaming, 170
routing protocol for low power and lossy network (RPL), 138
RPL (routing protocol for low-power and lossy networks), 292
RPL (Routing protocol for low-power and lossy networks), 328

self-organizing medium access control (SMACS) protocol, 288
sensor, 310
sensor and actuator network (SANET), 7, 303
sensor-MAC (S-MAC) protocol, 287
service-oriented architecture (SOA), 15
service-oriented network (SON), 315
simple network management protocol (SNMP), 210
smart building, 157
smart city, 156, 157
smart grid, 163
smart grid interoperability panel (SGIP), 458
smart meter (SM), 425
smart utility network (SUN), 153
SmartHouse project, 469, 475
SONET/SDH technology, 213
SPIN (sensor protocols for information via negotiation) protocol, 291
spinning reserves, 167
stand-alone mode, 91
standard bodies, 152
subscriber identity module (SIM), 158
supervisory control and data acquisition (SCADA), 9, 37, 150, 167, 284, 308, 353, 388, 441
synchronous digital hierarchy (SDH), 209
synchrophasors, 43, 205, 279

TCP (transmission control protocol), 295
TEEN (threshold-sensitive energy efficient sensor network) protocol, 291
traffic-adaptive medium access (TRAMA) protocol, 288
transmission control protocol (TCP), 17
transmission network, 163, 165

universal mobile telecommunications systems (UMTS), 155
user datagram protocol (UDP), 17
utility function, 77

vehicle-to-grid (V2G) systems, 91, 170
Vickrey–Clarke–Groves (VCG) approach, 84
voltage support, 167

web-service description language (WSDL), 15
Wi-Fi, 132, 153
wide-area measurement system (WAMS), 7, 46, 205
wide-area network (WAN), 442
wide-area situational awareness, 165
WiMAX, 17, 123, 133, 155
wireless jamming, 377

wireless mesh network (WMN), 123
wireless sensor network, 139
Wireless World Research Forum, 147
World Wide Web Consortium (W3C), 332

Z-Wave, 100
ZigBee, 99, 131, 153, 156